Wood
Structure
and
Composition

INTERNATIONAL FIBER SCIENCE AND TECHNOLOGY SERIES

Wood
Structure
and
Composition

edited by

Menachem Lewin
Hebrew University of Jerusalem
Jerusalem, Israel
and
Polytechnic University
Brooklyn, New York

Irving S. Goldstein
North Carolina State University
Raleigh, North Carolina

Marcel Dekker, Inc.   New York • Basel • Hong Kong

Library of Congress Cataloging-in-Publication Data

Wood structure and composition / edited by Menachem Lewin, Irving S.
  Goldstein.
    p.   cm. -- (International fiber science and technology series; v. 11)
    Includes bibliographical references and index.
    ISBN 0-8247-8233-X
    1. Wood. 2. Wood-Chemistry. 3. Wood Anatomy. I. Lewin,
Menachem.II. Goldstein, Irving S.
III. Series.
TA419.W833 1991
620.1'2--dc20   91-15739
                                            CIP

This book is printed on acid-free paper.

MARCEL DEKKER, INC.
270 Madison Avenue, New York, New York 10016

Current printing (last digit):
10 9 8 7 6 5 4 3 2 1

PRINTED IN THE UNITED STATES OF AMERICA

About the Series

When human life began on this earth *food* and *shelter* were the two most important necessities. Immediately thereafter, however, came *clothing*. The first materials used for it were fur, hide, skin and leaves—all of them sheetlike, two-dimensional structures not too abundantly available and somewhat awkward to handle. It was then—quite a few thousand years ago—that a very important invention was made: the *manufacture* of two-dimensional systems—fabrics—from simple monodimensional elements—fibers; it was the birth of the textile industry based on fiber science and technology. Fibers were readily available everywhere; they came from animals (wool, hair, and silk) or from plants (cotton, flax, hemp, and reeds). Even though their chemical composition and mechanical properties were very different, yarns were made of the fibers by spinning and fabrics were produced from the yarns by weaving and knitting. An elaborate, widespread, and highly sophisticated art developed in the course of many centuries at locations all over the globe virtually independent from each other. The fibers had to be gained from their natural sources, purified and extracted, drawn out into yarns of uniform diameter and texture, and converted into textile goods of many kinds. It was all done by hand using rather simple and self-made equipment and it was all based on empirical craftsmanship using only the most necessary quantitative measurements. It was also performed with no knowledge of the chemical composition, let alone the molecular structure, of the individual fibers. Yet, by ingenuity, taste, and patience, myriads of products of breathtaking beauty, remarkable utility, and surprising durability were obtained in many cases. *This first era* started at the very beginning of civilization and extended into the twentieth century when steam-driven machinery invaded the mechanical operations and some empirical procedures—mercerization of cotton, moth-proofing of wool, and loading of silk—started to introduce some chemistry into the processing.

The second phase in the utilization of materials for the preparation and production of fibers and textiles was ushered in by an accidental discovery which Christian Friedrich Schoenbein, chemistry professor at the University of Basel in Switzerland, made in 1846. He observed that cotton may be converted into a soluble and plastic substance by the action of a mixture of nitric and sulfuric acid; this substance or its solution was extruded into fine filaments by Hilaire de Chardonnet in 1884.

Organic chemistry, which was a highly developed scientific discipline by that time, gave the correct interpretation of this phenomenon: The action of the acids on cellulose—a natural fiber former—converted it into a *derivative*, in this case into a cellulose nitrate, which was soluble and, therefore, spinnable. The intriguing possibility of manipulating natural products (cellulose, proteins, chitin, and others) by chemical action and thereby rendering them soluble resulted in additional efforts which led to the discovery and preparation of several cellulose esters, notably the cellulose xanthate and cellulose acetate. Early in the twentieth century each compound became the basis of a large industry: viscose rayon and acetate rayon. In each case special processes has to be designed for the conversion of these two compounds into a fiber, but once this was done, the entire mechanical technology of yarn and fabric production which had been developed for the natural fibers was available for the use of the new ones. In this manner new textile goods of remarkable quality were produced, ranging from very sheer and beautiful dresses to tough and durable tire cords and transport belts. Fundamentally these materials were not truly "synthetic" because a known natural fiber former—cellulose or protein—was used as a base; the new products were "artificial" or "man-made." In the 1920s, when viscose and acetate rayon became important commercial items, polymer science had started to emerge from its infancy and now provided the chance to make *new fiber formers* directly by the polymerization of the respective monomers. Fibers made out of these polymers would therefore be "truly synthetic" and represent additional, extremely numerous ways to arrive at new textile goods. Now started the *third era* of fiber science and technology. First the basic characteristics of a good synthetic fiber former had to be established. They were: ready spinnability from melt or solution; resistance against standard organic solvents, acids, and bases; high softening range (preferably above 220°C); and the capacity to be drawn into molecularly oriented fine filaments of high strength and great resilience. There exist literally many hundreds of polymers or copolymers which, to a certain extent, fulfill the above requirements. The first commercially successful class were the *polyamides*, simultaneously developed in the United States by W. H. Carothers of Du Pont and by Paul Schlack of I. G. Farben of Germany. The *nylons*, as they are called, commercially, are still a very important class of textile fibers covering a remarkably wide range of properties and uses. They were soon (in the 1940s) followed by the *polyesters*, *polyacrylics*, and *polyvinyls*, and somewhat later (in the 1950s) there were added the *polyolefins* and *polyurethanes*. Naturally, the existence of so many fiber formers of different chemical composition initiated successful research on the molecular and supermolecular structure of these systems and on the dependence of the ultimate technical properties on such structures.

As time went on (in the 1960s), a large body of sound knowledge on structure-property relationships was accumulated. It permitted embarkation on the reverse approach: "Tell me what properties you want and I shall *tailor-make* you the fiber former." Many different techniques exist for the "tailor-making": graft and block copolymers, surface treatments, polyblends, two-component fiber spinning, and cross-section modification. The systematic use of this "macromolecular engineering" has led to a very large number of *specialty* fibers in each of the main classes; in some cases they have properties which none of the prior materials—natural and "man-made"—had, such as high elasticity, heat setting, and moisture repellency. An important result was that the new fibers were not content to fit into the existing textile machinery, but they suggested and introduced substantial modifications and innovations such as modern high-speed spinning, weaving, and knitting, and several new technologies of texturing and crimping fibers and yarns.

This third phase of fiber science and engineering is presently far from being complete, but already a *fourth era* has begun to make its appearance, namely in fibers for uses *outside* the domain of the classical textile industry. Such new applications involve fibers for the reinforcement of thermoplastics and duroplastics to be used in the construction of spacecraft, airplanes, buses, trucks, cars, boats, and buildings; optical fibers for light telephony; and fibrous materials for a large array of applications in medicine and hygiene. This phase is still in its infancy but offers many opportunities to create entirely new polymeric systems adapted by their structure to the novel applications outside the textile fields.

This series on fiber science and technology intends to present, review, and summarize the present state in this vast area of human activities and give a balanced picture of it. The emphasis will have to be properly distributed on synthesis, characterization, structure, properties, and applications.

It is hoped that this series serves the scientific and technical community by presenting a source of organized information, by focusing attention on the various aspects of the fascinating field of fiber science and technology, and by facilitating interaction and mutual fertilization between this field and other disciplines, thus paving the way to new creative developments.

Herman F. Mark

Preface

During the past two decades, increasing interest in the use of renewable resources to ameliorate both the environmental and energy problems confronting our global society has developed. Wood, of course, is our most abundant renewable resource.

It is not surprising then that interest in wood chemistry has also seen a resurgence, with a proliferation of papers, international conferences, and even journals. (*The Journal of Wood Chemistry and Technology* published by Marcel Dekker, Inc. since 1981 is a notable example.)

Beginning students of wood chemistry are served by introductory textbooks and experienced practitioners by the voluminous literature. However, professional chemists who wish to become familiar with various aspects of modern wood chemistry have at present no alternative but to dive into the deep water of specialized journal articles. It is for this audience that this book is intended. We have attempted to present the most important aspects of wood structure and composition at a level of treatment intermediate between that suitable for an undergraduate and that for a specialist in the field. This should enable a professional chemist to become oriented quickly in the topic of interest.

Included are chapters on wood anatomy, wood analysis, and the composite nature of wood, as well as an overview of the chemical composition of wood and chapters on the structure and properties of cellulose, lignin, hemicelluloses, extractives, and bark.

Multiple authorship assures expert treatment of the topics, at the cost of a lack of uniformity of style and exposition. We hope that this volume will encourage the increased involvement of chemical professionals in aspects of wood chemistry, in addition to broadening their understanding of it.

Menachem Lewin
Irving S. Goldstein

Contents

Contributors

Harald Berndt   University of California, Berkeley, California

Hou-min Chang   North Carolina State University, Raleigh, North Carolina

Chen-Loung Chen   North Carolina State University, Raleigh, North Carolina

Chyi-Cheng Chen   Hoffmann-La Roche, Inc., Nutley, New Jersey

Laurence Cool   University of California, Richmond, California

Dwight B. Easty*   The Institute of Paper Chemistry, Appleton, Wisconsin

Irving S. Goldstein   North Carolina State University, Raleigh, North Carolina

Murray L. Laver   Oregon State University, Corvallis, Oregon

G. D. McGinnis†   Mississippi State University, Mississippi State, Mississippi

Current affiliation:
*James River Corporation, Camas, Washington
†Michigan Technological University, Houghton, Michigan

Arno P. Schniewind   University of California, Berkeley, California

Ronald Sederoff   North Carolina State University, Raleigh, North Carolina

F. Shafizadeh[*]   University of Montana, Missoula, Montana

Richard J. Thomas   North Carolina State University, Raleigh, North Carolina

Norman S. Thompson   Institute of Paper Science and Technology, Atlanta, Georgia

Roy L. Whistler   Purdue University , West Lafayette, Indiana

Eugene Zavarin   University of California, Richmond, California

---

[*] Deceased.

Wood
Structure
and
Composition

1

Overview of the Chemical Composition of Wood

Irving S. Goldstein

North Carolina State University, Raleigh, North Carolina

I. INTRODUCTION

The purpose of this brief introductory chapter is to provide an overview of the composition of wood as a frame of reference to which each of the following chapters, which treat specific components or aspects of wood in depth, can be related. Armed with this background, a newcomer to wood chemistry can approach the detailed chapters in any order while retaining a perspective on their place within the overall framework.

Webster's defines wood as the hard fibrous substance, basically xylem, that makes up the greater part of the stems and branches of trees and shrubs beneath the bark. Woody plants are vascular, possessing conducting tissues consisting of wood (xylem) and bark (phloem); perennial; and exhibit stem thickening, the division of a growing layer (cambium) that annually forms new wood and new bark which are inserted between the older wood and bark.

The wood and bark are made up of individual cells that together determine their morphology. Cell walls may comprise as much as 95% of the mass of the woody plants. The chief components of woody plant cell walls are cellulose, hemicelluloses, and lignin. However, in bark these structural components may be overshadowed by suberins and phenolic acids. Wood components that are more variable in structure or quantity are extractives, the soluble compounds that are often species- or genus-specific.

1

II. CELLULOSE

Cellulose, the chief cell wall component (40–45%) and skeletal polysaccharide, is a long-chain polymer of β-D-glucose in the pyranose form linked together by (1→4) glycosidic bonds to form cellobiose residues that are the repeating units in the cellulose chain. The β configuration imposes a rotation of 180° on alternating glucose units. As a result, the linear cellulose chains are stiff and straight in contrast to the helical conformation of the α-linked amylose fraction of starch.

This cellulose structure favors the organization of the individual cellulose chains into bundles with crystalline order held together by hydrogen bonds, leading to a fibrous state. X-ray diffraction studies have shown that the small crystallites are oriented parallel to the fiber axis, with the length of the unit cell the cellobiose repeating unit. Cellulose is polymorphic. Four different forms have been reported, depending on the previous treatment of the samples. Native cellulose (cellulose I) is changed by strong alkali as in mercerization or regeneration to cellulose II, which has the same fiber axis repeating unit of 10.3Å, but with changes in the other dimensions of the unit cell.

Cellulose consists not only of these highly ordered crystalline regions, but contains disordered or amorphous regions as well. The degree of crystallinity depends on the origin and history of the cellulose, and decreases in the order cotton, wood pulp, mercerized cellulose, and regenerated cellulose.

Degree of polymerization (DP) as determined by various solution techniques also varies with origin and history, since the separation of the cellulose from accompanying polymers in its native state and its subsequent dissolution inevitably bring about some depolymerization. DP values can range from less than 1000 for regenerated cellulose to 7–10,000 for wood pulp, and as high as 15,000 for cotton.

Despite the fact that it is a polymer of glucose, cellulose is insoluble in water. Hydrogen bonding between the cellulose chains is so strong that water cannot disrupt it by complexing with the hydroxyl groups. Other reagents, however, including strong acids and bases, concentrated salt solutions, and various metal complexes are able to dissolve cellulose.

The presence of the three hydroxyl groups on each anhydroglucose residue in the cellulose chain does make cellulose very hygroscopic, and it readily adsorbs and desorbs water in the amorphous regions where the hydroxyls are not involved in interchain bonding. Reagents that interact with the hydroxyl groups must first penetrate the structure, so accessibility or availability of the hydroxyl groups is an important factor in all cellulose reactions. Cellulose derivatives may be prepared by esterification, etherification, xanthation, and grafting.

Some reactions of cellulose involve degradation, which can be desirable in controlling the viscosity and solubility of cellulose derivatives or producing sugars, or can be undesirable when the mechanical properties are adversely affected. Hydrolysis by acids or enzymes breaks the acetal links between the glucose units, and when carried to completion, yields glucose. The biodegradation of cellulose by microorganisms is an intrinsic part of the carbon cycle.

III. HEMICELLULOSES

Associated with the cellulose in the cell wall are carbohydrate polymers known as hemicelluloses. They consist, for the most part, of sugars other than glucose, both pentoses and hexoses, and are usually branched with DP ranging from less than 100 to about 200 sugar units.

In softwoods, galactoglucomannans are the principal hemicelluloses (about 20%) with smaller amounts (5–10%) of arabinoglucuronoxylan also present. The backbone of the glucomannans is a chain of (1→4) linked β-D-glucopyranose and β-D-mannopyranose units. To this are linked by (1→6) bonds α-D-galactopyranose residues. The C-2 and C-3 positions in the mannose and glucose units are partially substituted by acetyl groups. The arabinoglucuronoxylan consists of a backbone chain of (1→4) linked β-D-xylopyranose units partially substituted at C-2 by 4-O-methyl α-D-glucuronic acid groups and at C-3 by α-L-arabinofuranose units. The principal constituent sugars in softwood hemicelluloses in decreasing abundance are thus mannose, xylose, glucose, galactose, and arabinose.

In hardwoods, glucuronoxylan is most abundant (15–30%) with a much smaller quantity (2–5%) of glucomannan. The glucuronoxylan backbone consists of β-D-xylopyranose units linked by (1→4) bonds. The xylose units are partially substituted by (1→2)) 4-O-methyl α-D-glucuronic acid residues and acetyl groups at the C-2 or C-3 positions. The glucomannan is composed of β-D-glucopyranose and β-D-mannopyranose linked by (1→4) bonds. In hardwood hemicelluloses, the principal constituent sugars in decreasing abundance are xylose, mannose, glucose, and galactose, with minor amounts of arabinose and rhamnose.

Although the hemicelluloses are with certain exceptions insoluble in water, they can be dissolved in strong alkali. Hemicelluloses are also more readily hydrolyzed by acid than cellulose. Their greater solubility and susceptibility to hydrolysis than cellulose result from their amorphous structures and low molecular weights.

IV. LIGNIN

Lignin is the third major wood cell wall component (20–30%). It serves as a cement between wood fibers, as a stiffening agent within fibers, and as a barrier to the enzymatic degradation of the cell wall.

Lignins are three-dimensional network polymers of phenylpropane units with many different linkages between the monomers leading to a complicated structure that can only be defined by the frequency of occurrence of the various linkages. This random structure arises from an enzymatically initiated free radical polymerization of lignin precursors in the form of p-hydroxycinnamyl alcohols. In conifers, guaiacyl lignin is formed from coniferyl alcohol (3-methoxy-4-hydroxy-cinnamyl alcohol). In hardwoods, guaiacyl-syringyl lignins are formed from coniferyl alcohol and sinapyl alcohol (3,5-dimethoxy-4-hydroxycinnamyl alcohol). Thus, softwood and hardwood lignins differ in methoxyl content and in the degree of cross-linking. Hardwood lignins have more methoxyls, but these groups block potential reactive sites and reduce cross-linking.

The aromatic and phenolic character of lignin stems from its origin, as does its methoxyl content. More than two-thirds of the phenylpropane units in lignin are linked by ether bonds, the rest by carbon–carbon bonds. This explains the stringent conditions necessary for its depolymerization and the inability to bring about reversion to monomers. The most abundant bond type is the arylglycerol-β-aryl ether (48% in spruce, 60% in birch). Biphenyl linkages are more important in spruce (9.5–11%) than in birch (4.5%); diaryl ether more important in birch (6.5%) than in spruce (3.5–4%); phenylcoumaran more prevalent in spruce (9–12%) than in birch (6%); and noncyclic benzyl aryl ether and 1,2 diarylpropane bonds the same in both species (about 7%). There is evidence that covalent linkages exist between lignin and the hemicelluloses.

Since it is not possible to isolate lignin or remove it from wood without some degradation, the molecular weight of any lignin sample studied will depend on its previous history. Soluble lignin preparations exhibit molecular weight variations ranging from below a thousand to over a million daltons and are polydisperse within the same preparation. The chemical properties of lignins will also vary with their method of isolation.

V. CELL WALL STRUCTURE

Dissolution of either the lignin or carbohydrate portions of the wood cell wall leaving a carbohydrate or lignin skeleton has shown the intimate association of lignin and carbohydrate within the cell wall at the ultramicroscopic if not the molecular level. Cellulose occurs in microfibrils that are surrounded by the hemicelluloses and lignin. The cell wall may be likened to a fiber-reinforced plastic with crystalline cellulose fibers embedded in an amorphous matrix of hemicelluloses and lignin.

The composition is not uniform across the cell wall. Wood cells are bound together by a region called the middle lamella that consists principally of lignin. However, only about 25% of the total lignin is accounted for by this layer since it is very thin. The primary wall is also a thin layer containing randomly oriented cellulose microfibrils and is also highly lignified. The secondary wall contains most of the wood substance and about 75% of the lignin. It consists of three parts: thin inner and outer layers with helical cellulose microfibrils oriented almost perpendicular to the fiber axis, and a thick central layer with helical cellulose microfibrils oriented nearly parallel to the fiber axis.

The intimate association of the polymers in the cell wall influences their properties and response to processing. Separation of the components is difficult and expensive.

VI. BARK

Depending on the species, bark may comprise 10–15% of the total mass of the tree. Bark fibers are chemically similar to wood fibers and contain cellulose, hemicelluloses, and lignin, but in much lower concentrations. Because of the presence of extractives and the two other components unique to bark (suberins and phenolic acids), the cellulose content may be as low as 20–30%.

Suberins are found in the bark cork cells. They consist of long-chain (16–22 carbon atoms) ω-hydroxy monobasic acid esters. In addition, they contain α, β-dibasic acids esterified with diols as well as with monomeric phenylpropane phenol acids. The suberin content varies from 2–8% in the barks of pine, fir, aspen, and oak, whereas it is much more abundant in birch (20–40%) and cork oak (35–40%).

Phenolic acids are high-molecular-weight phenols that are often confused with lignin because of their insolubility in 72% sulfuric acid, but they differ from lignin in their lower molecular weight, alkali solubility resulting from a high carboxyl content, and lower methoxyl content. They are easily extracted from bark by alkali and may comprise almost 50% of the weight of the bark of conifers.

VII. EXTRACTIVES

Extractives are the extraneous wood components that may be separated from the insoluble cell wall material by their solubility in water or organic solvents. They are often genus- or species-specific, and it is possible to use their presence and abundance in taxonomic schemes based on chemical composition. Classification of extractives is made difficult by their great variety. Chemical classification makes use of the similarity of entire chemical structures to sort them into categories, but within these categories are found an extraordinarily large number of individual compounds. The location of the extractives may be in the heartwood, the resin canals of conifers, or as reserve materials in the living portion of the wood (sapwood).

Major categories of extractives include volatile oils, terpenes (turpentines, resin acids, sterols), fatty acids and their esters, waxes, polyhydric alcohols, mono- and polysaccharides, alkaloids, and aromatic compounds (acids, aldehydes, alcohols, phenylpropane dimers, stilbenes, flavonoids, tannins, and quinones).

2

Wood: Formation and Morphology

Richard J. Thomas

North Carolina State University, Raleigh, North Carolina

I. INTRODUCTION

The main stem of a tree is composed of wood and bark. Figure 1, which illustrates the end of the main stem, reveals that onsiderably more wood is present than bark. This results from the fact that more wood than bark cells are produced and also because wood cells are retained, whereas bark cells are constantly lost from the outermost portion of the bark.

In the living tree, wood performs the functions of conduction, support, and storage. The conduction role involves the movement of water from the roots up the woody stem to the leaves. The heights of trees and the weight not only of the stem but associated branches and foliage, as well as external loading (wind, ice, etc.), require a strong support system. The third main function of wood is the storage of food in the form of starch. Cells that perform the conduction and support roles are dead, whereas only living cells carry out the storage role. The death of the supporting and conducting cells usually occurs from 14 to 21 days after cell formation. Food-storing cells, on the other hand, remain alive from a few to many years, depending on species. Because of the close relationship between "form" and "function," an awareness of the functions of wood assists one in the study of wood anatomy.

7

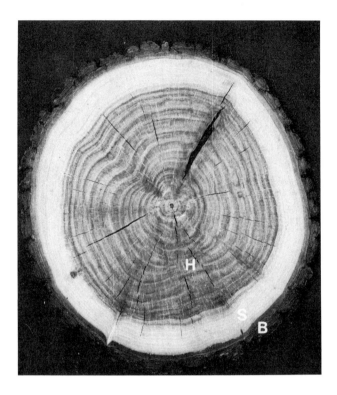

Figure 1   End view of a tree stem showing bark (B) and wood (H and S). The central darker-colored portion is called heartwood (H) and the outer light-color wood is termed sapwood (S).

II. GROSS MORPHOLOGICAL FEATURES

A. Sapwood and Heartwood

The end view of the tree stem depicted in Fig. 1 reveals that the innermost wood is considerably darker in color than the portion adjacent to the bark. The light-colored area is termed sapwood and the dark-colored portion heartwood. The sapwood, the most recently formed wood, performs the roles of support, conduction, and food storage. After some time elapses, the amount of which is species-dependent, the food-storing cells die. The death of these cells is accompanied by the secretion of oxidized phenols and in those woods that have a dark-colored heartwood, pigments. These secreted materials are termed extractives. In some species, heartwood formation is not accompanied by a change in color as food-storing cells die. However, because all the cells are physiologically dead, the area is technically heartwood. Since only the very outermost portion of the wood conducts and the food-storing cells are dead, heartwood functions in the support role only.

In many species, the extractives are toxic to decay organisms and substantially increase heartwood decay resistance. However, heartwood decay resistance is not related to the intensity of the color of heartwood. For example, the dark-colored heartwood of redwood and black locust is quite durable, whereas the dark-colored heartwood of sweetgum is not. On the other hand, the light-colored heartwood of white cedars is durable. Thus, decay resistance is directly linked to extractive toxicity rather than depth of color.

As new wood, which is sapwood, is formed between the existing wood and bark, additional interior sapwood adjacent to the heartwood zone is converted to heartwood. The proportion of sapwood to heartwood is variable. Some species are composed almost entirely of heartwood with only a very narrow band of sapwood, whereas others possess a small amount of heartwood.

Since sapwood contains the only living cells found in mature wood and they constitute, depending on species, 10–40% of the sapwood volume, it is obvious that the vast majority of cells that compose the wood portion of a living tree are dead. As more heartwood is formed and living food storage cells die, the percentage of living cells in the woody stem continues to decrease.

B. Softwoods and Hardwoods

Trees are classified into softwoods (gymnosperms) and hardwoods (angiosperms). The wood of hardwoods contains vessels that are responsible for water conduction and fibers that perform the support role. Softwoods contain cells called longitudinal tracheids that have the dual roles of conduction and support. Another classification is based on the retention of leaves by most softwoods, as opposed to the annual leaf shedding by most hardwoods. Thus, softwoods are called evergreen trees and hardwoods deciduous trees.

Although the average specific gravity of commercially important domestic hardwoods (0.50) is greater than it is for softwoods (0.41), the softwood–hardwood classification should not be taken exclusively as a measure of hardness as considerable overlap occurs in the range of specific gravities (softwoods 0.29–0.60; hardwoods 0.32–0.81). Thus, some softwoods are harder than some hardwoods and some hardwoods softer than some softwoods.

C. Growth Rings

Each growing season, trees insert a new layer of wood between existing wood and bark over the entire stem, branches, and roots. This new addition of wood is termed a growth ring or annual increment. The growth rings shown in Fig. 2 are easily detectable because of differences in the wood produced early as opposed to late in the growing season. The lighter-color portion of the growth ring, produced first, is called earlywood or springwood. The darker part, produced last, is termed latewood

Figure 2   End view of a softwood stem showing growth rings. Each growth ring or annual increment consists of a light area called earlywood (E) and a dark area termed latewood (L).

or summerwood. The color difference is due to variations in the structure of earlywood and latewood cells. Note that in Fig. 3, individual cells that constitute the earlywood zone are easily seen; however, it is difficult to detect individual cells in the latewood. At a higher magnification (Fig. 4), individual cells in both zones can be seen. Note that the earlywood cells have a large cross-sectional area, thin cell walls (±2 μm) and a large, open center called the lumen. The large lumen provides an efficient pathway for water conduction. The latewood cells have a smaller cross-sectional diameter, thicker cell walls (up to 10 μm), and a narrower lumen than earlywood cells. The thick cell walls provide substantial support for the tree, but the small lumen indicates less efficient conduction than provided by earlywood cells.

In some softwood species, the latewood zones are very narrow (Fig. 5). In these situations, it is much more difficult to distinguish growth rings since the contrast between the two zones is considerably reduced. Obviously, these woods have a more uniform structure.

Vessels, or pores that are characteristic of hardwoods, are composed of individual vessel segments or cells that possess a large diameter (up to 300 μm) and thin cell walls. Because of their large diameter, vessels in many hardwoods can be easily

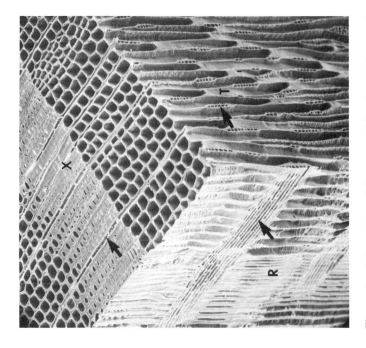

Figure 4 Softwood growth increments. Note the wide latewood zone in cross-sectional view (X). Individual latewood cells with thick walls and small radial diameters are visible. Arrows indicate wood rays in all three planes of study. (R, radial; T, tangential). (Courtesy of N.C. Brown Center for Ultrastructural Studies, SUNY College of Environmental Science and Forestry.)

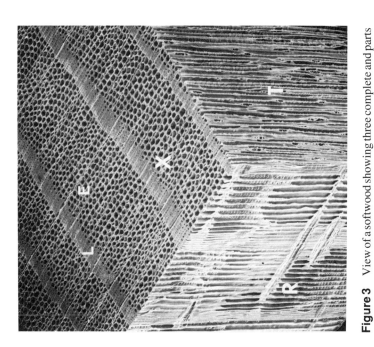

Figure 3 View of a softwood showing three complete and parts of two additional growth rings in the cross-sectional view (X). Individual cells are easily seen in earlywood (E) and difficult to detect in latewood (L). The two longitudinal surfaces (R, radial; T, tangential) are also revealed. (Courtesy of N.C. Brown Center for Ultrastructural Studies, SUNY College of Environmental Science and Forestry.)

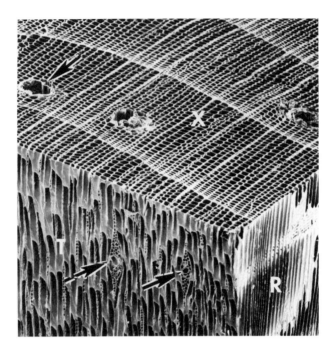

Figure 5 Softwood growth rings with narrow latewood zones in cross-sectional view (X). Arrow in cross-sectional plane (X) indicates a longitudinal resin canal. Fusiform rays (arrows) with embedded transverse resin canals are visible in tangential plane (T). Micrometer bar 100 μm. (Reprinted with permission from R. A. Parham and R. L. Gray, "The Practical Identification of Wood Pulp Fibers," *TAPPI*, Atlanta, Ga. 1982.)

seen by the unaided eye. Hardwoods are classified as ring-porous or diffuse-porous woods, depending on the vessel or pore arrangement within a growth ring. In a ring-porous wood (Fig. 6), the pores formed in the earlywood zone have a diameter considerably larger than the latewood pores. Diffuse porous woods (Fig. 7) reveal uniform pore diameters across the entire growth ring. As a result of this more uniform structure, it is often difficult to detect growth rings in diffuse porous woods. Some hardwoods show a gradual decrease in pore diameter across the growth ring and are termed either semiring or semidiffuse porous woods.

D. Wood Rays

Most wood cells have their long axis oriented parallel to the long axis of the tree stem. However, some cells have their long axis oriented perpendicular to the tree stem. Aggregations of these transversely oriented cells are called rays. At low magnifications, they appear as light-colored lines of varying width traversing

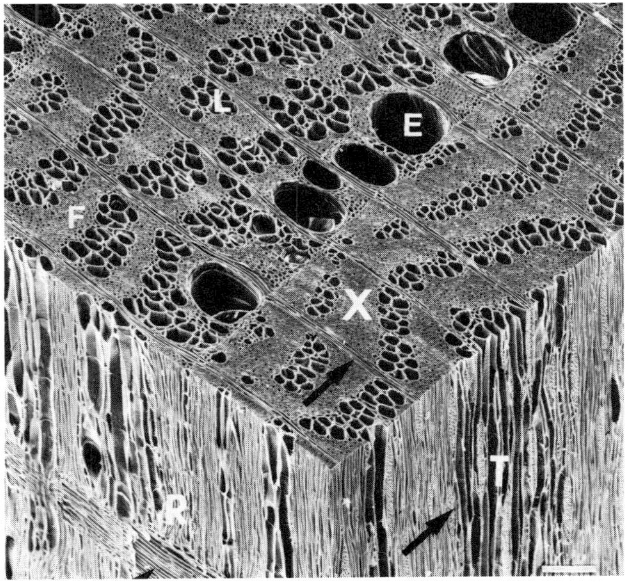

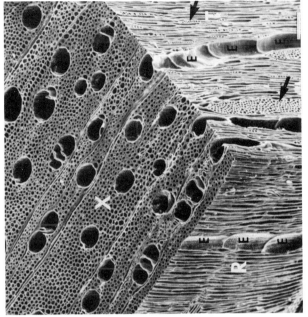

Figure 7 Diffuse-porous hardwood with fairly uniform vessel diameters across entire growth ring. Formation of vessels from individual vessel elements (E) shown in both radial (R) and tangential (T) views. Note presence of one-cell wide and multicell-wide rays in tangential view (arrows). Micrometer, bar 100 μm. (Courtesy of N.C. Brown Center for Ultrastructural Studies, SUNY College of Environmental Science and Forestry.) Micrometer, bar 100 μm.

Figure 6 Ring-porous hardwood illustrating abrupt change in earlywood (E) and latewood (L) vessel diameters in cross-sectional view (X). Fibers (F) can be seen between latewood vessel zones. Wood rays are apparent on all three surfaces (arrows). Micrometer, bar 200 μm. (Courtesy of N.C. Brown Center for Ultrastructural Studies, SUNY College of Environmental Science and Forestry.)

growth rings. In softwood species, the rays are mostly one cell in width, whereas in hardwoods, they vary from one to many cells. Figures 3 through 7 clearly depict the transverse orientation of ray cells in softwoods and hardwoods.

E. Planes of Study

Wood is an anistropic material, that is, it does not have the same properties in all directions. Also, since different aspects of the cellular structure are revealed in different directions, the study of wood anatomy requires knowledge of the three different planes.

Cross-sectional or transverse views are exposed by cuts made at right angles to the long axis of the tree stem. The end of a log (Figs. 1 and 2), or the end of a board, reveals, cross-sectional views. Since most wood cells have their longitudinal axis parallel to the tree stem, the transverse view reveals cells in cross section (Figs. 3, 4, and 5).

The remaining two major planes of study are longitudinal. A longitudinal radial view is exposed when the plane of cut is at right angles to the growth rings and parallel to the rays as seen in the cross-sectional view (Figs. 3, 4, and 5). If the plane of cut is extended across the entire stem, it will pass through the center.

The longitudinal tangential plane is revealed by a plane of cut parallel to growth rings and perpendicular to rays as seen in the cross sectional view (Figs. 3, 4, and 5). The tangential plane is at right angles to the radial plane and also is parallel and some distance from longitudinal plane through the center of the stem. Also note that in the tangential plane, the ray cells are revealed in cross section since their long axes are at right angles to the long axis of the tree stem.

An alternative method of identifying the planes of study utilizes the three orthotropic axes of wood. They are (1) longitudinal (L), parallel to the long axes of longitudinally oriented cells; (2) radial (R), perpendicular to the longitudinal cells and parallel to the rays; and (3) tangential (T), perpendicular to both longitudinal cells and rays. Based on the orthotropic axes, the three planes of study are referred to as the RT (cross-sectional), LR (radial), and LT (tangential) planes.

III. WOOD FORMATION

The increase in both the diameter and length of tree stems and branches is the result of the production of new cells by tissues termed meristems. Meristems consist of cells that are undifferentiated and retain, throughout their life, the ability to divide and produce new cells. After each cell division, one cell remains meristematic and the other eventually differentiates into a mature cell. The increase in the height of the main tree stem as well as the length of all branches and roots is designated as primary growth and is the result of cell production by apical meristems located at the tip or apex of the main stem, branches, and roots. Just below the apical meristems, some of the cells produced differentiate and form a lateral meristem called the cambium. The newly formed cambium is continuous with and thus part of the cambium formed during previous growing seasons. Cell division by the cambium, located between the wood and bark, inserts new wood and bark cells between existing wood and bark cells. This activity increases the diameter of the stem, branches, and roots and is referred to as secondary growth.

Fig. 8 shows the superimposed, cone-shaped layers of wood produced each year by a growing tree. Note the manner in which the main stem or trunk increases in diameter over a period of six years. The end of each cone reveals the location of the apical meristem at the end of the preceding growing season. Also shown is that the growth of branches, which except for the fact that they for the most part grow horizontally rather than vertically, is the same as growth in the main stem.

Wood consists of two interpenetrating systems of cells: one with the long axes of the cells oriented logitudinally to the main tree stem and the other transversely. Longitudinal cells are produced by cambial cells designated as fusiform initials and transverse cells by ray initials.

The phases of development for wood cells are as follows:

1.   cell division
2.   cell enlargement
3.   cell wall thickening
4.   lignification
5.   death

The first phase, cell division, occurs when a cambial initial divides, forming two cells. If the outermost cell (adjacent to the bark) remains meristematic, the innermost cell will develop into a mature wood cell as it passes through the remaining four phases of development. On the other hand, if the innermost cell remains meristematic, the outermost cell will differentiate into a bark cell. The production of wood cells and their subsequent enlargement move the cambium further away from the center of the tree (Fig. 9).

During the cell enlargement phase, the developing cell grows in length and diameter. Softwood cells increase their diameter primarily in the radial direction only and therefore maintain a tangential diameter essentially the same as the cambial initial from which it was formed. As a result, a rather uniform alignment of cells in the radial direction occurs (Fig. 5). In hardwoods, the transverse enlargement of vessels occurs in both the radial and tangential directions. Growth in the tangential direction pushes adjacent cells aside and disrupts the radial alignment of the cells (compare Fig. 5 with Figs. 7 and 27).

After cell division and during the enlargement phase, a very thin and plastic cell wall, called the primary wall, encases the protoplasm. During the cell wall thickening phase, wall thickness is increased by the addition of the secondary wall. The amount of cell wall thickening depends on the time of year the cell is formed

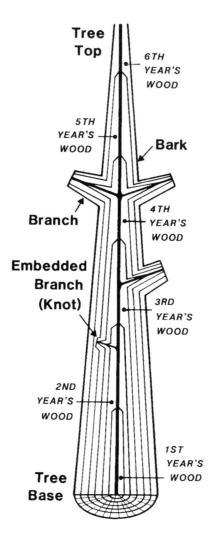

Figure 8 Diagram of growth increments in a tree illustrating their deposition as a series of inverted hollow cones. This diagram represents a longitudinal section of a six-year-old stem. Reprinted with permission from R.A. Parham and R. L. Gray, "The Practical Identification of Wood and Pulp Fibers" *TAPPI*, Atlanta, Ga, 1982.)

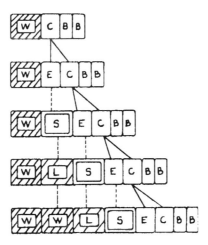

Figure 9 Sequence of wood cell development. B, bark cell; C, meristematic cambium cell; E, enlarging wood cell; S, thickening of cell wall; L, lignification of wood cell wall, W, mature cell. Note the displacement of cambium and bark cells as new wood cells inserted between existing wood and cambium.

(earlywood or latewood) or the function of the cell, or both. Once cell wall thickening occurs, no further growth of the cell is possible.

The last cell development phase, called lignification, involves the formation of lignin between the newly formed cells and within their cell walls. Early and rapid lignification of the areas between the cells, designated middle lamella, and the primary wall takes place (1, 2). Secondary wall lignification is a more gradual process, initiated when the middle lamella lignin concentration is approximately 50% of the maximum (3) and moves into the secondary wall, lagging just behind secondary wall formation.

For most wood cells, death occurs immediately after lignification. However, for those wood cells that perform the function of storage, this step is postponed for an indefinite period. Cells performing the support and conduction roles usually pass through the five phases of cell development in 14 to 21 days.

Although the development of wood cells is classified into distinct phases, considerable overlap, especially among the enlargement, wall thickening, and lignification phases takes place. For example, cell enlargement at the ends of the cell and secondary thickening in the middle region may occur simultaneously. Also, as noted above, lignification of the secondary wall is initiated in the outermost layer of the wall while secondary wall formation is still underway adjacent to the lumen. Wood and bark cell production as described above increases the diameter of the stem. Obviously, as stem diameter increases, the circumference must also increase and the cambium must make adjustments to provide for this enlargement. When the fusiform initial produces new wood or bark cells, the new wall formed during cell division is in the tangential plane and therefore parallel to the bark (periclinal division). To increase the circumference, the fusiform initial will divide by creating a new cell wall in the radial plane, that is, perpendicular to the bark (anticlinal division), forming two fusiform initials. As these cells begin to form new wood cells through periclinal divisions, an additional radial file of wood cells is created. This action on the part of the cambium accommodates the increase in circumference and prevents radial cracks from forming.

IV.   WOOD MICROSTRUCTURE

An obvious difference revealed at the gross structural level between softwoods and hardwoods is the presence of vessels in hardwoods. In some hardwood species, vessels can be seen by the unaided eye. At the microstructure level, more differences are detectable. For example, the number of different cell types is greater for hardwoods than for softwoods and leads to a more complex and varied structure. Because of their differences, the anatomy of softwoods and that of hardwoods are described separately.

A.   Softwood Microstructure

Although seven different softwood cell types exist, most softwood species contain only two types, longitudinal tracheids and ray parenchyma cells, and none contain all seven cell types. Panshin and deZeeuw (4) classified softwood cell types by orientation and function of the cells. A modification of their classification system is presented in Table 1.

Cells with their long axis oriented parallel to the tree stem are classified as longitudinal cells, whereas those oriented perpendicular to the stem are classified as transverse cells. In addition within each orientation group, the cells are further classified with regard to their function in the living tree.

1.   Longitudinal Cells

Of the four longitudinal cell types listed in Table 1, the longitudinal tracheid is the most important as it is found in all softwood species, occupies the greatest volume, and is the largest cell. Two of the remaining three cell types, strand tracheids and longitudinal tracheids, are not found in most species. When present, they occur in small amounts. Longitudinal epithelial cells are associated with longitudinal resin canals and are found only in pines, spruces, larches and Douglas fir.

Table 1 Softwood Cell Types

Longitudinal	Transverse
A. Support, conduction, or both	A. Support, conduction, or both
1. *longitudinal tracheid*	1. ray tracheids
2. strand tracheid	
B. Storage or secretion	B. Storage or secretion
1. longitudinal parenchyma	1. *ray parenchyma*
2. epithelial	2. epithelial

a. The Longitudinal Tracheid

(1) Volume, Size, and Shape. The longitudinal tracheid is a general-purpose cell in that it performs the dual roles of support and conduction. The proportion of wood volume occupied by longitudinal tracheids varies from 90–94% (4). Figures 3 through 5 illustrate the high volume of wood composed of longitudinal tracheids. Note also, particularly in Fig. 3, the very long length of the tracheids with respect to their width. As a general rule, longitudinal tracheids are 100 or more times longer than they are wide. For commercially important softwoods, the average tracheid length varies from 3.0–5.0 mm. The variation is species-dependent. Average tracheid diameters, as measured in the tangential direction, range from 15–65 μm (4). Redwood may have individual tracheids up to 80 μm in diameter, whereas the spruces and cedars will not exceed 35 μm.

The shape of longitudinal tracheids depends on the plane within which they are viewed and as to whether they are earlywood or latewood tracheids. In the cross-sectional, view earlywood tracheids are generally hexagonal and latewood tracheids rectangular (Fig. 4). Also note the greater cell wall thickness and smaller lumen diameter in latewood tracheids vs. springwood tracheids. In longitudinal views, earlywood tracheids are rounded in the radial plane and pointed in the tangential plane, whereas latewood tracheid ends are pointed in both the radial and tangential planes (Fig. 10). Note also that the tangential diameters of earlywood and latewood cells are essentially the same, but the radial diameter is smaller for latewood cells than for earlywood cells (Figs. 4 and 5).

In summary, the longitudinal tracheid is a cell many times longer than wide, with an open space in the middle for efficient water conduction up the tree stem. The open space, termed the lumen, is surrounded by rigid cell walls that provide support for the stem.

(2) Cell Wall Features. The most prevalent and obvious features of the tracheid cell wall are pits. Since tracheid lengths are insignificant when compared to total tree height, a pathway for liquid flow must exist among tracheids if water is to be conducted up the tree.

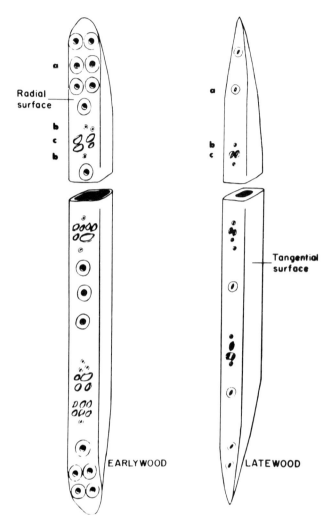

Figure 10 Drawing of isolated earlywood and latewood longitudinal tracheids that resemble elongated cylindrical tubes. Entire tracheids not shown as tracheids are at least 100X longer than wide. (a) bordered pits to adjacent longitudinal tracheids; (b and c) pits to adjacent ray cells. From E. T. Howard and F. G. Manwiller, *Wood Science*, 2:77. (1969).

Pits that provide the pathway are defined as a gap or a recess in the cell wall, open internally to the lumen and bounded externally by a membrane.

(a) Bordered Pit Pairs.   Note the numerous dome like structures on the radial walls of the longitudinal tracheids in Fig. 11. Although bordered pits are distributed

throughout the tracheid length, a greater concentration occurs near the ends of tracheids. Also fewer and smaller pits are characteristic of latewood tracheids (Fig. 10).

Fig. 12 depicts a bordered pit as viewed from the lumen of a tracheid. The domelike structure is called the pit border and the circular opening is termed the pit aperture. Removal of the pit border exposes the pit membrane (Fig. 13). The central, nonperforated area is called the torus, and the outer perforated region the margo. Liquid flow occurs readily through the margo openings. Since each bordered pit usually has a complimentary pit in the contiguous cell, liquid flows from the lumen of the tracheid through the aperture and the margo region and out the aperture into the lumen of the adjacent tracheid. Cross-sectional and longitudinal views of bordered pit pairs connecting tracheids are illustrated in Fig. 14. A single bordered pit pair as seen in cross section is depicted in Fig. 15. Note the thickness of the torus relative to the margo region. When the pit membrane is centrally located in the pit

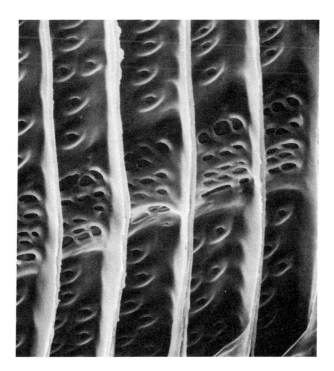

Figure 11   Longitudinal view of a portion of earlywood longitudinal tracheids as seen from their lumens. The circular domelike structures are bordered pits that allow liquid flow between contiguous tracheids. The smaller elliptical-shaped pits lead from the longitudinal tracheids to adjacent, transversely oriented ray cells. (Courtesy of N.C. Brown Center for Ultrastructural Studies, SUNY College of Environmental Science and Forestry.)

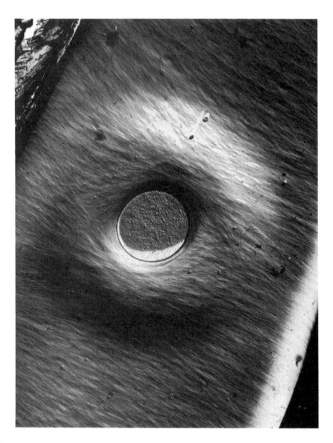

Figure 12   Bordered pit as seen from cell lumen. The domelike structure is called the pit border and the circular opening in the border is termed the pit aperture. The stringlike microfibrils that constitute the innermost wall layer (S_3) are oriented approximately 90° to the long axis of the cell. Micrometer bar 1 μm.

chamber (Fig. 15), the pit is in the nonaspirated condition and provides a pathway for liquid flow. However, during wood drying, as air–water menisci are pulled through the small openings in the margo, the pit membrane is displaced to one side of the pit chamber, effectively sealing the pit aperture with the torus (Fig. 16). Hart and Thomas (5) and Thomas and Kringstad (6) have provided a complete description of the mechanism of pit aspiration. Comparison of an aspirated pit pair (Fig. 16 and 17) with a nonaspirated pit pair (Figs. 13 and 15) reveals pit membrane displacement and the very tight seal between the torus and pit border in the vicinity of the pit aperture. In the event a living tree is damaged such that air enters the water-conducting pathways, pit aspiration confines the air embolism to a small area. Thus,

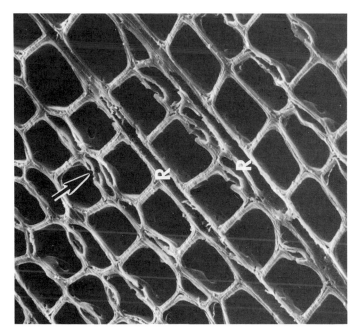

Figure 14 Cross-sectional view of earlywood longitudinal tracheids revealing both cross-sectional (arrow) and longitudinal views of bordered pits. Note the narrow, radially elongated, transversely oriented parenchyma cells that constitute rays (R). Small pits connecting ray cells and longitudinal tracheids are visible on radial walls of the longitudinal tracheids. (Courtesy of N.C. Brown Center for Ultrastructural Studies, SUNY College of Environmental Science and Forestry.)

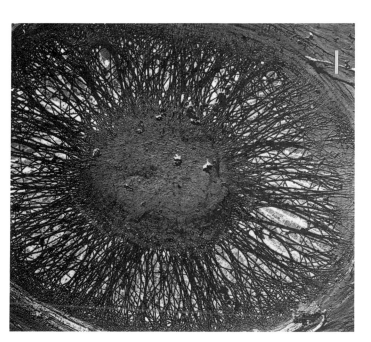

Figure 13 Bordered pit membrane as seen after removal of the pit border. The central, nonperforated region is the torus. The perforated outermost area is the margo. Water flows freely from cell to cell through the openings in the margo. The inside of the pit border from the adjacent cell can be seen through the margo openings.

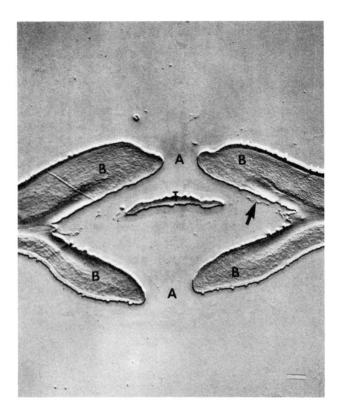

Figure 15 Cross-sectional view of a bordered pit pair. A, pit aperture; B, pit borders; T, torus. Most of the thin margo was disrupted during specimen preparation, some portions are visible (arrow). Liquid flows through the aperture around the torus through the margo, and out the other aperture into the adjacent cell. Micrometer bar 1 μm. Reprinted with permission from R. J. Thomas, in *Wood Structure and Chemical Composition in Wood Technology Chemical Aspects* (I. S. Goldstein, ed.), ACS Symposium Series 43, 1977.

water conduction is not interrupted throughout the entire stem. However, wood drying causes the vast majority of pit membranes to aspirate. Any subsequent process involving the penetration of liquids is obviously more difficult to perform after the wood is dried due to the reduction of liquid flow pathways resulting from pit aspiration.

(b) Pitting from Tracheids to Ray Parenchyma. Also obvious on longitudinal tracheid walls are pits that connect longitudinal tracheids to the transversely oriented ray parenchyma cells (Figs. 11 and 18). The function of these pits is not clear. Note that the membrane is, as expected, not perforated in order to provide protection to the protoplasm of the living parenchyma cells from the harsh

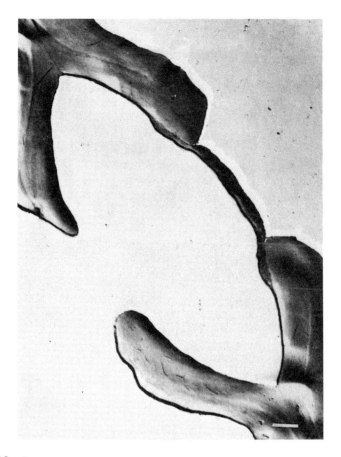

Figure 16   Cross-sectional view of an aspirated bordered pit pair. The pit membrane has moved to the border and sealed the aperture with the torus. In this condition, free liquid flow no longer occurs between contiguous cells. Micrometer bar 1 μm. Reprinted with permission from R. J. Thomas, in *Wood Structure and Chemical Composition in Wood Technology Chemical Aspects* (I. S. Goldstein, ed.), ACS Symposium Series 43, 1977.

environment of a tracheid lumen. The rather thick and nonperforated pit membrane precludes free liquid flow between ray parenchyma and longitudinal tracheids.

Due to the diverse structure of longitudinal to ray parenchyma pit pairs among species, they are useful for wood identification purposes. However, since wood identification is beyond the scope of this review, a description of the various types is not included. Readers interested in this topic should see Panshin and deZeeuw (4) or Phillips (7), or both.

Although additional cell wall features such as spiral thickenings, trabeculae, and

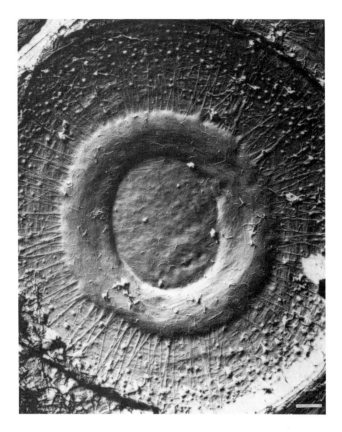

Figure 17 Longitudinal view of an aspirated bordered pit membrane. Note the imprint of the aperture through the torus as a result of the extremely tight seal between the torus and border. Micrometer bar 1 µm.

others are present in softwoods, their occurrence is not of sufficient magnitude to be considered. Descriptions of these features and others can be found in Panshin and deZeeuw (4).

b. Other Longitudinal Cells.   Of the three remaining longitudinal cells listed in Table 1, the strand tracheid and longitudinal parenchyma are never very abundant. A complete description of these cell types and their distribution can be found in Panshin and deZeeuw (4).

The longitudinal epithelial cells are secreting cells that surround longitudinal resin canals. Since longitudinal resin canals, in reality intercellular spaces, comprise 1% or less of softwoods, it is obvious that only a small amount of longitudinal epithelial cells is present. The description of transverse epithelial cells that follows

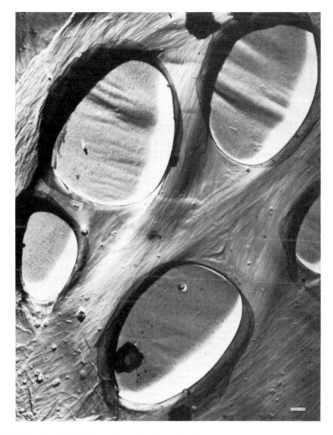

Figure 18 Longitudinal view from lumen of a longitudinal tracheid showing pits that connect a longitudinal tracheid to a ray cell. The pit membranes do not reveal any openings. Micrometer bar 1 µm. Reprinted with permission from R. J. Thomas, in *Wood Structure and Chemical Composition in Wood Technology Chemical Aspects* (I. S. Goldstein, ed.), ACS Symposium Series 43, 1977.

can be applied to longitudinal epithelial cells.

2. Transverse Cells

From the standpoint of volume of wood occupied and their presence in all softwood species, the most important transversely oriented cell type is the ray parenchyma. The transverse epithelial cells are found only in those groups possessing resin canals (pines, spruces, larches, and Douglas fir) and comprise a very small volume of wood. Ray tracheids are present in an even smaller volume and are found consistently in pines, spruces, larches, hemlocks, Douglas fir and Alaska yellow cedar (4).

a. Rays.   Transversely oriented cells form structures called rays. Softwoods may contain only uniseriate rays or uniseriate and fusiform rays. A uniseriate ray is one cell in width and does not contain a resin canal (Fig. 4). Fusiform rays contain a resin canal and are characterized by a fusiform shape as seen in the tangential plane (Fig. 5). As fusiform rays contain a resin canal, they are found only in those woods possessing resin canals as a normal feature (pines, spruces, larches, and Douglas fir).

The average ray volume for commercially important softwoods found in the United States is 7.0%, varying from 3.4–11.7% (4). Fusiform rays constitute less than 1% of these values (4).

(1) Uniseriate Rays.   With the exception of Alaska yellow cedar, uniseriate rays are composed either entirely of ray parenchyma or both ray parenchyma and ray tracheids. When both types of cells are present in the same ray, the ray tracheids are generally confined to the top and bottom of the ray. In Alaska yellow cedar, however, some uniseriate rays are comprised of ray tracheids only.

The variation in ray height is of some value in differentiating between softwoods. In some species ray heights are in excess of 60 cells, whereas in others ray heights average only 6 cells. As an overall average, 10–15 cells has been cited (4).

(2) Fusiform Rays.   Fusiform rays are composed of ray tracheids, ray parenchyma, and epithelial cells. Ray tracheids, as in uniseriate rays, are generally confined to the top and bottom of the fusiform ray, whereas ray parenchyma are found in both the narrow and wide portions of the ray. Also, the resin canal surrounded by epithelial cells is located in the wide portion.

b. Ray Parenchyma.   As viewed in the cross-sectional and radial plane, ray parenchyma appear as rectangular-shaped cells. They are approximately eight times longer than wide, with typical lengths ranging from 0.3 mm–0.8 mm, a size considerably less than that for longitudinal tracheids. Thin cell walls through which simple pit pairs provide connections characterize ray parenchyma. Simple pit pairs show little or no change in the diameter of the pit chamber from the membrane to the aperture. Recall that for bordered pit pairs a drastic decrease occurs, resulting in the formation of borders overarching the pit chamber. The pit membrane in simple pits contains no detectable openings and therefore does not appear to provide a passageway for liquid flow.

c. Ray Tracheids.   Ray tracheids are more or less rectangular-shaped cells shorter than ray parenchyma cells, with lengths from 0.1–0.2 mm. They possess bordered pits considerably smaller than longitudinal tracheid bordered pits. The pit membranes are characterized by a torus and a margo with very small openings (Fig. 19).

d. Epithelial Cells.   Epithelial cells line the walls of resin canals and secrete oleoresin into the canals. As viewed longitudinally, epithelial cells range from square to hexagonal. In pines, epithelial cells are thin-walled and apparently unpitted, whereas in the spruces, larches, and Douglas fir, the epithelial cells are

Figure 19 Longitudinal tracheid to ray tracheid pit membrane. Note the very dense network of microfibrils in the margo, resulting in a considerable reduction of margo openings. Micrometer bar 1 μm.

thick-walled and reveal pits. Because pines have the largest and most numerous resin canals, they have the highest volume of epithelial cells.

3. Resin Canals

Resin canals are tubular, intercellular spaces surrounded by epithelial cells (Fig. 5). Epithelial cells, derived from fusiform initials in the case of longitudinal canals and from ray initials for transverse canals, separate, leaving a resin canal in their midst.

Resin canals that are a normal feature of wood are found in both the longitudinal and transverse directions. As noted earlier, transverse canals are always embedded in fusiform rays. The pines, spruces, larches, and Douglas fir are species within which resin canals are normal features. Resins canals are larger, more abundant, and more evenly distributed in pines than in the other groups. Also, longitudinal resin canals are normally larger in diameter than transverse canals (Fig. 5).

Some softwoods form resin canals only as a result of injury. These canals are called traumatic resin canals. Longitudinal and transverse traumatic canals do not occur in the same sample. The epithelial cells of traumatic canals are normally thick-walled and appear to be lignified.

4. Summary: Softwood Microstructure

Softwoods consist primarily of longitudinal tracheids. These cells, in conjunction with a small number of other cells (strand tracheids, longitudinal parenchyma, and epithelial) that may or may not be present, comprise the vertical or longitudinal system of cells. The horizontal or transverse system consists primarily of ray parenchyma and a small amount of ray tracheids and epithelial cells found in some species. In simplified form, softwoods consist of two interpenetrating systems, that is, a longitudinal system composed of nonliving tracheids and a transverse system of rays composed of living ray parenchyma cells.

Earlywood longitudinal tracheids are larger in diameter and have thinner walls and a larger lumen than latewood tracheids. The walls of both are characterized by the presence of bordered pits, structures that permit liquid flow from tracheid to tracheid. With their long length, overlapping ends through which bordered pits provide a passageway for liquid flow, and their rigid cell walls, longitudinal tracheids are very effective in carrying out their dual roles of conduction and support in the living tree.

The living parenchyma cells, which are considerably smaller than longitudinal tracheids, perform the food-storing role. Their thin cell walls contain simple pits with unperforated membranes.

B. Hardwood Microstructure

Hardwoods are more complex than softwoods since they not only contain more cell types but show considerable variation in size, shape, and arrangement of the individual cell types. The cell types that occur in most hardwood of the northern temperate zone are listed in Table 2. All hardwoods contain, in varying amounts, vessel elements, fibers, longitudinal parenchyma, and ray parenchyma. Thus, differing amounts of four cell types constitute the largest volume of hardwoods. Recall that only two cell types, longitudinal tracheids and ray parenchyma, formed the greatest volume of softwoods.

1. Longitudinal Cells

Cells composing the longitudinal system consist of vessels, fibers, tracheids, and parenchyma (Table 2). Note that in contrast to softwoods, five different cell types (vessel elements, fibers, and tracheids) carry out the roles of conduction or support,

Table 2 Hardwood Cell Types

Longitudinal	Transverse

A. Support, conduction, or both
 1. *vessel elements*
 2. *fibers*
  a. fiber tracheids
  b. libriform fibers
 3. tracheids
  a. vascular tracheids
  b. vasicentric tracheids
	A. Support conduction, or both
	none

B. Storage *parenchyma*	B. Storage *ray parenchyma*

or both. In addition, longitudinal parenchyma, rare in softwoods, may constitute from 1–23% of the total cell volume of a hardwood (4).

a. Vessel Elements. Vessels or pores are structures designed to perform the conducting role. A vessel consists of a longitudinal series of cells termed vessel elements that have coalesced to form a tubelike structure (Fig. 7). In addition to following a longitudinal course, vessels deviate tangentially and radially (8). Also, vessels rarely end isolated, but rather terminate within a cluster of vessels. Water translocation from terminated vessels continues into adjacent vessels via intervessel pits. Vessel lengths vary considerably with lengths up to 11 m reported (9).

(1) Volume, Shape, and Size. The proportion of wood volume occupied by vessel elements is quite diverse, ranging from a low of 6% in hickory to 55% in basswood. Fig. 20 reveals the broad range of shapes exhibited by vessel elements. The very wide and short elements are characteristic of earlywood ring-porous vessel elements. The longer and narrower elements depicted are typical of diffuse-porous woods. In transverse section, they may be circular, oval or in some cases reveal an angular outline. In this view, they are often called pores.

Vessel elements vary greatly in length among species, ranging from 0.18 mm in black locust to 1.33 mm in black gum (4). Unlike other longitudinal cells, vessel elements do not increase significantly in length during the enlargement phase. Thus, they remain essentially the same length as the fusiform initials from which they were derived. In the case of ring-porous earlywood vessel elements, an actual decrease in length occurs as they increase in diameter. Since latewood vessel elements remain the same length as the fusiform initials and earlywood vessel elements decrease, considerable variation in vessel element length exists within ring-porous growth rings (compare Figs. 20A, and B with C).

The diameter of vessel elements varies enormously, ranging from 20–300 µm.

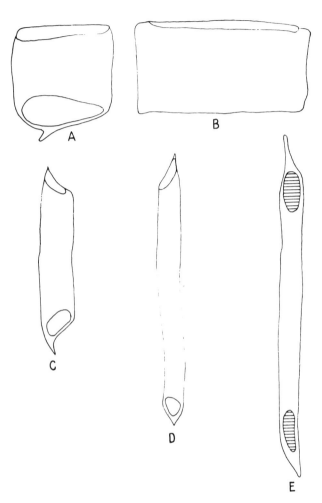

Figure 20 Types of vessel elements found in hardwoods. A and B, earlywood vessel elements from a ring-porous wood (note their short length relative to their diameter); C and D, vessel elements from diffuse-porous woods with simple perforation plates; E, diffuse-porous vessel element with scalariform perforation plates. Reprinted with permission from R. J. Thomas, in *Wood Structure and Chemical Composition in Wood Technology Chemical Aspects* (I. S. Goldstein, ed.), ACS Symposium Series 43, 1977.

For some ring-porous earlywood vessel elements, the width is greater than the length. The larger-diameter vessel elements are found in the earlywood zone of ring-porous woods. In oak, earlywood pores reach 300 μm in diameter and only 35 μm in latewood.

(2) Cell Wall Features.   The predominant cell wall features of vessel elements are those related to water conduction, namely, perforation plates and bordered pit pairs. Perforation plates provide a liquid flow pathway from vessel element to vessel element and intervessel pit pairs from vessel to vessel. Recall that vessels (pores) are structures comprised of individual vessel elements (Fig. 7).

(a) Perforation Plates.   At each end of a vessel element, openings, termed perforation plates, merge the individual vessel elements into a single structure called a vessel or pore. As a result of this merger, an elongated tubelike structure of considerable length and highly suitable for the longitudinal translocation of water is created. Two types of perforation plates are found in the northern temperate zone hardwoods. One type, called scalariform, consists of a number of parallel openings (Figs. 20E and 21) and the other, termed simple, consists of a single large opening (Figs. 20A through D and Fig. 22). Generally, the end walls of springwood ring-porous vessels are more or less transverse and their removal creates a perforation plate oriented transversely with no overlapping of the vessel elements. In diffuse-

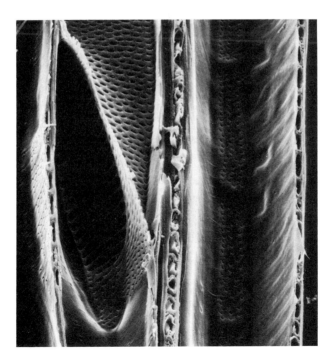

Figure 21   End of a vessel element with a scalariform perforation plate. Note the large number of intervessel bordered pits on the vessel element wall. (Courtesy of N.C. Brown Center for Ultrastructural Studies, SUNY College of Environmental Science and Forestry.)

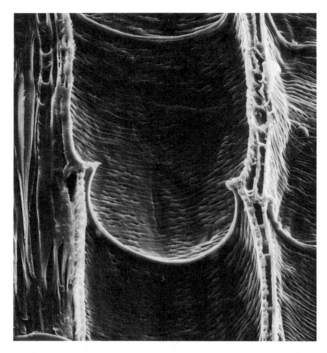

Figure 22 View of vessel elements and connecting simple perforation plate. (Courtesy of N.C. Brown Center for Ultrastructural Studies, SUNY College of Environmental Science and Forestry.)

porous woods and in the latewood of ring-porous woods, the end walls tend to be oblique and the perforation plate is formed on the longitudinal wall near the end of the cell. In this case, some overlapping of vessel elements occurs. In either case, free liquid flow between vessel elements is established through perforation plates.

(b) Intervessel Pitting.   As indicated earlier, when vessels end, they do so among a group of vessels. Water translocation continues into adjacent vessels through intervessel bordered pit pairs. Intervessel pitting is very obvious and is most often found on the tangential walls of vessel elements (Fig. 21). In many species, the entire wall is covered with pits to such an extent that most of the tangential wall consists of pit membranes. These pits differ from softwood bordered pits in that the pit membrane lacks a torus and easily detected openings (Fig. 23). Although in some diffuse porous species minute openings have been shown (10), detectable pit membrane openings are not characteristic of most species. Apparently, the membrane acts similarly to filter paper in that free liquid flow occurs, but openings are not easily seen because of their circuitous path through the membrane. Note the rather open and loose texture of the pit membrane shown in Fig. 24. With time, and

Figure 23 Pit membrane from an intervessel bordered pit pair. Comparison with a softwood bordered pit pair (Fig. 13) reveals the lack of a torus and the large openings characteristic of softwood bordered pit membranes. Micrometer bar 1 μm.

particularly during heartwood formation, the pit membranes become obstructed (Fig. 25) and in all likelihood prevent free liquid flow.

(3) Tyloses.   In many species, when the lumens of vessels become air-filled as sapwood is converted to heartwood, or as the result of an injury, adjacent, living, parenchyma cells form outgrowths through pit cavities into the lumens of vessels. These outgrowths, which are parenchyma cell wall, are termed tyloses (Fig. 26). In early growth, they are rounded ballonlike extensions of the parenchyma cell wall, but as growth continues, they press against each other and are compressed into more angular shapes that completely block vessel lumens (Fig. 26). Often, the parenchyma walls forming the tyloses thicken and pit pairs form at places where the cell walls of different parenchyma cells are in contact.

The amount of tyloses formed is species-dependent. Some woods such as black

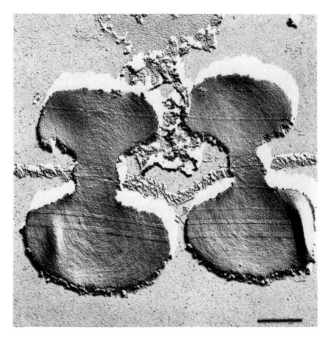

Figure 24 Cross-sectional view of intervessel bordered pit pairs. Note the loosely arranged microfibrils in the pit presumably through which liquid flow occurs. Micrometer bar 1 μm. From L. D. Bonner and R.J. Thomas, *Wood Science*, *3*:193 (1974).

locust and the white oaks form tyloses that completely fill vessel lumens. Others such as sweetgum or the red oaks form little or none. Obviously, wood permeability is considerably reduced when vessels are occluded with tyloses.

b. Tracheids.   Two types of tracheids, vascular and vasicentric, are found in hardwoods. Vascular tracheids resemble small-vessel elements except they lack perforation plates. In fact, in cross-sectional view one cannot distinguish between vascular tracheids and vessels.

Vasicentric tracheids are elongated cells, with rounded ends and heavily pitted walls. They are found in association with earlywood vessels of ring-porous woods. In the oaks, vasicentric tracheids are responsible for the movement of water from vessel to vessel as intervessel pits are lacking (11).

c. Fibers.   The longitudinal cells responsible for the support role in hardwoods are fibers. Fibers are defined as elongated cells with closed pointed ends and thick cell walls. Although two types of fibers, fiber tracheids and libriform fibers, are recognized, for all practical purposes the distinction based on the type of pitting is not easily ascertained due to the small size of the pits. Libriform fibers are characterized by simple pits and fiber tracheids by bordered pits.

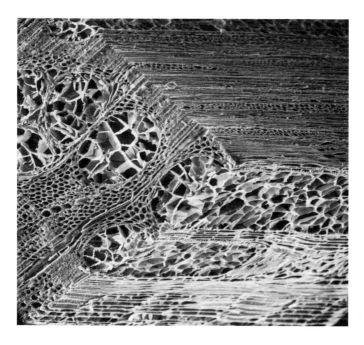

Figure 26 Vessels with lumens completely occluded with tyloses. (Courtesy of N.C. Brown Center for Ultrastructural Studies, SUNY College of Environmental Science and Forestry.)

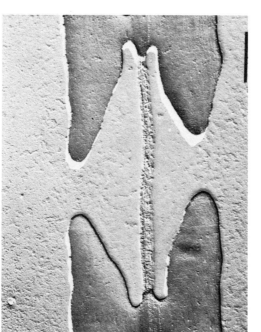

Figure 25 Cross-sectional view of an intervessel bordered pit pair from nonconducting region. Note the dense membrane (compare to, the membrane in Fig. 24) through which liquid flow would be greatly restricted. Micrometer bar 1 μm.

The amount of fiber cells on a volume basis varies from 26% in sweetgum to 75% in paper birch (4). As a general rule, the higher the fiber content, the more wood substance present and the higher the specific gravity.

Fiber lengths range from approximately 0.8 mm in big leaf maple to 2.3 mm in black gum, with an overall average slightly less than 2 mm. In diameter, fibers average about 20 μm.

Figs. 27 and 28 illustrate the obvious differences in relative size, shape, and wall thickness between fibers and other hardwood cell types.

d. Longitudinal Parenchyma.   Although some hardwoods, like softwoods, contain very little longitudinal parenchyma, most contain a substantial amount. In northern temperate regions, hardwood longitudinal parenchyma content ranges from 0.1–24% of the total wood volume (4).

The many different arrangements of longitudinal parenchyma, as viewed in cross section, provide a useful diagnostic feature for wood identification. A detailed description of the various arrangements is contained in Panshin and deZeeuw (4).

The longitudinal parenchyma cells are comparatively short, brick-shaped cells

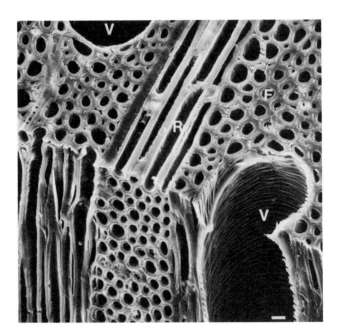

Figure 27   Hardwood showing vessels (V), fibers (F), and ray cells (R). Note the relative differences in size, shape, and wall thickness of the various cell types. Micrometer bar 10 μm. (Courtesy of N.C. Brown Center for Ultrastructural Studies, SUNY College of Environmental Science and Forestry.)

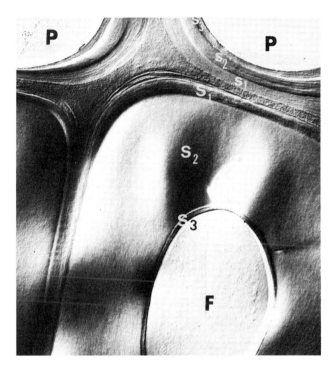

Figure 28 Fiber (F) and adjacent longitudinal parenchyma (P) cell walls in cross-sectional view. Cell wall layering is clearly revealed. Comparison of the thick fiber cell wall with the thin parenchyma cell walls reveals that the thick fiber wall is the result of a substantial change in the S_2 layer and little or none of the S_1 and S_3 layers. Micrometer bar 1 μm. From R. J. Thomas, *Journal of Educational Modules for Materials Science and Engineering, 1*:53 (1980).

with simple pits. Since their role is one of food storage, they remain alive with a functioning protoplasm providing a mechanism for the storage of carbohydrates in the form of starch. Barring abnormal occurrences, longitudinal parenchyma function until heartwood formation takes place.

2. Transverse Cells

As with softwoods, the transverse cells of hardwood constitute rays. However, hardwoods do not contain any transverse cells with a conduction role; thus, their rays are composed entirely of ray parenchyma.

a. Rays. Hardwood ray width varies considerably more than softwood rays. Recall that all softwoods contain uniseriate rays and some also have fusiform rays. A few hardwoods contain exclusively uniseriate rays; however, most possess rays varying from two to many seriate. Fig. 27 illustrates a ray seven cells in width. In domestic hardwoods, the widest rays, 30 or more cells, are found in the oaks. In a number of hardwoods, rays of two distinct sizes are common, with the smaller of the two usually uniseriate (Fig. 7).

In height, rays range from 1 to more than 2900 cells. The tallest rays, 2 or more in. in height, are found in the oaks.

As expected, the greatest ray volume occurs in the oaks, reaching approximately 30% of the wood volume. Most hardwoods have ray volumes between 10 and 20%, with an average of about 17% (4).

b. Ray Parenchyma. Although hardwood rays consist only of ray parenchyma cells, they may be composed of more than one type. The most frequent type is the radially elongated cell, termed procumbent. Cells that are as long, or even longer, in the longitudinal rather than in the radial direction are called upright cells. Upright cells are most often found at the top and bottom of a ray.

Note in Fig. 27 the relatively thick cell walls shown by the ray parenchyma. Obviously, increasing the ray parenchyma volume contributes significantly to wood-specific gravity.

3. Summary: Hardwood Microstructure

Hardwood structure is characterized by the presence of vessel elements, tracheids, fibers, longitudinal parenchyma, and ray parenchyma. The vessel elements, which vary greatly in size in different species, along with tracheids found in a few hardwoods, perform the conduction role. Fibers, which may constitute up to 60% of the wood volume, with their thick cell walls provide mechanical support. Parenchyma, both longitudinal and ray which can comprise up to 40% of the wood volume, perform the food storage function.

As a result of having more cell types and greater variation in the relative proportion of wood volume comprised of these cell types, hardwoods have a more complex structure and more variable wood properties than softwoods.

V.   CELL WALL STRUCTURE

Because cell wall structure is the product of the arrangement and type of chemical constituents that make up the wall, some knowledge of the chemical constituents and arrangement within the cell wall is useful. Therefore, prior to describing cell wall structure, a brief review of the major chemical constituents found in wood is presented.

A.   Chemical Constituents

The major chemical components of wood are cellulose, hemicellulose, and lignin. In addition, in many species, other chemicals, collectively called extractives, are deposited in cell walls during heartwood formation. Table 3 indicates the percentage of composition, polymeric nature, and role of the chemical constituents.

1. Cellulose

Cellulose is a linear polymer composed of anhydro-D-glucopyranose units linked by β-1-4 glycosidic bonds. The covalent bonding within and between glucose units results in a straight and stiff molecule with very high tensile strength. Lateral bonding of the cellulose molecules into linear bundles is primarily the result of hydrogen bonding. The large number of hydrogen bonds results in a relatively strong lateral association of cellulose molecules, which gives rise to crystalline regions in the cell wall. The crystalline regions, 60 nm in length, are interrupted by noncrystalline areas. The degree of crystallinity for wood ranges from 67–90% (12). The cellulose molecule, which ranges from 2500–5000 nm in length, passes through several crystalline and amorphous regions.

The affinity of water for cellulose is the result of the large number of OH groups present.

2. Hemocellulose, Lignin, and Extractives

Hemicellulose, like cellulose, is a polymer of sugar units. It differs from cellulose in that it is a smaller polymer, branched, usually contains more than one sugar type within the molecule, and does not form crystalline regions. Some typical sugars are glucose, galactose, mannose, xylose, and arabinose.

Lignin is the major noncarbohydrate portion of the cell wall. Chemically, it is quite different from cellulose and hemicellulose in that it is a very complex, crosslinked, three-dimensional polymer formed from phenolic units. The aromatic nature of the phenolic units makes lignin hydrophobic and the three-dimensional

Table 3 Chemical Constituents of Wood

	% Composition	Polymeric nature	Degree of polymerization	Molecular building blocks	Role
Cellulose	45–50	linear molecule crystalline	5,000 –10,000	glucose	framework
Hemicellulose	20-25	branched molecule amorphous	150-200	primarily nonglucose sugars	matrix
Lignin	20-30	three-dimensional molecule	100-1,000	phenolpropane	matrix
Extractives	0-10	polymeric	—	polyphenols	encrusting

provides rigidity for the cell wall. Without the presence of lignin, wood would be unable to carry out the support role.

Extractives is a term of convenience that includes a number of different chemical types such as terpenes and related compounds, fatty acids, aromatic compounds, and volatile oils. Extractives are found in varying amounts in the heartwood zone.

B. Cell Wall Layering

As indicated earlier, each cell has a primary wall formed during cell division and a secondary wall deposited after the cell enlargement phase has been completed. Conventional light microscopy, particularly with sections stained for lignin, reveals a lignin-rich area between contiguous cells. This area composed of the middle lamella and primary walls of the contiguous cells is called the compound middle lamella. The lignin in the middle lamella or intercellular layer bonds the cells together. Since the primary wall also contains a high amount of lignin and is very thin, it is very difficult to detect; thus, the use of the term compound middle lamella. Although the secondary wall appears as a rather homogeneous layer with conventional light microscopy, polarizing light microscopy reveals a three-layer structure. This differentiation of the secondary wall is due to the different orientation of the crystallite regions within the three layers.

Electron microscopy clearly reveals secondary wall layering (Fig. 28). Note the designation of the layers with the conventional notation of S_1, outermost layer; S_2, middle layer; and S_3, innermost layer.

Total cell wall thickness is largely controlled by the S_2 layer. Cells with thick walls contain a large S_2 layer, whereas thin-wall cells have a small S_2 layer. For example, note in Fig. 28, which depicts adjacent fiber and parenchyma cell walls, the wide S_2 in the thick fiber wall and the narrow S_2 in the parenchyma wall. Also note that little difference exists between the S_1 and S_3 layers of the two cell types. The average relative size of the cell wall layers is indicated in Table 4.

Electron microscopy studies have also revealed the presence of microfibrils (Fig. 29). These stringlike structures vary from 3–30 nm in width, are about half as thick as wide, and are of indefinite length. One explanation for the wide variation in width is based on the assumption that the microfibrils are formed through crystallization, thus, probably no uniform fibrils will be produced (13). Investigators (14, 15) have also shown that microfibrils are always composed of smaller strands called elementary fibrils with an average width of 3.5 nm. Later work (16) revealed the existence of cellulose fibrils 1.5–2.0 nm in width. These subelementary fibrils were found to be most abundant in the differentiating cell wall and to a much lesser degree in mature cell walls. Although the exact size of the basic cellulose fibril may not yet be known, it is clear that crystalline regions do exist within the microfibril and the long axes of crystallites are parallel to the long axes of microfibrils. The secondary wall layering revealed by electron microscopy is due to the differences in the angle

Table 4   Thickness of Various Cell Wall Layers and
Microfibril Angle Within the Layers

Wall layer	Relative thickness (%)	Average angle of microfibrils
PW	±1	random
S1	10–22	50–70°
S2	40–90	10–30°
S3	2–8	60–90°

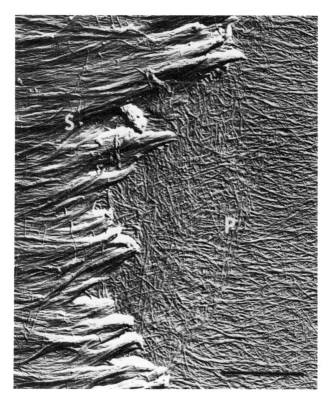

Figure 29   Microfibrils in the primary wall (P) and the S1 layer (S1) of the secondary wall. The microfibrils are loosely packed and randomly arranged in the primary wall. A tightly packed and parallel arrangement is evident in the S1 layer. Micrometer bar 1 μm. Reprinted with permission from R. J. Thomas, in *Wood Structure and Chemical Composition in Wood Technology Chemical Aspects* (I. S. Goldstein, ed.), ACS Symposium Series 43, 1977.

at which microfibrils are oriented within each layer. Table 4 indicates the average microfibril angle, measured from the long axis of the cell, which is typical for the different wall layers. Within the primary wall, the microfibrils are loosely packed and arranged in a random pattern with no evidence of lamination and in the secondary wall are closely packed and exhibit a high degree of parallelism (Fig. 29).

The outermost secondary wall layer (S_1) contains microfibrils with an average angle of 50°–70°. Both a left-hand helix (S) and a righthand helix (Z) are present in the S_1 layer (Fig. 30). Usually, the S helix is prominent.

The S_2 layer shows a Z helix of 10°–30°. Also, considerable evidence exists that the S_2, as well as the S_1 and S_3, are composed of numerous lamellae (17).

Microfibrils arranged in a Z helix with an angle of 60°–90° make up the S_3 layer (Figs. 12 and 30).

The foregoing discussion of microfibril angles implies an abrupt change in angle

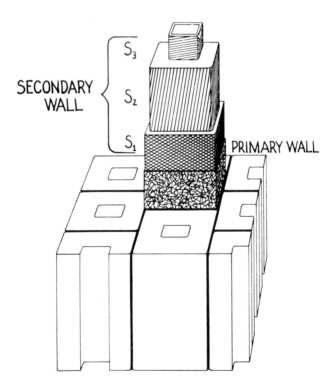

Figure 30 Idealized drawing of cell wall layering showing microfibril orientation and relative size of the various layers. Reprinted with permission from R. J. Thomas, in *Wood Structure and Chemical Composition in Wood Technology Chemical Aspects* (I. S. Goldstein, ed.), ACS Symposium Series 43, 1977.

from layer to layer. However, the angles presented are average values that include microfibril angles in transition lamellae between layers. Fig. 30 illustrates an idealized drawing of cell wall layering showing the relative size of each layer and the average microfibril orientation within each layer.

VI. SOME RELATIONSHIPS BETWEEN STRUCTURE AND PROPERTIES

Wood is an anisotropic material, that is, it has distinctly different properties in the three orthotropic axes of wood. These differences as well as other physical properties of wood are the direct result of the structure of the cell wall, cell orientation, the types of cells present, their distribution, and the relative proportions in which they are present. Examples of wood anisotropy are the following: The strength of wood in tension along the grain (longitudinal direction) is many times greater than across the grain (transverse direction); dimensional changes with the removal or addition of water is 10–15% across the grain and 0.1% along the grain; and the larger permeability of wood to liquid flow along as opposed to across the grain. Brief descriptions of some of the relationships between structure and properties follow.

The very high tensile strength of clear wood in the longitudinal direction is due to the structure of the cellulose molecule and the orientation of the microfibrils. Very strong covalent bonding both within and between glucose molecules creates a cellulose molecule with high tensile strength parallel to the long axis of the molecule. Furthermore, the arrangement of cellulose molecules within the micro-fibrils is parallel to the long axis of the microfibrils, which in the S_2 layer (the thickest layer) are almost parallel ($10°$–$30°$) to the longitudinal cell axis. This series of parallel longitudinal arrangements from the cellulose molecule to the longitudinal cells accounts for the high longitudinal tensile strength of wood. The very low tensile strength across the grain is expected since the cellulose molecules and microfibrils are bonded laterally with relatively weak hydrogen bonds.

The very high shrinking and swelling of wood across the grain (10–15% total transverse shrinking and swelling vs. 0.1% longitudinal) are also the result of cellulose molecule and microfibril orientation within the cell wall. As mentioned earlier, water has a high affinity for cellulose due to the larger number of OH groups present. However, water cannot enter the crystalline regions, but is confined to the amorphous regions within and between microfibrils. Since the long axis of crystalline regions is oriented parallel to the long axis of the microfibril, and water cannot enter crystalline regions, insertion of water swells the exterior of the microfibrils and pushes the crystalline regions as well as microfibrils further apart. Since most microfibrils are oriented almost parallel to the longitudinal cell axis, the cell wall swells more across rather than along the wall. Inasmuch as the long axis of most cells is parallel to the long axis of the tree stem, wood shrinks and swells much more transversely than longitudinally.

The vastly superior longitudinal permeability of wood compared to transverse permeability is not surprising since one of the functions of wood in the living tree is the conduction of water longitudinally from the roots to the leaves. Cell structure and orientation control permeability. The major structural aspects are the extreme length of cells compared to their width (100 times longer than wide in softwoods), the large lumen area of many cells relative to cell wall area, and the specialized conducting structures, vessels, formed from the merging of vessel elements. The fact that the cells have their long axes oriented in the longitudinal direction provides more uninterrupted flow longitudinally than transversely. Important cell wall structures are bordered pits and perforation plates that provide for liquid flow from cell to cell and thus permit the flow of water beyond the finite length of Individual cells.

The influence of the relative proportion of cell types on wood physical properties is illustrated in Table 5. Mark the trend of increasing specific gravity with decreasing vessel volume. The replacement of thin-walled, large lumen vessels with thick-walled, small lumen fibers increases the amount of cell wall material and thus wood-specific gravity increases. Also note, that the specific gravities of basswood and sweet gum are quite different, although their vessel volume is essentially the same. This is due to the lower parenchyma content and the presence of thinner-walled fibers in the basswood. Parenchyma cells, with their relatively thick walls and the presence of numerous cross walls because of their short length, significantly increase wood-specific gravity as their proportion of the wood volume increases.

Table 5 Relationship Between Specific Gravity and the Relative Proportions of Hardwood Cell Types

Species	% Vessels	% Fibers	% Parenchyma	Specific gravity
Basswood	56	36	8	0.32
Sweetgum	54	26	20	0.46
Birch	21	64	15	0.50
Hickory	6	67	22	0.64

REFERENCES

1. A. B. Wardrop, *Tappi*, *40*:225 (1957).
2. H. Imagawa, K. Fukazawa, and S. Ishida, *Gord*; *Bull. Hokkaido Univ. For.*, *33*:127 (1976).
3. S. Saka and R. J. Thomas, *Wood Sci. and Tech.*, *16*:1 (1982).
4. A. J. Panshin and Carl deZeeuw, *Textbook of Wood Technology*, McGraw-Hill, New York, 1980.
5. C. A. Hart and R. J. Thomas, *For. Prod. J.*, *17*(11):61 (1941).
6. R. J. Thomas and K. P. Kringstad, *Holzforschung*, *25*(5):143 (1971).
7 E. W. J. Phillips, J. Linn. Soc. London, *Bot.*, *52* (343):259.
8. M. H. Zimmermann and P. B. Tomlinson, *Int. Assoc. Wood Anat. Bull.*, (1):2 (1967).
9. D. S. Skene and V. Balodis, *J. Exptl. Bot.*, *19*:825 (1968).
10. L. D. Bonner and R. J. Thomas, *Wood Sci. and Tech.*, *6*(3):196 (1972).
11. E. A. Wheeler and R. J. Thomas, *Wood and Fiber*, *13*(3):169 (1981).
12. J. M. Dinwoodie, *Microscopy*, *104*:3 (1975).
13. K. Muhlethaler, in *Cellular Ultrastructure of Woody Plants*,(W. A. Côte, Jr., ed.), Syracuse University Press, Syracuse, N.Y. 1965, p.191.
14. K. Muhlethaler, *Z. Schweiz. Forstu.*, *30*:55 (1960).
15. A. Frey-Wyssling and K. Muhlethaler, *Makromol. Chem.*, *62*:25 (1963).
16. R. B. Hanna and W. A. Côte, Jr., *Cytobiologic*, *10*:102 (1974).
17. C. E. Dunning, *Wood Sci.*, *1*:65 (1968).

3

Wood Analysis

Dwight B. Easty[*]

The Institute of Paper Chemistry, Appleton, Wisconsin

Norman S. Thompson

Institute of Paper Science and Technology, Atlanta, Georgia

I. INTRODUCTION AND SCOPE

An array of classical, wet chemical procedures and a growing number of instrumental methods are presently available for the chemical analysis of wood. The wet chemical procedures typically require samples amounting to a gram or more and consist of separating the wood into macroscopic chemical components, for example, holocellulose, lignin, extractives. Because of the empirical nature of these separations, wood components are frequently described in terms of their method of isolation, such as Klason lignin or alcohol-benzene extractives, rather than by their exact chemical structure. By using these empirical separations, analysts have been able to acquire data on the gross composition of wood in spite of wood's structural and chemical complexity.

The advent of instrumental methods has greatly increased the specificity and convenience of wood analysis. Initially, chemical specificity was achieved on a macroscopic scale, as in the chromatographic separation and determination of the individual sugars in a wood hydrolyzate. More recently, the ultraviolet microscope and the electron microscope coupled with X-ray analysis have provided the topochemical specificity needed to describe the distribution of chemical constituents in individual fiber walls. Future instrumental developments should permit the chemical analysis of wood on any desired scale and, therefore, make gross preliminary separations of wood samples unnecessary.

The purpose of this chapter is to inform the reader about the methods available

[*]Current Affiliation: James River Corporation, Camas, Washington.

for the chemical analysis of wood. Analyses of pulp, pulping liquors, and cellulose derivatives are not included. Although lignosulfonates and alkali lignins are excluded, milled wood lignins and other preparations that approximate lignin in vivo are considered. Analyses of materials isolated from wood are discussed only when results may be directly related to wood.

Space does not permit the review of all methods available for wood analysis nor the presentation of procedures in sufficient detail to be followed in the laboratory. Those methods now used extensively or having the greatest potential for future use are emphasized. Possible difficulties in the procedures are noted to aid the analyst in selecting and applying a method. For detailed laboratory procedures, the reader is advised to consult the original literature or Browning's two-volume work, *Methods of Wood Chemistry* (1). Also available in the United States are test methods supplied by the Technical Association of the Pulp and Paper Industry (TAPPI) (2) and the American Society for Testing and Materials (ASTM) (3). Analytical methods are provided by similar associations in other countries.

II. SAMPLING AND PREPARATION OF SAMPLES FOR ANALYSIS

A. Sampling Problems and Objectives

No analysis is better than the sample on which it is based. Extensive analysis performed on a nonrepresentative or carelessly prepared sample is a gross misappropriation of resources. Unfortunately, the analyst is seldom consulted regarding this important initial step in wood characterization; his or her involvement most commonly begins when the collected sample is presented for analysis.

The objective of most sampling protocols is to obtain a sample that is representative of the population of interest. Results of subsequent analyses will then be significant for that population. Thus, the magnitude of the sampling needed for general characterization of a species would be quite different from that needed to evaluate the trees in a specified stand. For a given species, wood composition may vary with the tree's geographical location, because it is affected by soil, climate, and topography. Within a single tree, composition differs between roots, bole, and branches, heartwood and sapwood, springwood and summerwood. All of these factors must be considered in collecting a sample for analysis.

Some analyses do not require a representative sample. As Browning (4) has noted, "In investigations dealing solely with comparisons of composition, techniques, and methods, the only requirement is that the prepared sample be uniform, but when it is desired that results be characteristic of a species, proper selection of sample is essential."

B. Sample Collection

Guidance in sample collection is provided by TAPPI Test Method T257 os-76, "Sampling and Preparing Wood for Analysis" (5). The test method is appropriate for wood in all forms, that is, logs, chips, or sawdust. In an *engineered sampling plan*, the shipment of logs or chips is subdivided into approximately equal quantities by carloads, truckloads, or cords, and each subdivision is identified by a number. A random number table is used to indicate from which subdivisions samples are to be taken. Equal numbers of logs or amounts of chips are then taken at random from each of the indicated subdivisions.

To use a *probability sampling plan*, the standard deviation of measurements of the property of interest must be known from prior experience. One may then compute the number of subsamples required to assure within a defined confidence level that the average quality of a shipment lies within specified limits of the mean of the determinations. Details are given in T257.

After the sample logs are obtained, about one-third of the length is cut off one or both ends of each log. Then a power-driven saw having a guide to permit cuts across the end of a log the width of the saw teeth is used to make one or more such cuts across the ends of each sample log. Sawdust from these cuts is collected and used for analysis. An alternate procedure not in T257 involves chipping all or an equal portion of each of the logs. Samples of chips or sawdust are reduced by quartering to the amount needed for analysis (6).

Chips being moved by a conveyer are sampled by using a scoop to remove a portion of the chips at regular intervals (7). The chips are placed in a closed container from which a composite is taken for analysis.

An increment borer is used to take samples from living trees. The bore is made 4.5 ft (1.4 m) from the ground. Core samples are not representative, because a disproportionate amount of sample is taken near the center of the stem.

C. Size Reduction

After sawdust or chip samples have been allowed to air dry, they are ground in a mill of the Wiley type (Fig. 1). Wiley mills are available in several sizes. The larger sizes will accept whole chips, but chips must be subdivided manually or with a laboratory refiner before being fed into the smallest mill.

Samples are screened after grinding, and material that passes through a 40-mesh (0.40 mm) sieve is normally used for analysis. Delivery tubes with integral 40-mesh screens are available for the small Wiley mill. These permit grinding and screening in one operation. Fines differ from larger wood particles in composition, reactivity, and lignin content (8). Therefore, regrinding coarse material or discarding fines is not recommended. Following grinding, the sample is placed in an airtight container.

Subdivision of chips by cryogenic techniques has been used in attempts at minimizing degradation of wood polysaccharides. One procedure involved freez-

Figure 1   Wiley® mill. (Courtesy Arthur H. Thomas Company.)

ing the chips in liquid nitrogen, putting them through an Abbe mill, processing the Abbe-milled material through a liquid nitrogen-cooled hammermill, and finally processing the material in liquid nitrogen in an attritor (a process similar to ball milling) (9). Simplified versions of this procedure employed freezing the chips in liquid nitrogen or acetone-dry ice, splintering with a hammer, refreezing, and fiberizing with a Waring Blendor while frozen (10). Although these processes yielded products with excellent physical appearance, reduction of the polysaccharide degree of polymerization was not entirely avoided (10).

D.   Drying and Storage

After air drying, many chemical properties of wood and chips should be stable indefinitely. To avoid changes in reactivity and content of volatiles, wood samples collected for later analysis should not be oven-dried. Wood that must remain moist is usually placed in cold storage at 0–4°C; an environmentally acceptable biocide may be added to inhibit the growth of mold. Moist wood may be frozen to -5 or -10°C without producing excessive changes in structure.

E.   Removal of Extractives

Some determinations performed on wood require samples from which extractives have been removed. Preparation of extractive-free wood formerly involved successive extractions with ethanol-benzene (1:2), 95% ethanol, and distilled water (11). Ethanol-benzene is now considered adequate for removing most materials that interfere with analyses (12). The analyst should avoid inhalation of benzene vapor and any contact of benzene with the skin.

III. DETERMINATION OF WATER

Because of wood's hygroscopic nature, the moisture contents of wood samples and of materials made from wood are affected by the environment to which they are exposed. Moisture in these materials affects product performance (13) and the amount of salable product. Thus, commercial considerations require that accurate methods of moisture determination be available.

Chemical analyses of wood are almost always performed on air-dried samples, but results are reported on an oven-dry (o.d.) basis. A moisture determination must therefore be run on nearly every sample submitted for analysis.

A. Oven, Infrared, Microwave, and Vacuum Drying

The most commonly used method of determining the moisture content of a wood sample involves weighing the sample and then drying it to constant weight in an oven at 105°C (12, 14). TAPPI T264 specifies that constant weight is attained when successive weighings of a 2-g sawdust sample do not change by more than 0.002 g following 1-hr heating periods (12). Note that a moisture determination employing oven drying is in error to the extent that volatile substances other than water are removed from the sample during drying.

A moisture balance consists essentially of an infrared lamp mounted above the pan of a top-loading electronic balance. The energy input and drying time are controlled by the operator, and settings are normally based on previous experience with similar samples. Instrument readout indicates weight changes as they occur, expressed in grams or percentage of moisture lost. Potential advantages of a moisture balance are simplicity of operation and the reduced time required for a moisture determination.

Rapid moisture determinations in a variety of materials are possible through the use of a microwave oven (15). This technique, however, is not without potential problems. Some substances that absorb microwaves, for example, bark, are susceptible to inadvertent ignition. This may be avoided if the oven design includes a water loop and heat exchanger to absorb and remove excess energy. With other materials, the sample becomes more transparent to microwave radiation as moisture is evaporated, making the last traces of moisture difficult to remove. This problem is overcome by adding a weighed amount of a microwave absorber, such as ferrous oxide, to the ground sample (15).

A microwave oven with a built-in electric balance is also available (15). In an automatic operational mode, this device will turn itself off when the sample weight changes less than 0.01 g in 30 sec. The utility of this system for determining moisture in wood remains to be evaluated.

Vacuum drying may be used to determine moisture in wood and wood-derived materials. It is used frequently when other analyses are to be performed on a sample of which there is an insufficient quantity for a separate moisture determination. In that case, the sample may be dried completely in a vacuum desiccator over P_2O_5 without suffering undesirable changes that might occur in an oven at 105°C.

B. Azeotropic Distillation

Water in wood may be determined by azeotropic distillation with a water-immiscible solvent. Toluene (16) and xylene are solvents commonly used for this determination. The apparatus is shown in Fig. 2. Water from the sample distills with the solvent. Upon condensation, the water and immiscible solvent separate and flow into the trap below the condenser. Excess solvent overflows the trap and runs back into the distilling flask. The water is measured by observing the meniscus at the water-solvent interface at the conclusion of the test.

Azeotropic distillation is said to provide a better measure of true water content than oven drying (16). This method can completely remove water from wood samples, and its results are not affected by volatile substances other than water in the sample for analysis.

C. Karl Fischer Titration

The Karl Fischer method is included in the ASTM standard method for determining moisture in cellulose (17), and it has been used successfully for determining the moisture content of wood (18). The wood is ground, and its water is displaced by dry methanol during a 20-min extraction time. Water in the methanol is then measured by Karl Fischer titration. Unlike oven-drying methods, the Karl Fischer titration is not adversely affected by volatile extractives in the wood (18). In analysis of bark samples, the most accurate way to determine moisture employed drying over P_2O_5 followed by Karl Fischer titration to measure any water remaining in the bark (19). The Karl Fischer reagent contains iodine, sulfur dioxide, pyridine, and methanol. It reacts with water as shown below:

$$C_6H_5N \cdot I_2 + C_6H_5N \cdot SO_2 + C_6H_5N + H_2O \rightarrow$$
$$2\,C_6H_5N \cdot HI + C_6H_5N \cdot SO_3 \qquad (1)$$
$$C_6H_5N \cdot SO_3 + CH_3OH \rightarrow C_6H_5N(H)SO_4CH_3 \qquad (2)$$

In the second reaction step, the pyridine-sulfur trioxide complex, which is also capable of consuming water, is removed by reaction with excess methanol. All components of the reagent are present in excess except iodine, which controls the

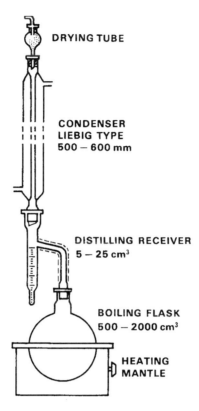

Figure 2 Apparatus for moisture determination by toluene distillation. (Reprinted from TAPPI Test Method T208 os-78, "Moisture in Wood, Pulp, Paper, and Paperboard by Toluene Distillation," Technical Association of the Pulp and Paper Industry, Atlanta, Ga., © 1978, with permission. Copies available from TAPPI, Technology Park/Atlanta, P.O. Box 105113, Atlanta, GA 30348.)

reagent's combining capacity for water.

Although the end point of the Karl Fischer titration can be perceived visually, electrometric methods of end-point detection are preferred. A widely used technique involves current measurement with two polarized electrodes (20). There are two types of automatic Karl Fischer titrators commercially available, both with electrometric end-point detection (21): titrators with motor-driven burets and titrators with coulometric generation of iodine (22).

D. Spectrometry

I. Infrared

Water absorbs in the near infrared at 1.45, 1.55, 1.77, and 1.93 μm. The 1.93 μm band is the strongest. Moisture has been determined in kraft black liquor, molasses, and pulp dissolved or dispersed in dimethylsulfoxide (DMSO) by measuring the absorbance at 1.93 μm (23). Although this technique has apparently not been used to determine moisture in wood, displacement of the water from the wood by DMSO followed by measurement of infrared absorbance should be feasible.

Several infrared techniques for determining moisture in wood particles and chips have been reported (24–26) and attempts have been made to apply the techniques to continuous operation (27).

2. Nuclear Magnetic Resonance

Nuclear Magnetic Resonance (NMR) spectroscopy is capable of distinguishing between protons existing in different magnetic environments. Therefore, the proton NMR spectrum of free water is different from that of the bound, nonfreezing water in cellulose, which is also different from that of cellulose itself (28). Utilizing this concept, wide-line NMR spectrometers were used successfully to determine moisture in wood samples (29, 30). An NMR technique combined with weight measurements of the wood samples provided moisture content data as accurate as those from oven drying (31). A concern about these and other NMR methods of moisture measurement is the possible nonrepresentative nature of wood core samples used in the determination.

Pulsed NMR techniques have been proposed as improvements over the older steady-state NMR methods for measuring water in wood (32). A short, intense burst (pulse) of a magnetic field at the ^1H resonance frequency is applied at right angles to the static magnetic field. Measurements of the induced voltage immediately after the pulse is applied give a measure of the relative numbers of ^1H nuclei in different samples. Although these measurements cannot be made at precisely zero time after the pulse, analysis of the free-induction decay provides the desired information. The free-induction decay amplitude measured 50 μsec after the pulse was proportional to all of the water in wood samples that could be removed by oven drying (33). The pulsed NMR technique has been used to measure moisture content in sapwood ranging from 0–176%. Relationships between the water content of wood and transverse relaxation times have been studied (34, 35).

Low-cost, low-resolution NMR instrumentation is available that should be useful for routine moisture determinations of this type (32).

E. Other Methods

Nondestructive methods of determining moisture in wood may be based on measurement of electrical resistivity. As wood moisture content is increased, resistivity decreases. There is a linear relationship between the logarithm of resistivity and moisture content up to 7% moisture (36). From 7% to fiber saturation, the logarithm of the resistivity is related linearly to the logarithm of the moisture content. Above fiber saturation, the change in resistivity with wood moisture content is relatively small.

At a given temperature and frequency, the dielectric constant of wood and cellulose increases with moisture content (37). The increase is caused by the higher dielectric constant of water compared with that of cellulose. Although this phenomenon should provide the basis for a useful method of determining moisture, dielectric methods have been described as unsatisfactory for hard solids like moist wood (28). They are adversely affected by the voids, the dielectric anisotropy, and the ionic nature of dissolved impurities. Despite these potential problems, commercial dielectric-type meters for measuring moisture in wood are available. The ASTM standard that describes their use stresses the importance of following the operating instructions provided with each instrument (38).

An additional principle on which wood moisture content measurements may be based is attenuation of β or γ radiation. Readout from such a method must be corrected for the density of the wood or chips being monitored.

A typical wood chip moisture-measuring system is shown in Fig. 3 (39). This device is based on the microwave power absorption technique.

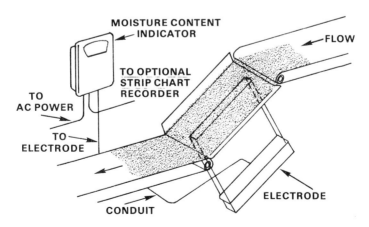

Figure 3   Microwave absorption wood chip moisture measurement system. (Reprinted with permission from Ref. 39.)

IV. CELLULOSE AND HEMICELLULOSES

A. The Problem of Separating Wood Components

Cellulose, hemicellulose, and lignin exist in wood as interpenetrating systems. Separation of all of these components, unchanged and free of contamination, has not yet been achieved. Part of the difficulty in accomplishing this separation has been attributed to bonding between lignin and carbohydrates. In spite of these problems, separation and analysis techniques have been devised that yield considerable information about the chemical composition of wood samples.

B. Isolation and Determination of Cellulose

1. Hydrolysis/Chromatographic Method for Estimation of Cellulose

Acid hydrolysis of the carbohydrates in a wood sample, followed by separation and quantitation of the resulting monosaccharides by a chromatographic technique, provides data from which cellulose can be estimated. Cellulose is assumed to be equal to the total glucan less the glucan associated with the glucomannan and galactoglucomannan in the hemicelluloses (40–42).

Immature woody tissues and cell exudates contain xyloglucan, a cellulose derivative (43). Softwood hemicelluloses have been reported to contain a species-dependent amount of glucan ranging from one-fourth to over one-third of the mannan (44). In hardwoods, the ratio between glucan and mannan is 1:2 except for birch, in which it is 1:1 (40).

A widely used hydrolysis procedure employs 72% sulfuric acid for 1 hr at 30°C, followed by dilution and an additional hour in an autoclave at 120°C (45, 46). Analysis by gas chromatography requires about 2 hr for preparation of the volatile alditol-acetate derivative of the sugars (46). Liquid chromatography does not require derivatization. Chromatographic separations are accomplished in 35–45 min. Further discussion of these methods is included in the section of this chapter on the determination of sugars.

2. Holocellulose Preparation

Ideally, a holocellulose preparation should contain all of the cellulose and hemicellulose and none of the lignin and any other components originally present in a wood sample. In practice, preparation of a holocellulose always involves some loss of carbohydrates and some retention of lignin. Procedures most commonly used for preparing a holocellulose involve treating the ground, extractive-free wood with either chlorine gas (47), an acid solution of sodium chlorite (48), or peracetic acid (49). Holopulp is a term coined to describe fibrous holocellulose modified in such a manner as to provide a papermaking furnish (50).

Details of the chlorination method are provided in TAPPI Test Method T9 wd-

75 and in TAPPI Useful Method 249 (51). Chlorine gas is drawn through the sample contained in the apparatus shown in Fig. 4. Ice water surrounds the filter crucible that holds the sample. After 5 min chlorination is stopped, the ice water drained, and the sample washed twice with hot monoethanolamine in ethanol. Chlorination and washing are repeated until the sample is white or does not become lighter upon further chlorination. The holocellulose is dried at 105°C and weighed, although drying must be at 60°C in a vacuum oven if the sample is to be used for subsequent analyses.

Several modifications to the chlorine holocellulose method have been proposed to minimize the degradation of carbohydrates (52–54). The chlorination method is especially useful for the study of hardwoods. Small amounts of acetyl groups originally present in wood are lost in this procedure (52), especially if alkaline interactions are employed to assist delignification.

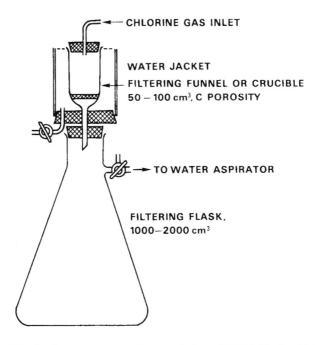

Figure 4   Chlorination apparatus. (Reprinted from TAPPI Useful Method 249, "Delignification of Unbleached Pulp (Chlorine Gas Method)," Technical Association of the Pulp and Paper Industry, Atlanta, Ga., with permission. Copies available from TAPPI, Technology Park/Atlanta, P.O. Box 105113, Atlanta, GA 30348.)

The acid chlorite procedure was devised as an alternate method for the preparation of holocellulose. In this procedure (55), water (160 mL), glacial acetic acid (0.5 mL), and sodium chlorite (1.5 g) are added to the flask containing the sample of wood meal (5 g). The reaction, maintained for 1 hr at 70°C with a hot plate or water bath, is conducted in a fume hood to remove toxic chlorine dioxide gas. Without cooling the sample, additional glacial acetic acid and sodium chlorite are added and allowed to react for a second hour. A third chloriting step is used for hardwoods; softwoods may require four chlorite treatments. Because this technique is quite species-dependent, a preliminary experiment may be necessary to determine the extent of chloriting required by a given wood sample. The reaction must be stopped with 2–4% lignin remaining in the holocellulose to avoid excessive carbohydrate degradation. After cooling, the holocellulose is filtered, washed on the filter with water and alcohol, and dried. Air drying or freeze drying are used if further analyses are to be performed on the holocellulose.

Modifications to the chlorite procedure for holocellulose preparation employing additional buffering conditions (56), pretreatments for specialized purposes (57, 58), and lower temperatures and longer reaction times (59, 60) have been devised to achieve more complete delignification with minimized carbohydrate loss. Reaction times for beechwood involved chloriting at 30°C for 4 days (61), whereas extractive-free southern pine wood required up to 14 days delignification at room temperature (57).

Considerable controversy exists about the relative effectiveness of the chlorination, peroxyacetic acid, and chlorite techniques of holocellulose preparation (57, 60, 62–64). It is likely that all processes can yield roughly equivalent cellulose components if careful experimental precautions are taken. The chlorination process is less destructive of the hemicellulose components, but the chlorite technique is more convenient for preparing large quantities of hemicellulose and is the only process for delignifying complete woody structures (65). Many of the reactions of chlorine dioxide with lignin and phenols are quite similar to those that occur during chloriting (66). Chlorine dioxide is a much less convenient method of preparing holocellulose, however (67).

3. Other Cellulose Preparations

a. Cross and Bevan Cellulose.   The chlorination procedure originally applied by Cross and Bevan (68) yields a preparation consisting largely of cellulose but also containing a significant amount of hemicelluloses (69). This technique is now considered to be principally of historical interest. Details of the method are provided by Browning (69).

b. α-, β-, and γ-Cellulose.   α-Cellulose is defined as the residue that is insoluble in 17.5% sodium hydroxide when the treatment is conducted under specified conditions. It is considered to represent the undegraded, higher-molecular-weight cellulose in the sample. β-Cellulose is that portion of the sample that dissolves in the caustic but is reprecipitated when the solution is neutralized, and γ-cellulose is that part of the dissolved material that remains soluble in neutral or acidic solutions. β- and γ-cellulose represent degraded cellulose and hemicelluloses. Before these determinations can be performed on a wood sample, the material must be delignified, normally by pulping and bleaching or by preparation of a holocellulose. Consequently, α-, β-, and γ-cellulose are more commonly determined on bleached pulp than wood.

Because these are empirical determinations, strict adherence to details of the procedure is essential. This is especially true for the washing process when the α-cellulose is determined gravimetrically. β- and γ-Cellulose are determined volumetrically by oxidation with potassium dichromate. Significant changes in the procedure were made in 1974 with the issuance of a revised TAPPI Test Method (70). Compared with the older procedure (71), the new method is run at 25°C rather than 20°C, requires dilution of the 17.5% caustic to 9.45% instead of 8.3%, and determines α-cellulose by difference rather than direct weighing. The new method gives α-cellulose values that are about 0.5% higher than those obtained by the older, gravimetric procedure (70).

Carbohydrate analysis by the hydrolysis-chromatographic method is replacing the α-, β-, and γ-cellulose determinations for analytical purposes but not rapid quality control.

4. Characterization of Cellulose Preparations

In terms of quantity and commercial value, wood pulp is the cellulose preparation having greatest importance. However, complete consideration of the chemical methods for evaluation of wood pulp is beyond the scope of this chapter. Methods for analysis of wood, wood pulp, and cellulose preparations were described in detail by Browning (1). Included were determinations of molecular weight and molecular weight distribution, solubility in alkaline solutions, functional groups, and sugar composition. Selected newer methods of interest are reviewed below. Part of the section on lignin in this chapter is devoted to functional group determinations. Methods for determining sugar composition are discussed in the portion of this chapter on hemicellulose structure.

a. Molecular Weight.   Estimates of cellulose average molecular weight are most commonly obtained from a determination of the viscosity of a solution of cellulose in an appropriate solvent: Problem areas in determining the molecular properties of cellulose have been reviewed by several researchers (72, 73). Solvents routinely used include cuprammonium hydroxide (74), cupriethylenediamine (cuene) (75), cadoxen (76), and iron sodium tartrate (EWNN or FeTNa) (77, 78). An empirical relationship, the G factor, has been developed to relate the cuene viscosities of unbleached alkaline pulps to pulping conditions (79).

More recently, dimethyl-sulfoxide-paraformaldehyde (DMSO/PF) (80) has been proposed for viscosity determination (81). Like cadoxen, DMSO/PF forms clear, colorless cellulose solutions. Kraft pulps (4% lignin) and pulps with high hemicellulose content (holopulps), which are apparently insoluble in cuene and cadoxen, dissolve in DMSO/PF (81). Of concern in the use of DMSO/PF is the variable composition of the methylol cellulose produced under different conditions for dissolving the sample. At lower solution temperature, 80° instead of 120°C, the method yields methylol cellulose of high molecular substitution unevenly distributed throughout the cellulose molecule (82).

Lignin in the cellulose preparation is a potential problem in viscosity determinations employing any of the above solvents. Even though the cellulose may dissolve, the undissolved lignin can influence the flow of the solution through a capillary viscometer. Removal of the suspended lignin by filtration is tedious and must be done quantitatively to permit adjustment of the sample weight. These problems can be reduced if the lignin is removed from the sample by a single mild chlorite treatment.

Nitration may be used to prepare samples for viscosity measurement in ethyl acetate without preliminary delignification and with little or no degradation (83, 84). However, cellulose nitrates are unstable, and nitrated xylans in holocellulose preparations are insoluble (84). The molecular weight distribution of cellulose nitrate may be measured by fractional precipitation from acetone solution, with water used as the nonsolvent.

A newer method of determining the molecular weight of cellulose preparations involves the synthesis of the carbanilate derivative (Fig. 5) (85–87). The main advantage of this derivative is that carbanilation proceeds to full substitution in one reaction step without degradation of the cellulose chain. Although this technique avoids the disadvantages of nitration, samples containing lignin must first be converted into holocellulose. The molecular weight distribution of the cellulose or holocellulose carbanilates is performed by gel permeation chromatography on a styrene divinylbenzene column using a tetrahydrofuran eluent. An on-line technique has been developed (88), as have improvements in the theoretical interpretation of GPC data (89).

b. Other Macromolecular Properties.   Instrumental methods that have received wide use in the characterization of cellulose include infrared (IR) spectrometry (90), X-ray diffraction, and electron diffraction (91). Differential scanning calorimetry has been used to determine cellulose accessibility (92).

The X-ray and electron diffraction techniques have been especially valuable in identifying the polymorphic forms (cellulose I-IV) which exist in a sample (93). X-ray diffraction is typically used to estimate cellulose crystallinity (94). Although nuclear magnetic resonance (NMR) spectroscopy has received some use in structural studies, its utility is limited because cellulose and its derivatives are insoluble in the usual NMR solvents (95).

The NMR technique by which spectra can be obtained on solid cellulose samples is cross polarization with magic angle spinning (CP/MAS) (96, 97).

Raman spectroscopy offers an alternative to IR spectroscopy as a consequence of

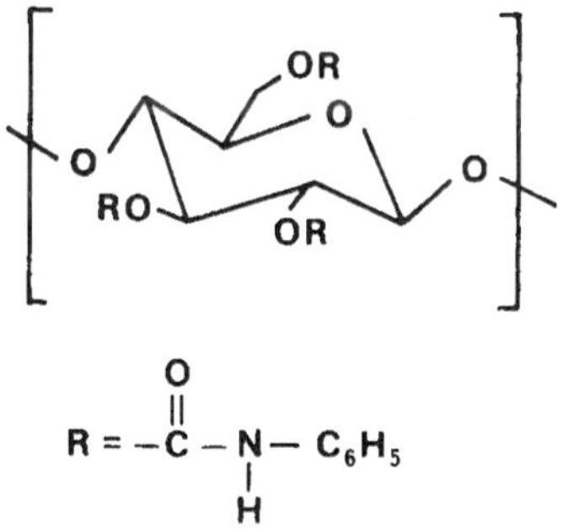

$$R = -\overset{\overset{\displaystyle O}{\|}}{C} - \underset{\underset{\displaystyle H}{|}}{N} - C_6H_5$$

Figure 5   Monomer unit of cellulose tricarbanilate. (Reprinted from L. R. Schroeder and F. C. Haigh, "Cellulose and Wood Pulp Polysaccharides: Gel Permeation Chromatographic Analysis," *Tappi*, 62 (10):103(1979), Technical Association of the Pulp and Paper Industry, Atlanta, with permission. Copies available from TAPPI, Technology Park/Atlanta, P.O. Box 105113, Atlanta, GA 30348.)

its dependence on the changes in the polarizability of vibrating bond systems rather than changes in the associated molecular dipoles of polymers (98, 99). The use of laser sources produces spectra sensitive to the skeletal vibrations of the cellulose molecule, whereas the mode of packing has only secondary effects. Such data, in combination with other spectral information such as solid-state ^{13}C (CP/MAS) NMR, can lead to the detection of subtle differences between celluloses from different biological sources as well as between technical celluloses (100, 101). In an extension of these studies, by using a Raman microprobe, it is possible to define the polarization of exciting and scattering radiation relative to the molecular orientation of cellulosic fibrils (102, 103) and to investigate the orientation of lignin in wood (104).

C.   Isolation and Characterization of Hemicelluloses

1.   Extraction and Fractionation

Hemicelluloses are commonly isolated from wood or from holocellulose prepara-tions by extraction with aqueous alkaline solutions. A large portion of the xylan in hardwoods may be removed by direct extraction of the wood with aqueous potassium hydroxide (40). This yields a hemicellulose fraction free of glucomannan, most of which cannot be removed until the wood is converted into holocellulose. With softwoods, a holocellulose preparation is usually essential before any signifi-cant hemicellulose fraction can be removed. An exception is the water-soluble arabinogalactan that, when present, is readily extracted from conifers, especially larch.

Potassium hydroxide solutions are used widely for the extraction of xylose-containing hemicelluloses, because the potassium acetate formed on neutralization with acetic acid is more soluble in the ethanol used to precipitate the isolated hemicellulose (105). Lithium, sodium, and potassium hydroxide solutions are about equal in their ability to remove xylose-containing polymers from a chlorite holocellulose (106). Sodium hydroxide has been found more effective in extracting the resistant mannose-containing hemicelluloses.

In a typical isolation procedure (105), the chlorite holocellulose in an Erlenmeyer flask purged with nitrogen is treated with 5% potassium hydroxide at 20°C for 2 hr. After filtering and washing of the fibrous residue, the hemicellulose in the filtrate is precipitated by the the the addition of ethanol. A 24% solution of potassium hydroxide is used in a second extraction of the fibrous residue. Following filtration, the hemicellulose in the filtrate from the second extraction is also precipitated by ethanol addition. The isolated hemicelluloses are separated from the ethanol by decantation and centrifugation and then dried.

Fractionation of the hemicelluloses may be achieved by successive extraction with a series of solvents. After preswelling with liquid ammonia, the holocellulose is subjected to extraction with dimethylsulfoxide or hot water to isolate the naturally acetylated xylan (107, 108). Later in a series of extractions with aqueous potassium or sodium hydroxide, like those described above, borate is added to aid in the removal of the glucomannans (109). Borate acts by complexing with the cis hydroxyl groups in mannan, and a detailed study of the effect of cations and borate on the extraction of pulps has been published (110).

The glucomannan is separated from contaminating xylan by precipitation with barium hydroxide (111).The addition of barium hydroxide at several points in a separation scheme followed by acidification with acetic acid and precipitation with ethanol yields glucomannan and galactoglucomannan fractions from softwood holocelluloses (112–114). Xylans and glucomannans dissolved in saturated aqueous alkaline earth halides may be fractionated with iodine according to their extent of branching by precipitating the less highly branched components as a blue complex (115).

An effective alternate procedure for extracting hemicelluloses from softwoods employs barium hydroxide added to the holocellulose to block dissolution of mannose-containing polysaccharides (116). Xylans are then readily extracted with 10% aqueous potassium hydroxide. After the removal of barium ions, sequential extraction with 1 and 15% sodium hydroxide removes galactoglucomannan and glucomannan, respectively.

More recently, the purification of hemicellulose extracts has employed ion-exchange chromatography on columns containing dethylaminoethyl (DEAE) cellulose. In one application (117), the hemicellulose was first extracted from a spruce holocellulose preparation with potassium hydroxide solution. After neutralization, a portion of the extract was applied to a DEAE cellulose column in the acetate form.

Fractions were collected as the concentration of eluent was increased from 0.1 to $1N$ NaOH. Most fractions contained principally arabinoxylan; some galactoglucomannan was found in two fractions. Unfortunately, elution of the column with alkali was accompanied by dissolution of the ion exchanger. In the case of cambial tissues (118), water and then $0.5M$ ammonium carbonate were used to elute polysaccharides from a DEAE cellulose column in the carbonate form.

A study of the cell wall polysaccharides of tobacco provides an excellent example of the combined use of classical and modern techniques for the separation of a wide range of hemicelluloses (119).

2. Determination of Structure

Methods for determination of carbohydrate structures have recently been reviewed by Lindberg (120). A more thorough discussion of polysaccharide degradation techniques is provided in a somewhat older review from the same laboratory (121). Newer techniques and instrumental methods of analysis permit structure determination on samples as small as 0.1 mg; this should allow characterization of the polysaccharides from different parts of the cell wall (120). Although these methods have been applied principally to nonwood polysaccharides (122), they should be directly applicable to the study of wood hemicelluloses.

Recent research has suggested a computerized approach to structural determination, called CASPER, that uses information from NMR spectroscopy and simple sugar analysis; this should alleviate the arduous task of structural determination (123).

a. Determination of Sugars. Determination of the sugar units comprising a polysaccharide involves hydrolysis to monosaccharides followed by, most commonly, separation of the monosaccharides by a chromatographic technique. Procedures for determining the total sugar content of hydrolyzates, the Munson and Walker method (124) and the Somogyi method (125), have been reviewed by Browning (126). Browning has also described in detail the separation technique employing paper chromatography (127, 128). Discussed below are hydrolysis procedures and separations by gas and liquid chromatography.

(1) Hydrolysis. Sulfuric acid is most commonly used for hydrolysis of polysaccharides because of the ease with which the acid is removed as insoluble barium sulfate. Complete hydrolysis of insoluble polysaccharides is accomplished in two steps. Primary hydrolysis involves treatment of the sample with 72% sulfuric acid for 1 hr at 30°C (45). Secondary hydrolysis requires dilution of the sulfuric acid to 4% and refluxing for 4 hr or heating for 1 hr in an autoclave at 120°C (15 psi). To avoid possible precipitation of higher-molecular-weight polysaccharides, primary hydrolysis in 77% sulfuric acid (129) and secondary hydrolysis under reflux in 3% sulfuric acid (130) have also been proposed.

Trifluoroacetic acid (TFA) has been proposed as a better reagent for the hydrolysis of polysaccharides than sulfuric acid (132, 133). Loss of monosaccharides is minimized, hydrolysis normally requires 1 hr, and the acid can be completely removed by evaporation. Water- or alkali-soluble polysaccharides are refluxed in 2N TFA for 1 hr. Cellulose-containing samples are steeped in concentrated TFA for 15 min and then successively refluxed for 15-min periods in concentrated and 80% TFA. A final 30-min reflux in 30% TFA completes the hydrolysis of cellulose. After repeated concentration on a vacuum rotary evaporator, with the addition of water after the first concentration, the sample is ready for analysis by liquid chromatography. A modification to this procedure employs hydrolysis with 80% TFA for 2 hr at 100°C in sealed tubes (134).

(2) Gas Chromatography.   Following hydrolysis, the sugars must be converted into volatile derivatives in order to be separated by gas chromatography (GC). Classical techniques such as methanolysis are continually being reevaluated for their applicability to this type of modern instrumentation (135). At present, volatile trimethylsilyl (TMS) derivatives are prepared by treating the sugar mixture in anhydrous pyridine with hexamethyldisilazane and trimethylchlorosilane (137). The derivatization reaction requires only 5 min. Because a separate derivative is formed from each anomeric form of each sugar in the mixture (136), complete separation of the peaks in a wood or pulp hydrolyzate is difficult by packed column GC (130). Lithium perchlorate has been suggested as a catalyst to equilibrate mixtures of anomers before GC (131).

As shown in Fig. 6, a single peak is observed for each sugar when sugars are chromatographed as their alditol-acetate derivatives (138). The monosaccharides in a hydrolyzate are reduced to alditols with sodium borohydride and then acetylated with acetic anhydride and sulfuric acid (46, 139). Suggested improvements to this procedure include an efficient system for concentrating samples and the use of a standard pulp for method calibration and quality assurance (140). A method for analysis of 10-mg or smaller samples has also been proposed (141).

A microscale derivatization has been developed for the study of polysaccharide samples as small as 20 ng (142). The alditols were converted to trifluoroacetates, which were then detected by the highly sensitive electron-capture detector.

Sugars in wood hydrolyzates have also been determined as acetates of aldononitriles (143). Following hydrolysis, neutralization, and evaporation, the dry residue was heated with pyridine and hydroxylamine hydrochloride. This was followed by acetylation with acetic anhydride.

Gas chromatographic resolution of sugar derivatives is greatly enhanced by the use of glass support-coated open tubular (SCOT) or wall-coated open tubular (WCOT) capillary columns. A SCOT column coated with methyl silicone has been used for separating TMS derivatives of monosaccharides, cyclitols, sucrose, and quinic and shikimic acids in 60% ethanol extracts from *Pinus radiata* tissues (144). Separation of silated (-)-2-butylglycosides (145) or acetylated (+)-2-octyl glycosides (146) on WCOT columns permitted determination of the D or L configuration of monosaccharides. High resolution and sensitivity have been combined by

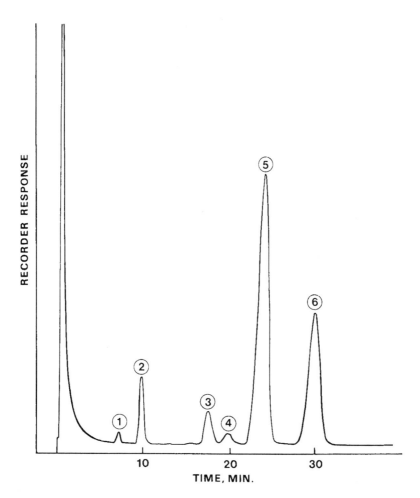

Figure 6 Gas chromatogram of sugar alditol acetates. Components: 1. Arabinose, 2. xylose, 3. mannose, 4. galactose, 5. glucose, 6. inositol (internal standard).

separating the trifluoroacetyl esters of sugars on a capillary column connected to an electron-capture detector (147). The separation of monosaccharide alditol-acetates was achieved on a chiral stationary phase on a WCOT column (148).

(3) Mass Spectrometry.   In routine sugar determinations by GC, the sugars are normally identified by their retention times. Identifications are more conclusive when the gas chromatograph is coupled with a mass spectrometer. The principles of mass spectrometry (MS) of carbohydrate derivatives have been the subject of

several reviews (149, 150). GC/MS analysis is not limited to derivatives of simple sugars. The technique may be used to identify the volatile peralkylated derivatives of disaccharides, most trisaccharides, and some tetrasaccharides prepared in the course of carbohydrate structural studies (151).

(4) Liquid Chromatography.   Early attempts to separate sugars by liquid chromatography (LC) employed strongly basic anion-exchange resins. One approach utilized the resin in the borate form and an aqueous borate solution as the eluting solvent (152). In this method, borate ions react with sugars to produce negatively charged sugar-borate complexes; the differences in the strengths of these complexes provide the basis for separation. In the other approach, the anion-exchange resin is in the sulfate form, and the eluting solvent is aqueous ethanol (153). Separations occur in this method because the strongly polar sugar solutes have a greater affinity for the solution inside the resin where the water concentration is higher compared with that in the external solution (154). Because both anion-exchange methods used postcolumn derivatization to detect the emerging sugars, automated procedures were developed using the anthrone (155) or orcinol (156, 157) colorimetric techniques.

Later developments in the anion-exchange separation of sugars as borate complexes achieved decreased separation time and allowed the application of sulfuric acid hydrolyzates directly to the column (158). A copper dye complex for postcolumn sugar detection was developed and applied (159, 160). This reagent is less corrosive, more stable, and gives better peak resolution than the viscous orcinol-sulfuric acid reagent (160). An LC system of this type, shown in Fig. 7, can separate mannose, arabinose, galactose, xylose, and glucose in wood and pulp hydrolyzates (161).

Carbohydrate mixtures can also be separated by partition chromatography on cation-exchange resins. The carbohydrates interact with a mono-, di-, or trivalent metal ionically bound to a sulfonated polystyrene-divinylbenzene copolymer. Separation is based on the stereochemistry of the carbohydrates, which are hypothesized to form complexes with the ionically bound metal via the configuration of their hydroxyl groups (162). Variation of the metal alters the behavior of the column toward the sugar molecules. Column packings of this type have been optimized for the separation of specific monosaccharides (163). Resins in the calcium and silver forms have been used for oligosaccharide separations (164–166), and complete resolution of the monosaccharides in wood hydrolyzates has been achieved on a resin in the lead form (167). Deionized, distilled, degassed water at 85°C is the most common eluent for ion-moderated partition, as this technique has been termed (169). A refractive index detector is normally used. Carbohydrates can also be determined using ion-exchange chromatography coupled with amperometric detection (168).

Liquid chromatographic separations of carbohydrates have also been performed on bonded-phase silica packings. In packings of this type, polar or nonpolar

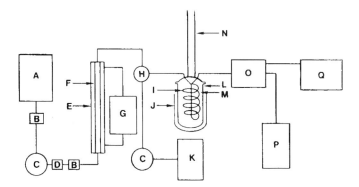

Figure 7   Liquid chromatograph for sugars A, eluent; B, filter; C, pump; D, injector; E,. jacket; F, column; G, thermostated circulator; H, mixing tee; I, reactor coil; J, heating mantle; K, copper dye complex; L, resin kettle; M, boiling water; N, air condenser; O, detector; P, waste; Q, recorder. (From Ref. 161.)

functional groups are covalently bound via silane bridges to the surface of microparticulate silica gel. A bonded phase containing the polar aminoalkyl functional group has been used for separating carbohydrates (170, 171). Gradient elution using water in acetonitrile separated monosaccharides, disaccharides, and trisaccharides; glucose and galactose were not completely resolved (171). A variable wavelength detector was used at 192 nm.

The HPLC analysis of wood sugars has been compared with paper chromatography, and the relative effectiveness of each has been described (172).

Fractionation of complex mixtures of peralkylated oligosaccharides was achieved by reverse-phase chromatography on an octadecylsilane bonded phase (151). The eluting solvent was water-acetonitrile (1:1).

(5) Determination of Pentosans.   In TAPPI Test Method T223 (173), pentosans in an extractive-free wood sample are hydrolyzed by boiling 3.85N hydrochloric acid to furfural, which is collected in the distillate and determined colorimetrically with orcinol-ferric chloride reagent (174). Ultraviolet absorption is used to measure the furfural in an alternate method (175). Another method utilizes harmine and L-cysteine, which produce an intense pink color with pentoses (176). However, with the development of efficient GC and LC methods for sugars in hydrolyzates, an effective pentosan determination should only require summation of the amounts of pentose sugars separated on the chromatograms. Values for the repeatability and reproducibility of the GC method for sugars (46) are much better than those for the colorimetric method for pentosans (173).

(6) Determination of Uronic Acids. The principal uronic acid in wood hemicelluloses is 4-O-methyl-D-glucuronic acid. In hardwoods, the uronic acid units exist as single, terminal side chains attached to the 2-position of a xylan polymer; the uronic acid:xylose ratio is about 1:10 (40). In softwoods, the ratio is about 1:5. The pectic materials in wood contain galacturonic acid units. However, because most pectic substances are dissolved during delignification, little galacturonic acid would be expected in a holocellulose preparation.

Most carboxylic acids in wood preparations can be detected visually or with electron microscopy by means of the complex between ferric ion and the hydroxamic acid derivative of the organic acid (177)

The determination of total uronic anhydride in a wood sample or isolated hemicellulose is based on decarboxylation with strong mineral acid (178). In this determination, the theoretical quantity of carbon dioxide evolved from the uronic acids in the sample is trapped and determined. An apparatus for performing the determination gravimetrically is shown in Fig. 8 (178). The sample is decarboxylated in boiling 12% hydrochloric acid. A stream of nitrogen carries the CO_2 through a silver phosphate trap to remove furfural, an anhydrone-filled tube to remove water, and into a tube containing ascarite where the CO_2 is absorbed. The weight gain of the ascarite tube, after corrections have been made, represents the CO_2 from the uronic acids in the sample. A preliminary heating of the sample, before uronic acid decarboxylation occurs, provides the correction for carbonates. At the conclusion of decarboxylation, which may require 4 hr, heating and periodic weighing are continued. Weight increases provide the correction for the constant evolution of CO_2 from the decomposition of nonuronic acid carbohydrates.

The gravimetric method for total uronic anhydride requires a special absorption train, a semimicro balance, several hours for each determination, and careful laboratory technique. Variations of this procedure have been reviewed by Browning (179) and have been compared with colorimetric methods (180).

A semimicro method for determining uronic acids was developed in which the sample was decarboxylated with hydriodic acid (181). Evolved CO_2 was absorbed in a barium hydroxide solution. The sigmoid-shaped recording of the solution's conductivity vs. time became linear at the end of the determination. Extrapolation of the linear increase in conductivity after the uronic acid CO_2 was evolved provided a correction for nonuronic acid CO_2. A complete determination required 45 min.

The carbazole-sulfuric acid colorimetric method (182) has been widely used to determine hexuronic acids. Development of the full color requires 2 hr. In a related method employing harmine, an intense pink color develops immediately (176). Meta-hydroxydiphenyl has been used to study aldonic acids and the mucopolysaccharides of wine (177). Use of the colorimetric reagent 3, 5-dimethylphenol for determining hexuronic acids has also been proposed (183). This reagent is selective for 5-formyl-2-furancarboxylic acid (5FF) formed from uronic acids in concentrated H_2SO_4 at 70°C. Prior to color development, water-insoluble polysaccharides

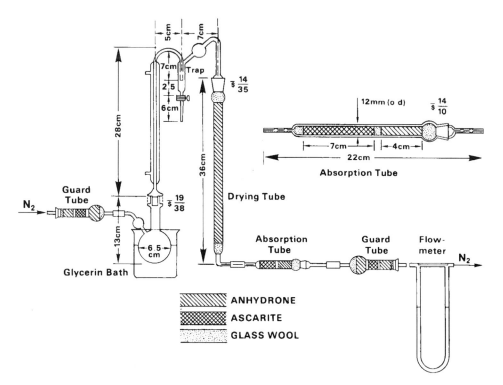

Figure 8 Uronic acid apparatus. (Reprinted from B. L. Browning, "Apparatus for Determination of Polyuronide Carboxyl," *Tappi*, *32*(3):119(1949), Technical Association of the Pulp and Paper Industry, Atlanta, Ga. ©1949 with permission. Copies available from TAPPI, Technology Park/Atlanta, P.O. Box 105113, Atlanta, GA 30348.)

are dissolved in 72% H_2SO_4 at 50°C for 10 min. After addition of NaCl and concentrated H_2SO_4, heating at 70°C for 10 min, and then cooling to room temperature, the 3, 5-dimethylphenol solution is added. Absorbance is read after allowing 10–15 min for color development. Galacturonic acid is used as a standard; its absorptivity is the same as that of 4-O-methyl glucuronic acid. The absorptivity of uronic acids in polymers was 12% higher than in monomers.

The rate and extent of formation of 5FF depended on the uronic acid configuration. Although galacturonic and 4-O-methyl glucuronic acids reacted quickly, borate addition was needed to develop the color from glucuronic acid. The yield of 5FF was also reduced by esterification of the uronic acid carboxyl group.

Interference from neutral sugars and lignin is reduced by measuring absorbances at two wavelengths, usually 450 and 400 nm. Determinations of uronic acids in wood and isolated hemicelluloses gave slightly lower results than those obtained by CO_2 evolution (183).

Difficulty was experienced in achieving a quantitative determination of the 4-O-methyl-D-glucuronic acid units in hemicellulose hydrolyzates, because the bond between the uronic acid and aldose sugars in the hemicellulose is more resistant to acid hydrolysis than the bonds between the neutral sugar units (184). Conditions necessary for complete cleavage of the glycosidic bonds also cause some decomposition of the uronic acid (185).

This problem may be avoided by reduction of the uronic acid group either before hydrolysis (186) or after a rehydrolysis of reduced partial hydrolysates (187). Analysis of the resulting sugars is usually accomplished by GC of alditol-acetate derivatives.

Unfortunately, however, packed column GC could not resolve 4-O-methyl glucitol pentaacetate from galactitol hexaacetate (148). Adequate separation has been achieved on a glass capillary column wall coated with N-propionyl-L-valine-tert-butylamide polysiloxane (148).

LC has also been used to separate uronic acids. A strong anion-exchange column in the borate form was eluted with a series of borate buffers to separate individual monosaccharides and mannuronic, guluronic, galacturonic, and glucuronic acids (188). Acidic sugar residues containing 4-O-methyl-D-glucuronic acid groups have been separated on anion-exchange resins in the acetate form with sequential elution using sodium acetate and acetic acid solutions (189).

(7) Determination of Acetyl Groups.   The hemicelluloses of hardwoods and softwoods are partially acetylated (40). Because of their ease of saponification, the O-acetyl groups are lost when the hemicelluloses are extracted from wood with alkaline solutions. The acetyl groups are largely retained during holocellulose preparation (190).

Several methods for determining acetyl in wood are based on transesterification. Older procedures based on this principle employ a titrimetric finish (191, 192). Wood meal is thoroughly dried in vacuo over P_2O_5. Sodium methoxide and methanol are added to the flask containing the sample, and the methyl acetate formed by transesterification is distilled. The collected distillate is refluxed with $0.1N$ NaOH to saponify the ester, and excess alkali is determined by titration with $0.1N$ HCl.

A simple semimicro method for the determination of acetyl groups in wood proposes hydrolysis with oxalic acid, followed by analysis for acetic acid using GC (193).

A newer, micromethod involves the gas chromatographic determination of benzyl acetate (194). After using an ultrasonic bath to disperse the ground and extracted wood sample in sodium ethoxide solution, the suspension is allowed to stand for 3 hr. The addition of water causes saponification of the ethyl acetate. After passing the solution through a cation-exchange column (H-form), tetrabutylammonium hydroxide is added to convert the acetic acid to its nonvolatile tetrabutylammonium salt. The sample is then concentrated to a syrup, dissolved in acetone, and benzyl bromide is added. Benzyl acetate is determined by GC.

Formyl, acetyl, and butyryl groups are readily removed from cellulose esters by treatment with pyrrolidine (195). This reaction has also been used to determine O-acyl groups in wood (196):

$$R-\overset{\overset{\displaystyle O}{\|}}{C}-OR' \quad + \quad \underset{N}{\overset{H}{\diagup}} \quad \longrightarrow \quad R-\overset{\overset{\displaystyle O}{\|}}{C}-N\diagup \quad + \quad R'OH \tag{3}$$

Wood is ground, extracted with acetone, and air-dried. Pyridine and pyrrolidine (1:1) containing 1-propionyl-pyrrolidine internal standard are added to the sample and allowed to stand at 80°C for 18 hr. After removal of solids by filtration, a portion of the filtrate is injected into the GC for determination of the acylpyrrolidine.

Acyl groups in wood have also been determined by IR spectrometry (197). The band at 1730 cm^{-1}, is useful for this purpose whereas the distribution of acetyl in willow xylan has been determined by ^{13}C NMR techniques (198).

b. Methylation.   Methylation analysis to identify hemicellulose structure involves complete methylation of all free hydroxyl groups on the polysaccharide, hydrolysis of the glycosidic bonds in the polymer, and identification of the methylated sugars. The free hydroxyl groups in the methylated sugars represent positions at which the sugars were involved in glycosidic bonds in the original polysaccharide. This technique does not, however, indicate the sequence in which the sugars were linked in the original polymer. Methylation techniques have recently been reviewed (199).

The older carbohydrate methylation procedures employing dimethyl sulfate (200), methyl iodide, and silver oxide (201, 202) have been largely replaced by the Hakomori procedure (203) or one of its modifications (204). In this method, methylsulfinyl carbanion is generated by dissolving sodium hydride in dimethyl sulfoxide. This is combined with the carbohydrate and stirred under nitrogen for 10 min at room temperature; then an excess of methyl iodide is added and stirred for 20 min. After diluting the reaction mixture with water, the methylated products are extracted with chloroform, washed with water, and the chloroform evaporated. More details of the procedure are provided by Sandford and Conrad (206). Complete methylation is achieved in one treatment, and the procedure requires from 1 to several hours, depending on the experimental details and material to be methylated.

The critical step in this process is generation of the sugar alkoxides catalyzed by methylsulfinyl carbanion. The carbanion has been prepared by potassium tert-butoxide in dimethyl sulfoxide, as well as by sodium hydride (207). Less background has been observed in subsequent GC analysis of the methylated products when potassium tert-butoxide is used (207).

A recently proposed modification to the Hakomori method involves pretreatment of the polysaccharide with a mixture of DMSO and 1, 1, 3, 3-tetramethylurea in a 1:1 ratio (208). With the polysaccharides tested, methylation was complete in 5 min.

It is advisable to prereduce alkali-labile polysaccharides with borohydride or borodeuteride to avoid "peeling" reactions that can occur under the alkaline conditions used for permethylation (199). For the study of polysaccharides containing uronic acid groups, an initial methylation has been recommended, then reduction of the uronic acid methyl ester to a hydroxyl group with lithium aluminum hydride, and finally remethylation (206).

The final steps of methylation analysis involve hydrolysis, conversion of the resulting partially methylated sugars into alditol acetates, and identification of the sugar units by GC/MS (199, 209).

Depolymerization of methylated polysaccharides can also be achieved using a reductive mixture of trimethylsilylmethylsulfonate and boron trifluoride (210).

c. Partial Acid Hydrolysis. Information regarding the sequences in which sugar units are arranged in polysaccharides is provided by results of partial acid hydrolysis studies. Steps in the method proposed by Albersheim et al. (151, 205) for such studies are outlined in Fig. 9.

In this procedure, the carbohydrates are methylated by a modification of the Hakomori procedure. The permethylated carbohydrate is then converted by partial formolysis or hydrolysis into a complex mixture of partially methylated oligosaccharides. Components of the oligosaccharide mixture are reduced with sodium borodeuteride, and the free hydroxyl groups are labeled by ethylation. The mixture of totally alkylated oligosaccharides is fractionated by reversed-phase LC on an octadecylsilane column, and the fractions are then collected. Structures of the oligosaccharides in the fractions are determined by total hydrolysis, conversion to alditol-acetate derivatives, and analysis by GC/MS. With the overlap of a sufficient number of oligosaccharides and data regarding their composition, the information is pieced together to determine the structure of the complex carbohydrate.

d. Periodate Oxidation. Periodate (and lead tetraacetate) oxidizes polysaccharide sugar units that possess vicinal free hydroxyl groups. Formaldehyde is produced from a primary alcohol group that has a hydroxyl group on the adjacent carbon. A secondary hydroxyl group adjacent to two other hydroxyl groups yields formic acid, as indicated in Fig. 10.

Periodate oxidation studies involve measurement of periodate consumed and formaldehyde and formic acid produced, and characterization of the oxidized polymer. The method employing periodate oxidation followed by borohydride reduction and mild acid hydrolysis of the polymer is called Smith degradation (211).

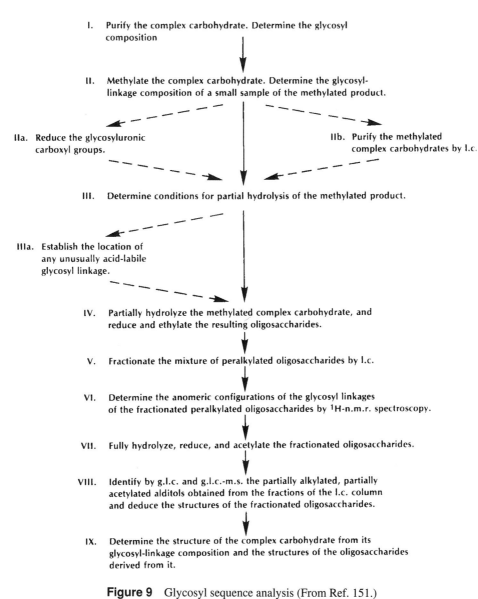

I.   Purify the complex carbohydrate. Determine the glycosyl composition

II. Methylate the complex carbohydrate. Determine the glycosyl-linkage composition of a small sample of the methylated product.

IIa. Reduce the glycosyluronic carboxyl groups.

IIb. Purify the methylated complex carbohydrates by l.c.

III. Determine conditions for partial hydrolysis of the methylated product.

IIIa. Establish the location of any unusually acid-labile glycosyl linkage.

IV. Partially hydrolyze the methylated complex carbohydrate, and reduce and ethylate the resulting oligosaccharides.

V.   Fractionate the mixture of peralkylated oligosaccharides by l.c.

VI. Determine the anomeric configurations of the glycosyl linkages of the fractionated peralkylated oligosaccharides by ^{1}H-n.m.r. spectroscopy.

VII. Fully hydrolyze, reduce, and acetylate the fractionated oligosaccharides.

VIII. Identify by g.l.c. and g.l.c.-m.s. the partially alkylated, partially acetylated alditols obtained from the fractions of the l.c. column and deduce the structures of the fractionated oligosaccharides.

IX. Determine the structure of the complex carbohydrate from its glycosyl-linkage composition and the structures of the oligosaccharides derived from it.

Figure 9   Glycosyl sequence analysis (From Ref. 151.)

When a sugar residue of a polysaccharide is cleaved by periodate and reduced, the resulting alcoholic derivative, a true acetal, is sensitive to acid hydrolysis. When a sugar unit that survives oxidation is joined to a unit that is cleaved, the surviving unit is a glycoside that is more stable to acid. Thus, the products of periodate oxidation, reduction, and mild hydrolysis are low-molecular-weight glycosides, the aglycons of which can provide structural information concerning the sugar residues from which they are derived (121).

Analysis may be performed by GC/MS after the low-molecular-weight glycosides are converted into volatile derivatives (121). These materials may also be studied by methylation analysis. Successive Smith degradations can be performed when the initial Smith degradation gives a polymeric residue.

Lead tetraacetate has been used to differentiate between cis and trans hydroxyl groups in glucomannan (212).

e. Degradation by β-Elimination.   During the Hakomori methylation, the carboxyl group of a uronic acid is converted into the methyl ester. On subsequent treatment with base, the substituent at C-4 is removed by β-elimination, leaving an acid-labile unsaturated uronic acid residue (120, 213), as shown in Eq. 4 (214). Mild acid hydrolysis degrades the uronic acid residue and releases its substituents, including the remainder of the polysaccharide of which the uronic acid was a part. Characterization of the polysaccharide involves etherification with trideuteriomethyl or ethyl groups, hydrolysis, conversion to alditol-acetates, and analysis by GC/MS.

Studies have indicated that the Hakomori methylation of compounds with 4-O-substituted uronic acid residues yields β-elimination products in addition to permethylated derivatives (215, 216). It was concluded that β-elimination occurred in the ionization step involving treatment with sodium methylsulfinylmethanide (217).

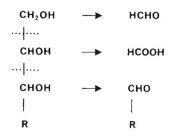

Figure 10   Reactions in periodate oxidation.

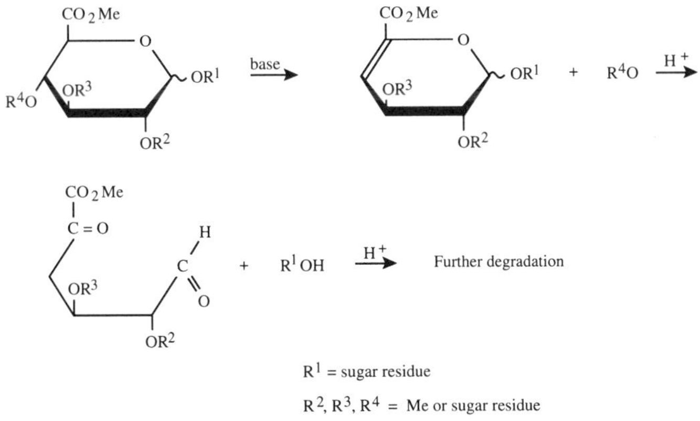

$$R^1 = \text{sugar residue}$$
$$R^2, R^3, R^4 = \text{Me or sugar residue}$$

(4)

3. Determination of Molecular Weight

Physical and chemical methods for determining the molecular weight of hemicelluloses were reviewed in detail by Browning (218). The physical methods were developed for the study of other polymers and have been adapted for the characterization of polysaccharides.

a. Osmometry. Osmotic pressure measurements are used to determine number-average molecular weights over the range 25,000–500,000 (218). If two solutions of different solute concentration are separated by a membrane permeable to solvent but impermeable to solute, the solvent will diffuse into the more concentrated solution. The extra pressure that must be applied to a solution in order just to prevent the flow of solvent through the membrane from the pure solvent is known as the osmotic pressure of the solution. Osmotic pressure is commonly indicated by the rise of solution in a capillary connected to the chamber of the osmometer containing the solution of higher-solute concentration. The chambers are usually separated by a cellophane membrane that has never been dried.

In using this procedure, the osmotic pressure π is measured for a series of concentrations c and a plot of π/c vs. c is extrapolated to zero concentration. The number-average molecular weight M_n is computed from:

$$(\pi/c)_{c \to 0} = RT/M_n \tag{5}$$

A similar equation, shown below, has been used to determine the molecular weight of a methylated or nitrate derivative of a glucomannan from red oak (219). The osmotic pressure is expressed as h, the osmotic height of the solution, and w is the concentration in grams per kilogram of solution.

$$M_n = 25{,}700/(h/w)_{w \to o} \tag{6}$$

In osmotic pressure measurements, all molecules count equally, irrespective of their size, so the number-average molecular weight is obtained (220):

$$M_n = \sum_i n_i M_i / \sum_i n_i \tag{7}$$

where n_i is the number of molecules of molecular weight M_i per liter. Other methods yield the weight-average molecular weight M_w:

$$M_w = \sum_i n_i M_i^2 / \sum_i n_i M_i \tag{8}$$

When all molecules have the same weight, $M_n = M_w$. When M_n and M_w differ, the ratio M_w/M_n indicates the polymolecularity of the sample.

Prior to osmotic pressure measurements, isolated hemicelluloses are frequently converted into derivatives soluble in organic solvents. Derivatization is essential for compounds containing uronic acid carboxyl groups, because osmotic pressure results are difficult to interpret for polyelectrolyte samples in aqueous solution. Representative derivatives and solvents for osmometry are shown in Table 1. Also included in the table are some water-soluble hemicelluloses that were run without derivatization.

b. Gel Permeation Chromatography. In gel permeation chromatography (GPC), molecules are separated according to their effective size in solution. Column packings are porous solids having various pore sizes. Separation is based on the selective diffusion of sample molecules into and out of the pores. Small molecules diffuse into all of the pores; molecules of intermediate size fit into some of the pores and are excluded from others. The largest molecules are excluded from all of the pores in the packing. Thus, the large molecules emerge from the column first,

Table 1   Osmometry of Hemicelluloses

Hemicellulose	Derivative	Solvent	Ref.
Laricinan (an	nitrate	acetone	(221)
acidic glucan)	methyl	toluene	
Glucomannan	nitrate	n-butyl acetate	
	methyl	chloroform-ethanol	
		(9:1)	(219)
4-O-Methylglucu-			
ronoxylan	acetate	chloroform-ethanol (9:1)	(223)
Xyloglucan	—	0.2M aqueous NaCl	(118)
Arabinogalactan	—	water-DMF (20:80)	(224)

followed by medium-sized molecules and, finally, the smallest molecules. Results reflect the molecular size or weight distribution of the sample.

Much of the analysis of hemicelluloses by GPC has been performed on soft gel column packings, especially cross-linked dextran resins. One example is the determination of the molecular weight distribution of a xyloglucan (118). Because these packings are fragile and cannot withstand high pressure (225), elution of the xyloglucan with sodium acetate-acetic acid buffer required 20–25 hr. Xyloglucan fractions were collected and quantitated by the phenol-sulfuric acid method (226). Dextrans and sugars of known molecular weight were used to calibrate the column. Also studied on cross-linked dextran packings were arabinogalactans (224, 227, 228), a 4-O-methylglucuronoxylan (229), and pectin (229) using water or sodium acetate buffer as the eluent.

Cross-linked polyacrylamide packings have been used for GPC analysis of water-soluble galactoglucomannan and an arabinose-, mannose-, and galactose-containing polymer from loblolly pine using water as the eluent (230). The same type of column was also used for the study of arabinogalactans from acacia (231) and larch (232). Separations required several hours, because these are also compressible, soft-gel packings.

In contrast, in about 1 hr. a granular porous glass packing separated larch arabinogalactan components having molecular weights of 100,000 and 18,000 (230); the rigid packing permitted high pressure and flow rate. A refractive index detector was used with water eluent. Another rigid packing consisting of a glycerylpropylsilyl layer covalently bonded to spherical silica gel particles has been used for GPC separations of pectins (233). The chromatogram required only 10 min.

Separations of this type are not used exclusively for the study of the native polymers. Products of Smith degradations generated in polysaccharide structural studies have also been analyzed by GPC (231).

GPC and other chromatographic techniques for the analysis of carbohydrates have been reviewed by Churms (234, 235).

c. *Viscosity.*   Measurement of the viscosity of hemicellulose in an appropriate solvent is a straightforward way to determine average molecular weight. These measurements are normally performed in a capillary viscometer. The relative viscosity is the viscosity of the sample solution relative to that of the solvent. Subtracting 1 from the relative viscosity yields the specific viscosity η_{sp}. Extrapolating a plot of η_{sp}/c vs. concentration to zero concentration yields the intrinsic viscosity $[\eta]$. The degree of polymerization (DP) of the polymer is obtained from the empirical relationship, $[\eta] = k_m \times DP$, where k_m is the Staudinger constant. Values of k_m for glucomannan in a NaOH solution and as the acetate in nonaqueous solvents have been published (236). Cupriethylenediamine is commonly used as the solvent for viscosity determinations on hemicelluloses (237).

Care must be taken in interpreting the viscometric analysis of hemicelluloses as their low molecular weights are frequently in a nonidealistic range (238).

Anhydrous dimethyl sulfoxide (DMSO) was used as the solvent for determination of the viscosity of laricinan, an acidic glucan from tamarack (221). A plot of η_{sp}/c vs. concentration showed a viscosity increase at low concentration. This behavior is characteristic of polyelectrolytes and confirmed the acidic nature of the polymer. The polyelectrolyte effect disappeared when the viscosity was determined in DMSO-water (4:1) containing $0.05M$ NaCl.

d. Other Methods.   A low-angle laser-light HPLC technique for the estimation of the molecular-weight distribution of amylose might be employed for the study of wood polysaccharides as well (222). Molecular-weight determinations employing light scattering, the ultracentrifuge, and end-group determinations have been discussed by Browning (218).

D. Distribution of Carbohydrates in Cell Walls

Methods for studying the distribution of wood components in the walls of woody cells have been reviewed (239). Those procedures that are most appropriate for studying the distribution of lignin are considered in a later section of this chapter.

Among techniques used in investigations of polysaccharide distribution are light and electron microscopy, selective staining (240), selective destruction of components by chemical and enzymatic methods (241), autoradiography (242, 243) and micromanipulation or beating of fibers followed by chemical analysis (244, 245).

Another technique involves the use of solvents and their relative rates of extraction of fibers (246). Many of these methods were designed to improve visualization of components in subsequent microscopic examination.

Recent studies have employed improved procedures for micromanipulation and analysis (247). Undamaged single fibers were isolated from thin spruce chips from which 10% of the lignin had been removed by a chlorite treatment. Separation of cell wall layers was performed by hand using specially sharpened tweezers under a stereomicroscope equipped with a polarizing accessory. The fibers were immersed in water during this procedure. Carbohydrate analysis of the isolated cell wall layers involved microscale hydrolysis and reduction to alditols, preparation of the trifluoroacetate derivatives, and analysis on a gas chromatograph with an electron capture detector (142).

V. LIGNIN

A. Detection of Lignin

Several color-forming reactions and instrumental analysis techniques are available for detecting lignin in wood. Stains and color reactions have been reviewed by Browning (248).

Two color reactions that are used widely are the phloroglucinol test and the Mäule reaction. The phloroglucinol stain is prepared by dissolving phloroglucinol in a mixture of methanol, concentrated hydrochloric acid, and water (249). It is sprayed on paper or pulp to indicate the presence of groundwood. The red or magenta color produced has been attributed to cinnamaldehyde-type groups in the lignin (250).

The Mäule test (251) is used to distinguish hardwood from softwood. A wood sample is treated with 1% aqueous potassium permanganate, washed, treated with 12% hydrochloric acid, washed again, and then moistened with 10% ammonium hydroxide. Hardwood shows a deep rose-red color, and softwood is stained pale yellow or brown. The red color has been attributed to the 3-methoxy-o-quinone type structure derived from hardwood (syringyl) lignin (252). The application of this test to the determination of hardwood/softwood proportions in wood chip blends has recently been described (253).

Lignin may also be detected by its characteristic ultraviolet and infrared spectra. An infrared spectrum of softwood lignin is shown in Fig. 11.

B. Determination of Lignin

1. Acid-Insoluble Lignin

The most widely accepted method for determining lignin in wood involves hydrolysis and solution of the polysaccharides with strong mineral acid followed by filtering, washing, drying, and weighing the lignin residue. The use of sulfuric acid for this purpose, originally proposed by Klason (254), has been incorporated in TAPPI (255) and ASTM (256) standard test methods.

Prior to treatment with sulfuric acid, a groundwood sample is preextracted with

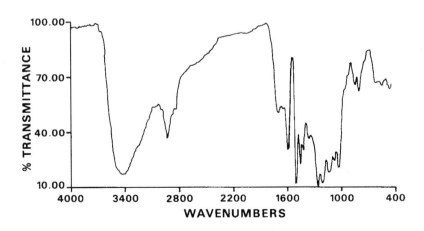

Figure 11   Infrared spectrum of lignin.

alcohol-benzene (1:2). The ASTM method recommends the use of a 95% ethanol extraction for woods with a high tannin content.

Following preextraction, the air-dried 1-g sample is treated with 72% sulfuric acid for 2 hr at $20 \pm 1°C$. The sample is then diluted with water to reduce the sulfuric acid concentration to 3% and boiled for 4 hr. After the lignin is allowed to settle, it is collected in a tared filtering crucible, washed free of acid, dried, and weighed. Because of condensation reactions occurring during the sulfuric acid treatment, this lignin is changed significantly from its original structure in the wood.

Results from a modified procedure for acid-insoluble lignin (257) have been found to correlate well with those from the standard method. The modified procedure requires only 200–300 mg of the sample and uses an autoclave at 120°C for 1 hr for secondary hydrolysis.

Correct preparation of 72% sulfuric acid is essential in the Klason lignin determination. The ASTM procedure specifies a concentration of $72 \pm 0.1\%$ (254). Errors resulting in a sulfuric acid concentration less than 72% are especially damaging, because they lead to erroneously high lignin values. Specific gravity measurements are recommended for assuring the correct acid concentration.

Other modifications of the Klason lignin determination have also been published (258).

2. Acid-Soluble Lignin

During the determination of acid-insoluble lignin, some of the lignin dissolves in the acid. This acid-soluble lignin amounts to 0.2–0.5% of softwoods and 3–5% of hardwoods (255). Determination of acid-soluble lignin involves measuring the ultraviolet absorbance of the filtrate from the acid-insoluble lignin isolation (259). The wavelength at which the measurement is made is 205 nm, and an absorptivity of 110 L/g·cm is used in the calculation (260–262). It is essential in this determination that 3% sulfuric acid be used as the reference solution in the UV measurement.

When available instrumentation permits, a wavelength scan in the 205-nm region is recommended. This assures that the absorbance is measured at the apex of the peak. It also confirms that a peak exists and that the absorbance is not rising continuously due to an interference.

3. Spectral Method for Total Lignin

Lignin in wood may be determined by dissolving the finely ground sample in acetyl bromide in acetic acid and then measuring the ultraviolet absorbance of the resulting solution at 280 nm (263). Accuracy and reproducibility are improved by the use of freshly prepared digestion mixture free from moisture contamination (264). Computation of lignin in wood requires the use of an absorptivity value measured separately. This may be determined by applying the same procedure to an isolated lignin or to a wood sample whose lignin content had been measured by the Klason method.

A technique for the determination of lignin in polyphenol-rich eucalyptus wood has been described (265). When this method is used, a lignin determination can be completed in an hour. Only small amounts of sample (10–25 mg) are required. Further discussion of this and other methods for determining lignin is provided by Browning (266).

C. Determination of Morphological Distribution of Lignin

Early studies of the distribution of lignin in wood fibers were summarized by Berlyn and Mark (267). They estimated that less than 40% (subsequently found to be less than 30%) of the total lignin is located in the middle lamella plus primary wall, and they pointed out the need for additional work in this area. The principal tools used in subsequent investigations have been light (especially ultraviolet) and electron microscopy.

1. Light Microscopy

Applications of the ultraviolet microscope for the quantitative determination of lignin distribution have been perfected by Goring et al. (268, 269). Although studies of lignin distribution approach the lower limit of resolution of the light microscope, the use of ultraviolet light for this task improved resolution by a factor of 2. Lignin is the only major component of the wood cell that possesses absorption maxima at 212 and 280 nm. Thus, measurements in these spectral regions provide the basis for the specific determination of lignin.

Development of an accurate ultraviolet microscopic method for lignin distribution required optimization of each stage of the procedure (268). Small wood chips were embedded in Epon 812, and ultrathin sections were cut with a diamond knife. Photomicrographs were taken at 280 nm for quantitation and at 240 nm for display purposes. The silver density of the developed negative was determined with a recording densitometer. A photomicrograph and densitometer tracing are shown in Fig. 12. Quantitation was based on the Beer–Lambert law and the absorptivity of sodium lignosulfonate solutions. Lignin quantitation using an ultraviolet microscopic image analyzer is believed to be more accurate than the indirect photographic method (270).

The lower limit of resolution of the microscope at 280 nm was about 0.16 μm. In transverse sections, this permitted resolution of the compound middle lamella but not its individual components, the primary walls of adjacent tracheids and the true middle lamella.

The ultraviolet microscope has been used to study the character as well as the distribution of cell wall lignin (271). Lignins rich in syringyl units exhibit a lower ratio between absorbances at 280 and 260 nm than is typical of softwood lignin (272, 273). Results from the use of this technique have suggested that middle lamella lignin in hardwoods is rich in guaiacyl units and that the ratio of syringyl to guaiacyl units is higher in the fiber walls.

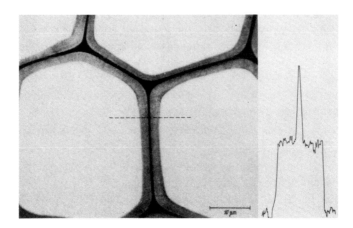

Figure 12 Cross section of Epon-embedded tracheids of black spruce earlywood photographed in ultraviolet light of wavelength 240 nm. The densitometer tracing was taken across the tracheid wall on the dotted line (From Ref. 269.)

Distribution of phenolic hydroxyl content in lignin has also been measured with the ultraviolet microscope (274). The difference between absorbance at pH 12 and pH 6, measured at 300 nm, is considered to be proportional to phenolic hydroxyl content (275).

In addition to the study of wood, the ultraviolet microscope has been used in investigations of the topochemistry of delignification (271, 276–280). Techniques are similar to those used on intact wood.

Lignin distribution has also been determined by interference microscopy (281). Volume fractions of lignin in different parts of the cell wall in spruce and pine wood were calculated from the refractive indices of these parts. Results were similar to those obtained by ultraviolet microscopy.

2. Electron Microscopy

The transmission electron microscope (TEM) and the scanning electron microscope (SEM) have both been used to determine lignin distribution. Early studies, which employed visual examination of photomicrographs, yielded only qualitative and semiquantitative information. Visualization of the lignin was aided by selective removal of carbohydrates with hydrofluoric acid (creating "lignin skeletons") (282, 283) or by application of potassium permanganate stain (284). More detailed qualitative information has recently been obtained by the use of the scanning transmission electron microscope (STEM) (285).

Combining the electron microscope with energy-dispersive X-ray analysis (EDXA) has provided quantitative data on lignin distribution in wood (286–290). When the sample in the microscope is bombarded with electrons, X rays are emitted with energies characteristic of the elements in the sample. Most studies have been performed with instrumentation that could not detect elements with atomic numbers less than 11 (sodium). Bromination has been used to incorporate in the lignin an element whose X rays could be detected (286). It has been assumed that lignin bromination takes place primarily by addition to double bonds in side-chain structures and by substitutions on phenolic nuclei. Prior to bromination, samples are treated with 2% NaOH, rinsed with water, dehydrated with methanol, and stirred in chloroform. To obtain absolute quantitative data, an X-ray emission count from an appropriate brominated lignin preparation is required as a standard (287).

Initial studies revealed that relative amounts of lignin in the middle lamella and the secondary wall determined by ultraviolet microscopy and the SEM-EDXA technique were not in agreement (288). This discrepancy was resolved when it was learned that secondary wall lignin was 1.7 times more reactive toward bromination than the middle lamella lignin (291).

In applying the SEM-EDXA technique, the size of the X-ray emitting area was larger than the morphological regions. Thus, it was not possible to study the lignin in the middle lamella, primary wall, and S_1 and S_3 layers as separate parts of the fiber cross section. However, because of its better resolution, the TEM-EDXA technique allowed the study of the various cell wall layers as separate entities (287). Values for the spatial resolution of the SEM-EDXA and TEM-EDXA techniques were reported to be 0.3 and 0.03 μm, respectively (288).

SEM and SEM-EDXA techniques have also been used to study the topochemistry of pulping (286, 292–294). In addition to bromine, chemical markers used to permit lignin to be detected by EDXA include sulfur from sodium sulfite pulping liquor (293) and chlorine (294, 295).

3.   Analysis of Separated Fiber Components

Separation of wood fibers into their components followed by analysis of the fractions provides another way to study lignin distribution. Separations have been achieved by disintegration of the wood, followed by sieving (296), differential sedimentation (297), or fiber-length classification of a mechanical pulp (298). The fractions isolated from spruce by sieving were (1) fibers, (2) ray cells, and (3) fines originating from the middle lamella and primary wall. Micro or semimicro chemical methods (methoxyl, nitrobenzene oxidation, spectrophotometric lignin determination, and methylation followed by permanganate oxidation) were used to characterize the lignin in the sieved fractions (296). Differential sedimentation yielded middle lamella and secondary wall tissue, which were characterized by elemental analyses and methoxyl, carbonyl, and carboxyl determinations (302). In other work, flakelike middle lamella fragments were studied by ultraviolet microscopy (299). Ultraviolet microscopy has also been used to study tissue fractions that were pulped by the kraft, acid sulfite, and acid chlorite methods (271).

A comprehensive study has recently described the guaicyl and syringyl composition of hardwood cell components by a variety of spectral and analytical techniques (300, 301).

D. Characterization of Lignin

Methods used to determine the basic structure and bonding in lignin have been reviewed by Lai and Sarkanen (303) and by Adler (304).

1. Isolation of Lignin

Studies to characterize lignin are performed on isolated lignin preparations or directly on wood. Lignin preparations resulting from the acid hydrolysis of polysaccharides, as in the Klason lignin determination, are not favored for these studies because of the formation of carbon–carbon bonds ("condensation reactions") in the strong acid medium. "Brauns lignins," soluble lignins originally isolated by ethanol extraction (305), represent only a small proportion of the total lignin in the wood; they presently receive little use in structural investigations.

In preparation of Bjorkman or milled wood lignins (MWL), wood is suspended in toluene and finely disintegrated in a vibratory ball mill (306, 307). After this treatment, much of the lignin in the wood becomes extractable with dioxane-water (9:1). The procedure may be shortened by milling with steel rather than porcelain balls and extracting in an ultrasonic generator (308). An alternate procedure involves dry milling in a rotary ball mill (309) and then extracting the MWL with 80% acetone (310). The yield of lignin is increased when ball milling is followed by treatment with cellulolytic enzymes (311, 312). Lignin is then extracted with aqueous dioxane. MWL and "cellulolytic enzyme lignins" contain small amounts of carbohydrates. Although these preparations are not identical with lignin in situ (313), they are considered to be the best lignin preparations currently available.

A shorter procedure for isolating lignin from extractive-free pulverized wood involves Soxhlet extraction with dioxane containing a small amount of HCl (314, 315). Although it was thought that this procedure did not produce major structural changes (316), the infrared spectra of this dioxane lignin and MWL showed some differences (315).

Details of these and other lignin isolation procedures have been described by Browning (317).

2. Functional Groups

a. Methoxyl.   Methoxyl groups in lignin as well as in polysaccharides and intact wood are usually determined by procedures based on the original method of Zeisel (318). The sample is boiled with constant boiling hydriodic acid; the methoxyl groups are converted to volatile methyl iodide, which is determined by gravimetric, volumetric, or gas chromatographic methods. Procedures have been reviewed by Browning (319).

ASTM D1166-60 (320) employs the apparatus shown in Fig. 13 for the determination of methoxyl groups. In this method, the methyl iodide is collected in an acetic acid solution containing potassium acetate and bromine. The following reactions occur:

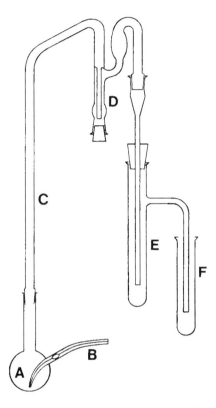

Figure 13 Apparatus for methoxyl determination. A, reaction flask; B, CO$_2$ inlet tube; C, vertical air-cooled condenser; D, trap or scrubber; E and F, absorption vessels or receivers. (From Ref. 320, American Society for Testing and Materials, Philadelphia, Pa. reprinted with permission.)

$$CH_3I + Br_2 \rightarrow CH_3Br + IBr \qquad\qquad (9)$$
$$IBr + 2\ Br_2 + 3\ H_2O \rightarrow HIO_3 + 5\ HBr \qquad\qquad (10)$$

Excess bromine is destroyed by the addition of formic acid. Iodic acid is then determined by titration of the iodine liberated in this reaction:

$$HIO_3 + 5\ HI \rightarrow 3\ I_2 + 3\ H_2O \qquad\qquad (11)$$

The ASTM procedure specifies the addition of phenol to the hydriodic acid in the reaction flask to promote complete conversion of methoxyl to methyl iodide. It was observed, however, that phenol tended to depress alkoxyl values in the analysis of vanillin and ethyl cellulose (321). Therefore, propionic anhydride was used in place of phenol in the TAPPI test method (now withdrawn) for methoxyl (322).

The Zeisel reaction has been coupled with the gas chromatographic determination of methyl iodide in a micromethod for methoxyl in wood tissue fractions (296). From 0.1 to 0.8 mg of sample was reacted with hydriodic acid and acetic anhydride in a sealed tube at 140°C . After 18 hr the tube was cooled, and the methyl iodide was extracted into toluene containing chloroform internal standard for GC analysis.

A modified method, which does not require hydriodic acid has been developed for determining methoxyl in lignin preparations (323). The demethylating agent was a mixture of potassium iodide and concentrated orthophosphoric acid.

b. Phenolic Hydroxyl.   Methods for determining free phenolic hydroxyl groups in lignins have been based on ionization difference spectra (324, 325), potentiometric titration in nonaqueous media (326, 327), conductometric titrations (328), pyrolytic gas chromatography (329), and other procedures (330). Experimental details of several methods have been reviewed by Browning (331). A specific method for phenolic groups in guaiacyl lignins is based on the finding that phenols with a methoxyl group in the o-position are oxidized by periodate to form the o-quinone and methanol (332). The methanol, formed in the reaction below, may be determined by GC.

$$R' \xrightarrow{\quad IO_4^- \quad} \quad + \quad MeOH \tag{12}$$

A ^{13}C NMR technique for the quantitative estimation of hydroxy groups in lignin has been described (333) as has a technique for the determination of phenolic and total hydroxyl groups in lignin (334).

C. Carbonyl.   Carbonyl groups in lignin may be determined by reaction with the common carbonyl reagents, such as phenylhydrazine (335) and hydroxylamine (336). A borohydride reduction procedure is used to determine carbonyl in materials as dissimilar as lignin (337) and bleached pulps (338).

In the borohyride method, the sample, borohydride, and sulfuric acid are placed separately in the three arms of a three-arm reaction vessel (337). The sample and borohydride are combined and allowed to react for several hours. Finally, the sulfuric acid is added to liberate hydrogen from the unreacted borohydride. The difference between the volumes of hydrogen liberated in a blank test and in the presence of the sample is equivalent to the borohydride that reacted with the carbonyl groups in the sample. Values for carbonyl in lignin preparations will be erroneously high if air is not excluded from the apparatus (339).

Carbonyl groups in lignin have also been measured by infrared analysis. It was assumed that the area under the 1670 cm^{-1} band was directly proportional to the carbonyl content of the sample (340). Comparison of this band in the spectra of isolated lignins with that in acetovanillone gave values for the carbonyl/methoxyl ratio; typical results were near 0.2.

Another method for determining carbonyl groups in lignin employs ultraviolet difference spectra (341). Spectra of borohydride-reduced lignin samples were subtracted from the spectra of alkaline solutions of the original samples. These spectra were compared with similar difference spectra of model compounds to give amounts of conjugated and nonconjugated carbonyl groups. Polarography has been considered for the analysis of lignin functional groups, especially carbonyl groups (342).

d. Benzyl Alcohol.   A method for determining benzyl alcohol groups was based on the finding that benzyl alcohols were readily oxidized by 2, 3-dichloro-5, 6-dicyano-benzoquinone to their corresponding α-carbonyl compounds (343).

3.   Structure and Bonding

a. Arylpropane Structural Units   The principal classical methods used to demonstrate the arylpropanoid structure of lignin include permanganate oxidation, nitrobenzene oxidation, hydrogenolysis, and ethanolysis to produce Hibbert ketones (304, 344).

In the permanganate oxidation technique (345, 346), spruce wood or lignin was first heated with 70% aqueous KOH to hydrolyze ether linkages. After protection of the liberated phenolic groups by methylation, permanganate oxidation yielded a mixture of mono- and polycarboxylic aromatic acids with methoxyl substituents. Results were consistent with the presence of guaiacyl propane and condensed structures and noncyclic ether bridges in lignin (304).

Alkaline nitrobenzene oxidation yields vanillin from softwoods (347) and vanillin and syringaldehyde from hardwoods (348). Comparable product distributions may be obtained by oxidation with cupric oxide (349, 350).

Hydrogenolysis of an isolated lignin produced propylcyclohexane derivatives in high yields (351). This was the first isolation of lignin degradation products that contained the complete C_6C_3 skeleton. Identical products have been obtained by the hydrogenation of wood (352). Studies in which hydrogenolysis is used to elucidate lignin structure have been reviewed by Hrutfiord (353).

When coniferous wood is refluxed with 2% ethanolic HCl, the so-called "Hibbert ketones" (Fig. 14) are produced (354). These were the first isolated lignin degradation products having guaiacylpropane structures.

In more recent investigations employing these classical techniques, products have been separated by packed (296, 355, 356) or capillary column (357) GC or automated methods (358) and identified by MS (359).

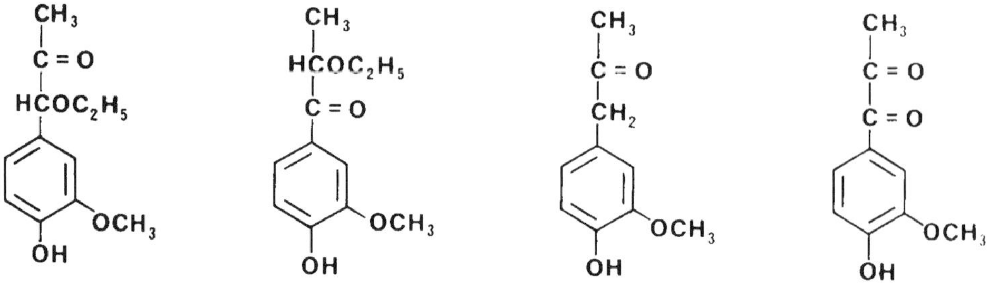

Figure 14   Hibbert ketones.

b. Dimeric Structures and Bonding.   Several degradation techniques have been developed to liberate dimeric structures from isolated lignins or wood. Identification and quantitation of the structures have revealed types and frequencies of the bonds existing in lignin. Mild treatments have been used to minimize condensation reactions. Current applications of these degradation methods use state-of-the-art chromatographic and spectrometric techniques for product analysis. Consequently, these analyses are more informative than those employed by the original investigators.

Acidolysis involves refluxing wood, isolated lignin, or model compounds with dioxane-water (9:1) containing the equivalent of 0.2$\underline{M}$ HCl (360). This revealed that the most abundant substructure in softwood lignin is arylglycerol-β-aryl ether. Other gentle hydrolysis procedures applied to wood and lignins involved percolation with water or 2% aqueous acetic acid at 100°C for several weeks (361) or treatment with dioxane-water at 180°C for 20 min (362–364).

Hydrogenolysis techniques, originally used to demonstrate the C_6C_3 skeleton, have been extended to show linkages in dimeric and trimeric structures (365–368).

Steps in thioacetolysis include treatment of wood with thioacetic acid and BF_3, saponification with 2$\underline{N}$ NaOH at 60°C, and treatment with Raney nickel and alkali at 115°C (369). Frequencies of the various bond types in beech lignin have been calculated on the basis of yields of degradation products from thioacetolysis, mild hydrolysis, and ^{13}C NMR spectroscopy (370).

One of the most informative degradation techniques is based on Freudenberg's permanganate oxidation (345, 346). The procedure is outlined in Fig. 15 (304). Wood or isolated lignin was methylated and oxidized to aromatic acids, which indicated the units in lignin with free phenolic hydroxyls. A preliminary kraft cook or cupric oxide oxidation (349) converted etherified units into phenolics; the increase in the yield of aromatic acids provided an estimate of etherified units. Alkaline peroxide degraded phenylglyoxylic acids to the corresponding benzoic acids, which were methylated and identified by GC and GC/MS (355, 359).

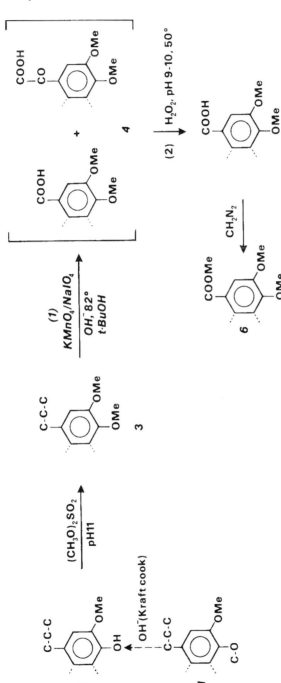

Figure 15   Oxidative degradation of lignin. (From Ref. 304.)

Combination of gel permeation chromatography with GC (371) provides an effective complement to the original GC procedure used to analyze oxidation products (372). Permanganate oxidation results were used as the basis for a calculation of the proportions of different types of bonds connecting the arylpropane units in spruce MWL (373).

An analysis technique, termed SIMREL, uses a computer to interpret primary analytical information from an isolated lignin preparation in terms of a definite chemical structure of lignin (374). Primary data include elemental analysis, contamination, functional groups, proton NMR spectra, and results from permanganate oxidation. The technique is shown conceptually in Fig. 16.

c. Bonding Between Lignin and Carbohydrates. Evidence for the existence of chemical bonds between lignin and carbohydrates has been obtained from studies of lignin-carbohydrate complexes (LCCs). In the isolation of LCC, wood meal is ball milled, MWL is extracted with 80% dioxane, and the LCC is then extracted from the residual wood meal with dimethyl formamide (375, 376). Other researchers use DMSO (377). The crude LCC may be fractionated by ion-exchange chromatography (DEAE Sephadex) using water, ammonium carbonate, and acetic acid to elute acetyl glucomannan, acidic polysaccharides, and a lignin-carbohydrate fraction, respectively (376). Further purification of the lignin-rich fraction involved treatment with β-glucosidase.

An alternate procedure used KOH extraction to isolate LCC from chlorite holocellulose (378, 379, 380). This was also followed by ion-exchange fractionation.

Techniques originally developed for the study of hemicellulose structures (described earlier in this chapter) have been used to reveal the monosaccharide units involved in lignin-carbohydrate bonds and also to indicate the positions at which bonding occurs. Results from Hakomori methylation (203) followed by hydrolysis and GC and MS analysis of methylated alditol acetates have indicated that galactose, arabinose, and xylose were involved in glycosidic linkages to lignin (376). Other studies have employed (1) treatment with 0.1M NaOH in the presence of borohydride followed by selective hydrolysis of furanosidic bonds, and (2) Smith degradation (211) (periodate oxidation followed by mild acid hydrolysis). From this work has come evidence for the existence of ether links (probably benzyl ether links) between lignin and C-2 or C-3 of arabinose and C-3 of galactose (318). Also indicated were ester bonds between lignin and 4-O-methylglucuronic acid units (381, 382).

An additional technique used to study LCCs is hydrophobic interaction chromatography on derivatized agarose gels (phenyl- and octyl-Sepharose) (383, 384). The affinity of LCC for the gels has been interpreted as proof of the existence of lignin-carbohydrate linkages in the LCCs.

Based on the results of an investigation of benzyl ether bond properties in model compounds, a method for the quantitative determination of benzyl ether bonds in a

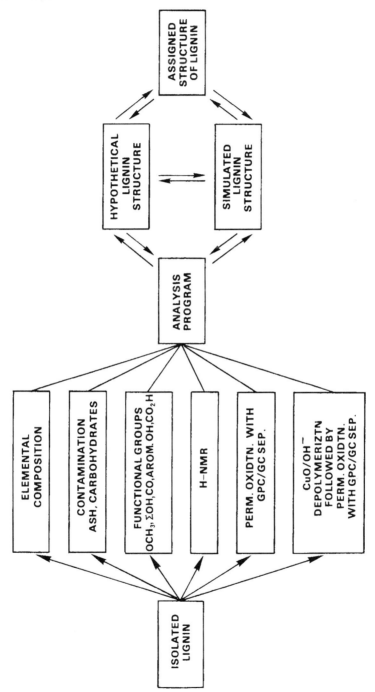

Figure 16 Conceptual presentation of SIMREL analysis technique. (From Ref. 374.)

spruce LCC has been developed (384). Total benzyl alcohol groups were determined by oxidation to the corresponding α-carbonyl compounds using 2,3-dichloro-
5,6-dicyano-1,4-benzoquinone. The p-hydroxyl benzyl alcohol groups were measured by reaction with quinone monochloroimide. The determination of benzyl
ether lignin-carbohydrate bonds employs these tests and the relative reactivities of
the various ether links in LCC under acidic and basic conditions.

The analysis of products from reactions of model compounds has confirmed that
benzyl ether bonds could be formed with all types of carbohydrate hydroxy groups
(385). Vanillyl alcohol was reacted with a range of sugars and polyhydroxy
compounds in aqueous solution. Reaction products were extracted, separated
chromatographically, and characterized by MS and NMR. Results suggest that the
5-position of arabinofuranose units and the 6-positions of 1,4- or 1,3-1inked
glucose, mannose, or galactose units should be favored for lignin-carbohydrate
bond formation (386).

4. Molecular Weight

Lignin must be removed from wood to measure its molecular weight. Because
structural changes are possible in all methods of lignin isolation, molecular weights
measured on lignin preparations provide only an approximation of the molecular
weight of the lignin in a sample of wood. Milled wood lignin (306) is regarded as
a preparation in which minimal degradation has occurred.

Methods available for measuring lignin molecular weight include osmometry,
light scattering, ultracentrifugation, viscosity and diffusion methods, and gel
filtration or gel permeation chromatography (387). The method presently receiving
greatest use is gel permeation chromatography (GPC). (See page 78) Results
provide an indication of the sample's molecular-weight distribution, or more
correctly, molecular-size distribution.

The early use of GPC to show qualitatively the molecular-weight distribution of
MWL employed a soft gel dextran (Sephadex G) swollen with DMSO (388).
Quantitative studies require that the GPC column be calibrated with substances of
known molecular weight. For that purpose, dehydrogenation polymers (DHP) of
coniferyl alcohol were prepared (389), fractionated by preparatory GPC, and their
molecular weights determined by sedimentation equilibrium ultracentrifugation
(390). An improved calibration method obtains the molecular weights of the DHP
fractions from the ultracentrifuge at zero concentration and zero field force (391,
393) Commercial polystyrene standards and dextrans are now commonly used for
calibration (394–396).

More recent GPC separations of lignins have employed cross-linked agarose gel
(Sepharose CL) (396, 397) and a cross-linked dextran gel (Sephadex G100) (398).
High-performance liquid chromatography (HPLC) has been conducted with rigid
or semirigid column packings such as porous glass (397) and polystyrene cross-
linked with divinylbenzene (μ Styragel) (396).

Gel permeation chromatograms of MWL have shown bimodal (399) or single symmetrical distributions (311, 394), depending on experimental conditions. The multiple maxima are believed to be due to the intermolecular association of lignin molecules (400), which can be eliminated by the addition of lithium chloride to the DMF eluent (396).

GPC data may be used to calculate the number average, weight average, and Z-average molecular weights and polydispersity of lignin samples (394, 401).

E. Spectral Analysis of Lignin

1. Ultraviolet

a. Origin of Spectra.   The absorption of ultraviolet (UV) light is caused principally by electronic excitation, the promotion of electrons from the ground state to higher energy states. In the transition to a higher electronic level, a molecule can go *from* any of a number of vibrational and rotational sublevels *to* any of a number of sublevels. As a result, a heterogeneous set of molecules would be expected to absorb energy over a wide range. Because the lignin polymer contains several different UV-absorbing arylpropane structures, the bands in its UV spectrum are especially broad (402).

Hexafluoropropanol has been proposed as a valuable solvent for lignin in UV and IR spectrography (392).

Methods used to interpret the UV spectrum of lignin have been reviewed in detail by Goldschmid (403). The principal tool has been UV difference spectroscopy (404). Ionization difference spectra were obtained from measurements in neutral and alkaline solution. Phenolic hydroxyl contents were determined by comparing the heights of the difference maxima of lignin preparations with those of model compounds (275). Carbonyl groups were determined from the difference in spectra produced before and after borohydride reduction (241) or ethanolysis (405). Borohydride reduction and ethanolysis difference spectra of lignins, like ionization difference spectra, were compared with those of a large number of model compounds. Similarly, hydrogenation difference spectra indicated the UV maxima due to α–β double bonds (406).

Computer techniques have been used to resolve the electronic absorption spectra of lignins into their components (407-409). The results of one of these approaches suggested that the electronic spectra of lignin preparations could be described by a combination of 13 Gaussian bands (408).

A technique currently in use for studying the UV spectra of lignin and related compounds is derivative spectroscopy (410, 411). Many modern UV-visible spectrometers have the capacity to generate up to second-derivative and in some cases fourth-derivative spectra (412). Broad, diffuse absorbance spectra of lignin model compounds are converted into well-resolved peaks when their second-derivative spectra are obtained (410). The peaks in derivative spectra have been reported to be more convenient to use for quantitative measurements (410).

b. Applications.   Applications of UV spectrometry described previously in this chapter include the determination of acid-soluble lignin (259), determination of total lignin in wood dissolved in acetyl bromide and acetic acid (263), and studies of the distribution of lignin in wood by means of the UV microscope (268, 269).

UV spectra, as well as spectra obtained by IR and NMR, are used routinely for the characterization of isolated lignins (413, 414). UV difference spectra have also been used for lignin characterization (415). The approach, involving ionization, hydrogenation, and acidolysis, was similar to that described above for interpreting the UV spectrum of lignin (404).

2.   Infrared

a. Interpretation of Spectra.   Assignment of infrared (IR) absorption bands to various functional groups in lignin has been achieved by studies of the band shifts produced by the derivatization of lignin and lignin model compounds. These procedures have been reviewed by Hergert (416) IR band assignments for softwood and hardwood lignins are shown in Table 2 (416).

b. Applications.   IR spectra are commonly used to characterize isolated lignins and to document changes produced in lignin in wood by various treatments.

Spectra are obtained on lignin samples in the form of KBr pellets, mulls in mineral oil, or as films cast on a salt plate by evaporation of the solvent from a lignin solution.

Samples of wood may be examined as thin sections or ground and formed into KBr pellets. Transmission IR spectra have been recorded with wood sections impregnated with mineral oil to reduce scatter or sandwiched in disks between layers of KBr or KCl (417, 418). Alternatively, the microtome sections may be clamped in contact with a KRS-5 crystal (a TlBr-TlI mixture); the wood is wetted with mineral oil to promote contact (419). The IR spectrum is then obtained by the attenuated total reflection (ATR) or multiple internal reflection (MIR) technique. Similar spectra are produced by the transmission and reflection methods. However, bands in the ATR spectrum appear deeper at longer wavelengths than in transmission spectra due to the wavelength dependence of the ATR effect (419).

IR spectra of wood components have been obtained after the other components, such as lignin or hemicellulose, were chemically removed (417). Similar information may be gained by placing appropriate materials in the reference beam of a double-beam spectrophotometer. The IR spectrum of lignin in wood has thereby been recorded with wood (ground and pressed into a KBr pellet) in the sample beam and holocellulose from the same wood in the reference beam (420, 421). Differential IR spectra have been used to study the effect of chlorine dioxide treatments on lignin in wood (422, 423). A characteristic of these differential spectra is the

Table 2 Assignment of Infrared Absorption Bands in Mildly Prepared Wood Lignins (416)

Position, cm^{-1}		
Guaiacyl lignin	Guaiacyl-syringyl Lignin	Band origin
3425-3400	3450-3400	OH-stretching (H-bonded)
2920	2940	OH-stretch in methyl and methylene groups
2875-2850	2880	same
2820	2845-2835	same
1715	1715-1710	carbonyl stretching , unconjugated ketone and carboxyl groups
1675-1660	1660	carbonyl stretching, para-substituted aryl ketone
1605	1595	aromatic skeletal vibrations
1515-1510	1505	same
1470	1470-1460	C-H deformations (asymmetric)
1460		same
1430	1425	aromatic skeletal vibrations
1370	1370-1365	C-H deformation (symmetric)
	1330-1325	syringyl ring breathing with CO-stretching
	1235-1230	same
1270	1275	guaiacyl ring breathing with CO-stretching
1230		same
1140	1145 shoulder	aromatic C-H in-plane deformation, guaiacyl-type
	1130	aromatic C-H in-plane deformation, syringyl-type
1085	1085	C-O deformation, secondary alcohol and aliphatic ether
1035	1030	aromatic C-H in-plane deformation, guaiacyl-type, and C-O deformation, primary alcohol
970	970	=CH out-of-plane deformation (trans)
	915	aromatic C-H out-of-plane deformations
855	860 shoulder	same
815	835	same
750-770 shoulder	750-770 shoulder	same

extreme care necessary to balance the energy levels of the sample and reference beams.

IR spectrometry has been revolutionized by the advent of Fourier transform infrared (FTIR) spectrometers (424). The heart of the instrument is an interferometer, which permits the FTIR system to measure a complete IR spectrum in the same time it takes a dispersive spectrometer to measure one resolution element. A computer is used to run the instrument, collect the raw data (interferogram), and calculate the Fourier transform to give the absorbance or transmittance spectrun. A single FTIR scan can be performed in less than a second. Thus, signal averaging of multiple FTIR scans can be used to obtain a useful spectrum with little noise from measurements involving low levels of IR energy. The FTIR has no energy-limiting slits and allows more energy (than dispersive IR) to reach the sample and subsequently the detector. This yields higher analytical sensitivity.

The effort required to generate differential spectra by dispersive IR is minimized by FTIR; the spectrum of the reference material is simply subtracted from that of the sample. FTIR difference spectra were used to measure changes in carboxyl, carbonyl, and phenolic hydroxyl in lignin when fiberized loblolly pine wood was treated with ozone (425, 426).

A consequence of the great sensitivity of FTIR is its ability to generate useful spectra from the surface of ground or powdered samples by diffuse reflectance. The lignin content of pulps has been measured in this manner (427). Lignin in wood had earlier been measured by dispersive IR with KBr pellets containing wood meal in the sample beam and lignin in the reference beam (420). Additional IR techniques for determining lignin in wood have been described (428, 429).

3. Nuclear Magnetic Resonance

a. Proton NMR.   Proton NMR (PMR) spectroscopy possesses unique features that make it attractive for the qualitative and quantitative analysis of organic materials. Its qualitative value is derived from its ability to differentiate between protons in different magnetic, and thus chemical, environments in a molecule. Quantitative data may be obtained by integrating the areas of the absorption signals produced by each of the types of protons in the sample.

Severe limitations are experienced when PMR is used to study complex, high-molecular-weight materials such as isolated lignins and lignin derivatives. The spectra consist of broad bands that represent many overlapping signals, as typified by Fig. 17 (430). Evident in the spectrum is the small amount of dimethylformamide added to aid in solubilization of the dioxane lignin and to thereby promote acetylation. The spectrum was obtained on a 100-MHz instrument.

In spite of its limitations, PMR has been used extensively for the study of lignins and lignin model compounds. Spectra run on model compounds have been used to assign chemical shifts in lignin spectra to characteristic proton environments (432). Typically, samples are run as 15% solutions in chloroform-*d* with tetramethylsilane

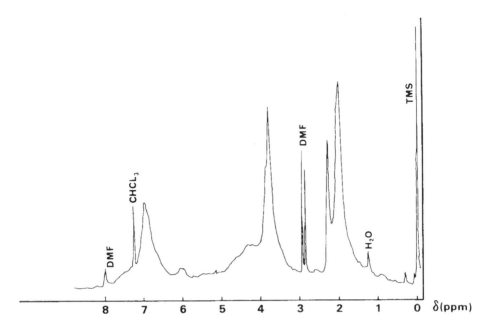

Figure 17   PMR spectrum of acetylated loblolly pine dioxane lignin. (From Ref. 430.)

(TMS) as an internal standard. Although lignin preparations are usually acetylated, PMR spectra have also been recorded on underivatized lignins in DMSO-d_6, thionyl chloride, trifluoroacetic acid (433), and dioxane-d_8-D_2O (5:1) (434).

Semiquantitative data on amounts of specific types of protons per C_9 unit in a lignin preparation can be computed from the integrated areas of the spectral ranges corresponding to those protons. Used in the determination are area percentages from the spectrum, carbon-hydrogen and methoxyl data, and the assumption that the isolated lignin is totally composed of guaiacylpropane units (430, 431, 433).

The determination of hydroxyl protons in MWL was facilitated by silation (435). The resulting large peaks due to the trimethylsilyl protons derived from hydroxyl groups appeared in a region of high magnetic shielding.

Early PMR studies of lignins utilized spectrometers operating at 60 MHz. Higher-frequency instruments produce spectra that are much more amenable to detailed interpretation. Extensive recent studies have employed a 270-MHz instrument operating in the pulse Fourier mode (436–439). Spectra run on model compounds (436–438) and on MWL from spruce (439) and birch (437) yielded more precise peak assignments and greatly increased amounts of structural information. The technique is also capable of the analysis of carbohydrates in lignin preparations (440).

PMR is currently receiving wide use in studies of changes incurred by lignin and lignin-related compounds during pulping (430, 431, 441–443).

b. Carbon-13 NMR. Carbon-13 NMR has several advantages over PMR for the analysis of complex high-molecular-weight materials such as lignin (444). The ^{13}C spectrum of lignin contains about 40 peaks over a 200-ppm range compared with 4 broad peaks over a 10-ppm range in the PMR spectrum. Carbon-13 nuclei have longer spin–spin relaxation times than protons; this leads to narrower line widths for the ^{13}C signals. Complete proton decoupling can produce a ^{13}C spectrum with a single, sharp peak for every type of carbon atom. Figure 18 shows the ^{13}C NMR spectrum of loblolly pine dioxane lignin (430).

Carbon-13 NMR spectrometry has some disadvantages. Because of the low natural abundance of ^{13}C, the ^{13}C NMR technique is inherently less sensitive than PMR. This deficiency may be largely overcome by the use of long accumulation times on modern Fourier transform ^{13}C NMR spectrometers. Carbon-13 NMR peak areas are not proportional to the number of nuclei producing the signal; thus, this

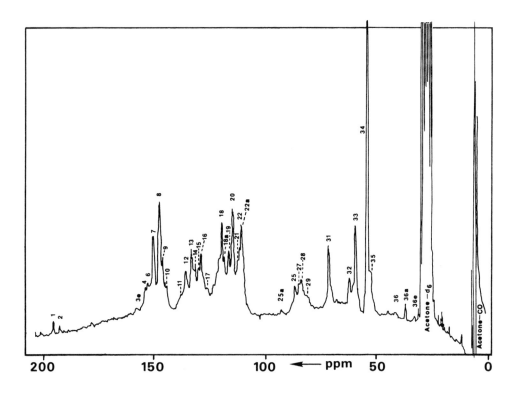

Figure 18 ^{13}C NMR spectrum of loblolly pine dioxane lignin. (From Ref. 430.)

technique is not normally used for quantitative measurements. In recent work, however, experimental conditions have been devised that may permit the quantitative analysis of lignins by ^{13}C NMR (445). Required conditions include an extended pulse delay time and an antigate decoupling sequence to cancel the nuclear Overhauser effect.

The solid-state ^{13}C NMR spectrum of wood has been described (446) as has a quantitative estimate of the hydroxyl groups in wood and wood products by ^{13}C NMR (447) and CP/MAS ^{13}C NMR (448) techniques.

To assign the approximately 40 peaks in the ^{13}C NMR spectrum of lignin, spectra of about 60 lignin model compounds were recorded (450–452). The ^{13}C NMR spectrum of lignin was found to be comprised of three main areas: 200–165 ppm, carbonyl carbon atoms; 165-100 ppm, aromatic and olefinic carbon atoms; and 100-50 ppm, aliphatic carbon atoms.

Because of the great amount of information provided, the ^{13}C NMR technique has been useful in elucidating the structural differences between lignins from different plant sources (449, 453). In recent work, ^{13}C NMR has been used to characterize changes in lignin and model compounds occurring in oxidation (454) and biodegradation (455–457), as well as soda pulping (458) and steam hydrolysis (459).

Conventional ^{13}C NMR studies on lignins suffer from the limitation that only soluble lignin preparations can be used. Wood lignins undoubtedly incur modifications as they are converted to soluble forms. It is therefore difficult to make conclusions regarding the structure of lignin in vivo from NMR studies of soluble lignins. Early attempts at obtaining (proton) NMR spectra on solid lignins yielded broad lines (460). The recent use of the combined techniques of cross-polarization, dipolar decoupling, and magic-angle sample spinning (CP/MAS) on ^{13}C NMR spectrometers has produced informative spectra on solid lignin, wood, and pulp samples (461–465). Significant structural differences were observed between solid lignins and lignins in solution (461).

4.   Other Spectrometric Techniques

a. Fluorescence.   A study of the fluorescence spectra of lignins indicates that lignin behaves as if it contains a single chromophore and not a mixture of compounds (466). Fluorescence has been reported to increase greatly following borohydride reduction.

b. Raman.   A krypton laser was used in obtaining the Raman spectrum of lignin (467). The lignin spectrum was recorded as the difference between loblolly pine groundwood before and after delignification with chlorite.

The use of a microprobe makes possible the study of lignin in the secondary wall of cells (102, 104).

c. Photoacoustic.   A UV photoacoustic spectrum of pine wood lignin in situ was obtained by subtracting the spectrum of microcrystalline cellulose from that of the wood (468). A dual-channel instrument allowed collection of the difference spectra (469). The spectrum of in situ lignin was similar to that of MWL, although significant differences were observed.

d. Electron Spectroscopy for Chemical Analysis. Electron Spectroscopy for Chemical Analysis (ESCA) spectra have been recorded for lignin, purified cellulose, and papers (470–472). The possibility of using ESCA to determine carbonyl groups in lignin has been suggested (472). ESCA has been used to estimate the relative amounts of lignin and polysaccharides on the surfaces of wood fibers separated by different pulping processes (473).

VI. Extraneous Organic Materials

Unlike cellulose, hemicelluloses, and lignin, the extraneous materials found in wood are nonpolymeric (except the pectins and condensed tannins) and can be studied in their original form without extensive degradation or derivatization. (Reactions that are used to facilitate analysis include saponification of high-molecular-weight esters and methylation of carboxylic acids.) Because most of the extraneous organic materials are commonly isolated from wood by solvent extraction, they are frequently called extractives. The sequential extraction of wood with petroleum ether, acetone, and water yields fractions whose principal contents are resins, phenolics, and carbohydrates, respectively (474).

In spite of the great diversity of the extraneous materials, their analysis employs many similar techniques, including (1) extraction (or distillation of volatiles), (2) chemical or chromatographic separation into groups of similar structure, and (3) instrumental chromatographic separation and analysis (GC, LC, GC/MS). Specific procedures used in the analysis of the extraneous materials are described below.

A. Volatile Materials

The volatile components of wood are termed volatile oils or essential oils. They are present in significant quantities in many softwoods; the amounts in hardwoods are negligible. The volatile essential oil of pines is called turpentine. Turpentine consists primarily of monoterpenes, such as α-pinene. Small amounts of aliphatic hydrocarbons, such as n-heptane, may also be present. Types of turpentine include gum turpentine, from distillation of oleoresin; wood turpentine, from distillation of wood extractives; and sulfate turpentine, obtained from condensates recovered during pulping of wood by the sulfate process.

1. Isolation of Volatiles

When preparing wood samples for the laboratory estimation of essential oils or turpentine, care must be exercised to avoid volatilization losses. Sawdust should be placed in an airtight container similar in size to the volume of the sample, and analysis should begin within 1 hr after sawing. An alternative would involve the use of frozen wood chips that have been milled to pass a 9-mm screen immediately prior to analysis (475).

The determination of volatile oils in wood typically requires distillation in a device like that shown in Fig. 19. The use of a resin kettle is more convenient than a distilling flask. Distillation may be conducted from a caustic solution (476) that may also contain ethylene glycol (477). The alkaline condition is necessary to prevent degradation and isomerization (477). The elevated temperature associated with the addition of ethylene glycol more closely simulates the kraft digester. To minimize bumping during distillation, ground wood is tied in a cheesecloth bag that is placed on a supporting metal screen in the resin kettle (478). Reflux is normally conducted for 2 hr, although the time may be extended to 3–4 hr to improve the recovery of high boiling compounds (479).

The recovery of volatiles may involve condensation techniques or the adsorption of vapors onto charcoal or other suitable materials (480).

Before distillation is started, the trap is partially filled with water. When distillation is complete, the water is drawn off through the stopcock of the trap until the oil-water meniscus is at zero on the graduated scale; the volume of oil liberated from the sample is then measured.

Sources of potentially toxic odorous emissions detected during turpentine distillation have been identified as 2-acetyl-5-methylfuran and 2-propionyl-5-methylfuran (479). The potential hazard from these compounds can be minimized by conducting the distillation in a hood.

An alternate procedure for determining turpentine yield involves the use of an internal standard and gas chromatographic analysis (475, 481). A known amount of the internal standard, tetradecane, is added to the ground wood sample prior to distillation. After distillation in the normal manner, a portion of the turpentine-tetradecane layer is taken from the trap and analyzed by GC. Yield is determined by recording the proportion of tetradecane to terpene components in the sample. With this method, the turpentine yield averages 5% greater than with the volumetric procedure.

2. Determination of Composition of Volatiles

Much of the recent analysis of turpentine has been undertaken to characterize changes resulting from the treatment of trees with the herbicide paraquat, which induces the accumulation of oleoresin in the wood (478). Lightwood refers to coniferous wood abounding in pitch, and wood in which oleoresin accumulation is stimulated by herbicide treatment is called induced lightwood.

Turpentine composition is commonly determined by GC on a column utilizing Carbowax 20M as the liquid phase (482). Temperature programming should be

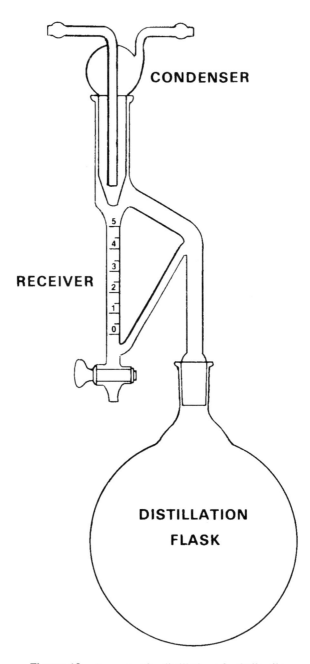

Figure 19   Apparatus for distillation of volatile oils.

continued until the sesquiterpenes are eluted from the column. Quantitation is usually based on the addition of decane or diethyl carbitol as internal standards. The composition of turpentine from a variety of conifers has been reported (477).

Special techniques have been used to characterize turpentine containing substantial amounts of low-boiling constituents, such as n-heptane (483). Because much of the n-heptane was lost during steam distillation, volatiles were isolated by shaking the sawdust with n-tetradecane for 6 hr at 5°C under nitrogen. The internal standard was n-undecane. On GC analysis, the tetradecane solvent peak emerged after the n-heptane and α-pinene. Less volatile constituents were extracted from a separate portion of the wood with pentane, which was concentrated before injection into the GC. This technique could not be used to determine heptane, because it was lost during evaporation of the solvent.

Volatile wood components have also been studied by sampling and analyzing the headspace in a glass vessel filled with chips (484). The familiar monoterpenes were predominant in the headspace, as they also were in distilled volatile oil. The identification of unknown terpenoids has been achieved by GC/MS with the matching of unknown and known mass spectra via a computer (485).

B.   Nonvolatile Resinous Extractives

The nonvolatile extractives in softwoods consist principally of resin acids, fatty acids, and nonsaponifiables. Hardwoods do not contain significant quantities of resin acids. Resin acids are di- and tri-cyclic diterpenes that have a carboxylic acid functionality. The most abundant resin acids are of the abietic acid and pimaric acid types (shown in Fig. 20). Abietic-type resin acids are susceptible to isomerization by heat and acids.

Fatty acids are straight-chain (C_{16}–C_{24}) saturated or unsaturated monocarboxylic acids. The fatty acids extracted from pine are not part of the oleoresin, but are part of the tree's food reserve system (478). In the tree the fatty acids exist in a free form or are esterified with glycerol or other alcohols and sterols.

Nonsaponifiables are neutral, water-insoluble hydrocarbons, alcohols, aldehydes, etc., existing in free form, as well as the alcohols produced during the saponification of esters. Sterols, especially sitosterol, make up a large part of the nonsaponifiables in pine.

1.   Isolation of Nonvolatile Extractives

Changes in the character and location of the extractives begin as soon as the tree is felled. It is therefore essential that samples for analysis be immediately ground and extracted. The freezing of wood samples minimizes degradation. Freeze drying has also been proposed (486).

Nonvolatile extractives are normally isolated by the use of a Soxhlet extractor. The solvent chosen for extraction depends on the purpose of the analysis. Maximum amounts of extractives are removed when ethanol-benzene (1:2) is used (12).

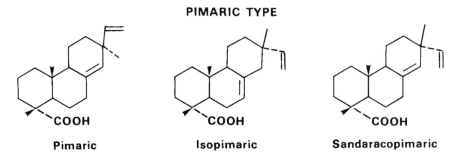

Figure 20   Resin acids, Abietic type.

Aliphatic and aromatic hydrocarbon solvents tend not to extract all of the oleoresins (478). Ethyl ether is the preferred solvent for the study of pine extractives; it provides essentially complete extraction of desired materials with minimal removal of extraneous hydrophilic extractives (478). Extraction temperature and possible thermal degradation of extractives are controlled by the boiling point of the solvent. The esterification of free acids is a potential problem with alcohol solvents. Soxhlet extraction is normally conducted for 8–16 hr.

2. Examination of Nonvolatile Extractives

A number of schemes for separating nonvolatile extractives into groups of components having similar properties have been outlined by Browning (487). The scheme shown in Fig. 21 has proven useful for determining amounts of unsaponifiables and free and combined fatty and resin acids in wood. After saponification and water addition, excess alcohol is removed on a steam bath. Selective esterification employs methanol-sulfuric acid to methylate fatty acids but not resin acids. Significant amounts of combined resin acids are seldom found in wood extracts.

An integrated analytical scheme for pine extractives (See Fig. 22) has been developed in which chromatographic techniques are used to separate the constitu-

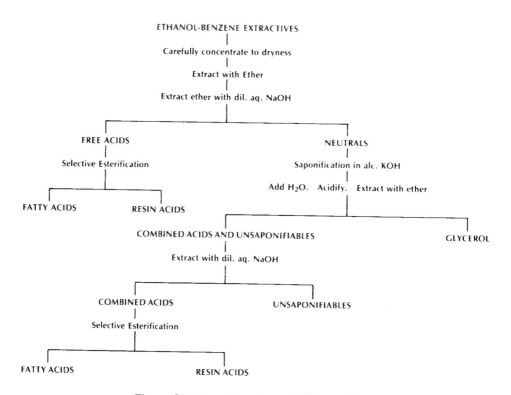

Figure 21   Separation of nonvolatile extractives.

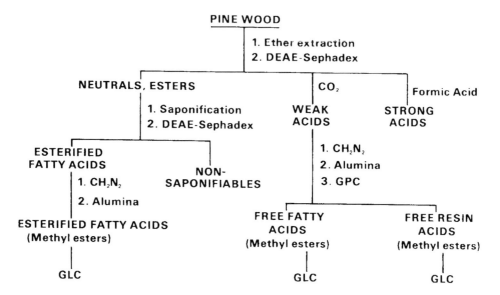

Figure 22   Analytical scheme for pine extractives. (Reprinted from D. F. Zinkel, "Tall Oil Precursors—An Integrated Analytical Scheme for Pine Extractives," *Tappi, 58*(1):109(1975), Technical Association of the Pulp and Paper Industry, Atlanta, Ga., ©1975, with permission. Copies available from TAPPI, Technology Park/Atlanta, P.O. Box 105113, Atlanta, GA 30348.)

ents (488). Ion-exchange chromatography is used to separate neutrals and free acids. This avoids the emulsions that form during extraction of soap solutions and isomerization of abietic-type resin acids on subsequent acidification. Diazomethane is used to prepare methyl esters of the fatty and resin acids for subsequent analysis by GC (489). GC retention times of resin acid methyl esters and other valuable data on resin acids are available (490). Prior to GC analysis, resin acid methyl esters as a group are separated from fatty acid methyl esters by GPC (491). GC on several different columns has been used to separate the different classes of compounds comprising the extractives and also to separate the compounds within the classes (492).

Because of the acids' structural similarity, no single liquid phase in a packed GC column can resolve all of the pine resin acids (493). Highly polar cyanosilicone liquid phases were able to achieve the difficult separation of methyl and tert-butyl palustrate and levopimarate.

Improved resolution of resin (494) and fatty acid esters (495) is achievable by GC with glass capillary columns. That technique has permitted the separation and quantitation of all the main fatty and resin acids in wood extractives and tall oil products (496, 497). Glass capillary GC has also been used to determine the fatty alcohols, triterpene alcohols, and sterols in the unsaponifiable fraction of the extractives (498).

Chemical transformations at elevated temperature and the need to prepare volatile derivatives are both avoided if LC is used to separate extractives. Fatty acids have been separated by reverse-phase high-performance LC on a bonded-phase packing (499). Fatty and resin acids and their methyl esters may be separated by argentation resin chromatography (500–502). In this technique, silver ions form π-electron complexes with the double bonds of the acids. A sulfonated polystyrene cation-exchange resin in the silver form is used with nonaqueous eluting solvents. In a related method, silver perchlorate is added to the methanol-water eluent on a reverse-phase C-18 column (503).

Limited characterization of nonvolatile extractives is possible without the use of chromatographic or extensive wet chemical separations. An extraction and titration procedure may be used to determine the acid content of a wood sample that has been subjected to treatment with aqueous caustic (504). When a typical acid number (mg KOH per g tall oil) is assumed, results can be used to estimate the amount of crude tall oil in the wood. Total resin acids in an extract are determined by titration with standard alcoholic KOH after the fatty acids have been selectively esterified with a solution containing butanol, toluene, and sulfuric acid (505). Methanolic HCl is also used for the selective esterification of fatty acids (506).The fatty acid content may then be computed from the difference between the acid number and the titrated resin acids.

A rapid column method has been proposed for determining unsaponifiable matter in oils and fatty acids (507). After the oil is ground with KOH, heated, mixed with Celite powder, and placed in a glass column, the unsaponifiables are eluted with dichloromethane.

The results of a series of traditional analytical methods are used as specifications for fatty acids or rosin when these materials are bought and sold. Some of the standard tests adopted by the American Oil Chemists' Society (AOCS) and the American Society for Testing and Materials (ASTM) are identified in Table 3. Similar methods for tall oil analysis are published by the Pulp Chemicals Association (504, 508). Traditional procedures for the analysis of rosin and crude tall oil are given in TAPPI T621 and T689, respectively (2). Further discussion of these analyses is provided by Browning (509) and Metcalfe (510).

C. Phenolic Extractives

Phenolic compounds extractable from wood, bark, and foliage range in complexity from simple phenolics, for example, vanillin, to polymeric condensed tannins. Structures of representative compounds are shown in Fig. 23. Larger amounts of the

Table 3 Wet Chemical Methods for Fatty Acids and Rosin

	Fatty acids		Rosin
Method	AOCS	ASTM	ASTM
Acid value	Te 1a	D1980	D465
Saponification value	Tl 1a	D1962	D464
Unsaponifiable matter	Tk 1a	D1965	D1065
Iodine value	Tg 1a	D1959	—
Ash	Tm 1a	D1951	D1063
Moisture, Karl Fischer	Tb 2	E203	—
Rosin acids	Ts 1a	D1240	—
Fatty acids	—	—	D1585
Sampling and grading	—	—	D509

phenolics are frequently found in the bark, foliage, and heartwood than in the sapwood.

Flavonoids are phenolics that have a C_6-C_3-C_6 carbon skeleton. They are found in the free, monomeric form, for example, taxifolin; combined with carbohydrates as glycosides, for example, quercitrin; or polymerized to varying degrees to form condensed tannins. Tannin is comprised of repeating flavan-3-ol units (511); flavan-3-ol dimers and higher oligomers have been termed procyanidins (512). Hydrolysis to compounds such as gallic acid, catechol, pyrogallol, and glucose (513) is characteristic of the so-called hydrolyzable tannins. Thus, they are considerably different from the condensed tannins.

Lignans, for example, pinoresinol, are compounds formed by the oxidative coupling of p-hydroxyphenylpropane (514). Extracts containing lignans have also been found to contain dihydrobenzofuran derivatives (515). Additional types of compounds that are classified as phenolics include hydroxystilbenes (516), for example, pinosylvin; and tropolones, for example, thujaplicin (517). Dilignol glycosides (see Fig. 23) as well as glycosides of other phenolics have been isolated from conifer needles (518, 519). Several phenolic glycosides, of which salireposide is an example, have been isolated from the hot water extracts of the leaves and bark of hardwoods (520, 521).

1. Isolation, Separation and Characterization of Phenolic Extractives

Wood samples from which phenolic extractives are to be isolated are normally dried (or freeze-dried), ground, and preextracted with petroleum ether (522, 523) or benzene (516) to remove fatty or resinous materials. The penolic substances are then extracted with alcohol (513, 514, 524–527), acetone (518), acetone-water (522,

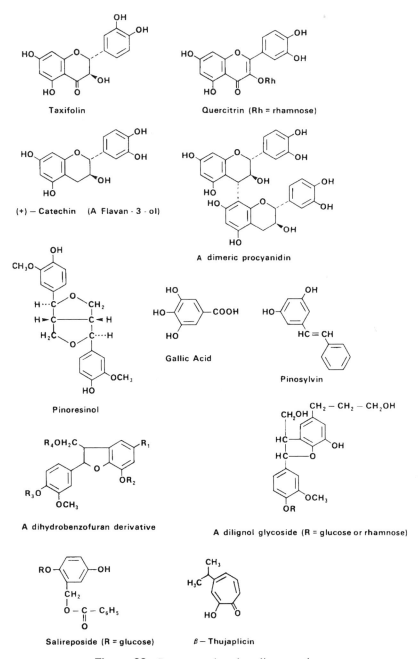

Figure 23   Representative phenolic extractives.

528), water (529) or dilute caustic (530, 531). Supercritical gas extraction with
acetone, tetrahydrofuran, dioxane, and toluene has also been used (532). Following
primary extraction, partition into a variety of solvents and fractionation by gel
permeation or silicic acid chromatography are used to obtain subfractions from
which individual compounds may be isolated.

Most of the present knowledge regarding the phenolic constituents of wood and
other plants has been gained by the use of paper chromatography (533), especially
two-dimensional paper chromatography. Developing solvents, sprays, and color-
forming reagents have been described in detail (534). Thin-layer chromatography
(TLC) has better resolution and speed than paper chromatography, but it suffers in
quantitative accuracy (535). TLC has been used for the isolation of lignans (535),
flavonoids (526) hydroxystilbenes (516), and phenolic glycosides (524). GC is fast
and accurate, but derivatives of many compounds of interest must be prepared to
impart volatility. Trimethylsilyl derivatives have been used in the GC analysis of
lignans (522) and tropolones (517) and for the structural studies of most phenolic
extractives by MS. Simple phenolics are run without derivatization (532). Thermal
degradation is a potential disadvantage of GC.

Currently, the method of choice for separating many phenolic extractives is high-
performance liquid chromatography (HPLC). Reverse-phase C_{18} column packings
have been used extensively for the HPLC analysis of phenolics. Phenolic acids and
glycosides, flavonoids, procyanidins, and anthocyanidins are among the types of
compounds studied by this technique (513, 527, 531, 533, 536–538). Size-exclusion
HPLC has also been used to separate polyphenolic extractives (531). To avoid the
strong association complexes that obstruct size exclusion chromatography in an
aqueous system, the compounds are methylated or acetylated and run with a
nonaqueous eluent (tetrahydrofuran) (531).

After a phenolic compound is isolated, it is characterized by UV, IR, NMR (539),
and MS (522).

D. Soluble Carbohydrates and Other Polar Extractives

Compounds in this category include monosaccharides, sucrose, arabinogalactans,
pectins, cyclitols, and low-molecular-weight carboxylic acids. Like the phenolics,
many of these materials tend to be present in the bark and foliage in higher
concentration than in the wood.

A typical procedure for isolating and detecting many of the low-molecular-
weight polar extractives employs extraction with a polar solvent, derivatization, and
gas chromatographic analysis on a capillary column (145). Plant material was
frozen in liquid nitrogen, freeze-dried, ground, dried further in vacuo, and extracted
with 60% ethanol at 60°C. The ethanol extract was reduced to dryness, dissolved in
water, and washed with diethyl ether to remove lipid materials. Interfering phenolics were removed by slurrying the aqueous solution with either activated charcoal

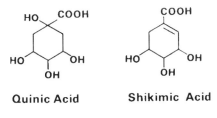

Quinic Acid   **Shikimic Acid**

Figure 24   Quinic and shikimic acids.

or insoluble polyvinylpyrrolidone. The sample was silated and analyzed by GC on a SCOT column coated with SE-30. Compounds identified included monosaccharides, cyclitols, O-methyl cyclitols, sugar alcohols, sucrose, quinic acid, and shikimic acid (Fig. 24).

A similar isolation procedure was used to determine the organic acids in pine needles (540). Acids and neutrals were separated on an anion-exchange column, and a packed GC column coated with XE-60 separated the individual silated acids. Quinic and shikimic acids were the main acidic constituents found.

Pectins are generally considered to be composed of polygalacturonic acids of varying methyl ester content. However, in wood the pectic substances also contain galactose and arabinose. The initial step in isolating pectins from willow bark involved extracting the bark with water at room temperature (541). Aqueous 5% cetyltrimethylammonium bromide was added dropwise with stirring to a solution of the pectin in water. The precipitate was collected by centrifugation, washed with water, dissolved in 5% acetic acid, and poured into ethanol. Then the precipitated polysaccharide mixture was collected, dissolved in water, and freeze-dried. Studies to determine the structure of the pectin employed partial acid hydrolysis, permethylation, and MS analysis of the methylated fragments (542). These procedures are described in the section of this chapter devoted to hemicelluloses.

Arabinogalactans are water-soluble polysaccharides found in large quantity (5–40%) in the lumen of heartwood tracheids of larch (224). After preextraction with nonpolar solvents (224, 230), the arabinogalactan is leached from groundwood or wood shavings with water (232). The aqueous solution is concentrated and poured into acetone or alcohol to precipitate the arabinogalactan.

Arabinogalactans have also been detected in cambial tissues (227). The addition of aqueous calcium chloride or trichloroacetic acid to the aqueous sample caused precipitation of noncarbohydrate, largely protein, contaminating material. GPC separated the high-molecular-weight arabinogalactan from the low-molecular-weight 4-O-methyl glucuronoxylan. Arabinogalactan structure determinations have involved the methylation, Smith degradation, and partial acid hydrolysis techniques described previously in this chapter. Analyses of the water-soluble polysaccharides comprising gum exudates are performed in a similar manner (231).

VII. INORGANIC CONSTITUENTS

Inorganic materials comprise from 0.2–0.9% of the dry weight of woods in the temperate zone of the United States and up to 4–5% of tropical woods (543). Natural amounts and types of inorganic elements represent those that are essential for plant growth and reflect the soil in which the tree has grown. Anthropogenic sources of inorganics in wood include forest fertilization, air pollution (especially affecting the foliage), and preservatives applied to lumber. The principal elements found in wood include (in addition to C, H, and O) calcium, magnesium, manganese, nitrogen, phosphorus, potassium, silicon, sodium, and sulfur. Trace levels of many other elements have been detected.

A. Determination of Ash

The procedure for determing ash in wood and pulp is specified in TAPPI Test Method T211 (544). The sample is placed in a tared platinum, porcelain, or silica crucible which is then put in an electric muffle furnace at a temperature no higher than 100°C. Raising the temperature gradually to $575 \pm 25°C$ carbonizes the sample without flaming. Ignition is continued at $575 \pm 25°C$ for 3 hr or longer until all of the carbon is burned away; this is indicated by the absence of black particles. The crucible is then covered, cooled in a desiccator, and weighed. Results are normally reported on the moisture-free basis. The sample is either oven-dried before ashing, or a moisture determination is performed on a separate portion of the sample.

The preparation of sulfated ash represents a variation on the normal ashing procedure. In this method, the alkali and alkaline earth metal compounds are converted to stable sulfates. Because the ash is not hygroscopic, it is easy to weigh. Ash contents measured by the sulfated ash procedure are somewhat heavier than those produced by the normal method, because sulfates are heavier than carbonates.

In the sulfated ash procedure, the sample is first heated at a low temperature to form a carbonaceous residue (545). One or two drops of dilute sulfuric acid are added to the residue. The sample is heated cautiously to fume off the excess sulfuric acid, and then ignition is completed at 700–800°C.

B. Determination of Metals

I. Destruction of Organic Matter

Destruction of organic matter is essential in preparing wood samples for analysis by most atomic spectrometric techniques, for example, atomic absorption, flame and arc emission, and plasma emission. Electrochemical, colorimetric, and gravimetric methods must also be preceded by the destruction of organics.

The two general procedures for destroying organic matter are dry ashing and wet digestion. Dry ashing is simpler, but some elements are lost in the process. Arsenic, selenium, mercury, cadmium, and silver are lost due to volatilization at 550°C. Lead will be quantitatively ashed at 450°C, but only in the absence of chloride. Copper may be lost by retention in the ashing vessel. Because of minimum retention in and contamination from the crucibles, platinum ware is generally preferred for dry ashing of samples for trace metal determinations. For some samples, magnesium sulfonate and potassium sulfonate ash aids reduce trace metal losses during dry ashing (546). Other ash aids include magnesium nitrate (547) and mixtures of nitric acid with potassium bisulfate or sulfuric acid (548).

Procedures for the destruction of organic matter by dry ashing are essentially as described above for determining the ash content of wood. An alternative to the muffle furnace is the low-temperature plasma. Samples in glass vessels are placed in the plasma tube in a stream of low-pressure oxygen. The tube is surrounded by an induction coil connected to a radio frequency generator. Loblolly pine wood was found to have a higher ash content, 0.335%, when determined by the plasma method than by the muffle procedure, 0.250% (549).

The ash from a wood sample is used directly for analysis by arc emission spectography, or it is dissolved in the dilute nitric or hydrochloric acids for analysis by flame emission, plasma emission, or atomic absorption. Ash which does not dissolve may be made soluble by treatment with flux reagents (550, 551). Other solvents are prescribed in the polarographic, colorimetric, and gravimetric procedures for specific metals.

Although wet digestion can be time-consuming and subject to contamination from acids, it is the method of choice for several metals (552). Mixtures of acids used for this purpose include nitric and perchloric (553) as well as nitric, sulfuric, and perchloric (554). Any analyst working with perchloric acid should be well acquainted with its hazardous properties and with techniques for its safe handling (555). Use of perchloric acid has been avoided by including hydrogen peroxide in digestion mixtures (547, 556–558). Samples such as plant tissue and paper are first digested in concentrated sulfuric or sulfuric plus nitric acids; then hydrogen peroxide is added dropwise.

Because of the high volatility of mercury, special wet digestion techniques are required. Pulp and paperboard have been digested in aqua regia (559); this technique should also be useful for wood. The cold vapor atomic absorption method (560) was used for the mercury measurements. When neutron activation analysis was used to determine mercury in wood chips, destruction of organic matter was not required (561).

Organic matter can also be destroyed by combustion in a Schöniger flask or Parr bomb (562). Because these devices are closed during combustion, loss of volatile elements is prevented. In contrast, chlorine may be lost during wet combustion as well as dry ashing. Therefore, samples containing organically bound Cl are usually burned in a Schöniger flask or oxygen bomb (563) to convert the chlorine to chloride for subsequent determination. Unfortunately, the combustion techniques are labor intensive and restricted to small samples.

2.   Atomic Spectrometry

Following destruction of organic matter, the metals in a sample may be determined by atomic absorption (AA) (564, 565), flame emission (566), arc emission (567), plasma emission (568, 569), or atomic fluorescence (570). Selection of the method depends upon the element(s) to be determined and the instrumentation available (571-575). Atomic absorption procedures are included in some TAPPI Test Methods, e.g., T438 (558). Detailed operating procedures are included in manuals supplied with individual instruments.

3.   Neutron Activation Analysis

Neutron activation analysis (NAA) has received appreciable use for the study of wood samples, because of its simultaneous multielement analysis capability. Most commonly, wood samples in solid form, sawdust, wood flour, or ash are irradiated in a high flux of thermal neutrons in a nuclear reactor. Many of the elements in the sample become radioactive and emit gamma rays having characteristic energies. A system employing a Ge(Li) or NaI(Tl) detector coupled to a multichannel gamma-ray spectrometer records the energies and the numbers of gamma rays emitted. The energies of the gamma rays indicate the elements in the sample, and the counts (number of gamma rays having a given energy) are used for quantification. Further discussion of NAA principles and application to wood is provided by Meyer and Langwig (576). Results comparing the use of NAA and AA for analysis of cottonwood sawdust are shown in Table 4 (577). For several elements the agreement between methods was poor or not feasible, because the elements could not be determined by both methods. Data obtained by NAA of wood and bark from other species are also available (578, 579).

4.   Other Methods

After a wood sample is prepared for analysis by dry ashing or wet digestion, the cations in solution may be determined by polarography (558), colorimetric techniques (580), and more recently, ion chromatography (581). The specific locations of elements in intact samples may be determined by an X-ray analysis attachment on a scanning electron microscope (582).

C. Determination of Nonmetals and Anions

Most methods for determining nonmetals in wood require that the organic matter be destroyed by one of the techniques previously noted. With the exception of nitrogen, which forms ammonium salts in Kjeldahl digestion, the other nonmetals

Table 4   Analysis of Cottonwood Sawdust (577)

Element	NAA (ppm)	AA (ppm)
Ca	1171 ± 314[a]	856 ± 64
Cl	30 ± 5	N.O.
Fe	N.O.[b]	110 ± 106
K	2581 ± 847	2270 ± 145
Mg	O.N.M.[c]	288 ± 41
Mn	4.7 ± 0.3	20 ± 1
Na	112 ± 29	940 ± 89
Zn	N.O.	30 ± 2

[a] ± 1 standard deviation.
[b] Not observed.
[c] Observed but not measured.

of interest (Si, P, S) form oxides in ash and anions in solution. The ash has been used in the determination of Si and P (as well as metals) in bark by arc emission spectography (583). Sulfur in solution as sulfate is commonly precipitated with barium ions and determined gravimetrically, turbidimetrically, nephelometrically, or by titration to a visual end point (584, 585). Sulfate is also determined by ion chromatography (581). An X-ray fluorescence technique, which did not require destruction of organic matter, indicated that pine needles and birch leaves contained 0.12% and 0.23% S, respectively (586). Approximately 0.05% S was found in alder and myrtle twigs by a titrimetric method (587).

Chloride may be determined by the familiar titrimetric procedures employing visual or electrometric end-point detection. An X-ray fluorescence method indicated about 0.1% Cl in pine and spruce needles (588). Methods compared for isolating chloride from eucalypt wood meal included oxygen bomb combustion, room temperature nitric acid extraction, and hot and cold water extraction (589). Potentiometric titration with silver nitrate yielded similar results from all of the isolation procedures; chloride amounted to 200 mg/kg o.d. wood or less.

VIII. SUMMATIVE ANALYSIS

Summative analysis represents the analyst's attempt to account for all of a sample. When this concept is applied rigorously, no constituents may be overlooked, there may be no overlapping determinations, and nothing is determined by difference. The summation, which should be close to 100%, then becomes a severe test of the analyst and the analytical methods. Possible reasons for failure to achieve a complete accounting of a wood sample have been reviewed by Browning (590).

Data in available collections of wood composition information (591–593) are rarely presented as detailed summative analyses. Included are results from many investigators who used older, empirical separation techniques, e.g., holocellulose, Cross and Bevan cellulose, and α-cellulose, to characterize the carbohydrate portions of the wood. Thus, these collections are of limited value.

Inclusion in summative analyses of a large amount of information about the carbohydrates in wood has been made possible by the development of chromatographic methods for carbohydrate analysis. Absolute determinations of the individual sugars are accomplished by the addition of an internal standard and by applying corrections for hydrolysis losses (46). One reservation about this method is that losses of sugars during hydrolysis of wood might not be identical to that of the pure sugars used to determine hydrolysis loss corrections.

An alternate approach is to determine ash, lignin, acetyl groups, and uronic anhydride and assume that the remainder of the wood is carbohydrate. Sugar analysis data are then adjusted so that their sum is equal to the total carbohydrates determined by difference. When this procedure is followed, summative analysis results will equal 100%. This practice is reflected in the data tabulated below.

Summative analysis data for three hardwoods and two softwoods, expressed on an extractive-free wood basis, are shown in Table 5 (42, 594). These data and component ratios in hemicelluloses (40, 594) were used to compute the chemical composition of the same woods in Table 6. Percentages in Table 6 are based on the ovendry weight of unextracted wood. Lignin values are comprised of acid-soluble as well as acid-insoluble lignin. Because the carbohydrates were hydrolyzed by a procedure which did not cleave aldobiouronic acids (129), corrections were made for incompletely hydrolyzed xylose units attached to 4-O-methyl glucuronic acid. The remainder of the uronic acid was considered to be pectin. Total hemicellulose is the sum of the individual hemicellulose fractions including pectin; it may also be estimated from the sum of the nonglucose sugars plus acetyl groups, uronic acids, and glucan in glucomannan and galactoglucomannan.

Caution is advised in using computed hemicellulose composition values like those shown in Table 6. Sugar ratios in the hemicellulose polymers may vary between species, individual trees, locations within a tree, and as a function of storage of the sample prior to analysis.

Table 5 Summative Analysis Data for Woods (42, 594)[a]

Component	Sweet gum	Hickory	White oak	Slash pine[b]	Longleaf pine[b]
Lignin[c]	26.0	25.3	26.0	26.8	26.1
Acetyl	3.9	3.1	3.7	1.4	1.3
Uronic acid	4.4	4.9	5.2	2.0	2.0
Arabinan	0.4	0.6	0.6	1.0	1.0
Xylan	17.5	21.3	16.1	6.5	6.4
Mannan	2.4	0.6	2.2	10.2	12.1
Galactan	0.6	1.4	1.1	2.3	2.7
Glucan	44.5	41.7	45.1	49.5	48.2
Ash	0.3	1.1	0.2	0.2	0.2

[a] All values as percentage of extractive-free wood.
[b] Sapwood, summerwood.
[c] Acid-soluble and insoluble lignin.

Table 6 Chemical Composition of Wood (42,594)[a]

Component	Sweet gum	Hickory	White oak	Slash pine[b]	Longleaf pine[b]
Extractives	1.1	9.0	5.3	0.9	1.4
Lignin	25.7	23.0	24.6	26.6	25.7
Cellulose	42.8	37.7	41.7	46.5	44.6
Hemicellulose	30.1	29.2	28.4	25.8	28.1
O-acetyl-4-O-methyl glucuronoxylan	23.6	24.9	21.0	—	—
Glucomannan	3.6	0.8	3.1	—	—
Arabinogalactan	1.0	1.8	1.6	1.6	1.8
Arabino-4-O-methyl glucuronoxylan	—	—	—	8.1	7.8
O-acetyl-galacto-glucomannan	—	—	—	15.3	17.7
Pectin	1.9	1.7	2.7	0.8	0.8
Ash	0.3	1.1	0.2	0.2	0.2

[a] Percentages based on ovendry weight of unextracted wood.
[b] Sapwood, summerwood.

ACKNOWLEDGMENTS

The authors gratefully acknowledge the assistance of the following colleagues who have reviewed all or part of this chapter: R.H. Atalla, L.G. Borchardt, D.R. Dimmel, W.F.W. Lonsky, E.W. Malcolm, and L.O. Sell. Support by The Institute of Paper Chemistry is also appreciated.

REFERENCES

1.   B. L. Browning, *Methods of Wood Chemistry*, Wiley-Interscience, New York, 1967.
2.   *TAPPI Test Methods*, Technical Association of the Pulp and Paper Industry, Atlanta, Ga.
3.   *ASTM Standards*, American Society for Testing and Materials, Philadelphia, Pa.
4.   B. L. Browning, *Methods of Wood Chemistry*, Vol. 1, Wiley-Interscience New York, 1967, p. 45.
5.   T257 os-76, in *TAPPI Test Methods*, Technical Association of the Pulp and Paper Industry, Atlanta, Ga.
6.   T605 om-82, in *TAPPI Test Methods*, Technical Association of the Pulp and Paper Industry, Atlanta, Ga.
7.   Useful Method 4, in *TAPPI Useful Methods*, Technical Association of the Pulp and Paper Industry, Atlanta, Ga.
8.   E. H. Harris, *Tappi*, *36*:402 (1953).
9.   Battelle-Columbus Laboratories, unpublished work, 1978.
10.   L. R. Schroeder and B. A. Wabers, unpublished work, Institute of Paper Chemistry, Appleton, Wis. 1979.
11.   T12 wd-82, in *TAPPI Test Methods*, Technical Association of the Pulp and Paper Industry, Atlanta, Ga.
12.   T264 om-82, in *TAPPI Test Methods*, Technical Association of the Pulp and Paper Industry, Atlanta, Ga.
13.   H. Tarkow, in *Wood: Its Structure and Properties* (F. F. Wangaard, ed.), Pennsylvania State University, University Park, Pa. 1979, p. 147.
14.   T258 os-76, in *TAPPI Test Methods*, Technical Association of the Pulp and Paper Industry, Atlanta, Ga.
15.   J. D. Pettinati, *J. Assoc. Offic. Anal. Chem.*, *58*:1188 (1975).
16.   T208 os-78, in *TAPPI Test Methods*, Technical Association of the Pulp and Paper Industry, Atlanta, Ga.
17.   D1348-61, in *ASTM Standards*, American Society for Testing and Materials, Philadelphia, Pa..
18.   G. Lohse and H. H. Dietrichs, *Holtz. Roh-Werkstoff*, *30*:468 (1972).
19.   J. D. Litvay and M.D. McKimmy, *Wood Sci.*, *7*:284 (1975).
20.   C.N. Reilley and R.W. Murray, in *Treatise on Analytical Chemistry* (I.M.Kolthoff and P.J. Elving, eds.), Part I, Vol. 4, Wiley, New York, 1963, p. 2189.

21.   F. E. Jones and C. S. Brickenkamp, *J. Assoc. Offic. Anal. Chem.*, *64*: 1277 (1981).
22.   A. S. Meyer, Jr. and C. M. Boyd, *Anal. Chem.*, *31*: 215 (1959).
23.   E. Spinner, *Anal. Chem.*, *47*:849 (1975).
24.   G. V. Shugliashvili, N. G. Charuev, Z. G. Anikashvili, E. S. Repin, and D. S. Shapto, *Derevoobrabat. Prom.*, *5*:11 (1979); *ABIPC*, *51*:A1549.
25.   M. D. Korsunskii and A.K. Veksler, *Derevoobrabat. Prom.*, *5*:9 (1983).
26.   W. L. James, Y. H. Yen, and R. J. King, U.S. Forest Service Note FPL-02050, (1985).
27.   H. Wunschmann, *Holztechnol.*, *26*(3):125 (1985).
28.   J. E. Carles and A. M. Scallan, *J. Appl. Polym. Sci.*, *17*:1855 (1973).
29.   A. J. Nanassy, *Wood Sci.*, *5*:187 (1973).
30.   A. J. Nanassy, *Wood Sci.*, *9*:104 (1976).
31.   H. Magnusson, L. Eriksson, and L.-O. Andersson, *Svensk Papperstid.*, *75*:619 (1972).
32.   A. R. Sharp, M. T. Riggin, R. Kaiser, and M. H. Schneider, *Wood and Fiber*, *10*:74 (1978).
33.   K. K. Brownstein, *J. Magn. Reson.*, *40*:505 (1980).
34.   H. Hsi, R. Hossfeld, and R. G. Bryant, *J. Coll. Interface Sci.*, *62*:389 (1977).
35.   E. Brosio, F. Conti, C. Lintas, and S. Sykora, *J. Food Technol.*, *13*:l07 (1978).
36.   R. T. Lin, *Forest Prod. J.*, *17*(7):54 (1967).
37.   R. T. Lin, *Forest Prod. J.*, *17*(7):61 (1967).
38.   D2016-74, in *ASTM Standards*, American Society for Testing and Materials, Philadelphia, Pa.
39.   J. R. Lavigne, *An Introduction to Paper Industry Instrumentation*, 2nd ed., Miller Freeman Publications, San Francisco, Calif., l977. p. 223.
40.   T. E. Timell, *Wood Sci. Technol.*, *1*:45 (1967).
41.   J. Janson, *Paperi Puu*, *52*(5):323 (1970).
42.   H. L. Hergert, J.D. Wilson, R. G. Rickey, and A. M. Hughes, in *Proceedings TAPPI Research and Development Division Conference*, Technical Association of the Pulp and Paper Industry, Atlanta, Ga., 1982, p. 43.
43.   L. T. Nealey, "Isolation, Characterization and Biological Testing of Xyloglucan from Suspension Cultured Loblolly Pine Cell Spent Medium," Ph.D.Dissertation, Institute of Paper Chemistry, Appleton, Wis., June 1987.
44.   T.E. Timell, *Tappi*, *44*:88 (1961).
45.   J. F. Seaman, W.E. Moore, R.L. Mitchell, and M.A. Millett, *Tappi*, *37*:336 (1954).
46.   T249 pm-75, in *TAPPI Test Methods*, Technical Association of the Pulp and Paper Industry, Atlanta, Ga.
47.   G. J. Ritter and E. F. Kurth, *Ind. Eng. Chem.*, *25*:1250 (1933).
48.   G. Jayme and G. Schwab, *Papierfabr.*, *40*:147 (1942).
49.   A. Poljak, *Angew. Chem.*, *60*:45 (1948).
50.   R. P. Whitney, N. S. Thompson, G. A. Nicholls, and S. T. Han, *Pulp Paper*, *44*(8):68 (1969).
51.   Useful Method 249, in *TAPPI Useful Methods*, Technical Association of the Pulp and Paper Industry, Atlanta, Ga.

52. G. W. Holmes and E. F. Kurth, *Tappi*, *42*:837 (1959).

53. T. E. Timell, *Pulp Paper Mag. Can.*, *60*:T26 (1959).

54. E. F. Kurth and A. A. Swelim, *Tappi*, *46*:591 (1963).

55. L. E. Wise, M. Murphy, and A. A. D'Addieco, *Paper Trade J.*, *122*(2):35 (1946).

56. E. Maekawa and T. Koshijima, *Tappi*, *66*(11):79 (1983).

57. N. S. Thompson and O.A. Kaustinen, *Tappi*, *47*(3):157 (1964).

58. N. S. Thompson and O.A. Kaustinen, *Tappi*, *53*(8):1502 (1970).

59. B. L. Browning, *Methods of Wood Chemistry*, Vol. 2, Wiley-Interscience, New York, 1967, p. 396

60. G. Wegener, *Das Papier*, *28*(11):478 (1974).

61. D. Fengel, H. Ucar, and G. Wegener, *Das Papier*, *33*:233 (1979).

62. B. Leopold, *Tappi*, *44*(3):230, 232 (1961).

63. E. Maekawa and T. Koshijima, *J. Japan Wood Res. Soc.*, *29*(10): (1983).

64. M. Kono, K. Sakai, and T. Kondo, *J. Jap. Tappi*, *19*(1): 27 (1965).

65. N. S. Thompson, *Holzforschung*, *22*(4):124 (1968).

66. B.O. Lindgren, *Svensk Papperstidn.*, *74*:57 (1987).

67. J. K. Hamilton and N. S. Thompson, *Pulp Paper Mag. Can.*, *59*(10):253 (1958).

68. C. F. Cross, E. J. Bevan, and C. Beadle, *Cellulose*, Longmans, Green, London, 1895.

69. B. L. Browning, *Methods of Wood Chemistry*, Vol. 2, Wiley-Interscience, New York, 1967, p. 403.

70. T203 os-74, in *TAPPI Test Methods*, Technical Association of the Pulp and Paper Industry, Atlanta, Ga.

71. B. L. Browning, *Methods of Wood Chemistry*, Vol. 2, Wiley-Interscience, New York, 1967, p. 418.

72. J. D. van Zyl, *Paperi Puu*, *68*(4):320 (1986).

73. H. U. Korner, D. Gottschalk, and J. Puls, *Das Papier*, *38*:255 (1984).

74. T206 wd-71, in *TAPPI Test Methods*, Technical Association of the Pulp and Paper Industry, Atlanta, Ga.

75. T254 pm-76, in *TAPPI Test Methods*, Technical Association of the Pulp and Paper Industry, Atlanta, Ga.

76. G. Jayme, *Tappi*, *44*:299 (1961).

77. G. Jayme and W. Bergmann, *Papier*, *10*:307 (1956).

78. L. Valtasaari, *Paperi Puu*, *39*:343 (1957).

79. G. J. Kubes, B. I. Fleming, J. M. McLeod, H. I. Bolker, and D.P. Werthemann, *J. Wood Chem. Technol.*, *3*(3):313 (1983).

80. D.C. Johnson, M. D. Nicholson, and F.C. Haigh, *Appl. Polym. Symp.*, *28*: 931 (1976).

81. H. A. Swenson, *Appl. Polym. Symp.*, *28*:945 (1976).

82. T. J. Baker, L. R. Schroeder, and D. C. Johnson, *Carbohyd. Res.*, *67*:C4 (1978).

83. B. L. Browning, *Methods of Wood Chemistry*, Vol. 2, Wiley-Interscience, New York, 1967, p. 545.

84. T. E. Timell, *Methods Carbohyd. Chem.*, *5*:100 (1965).

85. L. R. Schroeder and F. C. Haigh, *Tappi*, *62*(10):103 (1979).

86. J. J. Cael, D. J. Cietek, and F. S. Kolpac, *J. Appl. Polym. Sci.: Appl. Polym. Symp.*, *37*:509 (1983).

87. L. Valtasaari and K. Saarela, *Paperi Puu*, *57*:5 (1975).

88. J. M. Laureol, J. Comtat, P. Froment. F. Pla, and A. Robert, *Holzforschung*, *41*(3):165 (1985).

89. G. J. F. Ring, R. A. Stratton, and L. E. Schroeder, *J. Liquid Chromatogr.*, *6*(3):401 (1983).

90. J. Blackwell and R. H. Marchessault, in *Cellulose and Cellulose Derivatives* (N. M. Bikales and L. Segal, eds.), Part IV, Wiley-Interscience, New York, 1971, p. 1.

91. D. W. Jones, in *Cellulose and Cellulose Derivatives* (N. M. Bikales and L. Segal, eds.), Part IV, Wiley-Interscience, New York, 1971, p. 117.

92. M.S. Bertran and B. E. Dale, *J. Appl. Polym. Sci.*, 32(3):4241 (1986).

93. O. Ellefsen and B. A. Tonnesen, in *Cellulose and Cellulose Derivatives* (N. M. Bikales and L. Segal, eds.), Part IV, Wiley-Interscience, New York, 1971, p. 151.

94. L. Segal, J. J. Creely, A. E. Martin Jr., and C. M. Conrad, *Textile Res. J.*, *29*:786 (1959).

95. R. H. Barker and R.A. Pittman, in *Cellulose and Cellulose Derivatives* (N. M. Bikales and L. Segal, eds.), Part IV, Wiley-Interscience, New York, 1971,   p. 181.

96. R. H. Atalla, J. C. Gast, D. W. Sindorf, V. J. Bartuska, and G. E. Maciel, *J. Am. Chem. Soc.*, *102*:3249 (1980).

97. F. P. Miknis, V. J. Bartuska, and G. E. Maciel, *Am. Lab.*, *11*(11):19 (1979).

98. R. H. Atalla, *SPCI Internatl. Symposium on Wood and Pulping Chemistry*, Stockholm, June 9–12, 1981, Preprints, Vol. 1, p. 57.

99. R. H. Atalla, *J. Appl. Polym. Sci., Appl. Polym. Symp.*, *37*:295 (1983).

100. R. H. Atalla and D. L. VanderHart, *Sci.*, *223*:283 (1984).

101. R. H. Atalla, J. Tanua, and E. W. Malcolm, *Tappi*, *67*(2):96 (1984).

102. R. H. Atalla and U. P. Agarwal, *J. Raman Spectroscopy*, *17*:229 (1986).

103. J. H. Wiley, "Raman Spectra of Cellulose", Ph.D. Dissertation, Institute of Paper Chemistry, Appleton, Wis., 1986.

104. R. H. Atalla and U. P. Agarwal, *Sci.*, *227*:636 (1985).

105. B. L. Browning, *Methods of Wood Chemistry*, Vol. 2, Wiley-Interscience, New York, 1967, p. 567.

106. J. K. Hamilton and G. R. Quimby, *Tappi*, *40*:781 (1957).

107. E. Hagglund, B. Lindberg, and J. McPherson, *Acta Chem. Scand.*, *10*:1160 (1956).

108. P. D. Cafferty, C. P. J. Glaudemans, R. Coalson, and R. H. Marchessault, *Svensk Papperstid.*, *67*:845 (1964).

109. J. K. N. Jones, L. E. Wise, and J. P. Jappe, *Tappi*, *39*:139 (1956).

110. R. W. Scott, *J. Appl. Polym. Sci.*, *38*:907 (1989).

111. H. Meier, *Acta Chem. Scand.*, *12*:144 (158).

112. T. E. Timell, *Tappi*, *44*:88 (1961).

113. D. L. Brink and A. A. Pohlman, *Tappi*, *55*:380 (1972).

114. G. G. S. Dutton, B. I. Joseleau, and P. E. Reid, *Tappi*, *56*:168 (1973).

115. B. D. E. Gaillard, N. S. Thompon, and A. J. Morak, *Carbohydr. Res.*, *11*:509 (1969).

116.   A. Beelik, R. J. Conca, J. K. Hamilton, and E.V. Partlow, *Tappi*, *50*:78 (1967).
117.   D. Fengel and M. Przyklenk, *Svensk Papperstid.*, *78*:17 (1975).
118.   B. W. Simson and T. E. Timell, *Cellulose Chem. Technol.*, *12*:51 (1978).
119.   S. Eda, Y. Akiyama, M. Mori, and K. Kato, *International Symposium of Wood and Pulping Chemistry*, Tsukuba Science City, Japan, May 23–27 1983, Vol. 1, p. 138.
120.   B. Lindberg, *SPCI International Symposium on Wood and Pulping Chemistry*, Stockholm, June 9–12, 1981., Preprints, Vol. 4, p. 118.
121.   B. Lindberg, J. Lönngren, and S. Svensson, *Adv. Carbohyd. Chem.*, *31*:185 (1975).
122.   G. O. Aspinall, in *The Biochemistry of Plants., Vol. 3., Carbohydrates: Structure and Function* (J. Preiss, ed.), Academic Press, New York, 1980.
123.   P.E. Jansson, L. Kenne, and G. Widmalm. *Carbohydr. Res.*, *168*(1):67 (1987).
124.   L. S. Munson and P. H. Walker, *J. Am. Chem. Soc.*, *28*:663 (1906).
125.   M. Somogyi, *J. Biol. Chem.*, *195*:19 (1952).
126.   B. L. Browning, *Methods of Wood Chemistry*, Vol. 2, Wiley-Interscience, New York, 1967, p. 590.
127.   B. L. Browning, *Methods of Wood Chemistry*, Vol. 2, Wiley-Interscience, New York, 1967, p. 592.
128.   T 250 pm-75, in *TAPPI Test Methods*, Technical Association of the Pulp and Paper Industry, Atlanta, Ga.
129.   J. E. Jeffery, E. V. Partlow, and W. J. Polglase, *Anal. Chem.*, *32*:1774 (1960).
130.   M.L. Laver, D.F. Root, F. Shafizadeh, and J. C. Lowe, *TAPPI*, *50*:618 (1967).
131.   G.L. Cowie and J. L. Hedges, *Anal. Chem.*, *56*(3):497 (1987).
132.   P. Albersheim, D. J. Nevins, P. D. English, and A. Karr, *Carbohyd. Res.*, *5*:340 (1967).
133.   D. Fengel, G. Wegener, A. Heizmann, and M. Przyklenk, *Holzforschung*, *31*:65 (1977).
134.   M. G. Paice, L. Jurasek, and M. Desrochers, *Tappi*, *65*(7):103 (1982).
135.   E. J. Roberts, M. A. Marshall, M. A. Clarke, W. S. C. Tsang, and F. W. Parrish, *Carbohydr. Res.*, *168*(1):l03 (1987).
136.   K. Robards and M. Whitelaw, *J. Chromatogr.*, *73*(1):81 (l986).
137.   C. C. Sweeley, R. Bentley, M. Makita, and W. W. Wells, *J. Am. Chem.Soc.*, *85*:2497 (1963).
138.   E. P. Crowell and B. B. Burnett, *Anal. Chem.*, *39*:121 (1967).
139.   L. G. Borchardt and C. V. Piper, *Tappi*, *53*:257 (1970).
140.   L. G. Borchardt and D. B. Easty, *Tappi*, *65*(4):127 (1982).
141.   K. L. McDonald and A. C. Garby, *Tappi*, *66*(2):100 (1983).
142.   H. L. Hardell and U. Westermark, *SPCI International Symposium on Wood and Pulping Chemistry*, Stockholm, June 9–12, 1981, Preprints, Vol. 1, p.32.
143.   A. E. Mikel'son, T. E. Sharapova, and G. E. Domburg, *Khim. Drev. (Riga)*, *2*:94 (1980); *ABIPC*, *51*:A5338.
144.   A. M. Cranswick and J. A. Zabkiewicz, *J. Chromatogr.*, *171*:233 (1979).
145.   G. J. Gerwig, J. P. Kamerling, and J. F. G. Vliegenthart, *Carbohyd. Res.*, *62*:349 (1978).
146.   K. Leontein, B. Lindberg, and J. Lönngren, *Carbohyd. Res*, *62*:359 (1978).

147. G. Eklund, B. Josefsson, and C. Roos, *J. Chromatogr.*, *142*:575 (1977).

148. G. Holzer, J. Oro, S. J. Smith, and V. M. Doctor, *J. Chromatogr.*, *194*:410 (1980).

149. N. K. Kochetkov and O. S. Chizhov, *Adv. Carbohyd. Chem.*, *21*:39 (1966).

150. J. Lönngren and S. Svensson, *Adv. Carbohyd. Chem. Biochem.*, *29*:42 (1974).

151. B. S. Valent, A. G. Darvill, M. McNeil, B. K. Robertsen, and P. Albersheim, *Carbohyd. Res.*, *79*:165 (1980).

152. J.X. Khym and L. P. Zill, *J. Am. Chem. Soc.*, *74*:2090 (1952).

153. J.Dahlberg and O. Samuelson, *Acta Chem. Scand.*, *17*:2136 (1963).

154. P. Jandera and J. Churacek, *J. Chromatogr.*, *98*:55 (1974).

155. P. Jonsson and O. Samuelson, *J. Chromatogr.*, *26*:194 (1967).

156. R. B. Kesler, *Anal. Chem.*, *39*:1416 (1967).

157. K. Goel and A. M. Ayroud, *Pulp Paper Mag. Can.*, *73*:T146 (1972).

158. M. Sinner, M. H. Simatupang, and H. H. Dietrichs, *Wood Sci. Technol.*, *9*:307 (1975).

159. K. Mopper and E. M. Girdler, *Anal. Biochem.*, *56*:440 (1973).

160. M. Sinner and J. Puls, *J. Chromatogr.*, *156*:197 (1978).

161. T. S. Friberg, E. E. Barnes, and R. E. Meyers, unpublished work, Weyerhaeuser Co., Tacoma, Wa., 1981.

162. R. W.Goulding, *J. Chromatogr.*, *103*:229(1975).

163. F.E. Wentz, A.D. Marcy, and M.J. Gray, *J.Chromatogr. Sci.*, *20*:349 (1982).

164. H. D. Scobell, K. M. Brobst, and E. M. Steele, *Cereal Chem.*, *54*:905 (1977).

165. H. D. Scobell and K. M. Brobst, *J. Chromatogr.*, *212*:51 (1981).

166. K. M. Brobst and H. D. Scobell, *Starch*, *34*:117 (1982).

167. R. C. Pettersen and V. Schwandt, unpublished work, U.S. Forest Products Laboratory, Madison, Wi., 1982.

168. R. D. Rocklin and C. A. Pohl, *J. Liquid Chromtogr.*, *6*(9):1577 (1983).

169. T. Jupille, M. Gray, B. Black, and M. Gould, *Am. Lab.*, *13*(8):80 (1981).

170. R. E. Majors, *J. Assoc. Offic. Anal. Chem.*, *60*: 86 (1977).

171. R. E. Majors, *J. Chromatogr. Sci.*, *15*:334 (1977).

172. R. C. Pettersen, V. H. Schwandt, and M. J. Effland, *J. Chromatogr.*, *22*:478 (1984).

173. T223 os-78, in *TAPPI Test Methods*, Technical Association of the Pulp and Paper Industry, Atlanta, Ga.

174. V. Berzins, *Tappi*, *54*:1165 (1971).

175. R. W. Scott, W. E. Moore, M. J. Effland, and M. A. Millett, *Anal. Biochem.*, *21*:68 (1967).

176. A. H. Wardi, W. S. Allen, and R. Varma, *Anal. Biochem.*, *57*:268 (1974).

177. N. Blumenkrantz and G. Asboe-Hansen, *Anal. Biochem.*, *54*:484 1973).

178. B. L. Browning, *Tappi*, *32*:119 (1949).

179. B. L. Browning, *Methods of Wood Chemistry*, Vol. 2, Wiley-Interscience, New York, 1967, p. 635.

180. R. W. Scott. K. A. Libkie, and E. L. Springer, *J. Wood Chem. Technol.*, *4*(4):497 (1984).

181. M. Bylund and A. Donetzhuber, *Svensk Papperstid.*, *71*:505 (1968).

182. E. Dische, *J. Biol. Chem.*, *16/*:189 (1947).

183. R. W. Scott, *Anal. Chem.*, *51*:936 (1979).
184. T. E. Timell, *Adv. Carbohyd. Chem.*, *19*:271 (1964).
185. E. Maekawa and T. Koshijima, *J. Japan Wood Res. Soc.*, *34*(4):359 (1988).
186. R. L. Taylor and H. E. Conrad, *Biochem.*, *11*(8):1383 (1972).
187. E. Sjostrom, S. Juslin, and E. Sappala, *Acta Chem. Scand.*, *23* (10):3610 (1969).
188. M. H. Simatupang, *J. Chromatogr.*, 178:588 (1979).
189. K. Shimizu, M. Hashi, and K. Sakurai, *Carbohyd. Res.*, *62*:117 (1978).
190. B. L. Browning, *Methods of Wood Chemistry*, Vol. 2, Wiley-Interscience, New York, 1967, p. 653.
191. R. L. Whistler and A. Jeanes, *Ind. Eng. Chem. Anal., Ed.*, *15*:317 (1943).
192. T. E. Timell, *Svensk Papperstid.*, *60*:762 (1957).
193. R. Solar, F. Kacik, and I. Melcer, *Nordic Pulp Paper Res. J.*, *4*(2):139 (1987).
194. P. O. Bethge and K. Lindstrom, *Svensk Papperstid.*, *76*:645 (1973).
195. P. Mansson and L. Westfelt, *J. Appl. Polym. Sci.*, *25*:1533 (1980).
196. P. Mansson and B. Samuelsson, *Svensk Papperstid.*, *84*:R15 (1981).
197. V. B. Karklin', *Khim. Drev. (Riga)*, 6:24 (1976); *ABIPC*, *48*:A2639.
198. A. S. Karacronyi, J. Alfoldi, M. Kubackova, and L. Stupha, *Cellulose Chem. Technol.*, *17*:637 (1983).
199. H. Rauvala, J. Finne, T. Kruslus, J. Kärkkäinen, and J. Järnefelt, *Adv. Carbohyd. Chem. Biochem.*, *38*:389 (1981).
200. W. N. Haworth, *J. Chem. Soc.*, *107*:8 (1915).
201. T. Purdie and J. C. Irvine, *J. Chem. Soc.*, *83*:1021 (1903).
202. R. Kuhn, H. Trischmann, and I. Löw, *Angew. Chem.*, *67*:32 (1955).
203. S. Hakomori, *J. Biochem.* (Tokyo), *55*:205 (1964).
204. I. Ciucanu and F. Kerck, *Carbohydr. Res.*, *131*(2)209 (1984).
205. T. J. Waegbe, A. G. Darvill, M. McNeil, and P. Albersheim, *Carbohydr. Res.*, *123*(2):281 (1983).
206. P. A. Sandford and H. E. Conrad, *Biochem.*, *5*:1508 (1966).
207. J. Finne, T. Krusius, and H. Rauvala, *Carbohyd. Res.*, *80*:336 (1980).
208. T. Narui, K. Takahashi, M. Kobayashi, and S. Shibata, *Carbohyd. Res.*, *103*:293 (1982).
209. V. Kovacik and P. Kovac, *Carbohyd. Res.*, *105*:251 (1982).
210. J. G. Jun and G. R. Gray, *Carbohydr. Res.*, *163*(2):247 (1987).
211. I. J. Goldstein, G. W. Hay, B. A. Lewis, and F. Smith, *Methods Carbohyd. Chem.*, *5*:361 (1965).
212. J. M. Vaughn and E. E. Dickey, *J. Org. Chem.*, *29*(3):715 (1964).
213. J. Kiss, *Adv. Carbohyd. Chem. Biochem.*, *29*:229 (1974).
214. B. Lindberg, J. Lönngren, and J. L. Thompson, *Carbohyd. Res.*, *28*:351 (1973).
215. K. Shimizu, *J. Japan Wood Res. Soc.*, *22*:51 (1976)
216. K. Shimizu, *Carbohyd. Res.*, *92*:65 (1981)
217. K. Shimizu, *Carbohyd. Res.*, *92*:219 (1981).
218. B. L. Browning, *Methods of Wood Chemistry*, Vol. 2, Wiley-Interscience, New York, 1967, p. 703.

219. A. J. Mian and T. E. Timell, *Can. J. Chem.*, *38*:1511 (1960).
220. F. Daniels and R. A. Alberty, *Physical Chemistry*, Wiley, New York, 1955, p. 499.
221. G.C. Hoffman and T.E. Timell, *Svensk Papperstid.*, *75*:135 (1972).
222. S. Hizukuri and T. Takagi, *Carbohydr. Res.*, *134*(1):1(1984).
223. J. K. N. Jones, C. B. Purves, and T. E. Timell, *Can. J. Chem.*, *39*:1059 (1961).
224. B. M. Simson, W. A. Côté, Jr., and T. E. Timell, *Svensk Papperstid.*, *71*:699 (1968).
225. E. L. Johnson and R. Stevenson, *Basic Liquid Chromatography*, Varian Associates, Palo Alto, Ca., 1978.
226. M. Dubois, K. A. Gilles, J. K. Hamilton, P. A. Rebers, and F. Smith, *Anal. Chem.*, *28*:350 (1956).
227. B. W. Simson and T. E. Timell, *Cellulose Chem. Technol.*, *12*:63 (1978).
228. B. V. Ettling and M. F. Adams, *Tappi*, *51*:116 (1968).
229. B. W. Simson and T. E. Timell, *Cellulose Chem. Technol.*, *12*:79 (1978).
230. G. P. Belue and G. D. McGinnis, *J. Chromatogr.*, *97*:25 (1974).
231. S. C. Churms, E. H. Merrifield, and A. M. Stephen, *Carbohyd. Res.*, *55*:3 (1977).
232. S. C. Churms, E. H. Merrifield, and A. M. Stephen, *Carbohyd. Res.*, *64*:Cl (1978).
233. H. G. Barth, *J. Liquid Chromatogr.*, *3*:1481 (1980).
234. S. C. Churms, *Adv. Carbohyd. Chem.*, *25*:13 (1970).
235. S. C. Churms, *CRC Handbook of Chromatography, Carbohydrates*, Vol. 1, CRC Press, Boca Raton, Florida 1981.
236. J. K. Hamilton and H. W. Kircher, *J. Am. Chem. Soc.*, *80*:4703 (1958).
237. T. E. Timell, *J. Am. Chem. Soc.*, *82*:5211 (1960).
238. D. A. I. Goring, Discussion of paper by R. G. LeBel and D. A. I. Goring, *J. Polym. Sci.*, Part C (2):29 (1963).
239. N. Parameswaran and W. Liese, *SPCI International Symposium on Wood and Pulping Chemistry*, Stockholm, June 9–12, 1981, Preprints, Vol. 1, p. 16.
240. G. Cox and B. Juniper, *J. Microscopy*, 97:343 (1973).
241. M. Sinner, N. Parameswaran, H. H. Dietrichs, and W. Liese, *Holzforschung*, *27*:36 (1973).
242. R. H. Mullis, N. S. Thompson, and R. A. Parham, *Planta* (Berlin),*132*:241 (1976).
243. E. M. Byers, "Autoradiographic Study of Hemicellulose Distribution in the Walls of Pinus resinosa (Red Pine) Tracheids," Ph.D. Dissertation, Institute of Paper Chemistry, Appleton, Wis., 1989.
244. H. Meier, *Pure Appl. Chem.*, *5*:37 (1962).
245. R. P. Kibblewhite and D. Brookes, *Wood Sci. Technol.*, *10*:39 (1976).
246. R. W. Scott, *J. Wood Chem. Technol.*, *4*(2):199 (1984).
247. H.L. Hardell and U. Westermark, *SPCI International Symposium on Wood and Pulping Chemistry*, Stockholm, June 9–12, 1981, Preprints, Vol. 5, p. 17.
248. B. L. Browning, *Methods of Wood Chemistry*, Vol. 1, Wiley-Interscience, New York, 1967, p. 273.
249. T401 om-82, in *TAPPI Test Methods*, Technical Association of the Pulp and Paper Industry, Atlanta, Ga.

250. B. L. Browning, *Methods of Wood Chemistry*, Vol. 2, Wiley-Interscience, New York, 1967, p. 818.

251. C. Mäule, *Beitr. Wiss. Botanik*, *4*:166 (1900).

252. G. Meshitsuka and J. Nakano, *J. Japan Wood Res. Soc.*, *26*(8):576 (1980).

253. R. A. Parham, *Tappi*, *65*(4):127 (1982).

254. P. Klason, *Arkiv Kemi*, *3*(5):17 (1906).

255. T222 om-88, in *TAPPI Test Methods*, Technical Association of the Pulp and Paper Industry, Atlanta, Ga.

256. D1106-56, in *ASTM Standards*, American Society for Testing and Materials, Philadelphia, Pa.

257. M. T. Effland, *Tappi*, *60*(10):143 (1977).

258. K. Yoshihara, K. Kabayashi, T. Fujii, and I. Akamatsu, *J. Jap. Tappi*, *38*(4):86 (1984).

259. Useful Method 250, in *TAPPI Useful Methods*, Technical Association of the Pulp and Paper Industry, Atlanta, Ga.

260. V. Loras and F. Loschbrandt, *Norsk Skogind.*, *15*:302 (1961).

261. A. G. Schoning and G. Johansson, *Svensk Papperstid.*, *68*:607 (1965).

262. B. Swan, *Svensk Papperstid.*, *68*:791 (1965).

263. D. B. Johnson, W. E. Moore, and L. C. Zank, *Tappi*, *44*:793 (1961).

264. J. D. Zyl, *Wood Sci. Technol.*, *12*:251 (1978).

265. K. Iiyama and A. F. A. Wallis, *Appita*, *41*(6):442 (1988).

266. B. L. Browning, *Methods of Wood Chemistry*, Vol. 2, Wiley-Interscience, New York, 1967, p. 785.

267. G. P. Berlyn and R. E. Mark, *Forest Prod. J.*, *15*:140 (1965).

268. J. A. N. Scott, A. R. Procter, B. J. Fergus, and D. A. I. Goring, *Wood Sci. Technol.*, *3*:73 (1969).

269. B. J. Fergus, A. R. Procter, J. A. N. Scott, and D. A. I. Goring, *Wood Sci. Technol.*, *3*:117 (1969).

270. K. Fukazawa and H. Imagawa, *Wood Sci. Technol.*, *15*:45 (1981).

271. D. A. I. Goring, *SPCI International Symposium on Wood and Pulping Chemistry*, Stockholm, June 9–12, 1981, Preprints, Vol. 1, p. 3.

272. B. J. Fergus and D. A. I. Goring, *Holzforschung*, *24*:113 (1970).

273. Y. Musha and D. A. I. Goring, *Wood Sci. Technol.*, *9*:45 (1975).

274. J.M. Yang and D. A. I. Goring, *Trans. Tech. Sect., Can. Pulp Paper Assoc.*, *4*:2 (1978).

275. G. Aulin-Erdtman, *Svensk Papperstid.*, *57*:745 (1954).

276. A.R. Procter, W. Q. Yean, and D. A. I. Goring, *Pulp Paper Mag. Can.*, *68*:T445 (1967).

277. B. J. Fergus and D. A. I. Goring, *Pulp Paper Mag. Can.*, *70*:T314 (1969).

278. J. R. Wood, P. A. Ahlgren, and D. A. I. Goring, *Svensk Papperstid.*, *75*:15 (1972).

279. A. J. Kerr and D. A. I. Goring, *Svensk Papperstid.*, *79*:20 (1976).

280. L. Gädda, *Svensk Papperstid.*, *85*:R57 (1982).

281. B. Boutelje, *Svensk Papperstid.*, *75*:683 (1972).

282. W. A. Côté, A. C. Day, and T. E. Timell, *Wood Sci. Technol.*, *2*:13 (1968).

283. R. A. Parham and W. A. Côté, *Wood Sci. Technol.*, *5*:49 (1971).
284. D. E. Bland, R. C. Foster, and A. F. Logan, *Holzforschung*, *25*:137 (1971).
285. K. Ruel and F. Bernoud, *SPCI International Symposium on Wood and Pulping Chemistry*, Stockholm, June 9–12, 1981, Preprints, Vol. 1, p. 11.
286. S. Saka, R. J. Thomas, and J. S. Gratzl, *Tappi*, *61*(1):73 (1978).
287. S. Saka, R. J. Thomas, and J. S. Gratzl, in *Dietary Fibers: Chemistry and Nutrition* (G. E. Inglett and I. F. Falkehag, eds.), Academic Press, New York, 1979, p. 15.
288. S. Saka, R. J. Thomas, and J. S. Gratzl, *SPCI International Symposium on Wood and Pulping Chemistry*, Stockholm, June 9–12, 1981, Preprints, Vol. 1, p. 35.
289. S. Saka, S. Hosoya, and D. A. I. Goring, *International Symposium of Wood and Pulping Chemistry*, Tsukuba Science City, Japan, May 23–27, 1983, Vol. 1, p. 138.
290. S. Saka and R. J. Thomas, *Wood Sci. Technol.*, *16*:1 (1982).
291. S. Saka, P. Whiting, K. Fukazawa, and D. A. I. Goring, *Wood Sci. Technol.*, *16*:269 (1982) .
292. S. Saka, R. J. Thomas, and J. S. Gratzl, *Wood and Fiber*, *11*:99 (1979).
293. J. J. Kolar, B. O. Lindgren, and E. Treiber, *Svensk Papperstid.*, *85*:R21 (1982).
294. D. J. Gardner, J. M. Genco, R. Jagels, and G. L. Simard, *Tappi*, *65*(9):133 (1982).
295. S.J. Kuang, S. Saka, and D. A. I. Goring, Paper presented at *9th Cellulose Conference*, SUNY College of Environmental Science and Forestry, Syracuse, N.Y. May 24–27, 1982.
296. H.L. Hardell, G. J. Leary, M. Stoll, and U. Westermark, *Svensk Papperstid.*, *83*:44 (1980).
297. P. Whiting, B. D. Favis, F. G. T. St.-Germain, and D. A. I. Goring, *J. Wood Chem. Technol.*, *1*:29 (1981).
298. J. Savari, E. Sjostrom, A. Klemola, and J. E. Laine, *Wood Sci. Technol.*, *20*:35 (1986).
299. J. B. Boutelje and I. Eriksson, *Svensk Papperstid.*, *85*:R39 (1982).
300. J. R. Obst, *Holzforschung*, *36*(3):143 (1982).
301. J. R. Obst and J. Ralph, *Holzforschung*, *37*(6):297 (1983)
302. P. Whiting and D. A. I. Goring, *Wood Sci. Technol.*, *16*:261 (1982).
303. Y. Z. Lai and K. V. Sarkanen, in *Lignins* (K. V. Sarkanen and C. H. Ludwig, eds.), Wiley-Interscience, New York, 1971, p. 165.
304. E. Adler, *Wood Sci. Technol.*, *11*:169 (1977).
305. F. E. Brauns, *J. Am. Chem. Soc.*, *61*:2120 (1939).
306. A. Björkman, *Svensk Papperstid.*, *59*:477 (1956).
307. A. Björkman, *Svensk Papperstid.*, *60*:329 (1957).
308. G. Wegener and D. Fengel, *Tappi*, *62*(3):97 (1979).
309. H. H. Brownell, *Tappi*, *48*:513 (1965).
310. D. E. Bland and M. Menshun, *Appita*, *21*:17 (1967).
311. H.M. Chang, E. B. Cowling, W. Brown, E. Adler, and G. E. Miksche, *Holzforschung*, *29*:153 (1975).
312. J. Polcin and B. Bezuch, *Wood Sci. Technol.*, *12*:49 (1978).
313. W. Ziechmann and T. Weichelt, *Z. Pflanzenernahr. Bodenkunde*, *140*:645 (1977).

314. J. M. Pepper and M. Siddiqueullah, *Can. J. Chem.*, *39*:1454 (1961).
315. H. L. Hergert, *J. Org. Chem.*, *25*:405 (1960).
316. D. F. Arsenau and J. M. Pepper, *Pulp Paper Mag. Can.*, *66*:T415 (1965).
317. B. L. Browning, *Methods of Wood Chemistry*, Vol. 2, Wiley-Interscience, New York, 1967, p. 717.
318. S. Zeisel, *Monatsh.*, *6*:989 (1885); *7*:406 (1886).
319. B. L. Browning, *Methods of Wood Chemistry*, Vol. 2, Wiley-Interscience, New York, 1967, p. 660.
320. D1166-60, in *ASTM Standards*, American Society for Testing and Materials, Philadelphia, Pa.
321. E. P. Samsel and J. A. McHard, *Ind. Eng. Chem., Anal. Ed.*, *14*:750 (1942).
322. T209 wd-79, in *TAPPI Test Methods*, Technical Association of the Pulp and Paper Industry, Atlanta, Ga.
323. E. D. Gel'fand and L. F. Tushina, *Izv. VUZ Lesnoi Zh.*, *18*(5):163 (1975); *ABIPC*, *47*:A185.
324. G. Aulin-Erdtman, *Tappi*, *32*:160 (1949).
325. O. Goldschmid, *Anal. Chem.*, *26*:1421 (1954).
326. J. P. Butler and T. P. Czepiel, *Anal. Chem.*, *28*:1468 (1956).
327. K. Freudenberg, J. H. Harkin, and H. K. Wecker, *Chem. Ber.*, *97*:909 (1964).
328. K. Sarkanen and C. Schuerch, *Anal. Chem.*, *27*:1245 (1955).
329. P. Whiting and D. A. I. Goring, *Paperi Puu*, *64*:592 (1982).
330. Y. Z. Lai and K. V. Sarkanen, in *Lignins* (K. V. Sarkanen and C. H. Ludwig, eds.), Wiley-Interscience, New York, 1971, p. 201.
331. B. L. Browning, *Methods of Wood Chemistry*, Vol. 2, Wiley-Interscience, New York, 1967, p. 755.
332. E. Adler, S. Hernestam, and I. Wallden, *Svensk Papperstid.*, *61*:641 (1958).
333. G. Brunow and D. Robert, *International Symposium of Wood and Pulping Chemistry*, Tsukuba Science City, Japan, May 23–27, 1983, Vol. 1, p. 92.
334. P. Manssen, *Holzforschung*, *37*:143 (1983).
335. W. J. Schubert and F. F. Nord, *J. Am. Chem. Soc.*, *72*:977 (1950).
336. E. Adler and J. Gierer, *Acta Chem. Scand.*, *9*:84 (1955).
337. J. Gierer and S. Soderberg, *Acta Chem. Scand.*, *13*:127 (1959).
338. B. Lindberg and O. Theander, *Svensk Papperstid.*, *57*:83 (1954).
339. J. Marton, E. Adler, and K.I. Persson, *Acta Chem. Scand.*, *15*:384 (1961).
340. S. Kolboe and O. Ellefsen, *Tappi*, *45*:163 (1962).
341. E. Adler and J. Marton, *Acta Chem. Scand.*, *13*:75 (1959).
342. E. I. Evstigneev, L. V. Bronov, and V. M. Nikitin, *Khim. Drev. (Riga), No. 6*:82 (1979); ABIPC 51:A2559.
343. H.-D. Becker and E. Adler, *Acta Chem. Scand.*, *15*:218 (1961).
344. E. Sjöstrom, Wood Chemistry, Academic Press, New York, 1981, p. 71.
345. K. Freudenberg, A. Janson, E. Knopf, and A. Haag, *Ber. Deusch. Chem. Ges.*, *69*:1415 (1936).

346. K. Freudenberg, K. Engler, E. Flickinger, A. Sobek, and F. Klink, *Ber. Deusch. Chem. Ges., 71*:1810 (1938).

347. K. Freudenberg, W. Lautsch, and K. Engler, *Ber. Deusch. Chem. Ges., 73*:167 (1940).

348. R. H. J. Creighton, R. D. Gibbs, and H. Hibbert, J. Am. Chem. Soc., 66:32 (1944).

349. I. A. Pearl, *J. Am Chem. Soc., 64*:1429 (1942).

350. H.-M. Chang and G. G. Allan, in Lignins (K. V. Sarkanen and C. H. Ludwig, eds.), Wiley (Interscience), New York, 1971, p. 446.

351. E. E. Harris and H. Adkins, *Paper Trade J.* 107(20):38 (Nov. 17, 1938).

352. H. P. Godard, J. L. McCarthy, and H. Hibbert, *J. Am. Chem. Soc., 62*:988 (1940); 63:3061 (1941).

353. B. F. Hrutfiord, in Lignins (K. V. Sarkanen and C. H. Ludwig, eds.), Wiley (Interscience), New York, 1971, p. 487.

354. A. B. Cramer, J. M. Hunter, and H. Hibbert, *J. Am. Chem. Soc., 61*:509 (1939).

355. M. Erikson, S. Larsson, and G. E. Miksche, *Acta Chem. Scand., 27*:127 (1973).

356. T. P. Schultz, C. L. Chen, I. S. Goldstein, and F. P. Scaringelli, *J. Chromatog. Sci., 19*:235 (1981).

357. J. I. Hedges and J. R. Ertel, *Anal. Chem., 54*:174 (1982).

358. G. Alibert and A. Boudet, *Physiol. Veg. 17*:67 (1979).

359. S. Larsson and G. E. Miksche, *Acta Chem. Scand., 23*:917 (1969).

360. J. M. Pepper, P. E. T. Baylis, and E. Adler, *Can. J. Chem., 37*:1241 (1959).

361. H. Nimz, *Holzforschung, 20*:105 (1966).

362. A. Sakakibara and N. Nakayama, *J. Japan Wood Res. Soc., 8*:157 (1962).

363. S. Omori and A. Sakakibara, *J. Japan Wood Res. Soc., 25*:145 (1979).

364. M. Aoyama and A. Sakakibara, *J. Japan Wood Res. Soc., 25*:149 (1979).

365. S. Yasuda and A. Sakakibara, *J. Japan Wood Res. Soc., 21*:307 (1975).

366. K. Sudo and A. Sakakibara, *J. Japan Wood Res. Soc., 21*:164 (1975).

367. K. Sudo, B. H. Hwang, and A. Sakakibara, *J. Japan Wood Res. Soc., 25*:61 (1979).

368. B. H. Hwang, A. Sakakibara, and K. Miki, *Holzforschung, 35*:229 (1981).

369. H. Nimz, *Chem. Ber., 102*:799 (1969).

370. H. Nimz, *Angew. Chem. Intern. Ed., 13*:313 (1974).

371. N. Morohoshi and W. G. Glasser, *Wood Sci. Technol., 13*:249 (1979).

372. S. Larsson and G. E. Miksche, *Acta Chem. Scand., 21*:1970 (1967).

373. M. Erickson, S. Larsson, and G. E. Miksche, *Acta Chem. Scand., 27*:903 (1973).

374. W. G. Glasser and H. R. Glasser, *Paperi Puu, 63*:71 (1981).

375. A. Björkman, *Svensk Papperstid., 60*:243 (1957).

376. T. Koshijima, F. Yaku, and R. Tanaka, *J. Appl. Polym. Sci., Appl. Polym. Symp., 28*:1025 (1976).

377. G. Mishitsuka, Z. Z. Lee, J. Nakano, and S. Eda, *J. Wood Chem. Technol., 2*(3):251 (1982).

378. D. Fengel and M. Przyklenk, *Svensk Papperstid., 78*:617 (1975).

379. D. Fengel, *Cellulose Chem. Technol.*, *13*:279 (1979).

380. J. Feckl and D. Fengel, *Holzforschung*, *36*(5):233 (1982).

381. O. Eriksson, D. A. I. Goring, and B. O. Lindgren, *Wood Sci. Technol.*, *14*:267 (1980).

382. A. P. Lapan, V. B. Chekhovskaya, T. G. Paramonova, and Z. L. Kiseleva, *Khim. Drev.* (*Riga*), *1*:85 (1981); *ABIPC*, *52*:A4656.

383. F. Yaku, R. Tanaka, and T. Koshijima, *Holzforschung*, *35*:177 (1981).

384. S. Mukoyoshi, J. Azuma, and T. Koshijima, *Holzforschung*, *35*:233 (1981).

385. B. Kosikova, D. Joniak, and L. Kosakova, *Holzforschung*, *33*:11 (1979).

386. G. J. Leary, D. A. Sawtell, and H. Wong, *SPCI International Symposium on Wood and Pulping Chemistry*, Stockholm, June 9–12, 1981, Preprints, Vol. 1, p. 63.

387. D. A. I. Goring, in *Lignins* (K. V. Sarkanen and C. H. Ludwig, eds.), Wiley-Interscience, New York, 1971, p. 719.

388. E. Adler and B. Wesslen, *Acta Chem. Scand.*, *18*:1314 (1964).

389. K. Freudenberg, K. Jones, and H. Renner, *Chem. Ber.*, *96*:1847 (1964).

390. T. N. Soundararajan and M. Wayman, *J. Polym. Sci.*, Part C, *30*:521 (1970).

391. T. I. Obiaga and M. Wayman, *J. Appl. Polym. Sci.*, *18*:1943 (1974).

392. G. Wegener, M. Przyklenk, and D. Fengel, *Holzforschung*, *37*:303 (1983).

393. W. Q. Yean and D. A. I. Goring, *J. Appl. Polym. Sci.*, *14*:1115 (1970).

394. O. Faix, W. Lange, and E. C. Salud, *Holzforschung*, *35*:3 (1981).

395. W. J. Connors, *Holzforschung*, *32*:145 (1978).

396. W. J. Connors, S. Sarkanen, and J. L. McCarthy, *Holzforschung*, *34*:80 (1980).

397. A. Huttermann, *Holzforschung*, *32*:108 (1978).

398. S. Sarkanen, D. C. Teller, E. Abramowski, and J. L. McCarthy, *Macromols.*, *15*:1098 (1982).

399. G. Wegener and D. Fengel, *Wood Sci. Technol.*, 11:333 (1977).

400. S. Sarkanen, D. C. Teller, J. Hall, and J. L. McCarthy, *Macromols.*, *14*:426 (1981).

401. D. A. I. Goring, in *Lignins* (K. V. Sarkanen and C. H. Ludwig, eds.), Wiley-Interscience, New York, 1971, p. 698.

402. G. Aulin-Erdtman, A. Björkman, H. Erdtman, and S. E. Häglund, *Svensk Papperstid.*, *50* (11B):81 (1947).

403. O. Goldschmid, in *Lignins* (K. V. Sarkanen and C. H. Ludwig, eds.), Wiley-Interscience, New York, 1971, p. 241.

404. G. Aulin-Erdtman, *Svensk Kem. Tidskrift.*, *70*:145 (1958).

405. K. Sarkanen and C. Schuerch, *J. Am. Chem. Soc.*, *79*:4203 (1957).

406. J. Marton and E. Adler, *Acta Chem. Scand.*, *15*:370 (1961).

407. K. Iiyama, J. Nakano, and N. Migita, *J. Japan Wood Res. Soc.*, *13*:125 (1967).

408. H. Norrström and A. Teder, *Svensk Papperstid.*, *74*:85 (1971).

409. K. G. Bogolitsyn and I. M. Bokhovkin, *Zh. Prikl. Spektroskopii*, *31*:283 (1979); *ABIPC*, *51*:A238.

410. S. Y. Lin, in *Proceedings TAPPI Research and Development Division Conference*, Technical Association of the Pulp and Paper Industry, Atlanta, Ga. 1982, p. 119.

411. S. Y. Lin, *Svensk Papperstid.*, *85*:R162 (1982).

412. J. E. Cahill, *Am. Lab.*, *11*(11):79 (1979).

413. B. Bezuch and J. Polcin, *Cellulose Chem. Technol.*, *12*:473 (1978).

414. K. Lundquist and T. K. Kirk, *Tappi*, *63*(1):80 (1980).

415. M. Fiserova and L. Suty, *Cellulose Chem. Technol.*, *14*:243 (1980).

416. H. L. Hergert, in *Lignins* (K. V. Sarkanen and C. H. Ludwig, eds.), Wiley-Interscience, New York, 1971, p. 267.

417. C. Y. Liang, K. H. Bassett, E. A. McGinnes, and R. H. Marchessault, *Tappi*, *43*:1017 (1960).

418. K. J. Harrington, H. G. Higgins, and A. J. Michell, *Holzforschung*, *18*:108 (1964).

419. C.Y. Hse and B. S . Bryant, *J. Japan Wood Res. Soc.*, *12*:187 (1966).

420. S. Kolboe and O. Ellefsen, *Tappi*, *45*:163 (1962).

421. H. I. Bolker and N. G. Somerville, *Pulp Paper Mag. Can.*, *64*:T187 (1963).

422. N. G. Vander Linden, "Studies on Chlorine Dioxide Modification of Lignin in Wood," Ph.D. Dissertation, Institute of Paper Chemistry, Appleton, Wis., 1974.

423. N. G. Vander Linden and G. A. Nicholls, *Tappi*, *59*(11):110 (1976).

424. P. T. Griffiths, *Chemical Infrared Fourier Transform Spectroscopy*, Wiley-Interscience, New York, 1975.

425. T. E. Lyse, "A Study on Ozone Modification of Lignin in Alkali-Fiberized Wood," Ph.D. Dissertation, Institute of Paper Chemistry, Appleton, Wis., 1979.

426. R. D. McKelvey, N. S. Thompson, and T. E. Lyse, *Cell. Chem. Technol.*, *17*:335 (1983).

427. S. A. Berben, J. P. Rademacher, L. O. Sell, and D. B. Easty, *Tappi*, *70*(11):129 (1987).

428. V. B. Karklin´, Z. N. Kreitsberg, and M. Y. Ekabsone, *Khim. Drev.* (*Riga*), 2:53 (1975); *ABIPC*, *46*:A4638.

429. V. B. Karklin´, Y. A. Eidus, and Z. N. Kreitsberg, *Khim. Drev.* (*Riga*), 5:53 (1977); *ABIPC*, *49*:A142.

430. T. E. Crozier, "Oxygen—Alkali Degradation of Loblolly Pine Dioxane Lignin: Changes in Chemical Structure as a Function of Time of Oxidation," Ph. D. Dissertation, Institute of Paper Chemistry, Appleton, Wis., 1978, p. 48.

431. T. E. Crozier, D. C. Johnson, and N. S. Thompson, *Tappi*, *52*(9):107 (1979).

432. C. H. Ludwig, in *Lignins* (K. V. Sarkanen and C. H. Ludwig, eds.), Wiley-Interscience, New York, 1971, p. 299.

433. B. L. Lenz, *Tappi*, *51*:511 (1968).

434. K. Lundquist, *Acta Chem. Scand.*, *35B*:497 (1981).

435. A. Sato, T. Kitamura, and T. Higuchi, *Wood Res.* (*Kyoto*), *59/60*:93 (1976).

436. K. Lundquist and T. Olsson, *Acta Chem. Scand.*, *31B*:788 (1977).

437. K. Lundquist, *Acta Chem. Scand.*, *33B*:27 (1979).

438. K. Lundquist, *Acta Chem. Scand.*, *33B*:418 (1979).

439. K. Lundquist, *Acta Chem. Scand.*, *34B*:21 (1980).

440. K. Lundquist, R. Simonson, and K. Tingsvik, *Svensk Papperstid.*, *82*:272 (1979).

441. K. Lundquist, R. Simonson, and K. Tingsvik, *Paperi Puu*, *63*:709 (1981).

442. L. L. Landucci, *J. Wood Chem. Technol.*, *1*:61 (1981).

443. D. R. Dimmel, D. Shepard, and T. A. Brown, *J. Wood Chem. Technol.*, *1*:23 (1981).

444. H. H. Nimz, *Bull. Liaison Groupe Polyphenols* (Nancy), *8*:185 (1978).

445. D. Robert and D. Gagnaire, *SPCI International Symposium on Wood and Pulping Chemistry*, Stockholm, June 9–12, 1981, Preprints, Vol. 1, p. 86.

446. M. G. Taylor, Y. Deslandes, T. Bluhm, R. H. Marchessault, M. Vincendon, and J. Saint-Germain, *Tappi J.*, *66*(6):92 (1983).

447. D. R. Robert and G. Brunow, *Holzforschung*, *38*(2):85 (1985).

448. P. Tekeley and M. R. Vigna, *J. Wood Chem. Technol.*, *7*(2):215 (1987).

449. J. A. Hemmingson and R. F. H. Dekker, *J. Wood Chem. Technol.*, *7*(2):229 (1987).

450. H.-D. Lüdemann and H. Nimz, *Makromol. Chem.*, *175*:2393 (1974).

451. H. Nimz and H.-D. Lüdemann, *Makromol. Chem.*, *175*:2577 (1974).

452. H.-D. Lüdemann and H. Nimz, *Makromol. Chem.*, *175*:2409 (1974).

453. H. H. Nimz, D. Robert, O. Faix, and M. Nemr, *Holzforschung*, *35*:16 (1981).

454. H. H. Nimz, and H. Schwind, *SPCI International Symposium on Wood and Pulping Chemistry*, Stockholm, June 9–12, 1981, Preprints, Vol. 2, p. 105.

455. K. Haider, P.-C. Ellwardt, and L. Ernst, *SPCI International Symposium on Wood and Pulping Chemistry*, Stockholm, June 9–12, 1981, Preprints, Vol. 3, p. 93.

456. M. G. S. Chua, C.L. Chen, H.-M. Chang, and T. K. Kirk, *Holzforschung*, *36*:165 (1982).

457. C.L. Chen, M. G. S. Chua, J. Evans, and H.-M. Chang, *Holzforschung*, *36*:239 (1982).

458. L. L. Landucci, *J. Wood Chem. Technol.*, *4*(2):171 (1984).

459. J. F. Haw, G. E. Maciel, and C. J. Biermann, *Holzforschung*, *38*(6):327 (1984).

460. H. Hatakeyama and J. Nakano, *Rept. Progr. Polym. Phys. Japan*, *13*:351 (1970).

461. V. J. Bartuska, G. E. Maciel, H. I. Bolker, and B. I. Fleming, *Holzforschung*, *34*:214 (1980).

462. W. Kolodziejski, J. S. Frye, and G. E. Maciel, *Anal. Chem.*, *54*:1419 (1982).

463. D. L. W. Kwoh, S. S. Bhattacharjee, J. J. Cael, and S. L. Patt, in *Proceedings TAPPI Research and Development Division Conference*, Technical Association of the Pulp and Paper Industry, Atlanta, Ga., 1982, p. 113.

464. J. F. Haw, G. E. Maciel, and H. A. Schroeder, *Anal. Chem.*, *56*(8):1323 (1984).

465. J. F. Haw and T. P. Schultz, *Holzforschung*, *39*(5):289 (1985).

466. K. Lundquist, B. Josefsson, and G. Nyquist, *Holzforschung*, *32*:27 (1978).

467. R. H Atalla, unpublished work, Institute of Paper Chemistry, Appleton, Wis., 1979.

468. J. M. Gould, unpublished work, Northern Regional Research Center, U.S. Department of Agriculture, Peoria, Il., 1982.

469. R. E. Blank and T. Wakefield II, *Anal. Chem.*, *51*:50 (1979).

470. G. M. Dorris and D. G. Gray, *Cellulose Chem. Technol.*, *12*:9, 721 (1978).

471. S. Katz and D. G. Gray, *Svensk Papperstid.*, *83*:226 (1980)

472. K. Lundquist, *SPCI International Symposium on Wood and Pulping Chemistry*, Stockholm, June 9–12, 1981, Preprints, Vol. 1, p. 81.

473. A. N. Buckley and A. J. Michell, *Appita*, *36*:205 (1982).

474. R. G. Rickey, J. K. Hamilton, and H. L. Hergert, *Wood Fiber*, *6*:200 (1974).

475. J. W. Munson, in *Proceedings 6th Annual Lightwood Research Conference* (M. H. Esser ed.), Southeast Forest Experiment Station, Asheville, N.C. 1979, p. 120.

476. J. Drew, J. Russell, and H. W. Bajak, *Sulfate Turpentine Recovery*, Pulp Chemicals Association, New York, 1971.

477. J. Drew and G. D. Pylant, Jr., *Tappi*, *49*:430 (1966) .

478. D. F. Zinkel and C. R. McKibben, in *Proceedings 5th Annual Lightwood Research Conference* (M. H. Esser, ed.), Southeast Forest Experiment Station, Asheville, N.C. 1978, p. 133.

479. D. F. Zinkel, in *Proceedings 6th Annual Lightwood Research Conference* (M. H. Esser ed.), Southeast Forest Experiment Station, Asheville, N.C. 1979, p. 147.

480. M. Mayr, B. Hausmann, N. Zelmann, and K. Kratzyl, *Proceedings TAPPI Research and Development Division Conference*, Technical Association of the Pulp and Paper Industry, Atlanta, Ga., 1984, p. 155.

481. S. V. Kossuth and J. W. Munson, *Tappi*, *64*(3):174 (1981).

482. D3009-72, in *ASTM Standards*, American Society for Testing and Materials, Philadelphia, Pa.

483. E. Zavarin, Y. Wong, and D. F. Zinkel, *Proceedings 5th Annual Lightwood Research Conference* (M. H. Esser ed.), Southeast Forest Experiment Station, Asheville, N.C. 1978, p. 19.

484. K. Frodin and J. Andersson, *Eur. J. Forest Pathol.*, *7*(5):282 (1977).

485. R. P. Adams, M. Granat, L. R. Hogge, and E. Rudloff, *J. Chromatogr. Sci.*, 17:75 (1979).

486. P. J. Nelson, P. I. Murphy, and F. C. James, *Appita*, *30*:503 (1977).

487. B. L. Browning, *Methods of Wood Chemistry*, Vol. 1, Wiley-Interscience, New York, 1967, p. 115.

488. D. F. Zinkel, *Tappi*, *58*:109 (1975).

489. F. H. M. Nestler and D. F. Zinkel, *Anal. Chem.*, *35*:1747 (1963).

490. D. F. Zinkel, L. C. Zank, and M. F. Wesolowski, *Diterpene Resin Acids: A Compilation of Infrared, Mass, Nuclear Magnetic Resonance, Ultraviolet Spectra and Gas Chromatographic Retention Data* (*of the Methyl Esters*), USDA, Forest Service, Forest Products Laboratory, Madison, Wis. 1971.

491. D. F. Zinkel and L. C. Zank, *Anal. Chem.*, *40*:1144 (1968).

492. R. A. Chapman, H. M. Nugent, H. I. Bolker, D. F. Manchester, R. H. Lumsden, and W. A. Redmond, *CPPA Trans. Tech. Sect.*, *1*(4):113 (1975).

493. D. F. Zinkel and C. C. Engler, *J. Chromatogr.*, *136*:245 (1977).

494. M. Mayr, E. Lorbeer, and K. Kratzl, *J. Am. Oil Chem. Soc.*, *59*:52 (1982).

495. L. Sisfontes, G. Nyborg, L. Svensson, and R. Blomstrand, *J. Chromatogr.*, *216*:115 (1981).

496. B. Holmbon, *J. Am. Oil Chem. Soc.*, *54*:289 (1977).

497. G. M. Dorris, M. Douek, and L. H. Allen, *J. Am. Oil Chem. Soc.*, *59*:494 (1982).

498. K. Ukkonen, *Tappi*, *65*(2):71 (1982).

499. J. W. King, E. C. Adams, and B. A. Bidlingmeyer, *J. Liquid Chromatogr.*, *5*:275 (1982).

500. S. S. Curran and D. F. Zinkel, *J. Am. Oil Chem. Soc.*, *58*:980 (1981).

501. C. R. Scholfield, *J. Am. Oil Chem. Soc.*, *56*:511 (1979).

502. E. A. Emken, J. C. Hartman, and C. R. Turner, *J. Am. Oil Chem. Soc.*, *55*:561 (1978).
503. W. Kamutzki and T. Krause, *SPCI International Symposium on Wood and Pulping Chemistry*, Stockholm, June 9–12, 1981, Preprints, Vol. 5, p. 70.
504. J. Drew and M. Propst, *Tall Oil*, Pulp Chemicals Association, New York, 1981.
505. SCAN Testing Committee, *Svensk Papperstid.*, *82*:391 (1979).
506. E. E. Balder and E. P. Crowell, *J. Am. Oil Chem. Soc.*, *52*:14 (1975).
507. R. J. Maxwell and D. P. Schwartz, *J. Am. Oil Chem. Soc.*, *56*:634 (1979).
508. *Analytical Procedures for Tall Oil Products*, Pulp Chemicals Association, New York, 1976.
509. B. L. Browning, *Methods of Wood Chemistry*, Vol. 1, Wiley-Interscience, New York, 1967, pp. 132, 164.
510. L. D. Metcalfe, *J. Am. Oil Chem. Soc.*, *56*:786A (1979).
511. D. G. Roux, D. Ferreira, H. K. L. Hundt, and E. Malan, *J. Appl. Polym. Sci.:Appl. Polym. Symp.*, *28*:335 (1975).
512. R. S. Thompson, D. Jacques, E. Haslam, and R. J. N. Tanner, *J. Chem. Soc. Perkin Trans. I*:1387 (1972).
513. J. Parker, *J. Chem. Ecol.*, *3*:389 (1977).
514. R. S. McCredie, E. Ritchie, and W. C. Taylor, *Aust. J. Chem.*, *22*:1011 (1969).
515. T. Takehara and T. Sasaya, *J. Japan Wood Res. Soc.*, *25*:660 (1979).
516. T. W. Pearson, G. S. Kriz, Jr., and R. J. Taylor, *Wood Sci.*, *10*:93 (1977).
517. E. L. Johnson and A. J. Cserjesi, *J. Chromatogr.*, *107*:388 (1975).
518. T. Popoff and O. Theander, *J. Appl. Polym. Sci.: Appl. Polym. Symp. 28*:1341 (1976).
519. O. Theander, *SPCI International Symposium on Wood and Pulping Chemistry*, Stockholm, June 9–12, 1981, Preprints, Vol. 1, p. 89.
520. I. A. Pearl and S. F. Darling, *Phytochem.*, *10*:3161 (1971).
521. I. A. Pearl and S. F. Darling, *Phytochem.*, *7*:821 (1968).
522. R. Ekman, *Holzforschung*, *30*:79 (1976).
523. C.-L. Chen, H.-M. Chang, and E. B. Cowling, *Phytochem.*, *15*:547 (1976).
524. I. A. Pearl and S. F. Darling, *Phytochem.*, *9*:1277 (1970).
525. K. Miki, K. Ito, and T. Sasaya, *J. Japan Wood Res. Soc.*, *25*:665 (1979).
526. R. W. Spencer and E. T. Choong, *Holzforschung*, *31*:25 (1977).
527. M. Samejima and T. Yoshimoto, *J. Japan Wood Res. Soc.*, *25*:671 (1979).
528. J. J. Karchesy, P. M. Loveland, M. L. Laver, D. F. Barofsky, and E. Barofsky, *Phytochem.*, *15*:2009 (1976).
529. Y. Yazaki and W. E. Hillis, *Holzforschung*, *31*:20 (1977).
530. P. Fang and G. D. McGinnis, *J. Appl. Polym. Sci.: Appl. Polym. Symp.*, *28*:363 (1975).
531. E. Pulkkinen and S. Vaisanen, *Paperi Puu*, *64*:72 (1982).
532. A. Calimli and A. Olcay, *Holzforschung*, *32*:7 (1978).
533. L. W. Wulf and C. W. Nagel, *J. Chromatog.*, *116*:271 (1976).
534. B. L. Browning, *Methods of Wood Chemistry*, Vol. 1, Wiley-Interscience, New York, 1967, p. 223.

535. R. Andersson, T. Popoff, and O. Theander, *Acta Chem. Scand.*, *29B*:835 (1975).

536. W. A. Court, *J. Chromatogr.*, *130*:287 (1977).

537. P. Labosky, Jr. and J. A. Sellers, *Wood Sci.*, *13*:32 (1980).

538. M. Wilkinson, J. G. Sweeny, and G. A. Iacobucci, *J. Chromatogr.*, *132*:349 (1977).

539. S. F. Fonseca, L. T. Nielsen, and E. A. Ruveda, *Phytochem.*, *18*:1703 (1979).

540. S. K. Sarkar and S. S. Malhotra, *J. Chromatogr.*, *171*:227 (1979).

541. R. Toman, S. Karacsonyi, and M. Kubackova, *Carbohyd. Res.*, *43*:111 (1975).

542. R. Toman, S. Karacsonyi, and M. Kubackova, *Cellulose Chem. Technol.*, *10*:561 (1976).

543. B. L. Browning, *The Chemistry of Wood*, Wiley-Interscience, New York, 1963, p. 59.

544. T211 om-80, in *TAPPI Test Methods*, Technical Association of the Pulp and Paper Industry, Atlanta, Ga.

545. B. L. Browning, *Methods of Wood Chemistry*, Vol. 1, Wiley-Interscience, New York, 1967, p. 88.

546. M. S. Vigler, A. W. Varnes, and H. A. Strecker, *Am. Lab.*, *12*(8):31 (1980).

547. M. T. Friend, C. A. Smith, and D. Wishart, *At. Absorption Newslett.*, *16*:46 (1977).

548. D. L. Heanes, *Analyst*, *106*:172 (1981).

549. J. B. Zicherman and R. J. Thomas, *Holzforschung*, *26*:150 (1972).

550. R. Ganapathy, *Chemist-Analyst*, *66*(2):6 (1977).

551. C. O. Ingamells, *Anal. Chem. Acta*, *52*:323 (1970).

552. T. T. Gorsuch, *Analyst*, *84*:135 (1959).

553. Useful Method 243, in *TAPPI Useful Methods*, Technical Association of the Pulp and Paper Industry, Atlanta, Ga.

554. G. D. Christian, *Anal. Chem.*, *41*:24A (1969).

555. L. A. Muse, *J. Chem. Educ.*, *49*:A463 (1972).

556. J. P. Price, *Tappi*, *54*:1497 (1971).

557. N. M. Arafat and W. A. Glooschenko, *Analyst*, *106*:1174 (1981).

558. T438 om-82, in *TAPPI Test Methods*, Technical Association of the Pulp and Paper Industry, Atlanta, Ga.

559. D. C. Lee and W. C. Laufmann, *Anal. Chem.*, *43*:1127 (1971).

560. W. R. Hatch and W. L. Ott, *Anal. Chem.*, *40*:2085 (1968).

561. J. Marton and T. Marton, *Tappi*, *55*:1614 (1972).

562. R. Lammi, *Paperi Puu*, *63*:605 (1981).

563. E. King and O. Schalin, *Paperi Puu*, *57*:209 (1975).

564. H. L. Kahn, in *Trace Inorganics in Water*, Advances in Chemistry Series, No. 73, American Chemical Society, Washington, D.C. 1968, p. 183.

565. W. H. Smith, *Environ. Sci. Technol.*, *7*:631 (1973).

566. E. E. Pickett and S. R. Koirtyohann, *Anal. Chem.*, *41*(14):28A (1969).

567. L. H. Ahrens and S. R. Taylor, *Spectrochemical Analysis*, 2nd ed., Addison-Wesley, Reading, Mass., 1961.

568. V. A. Fassel and R. N. Knisely, *Anal. Chem.*, *46*:1110A (1974).

569. V. A. Fassel and R. N. Knisely, *Anal. Chem.*, *46*:1155A (1974).

570. J. D. Winefordner and R. C. Elser, *Anal. Chem.*, *43*(4): 24A (1971).

571. L. Laamanen, *Paperi Puu*, *66*(9):517 (1984).

572. S. Saka and D. A. I. Goring, *J. Japan Wood Res. Soc.*, *29*(10):648 (1983).

573. H. Nurmesniemi and H. Hayrynen, *Paperi Puu*, *65*(11):700 (1983).

574. W. F. DeGroot, *Carbohydr. Res.*, *142*(1):172 (1985).

575. C. F. Baes and S. B. McLaughlin, *Sci.*, *224*(4648):494 (1984)

576. J. A. Meyer and J. E. Langwig, *Wood Sci.*, *5*:270 (1973).

577. C. A. Osterhaus, J. E. Langwig, and J. A. Meyer, *Wood Sci.*, *8*:370 (1975)

578. T. Hattula and M. Johanson, *Radiochem. Radioanal. Lett.*, *32*:35 (1978).

579. H. E. Young and V. P. Guinn, *Tappi*, *49*(5):190 (1966).

580. E. B. Sandell, *Colorimetric Determination of Traces of Metals*, 3rd ed., Wiley-Interscience, New York, 1959.

581. H. Small, T. S. Stevens, and W. C. Bauman, *Anal. Chem.*, *47*:1801 (1975).

582. R. A. Parham, *Paperi Puu*, *55*:959 (1973).

583. M. L. Harder and D. W. Einspahr, *Tappi*, *63*(12):110 (1980).

584. K. L. McDonald, *Tappi*, *57*(3):163 (1974).

585. J. S. Fritz and S. S. Yamamura, *Anal. Chem.*, *27*:1461 (1955).

586. J. Turunen and A. Visapaa, *Paperi Puu*, *54*:59 (1972).

587. C. T. Garten, Jr., *Bull. Environ. Contam. Toxicol.*, *17*(2):127 (1977).

588. A. Visapaa, *Paperi Puu*, *55*:385 (1973).

589. W. R. Webb and P. J. Aylward, *Appita*, *36*:293 (1983).

590. B. L. Browning, *Methods of Wood Chemistry*, Vol. 1, Wiley-Interscience, New York, 1967, p. 12.

591. D. Fengel and D. Grosser, *Holz als Roh- und Werkstoff*, *33*:32 (1975).

592. I. H. Isenberg, M. L. Harder, and L. Louden, *Pulpwoods of the United States and Canada*, Vol. 1, Institute of Paper Chemistry, Appleton, Wis., 1980.

593. I. H. Isenberg, M. L. Harder, and L. Louden, *Pulpwoods of the United States and Canada*, Vol. 2, Institute of Paper Chemistry, Appleton, Wis., 1981.

594. H. L. Hergert, T. H. Sloan, J. P. Gray, and K. R. Sandberg, unpublished work, ITT Rayonier Inc., 1981.

4

Cellulose

G. D. McGinnis*

Mississippi State University, Mississippi State, Mississippi

F. Shafizadeh†

University of Montana, Missoula, Montana

I. HISTORICAL USE OF CELLULOSE

Cellulose is that ubiquitous material without which we could not have life as we know it on this planet. In nature, it is produced, consumed, and destroyed in tremendous quantities each day. Cellulose is the most abundant macromolecule on earth. Hess and others have estimated that 10^9–10^{11} tons of cellulose are produced annually (1).

The major function of cellulose in nature is as a structural component in plants. In higher plants (*Embryophyta*) and a majority of lower plants (*Thallophyta*), cellulose is the principal structural component of different layers and lamellas of the cell wall that are generally interlaced with a variety of matrix polysaccharides and often are impregnated with lignin.

Cellulose synthesis, however, is not limited to the plant kingdom but is also found in bacterial, fungal, and algal systems and certain animals (*Tunicate*). Cellulose also plays an important role in the carbon dioxide cycle on earth, and its stabilizing influence on the level of carbon dioxide in the atmosphere is of the same magnitude as that of the oceans. Cellulose has not only an important role in nature, but it has provided man with a ready fuel source for fire, food, clothing, and shelter

---

* Current affiliation: Michigan Technological University, Houghton, Michigan.
† Deceased.

and has had a profound influence in shaping the destiny of man and the development of science, technology, and economics.

Men's first use of cellulosic material was as a source of natural fuel. At the beginning of the Pleistocene era, the use of fire as well as stone tools and language probably distinguished mankind from other primates. One of the most momentous events in the development of mankind must have been the discovery of a method for generating fire by the ignition of cellulosic materials. The use of campfire for protection and supplementary heat by early man led to the arts of cooking, baking, pottery, and eventually production of charcoal and naval stores by the partial combustion and pyrolysis of wood. It is believed that the destructive distillation of wood has a long industrial history dating back to the ancient Chinese and Egyptians who used the tarry products for embalming.

Sheets of papyrus made by pressing the pith tissue of a sedge, *Cyperus papyrus*, were used as early as 3000 B.C. in Egypt for writing. In China, strips of bamboo or wood were used for drawing and writing until the discovery of paper, which is attributed to Ts'ai Lun in 105 A.D. The first paper was made in China from rags, bark fiber, and bamboo. Pieces of bamboo were soaked for more than 100 days and boiled in the milk of lime for approximately 8 days and nights to release the fibers. The art of papermaking finally reached Persia by 751 and from there spread to the Mediterranean countries, when the Moors brought the industry to Europe in the 12th century. The acceptance of paper and the development of the paper industry in Europe created a chronic shortage of rag, which was used as the raw material.

The shortage of paper eventually was relieved by the production of pulp from wood, which was responsible for literally raising the industry from rags to riches. The credit for starting this immense industry is obviously shared by many scientists and inventors, but it is interesting to note that the seeds for the idea of extracting the cellulosic fiber from abundant woody tissues not only could be found in the Chinese practice of boiling bamboo in the milk of lime, but also in a treatise submitted to the French Royal Academy by René Antoine Ferchault de Réaumur on November 15, 1719. In this treatise, Réaumur, a noted physicist and naturalist, observed that the American wasps form very fine paper, like ours; they extract the fibers of common wood of the countries where they live. They teach us that paper can be made from the fibers of plants without the use of rags and linen, and seem to invite us to try and make fine paper from the use of certain woods.

Another 120 years elapsed before the French chemist, Anselme Payen, demonstrated that a fibrous substance that he called cellulose (in 1893) could be isolated by the treatment of wood with nitric acid. The isolation of this substance opened the door for production of wood pulp by commercial methods of delignification, including the soda processes patented by Watt and Burgers (1853), the sulphite process invented by Tilgman (1866), the Kraft process developed by Eaton (1870) and Dahl (1884), and various bleaching methods. Numerous refinements and developments of these processes in the 20th century have led to the rapid growth of this industry and adaptation of paper not only for writing and printing, but also for wrapping, packaging, and a variety of disposable products.

The isolation of cellulose from plant tissues by the laboratory and industrial processes naturally stimulated the scientific community to determine the structure, composition, and properties of this material and how it is synthesized in the plant. It is interesting to note that, after more than a century of scientific investigation and numerous findings, controversies, and debates, the term cellulose (in pure form) still means different things to different groups. To the organic chemists, it means β-D-(1-4)-linked glucopyran. To the technologists, it means an asymptotic entity, often called α-cellulose, which represents the alkali-insoluble portion of the wood pulp. To the biologists, it means the fine microfibrils of plant cell walls that reach a high degree of purity and perfection in a group of green algae, including *Valonia*, *Cladophora* and *Chaetomorpha*. These groups have been concerned not only with the chemical structure of cellulose and its reactions, but also with its interrelated physical, morphological, and biological properties. Through various types of inquiries, they have tried to find how cellulose is crystallized and packed in fibrils to give it fibrous and physical properties; what the fibrils and microfibrils are and how they are organized within the layers and lamella of the cell wall; and whether the microfibrils are produced and organized by inanimate physical forces or by the direct biological influence of the living cell.

The isolation and chemical investigation of cellulose led to the production of cellulose nitrate, cellulose acetate, rayon, and cellophane, the forerunner of the modern plastic and polymer industry.

The efforts of John Wesley Hyatt to produce synthetic ivory for billiard balls led to the development of the first synthetic plastic, known as celluloid. In 1870, this material was produced from a mixture of partially nitrated cellulose, called pyroxylin, with camphor; pyroxylin has also been used in the manufacture of lacquers, films, and adhesives. Highly nitrated cellulose replaced the black powder and has been developed as a modern explosive and propellant.

Because of the flammability of cellulose nitrate, cellulose acetate was developed industrially as a safer substitute, first for coating wing fabrics of World War I airplanes, and subsequently, for photographic and motion picture films. Another major factor in the growth of the cellulose acetate industry was the development of the injection molding process. Cellulose acetate was one of the first materials to be used for manufacturing plastic objects by this process.

In 1892, Cross and Bevan discovered cellulose xanthate, which forms a viscous solution from which cellulose could be regenerated as a continuous fiber (rayon) or film (cellophane). The development of the industrial process for the continuous production of rayon for the textile industry and cellophane film for the packaging industry led by 1911 to the development of the dissolving pulp industry. Until that time, the pulp industry was concerned with the manufacturing of fibers for the paper industry, and chemical derivatives of cellulose were produced from purified rags or cotton linter called chemical cotton. The development of the dissolving pulp industry led to the mass production of relatively cheap and pure cellulose as a raw material for the expanding chemical industry. Large quantities of cellulose and rayon acctate are currently used for the production of synthetic fiber.

The rapid development of the pulp and paper industry provided a ready application for some of the sawmill residue, which was being burned for disposal or used as hog fuel. More and more of the chips and sawdust produced by the sawmill are being used as the raw material for the expanding pulp and paper industry. However, logging residues and tremendous amounts of waste paper that end up as municipal refuse, as well as the agricultural residues such as bagasse, straw, corn stalk, etc., which constitute a colossal source of cellulosic materials, are still wasted or inefficiently utilized.

All the above industrial developments involve the isolation, modification, or application of cellulose as a fibrous or structural polymer. However, from the early development of cellulose chemistry, the fact that enormous quantities of cellulosic wastes can be converted to simple sugar components as sources of food, alcohol, and industrial chemicals has created an industrial opportunity that has been repeatedly explored but only used within the controlled economy of the Soviet Union and Japan during the two World Wars.

Due to the increasingly demands for food, chemicals, and fuel and the limited supply of petroleum, the conservation and efficient utilization of cellulosic materials have become a more urgent problem, and investigation of the chemical conversion of cellulosic wastes to a wider variety of chemicals is gaining momentum. The current efforts are directed toward improving the older acid hydrolysis methods, as well as developing new processes utilizing enzymatic and pyrolytic procedures.

Currently, only a fraction of the annual harvest of cellulosic materials produced by forests and agricultural lands is utilized. The conversion of these materials to food and chemicals will provide a fundamental solution to the problems of supply and productivity.

II. SOURCE, ISOLATION, AND MOLECULAR STRUCTURE OF CELLULOSE

Cellulose is the most abundant and renewable natural resource available to man, but cellulose does not occur in pure form in any natural source. Even the seed hairs of cotton, the purest form of cellulose readily available in nature, contain about 6% by weight of noncellulosic polysaccharides, proteins, and mineral elements. In nature, cellulose always is associated with a number of other polysaccharides such as starch, pectin, and a variety of hemicelluloses. These hemicelluloses are mainly heteropolymers of galactose, mannose, xylose, arabinose, and other sugars and their uronic acids. In addition, almost all the cellulose in nature is deposited in an intimate physical admixture with lignin-,-a complex, three-dimensional, phenolic polymer that is quite distinct from both cellulose and the associated hemicelluloses. The heterogenous compositions of plant materials are illustrated in Table 1, which shows overall composition of a variety of plant materials.

A. Isolation of Cellulose

The general methods of isolating cellulose from a complex matrix such as wood or from some other plant involves removing the noncellulosic components. Extractive components can be removed by solvent extraction, whereas lignin and hemicellu-lose are removed by a combination of chemical or physical techniques that break down the noncellulosic materials, followed by extraction of the noncellulosic component with an appropriate solvent.

One widely used laboratory procedure for softwood fractionation is shown in Fig. 1 (2b). The extractives, in general, make up 1–5% of wood and can be removed by extraction with neutral organic solvents. Owing to the diversity of extractive substances that occur in plants, no single solvent is effective. Solvents that have been used to effectively remove the extractives include petroleum ether, ethyl ether, methanol or ethanol, acetone, benzene, ethanol-benzene (1:2), and water. The extraction, except for water extraction, is generally performed using a soxhlet-type extractor.

A minor portion of the hemicelluloses can be removed before delignification. However, in most cases it is necessary to remove most of the lignin before major

Table 1 Composition of Various Plant Materials

Source	Cellulose (%)	Lignin (%)	Hemicelluloses (%)
Hardwoods	40–55	18–25	24–40
Softwoods	45–50	25–35	25–35
Monocotyledons (e.g., grasses such as palms, bamboo, wheat, rice, and sugarcane)	25–40	10–30	25–50
Parenchyma cells of most leaves	15–20	—	80–85[a]
Fibers (e.g., seed hairs of cotton and bast fibers of flax)	80-95	—	5–20

[a] Most of these materials are pectins and other hemicelluloses.
Source: Ref. 2a.

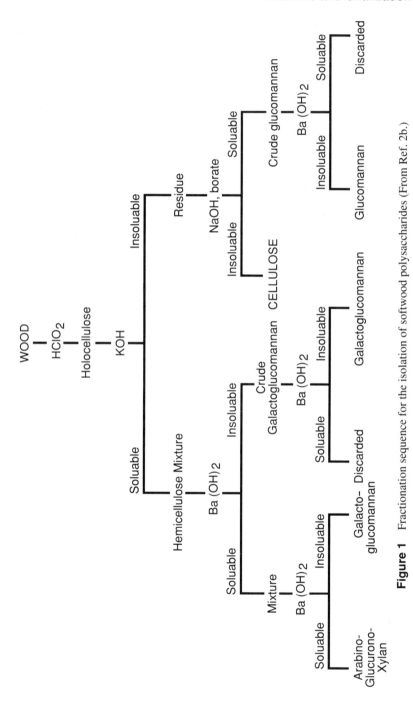

Figure 1 Fractionation sequence for the isolation of softwood polysaccharides (From Ref. 2b.)

portions of the hemicellulose or cellulose can be isolated. The process of selectively removing the lignin from plant materials is called "delignification" and the solid product remaining, which is mainly the plant carbohydrates, is called "holocellulose."

Laboratory processes to produce holocellulose are mostly based on the reaction of moist wood with chlorine gas or on digestion with an acidified solution of sodium chlorite. The classical method of Cross and Bevan (3) for the isolation of cellulose preparations comprises alternate chlorination and extraction with a hot, aqueous, sodium sulfite solution. Unfortunately, this procedure leads to the removal of a considerable part of the hemicelluloses along with the lignin. Currently, the most common laboratory method involves alternate chlorination, followed by extraction with a hot alcoholic solution of monoethanolamine (4). Another method (for preparation of "chlorite holocellulose") depends on the digestion of wood meal with an acidified solution of sodium chlorite (5). The temperature of treatment is varied from about 70–90°C. The extent of degradation of cellulose and holocellulose caused by these different methods of isolation has been compared and reviewed by several researchers (6,7).

The hemicelluloses are separated from cellulose by extraction, preferably of a holocellulose preparation, with aqueous alkaline solutions at room temperature. Effective extraction of the hemicelluloses and partial fractionation with extracting solvents are accomplished with solvents such as 5% sodium hydroxide, 16–18% sodium hydroxide, or 24% potassium hydroxide (8,9). The addition of boric acid or metaborates to the alkaline solutions enhances extraction of mannan-containing hemicelluloses.

For a quantitative isolation of a wood cellulose, the lignin is first removed with chlorous acid as described previously. The resulting holocellulose is exhaustively extracted with 24% aqueous sodium hydroxide containing 4% boric acid for elimination of the hemicelluloses. The cellulose thus obtained in almost quantitative yield normally contains only traces of mannose and xylose residues and on hydrolysis gives D-glucose. Its X-ray diagram is that of cellulose II and the polysaccharide is severely degraded, the original number-average degree of polymerization having been lowered from approximately 5,000–14,000 to 200–700 (10). The best procedure for the quantitative isolation of an undegraded cellulose is to directly nitrate with a nondegrading acid mixture in order to convert the cellulose into cellulose nitrate. Cellulose nitrate is very soluble in many organic solvents, and the molecular weight can be determined by viscometry or other techniques (11).

B.   The Molecular Structure of Cellulose

Cellulose was first isolated and recognized as a distinct chemical substance in the 1830s by the noted French agricultural chemist, Anselme Payen (12). Payen also concluded, more or less correctly, that cellulose and starch were isomeric substances because both had the same carbon and hydrogen analysis and, subjected to hydrolysis, both yielded D-glucose. However, nearly three-quarters of a century passed before the precise empirical formula of cellulose was established (13) $(C_6H_{10}O_5)_x$. Results from earlier studies on acetylation (14) and nitration (15, 16) had indicated that cellulose had three free hydroxyl groups per $(C_6H_{10}O_5)$ unit.

The next major step in unraveling the overall structure of cellulose was to determine if cellulose contained only D-glucose units or if it consisted mainly of D-glucose units with trace amounts of other sugars. When cellulose was hydrolyzed directly with acids, over a 90.7% yield of D-glucose was obtained (17). This was good evidence that the only repeating unit in the cellulose polymer was anhydro-D-glucose. Even better evidence was obtained when the cellulose was initially converted into cellulose acetate and then hydrolyzed to a mixture of methyl glucosides. With this procedure, over 95.5% yield of a crystalline mixture of methyl α- and β-D-glucosides was obtained (18, 19).

The next step in determining the overall structure of cellulose involved determining how the anhydro-D-glucose units were linked together. Samples of cellulose were methylated and then hydrolyzed to the individual monomeric units. The position of the methyl groups corresponds to the position of the free hydroxyl group within the cellulose molecule. When cellulose is methylated (20, 21) and hydrolyzed, the major product obtained was 2,3,6-tri-O-methyl D-glucose. This indicated that the free hydroxyl groups in cellulose were located at the 2, 3, and 6 positions and that the 1, 4, and 5 positions were linked by chemical bond. The two possibilities are a 4-O-substituted D-glucopyranose or a 5-O-substituted D-glucofuranose. Further structure studies by a variety of Investigators indicated that the pyranose structure was the correct form (22).

The final problem was to determine whether the linkage between the units was α or β. Evidence on this point came from experiment by the partial acid hydrolysis (23, 24) of cellulose. This treatment converted cellulose into a series of cellulose oligomers that include cellobiose and cellotriose. Structural studies on these two compounds indicated that the linkage connecting the individual units was β. Therefore, cellulose is a linear polysaccharide consisting of anhydro-D-glucopyranose units linked between the 1 and 4 position of adjacent sugar units by a β linkage.

In the determination of the average molecular weight of cellulose, the usual methods for polymers have been used, including osmometry, light scattering measurements, ultracentrifugation, gel permeation, and viscometric determinations (25–31). The degree of polymerization of cellulose that is polydispersed varies with the source and method of isolation. Of the various methods tried, direct nitration of the wood has been the most successful (32). Nitration is carried out in a mixture of nitric acid and phosphorus pentoxide, or nitric acid and acetic anhydride (33), whereby practically no degradation occurs.

Some of the more recent studies involving nitrate derivative give an average DP from viscometry of about 3,000–14,000 for higher plants (32–34). There are indications that native cellulose present in the secondary wall of plants is monodisperse (contains molecules all having identical molecular weights), whereas the cellulose in the primary wall has a lower molecular weight and a broad range of molecular weights (35–36). Most chemical and physical reactions of cellulose lead to a very rapid decrease in the molecular weight of the cellulose. More detailed information on how these processes affect molecular weight can be found in a later part of this chapter.

III. CELLULAR STRUCTURE OF CELLULOSE

It would be impossible to understand the chemical reactions of cellulose, account for the physical properties of wood or other plant materials, or even fully appreciate the complex process of biogenesis without some understanding of the arrangement of cellulose, hemicelluloses, and lignin in the cell wall. There are several excellent monographs and articles (37–42) in which the subject is explored in greater detail. The present discussion will briefly cover some features relevant to the topic of this chapter.

Plant cells are distinguished from animal cells by the presence of true cell walls containing polysaccharides as the major structural material. Cells originate at the tips of plants (meristems) by the division of existing cells. During the initial growth period, the wall (primary wall) is continually synthesized at or near the plasmalemma. During this period, the cell continues to increase in volume; the primary cell wall does not thicken, but it continues to expand. Eventually, the cell reaches its mature shape and size, but cell wall deposition does not cease; instead, expanding in area, the wall starts to thicken. This thickened layer constitutes the secondary walls. At some point during the thickening process, the cell dies and the cell wall remains as a hollow shell.

In this form, it has enormous commercial significance. All timbers and timber products, including paper, cotton, linen, hemp, ramie, and sisal fibers, are cell wall complexes, and many cell wall constituents are extracted and used as foods, chemicals, fiber, and many other products. Moreover, the cell wall is important in determining the consistence of, for instance, potatoes and apples, and is a feature of relevance even in commercial enterprises in which it is not actually used, as in the expression of oil from oil seed in which the wall needs to be fractured before the oil becomes accessible.

This simplified description of primary and secondary wall formation is still an area not completely understood. More detailed information about the processes of cell growth can be found in reviews by Preston (43), Albersheim (44), and others (45–46). One important point that should be made is that unique differences exist between the composition, molecular weight, molecular-weight distribution, and orientation of the polysaccharides in the primary and secondary walls, which indicates that the biological processes of biosynthesis of cell wall materials in these areas are uniquely different. This will become more evident in later sections of this chapter.

Cellulose is the major structural component of plant cell walls. It exists in the cell wall as long, threadlike fibers (microfibrils). The cellulose microfibrils in mature wood cells are embedded in a matrix composed mainly of hemicelluloses and lignin.

Electron microscopic studies of mature wood cells (Fig. 2) show that they consist of several layers of cell wall surrounded by amorphous, intercellular substance (47). A simplified picture of the organization of a typical softwood tracheid or hardwood fiber is seen in Fig. 3. Between the cells is a region called the true middle lamella that contains mainly lignin and pectin substances. The primary wall, which is only 0.1–0.2 μm in thickness, contains a randomly and loosely organized network of cellulose microfibrils embedded in a matrix that, until recently, was considered to

Figure 2   A replica of an earlywood tracheid showing the microfibrillar orientation in the S_2 and S_3 layers of Picea jezoensis (courtesy of Syracuse University Press).

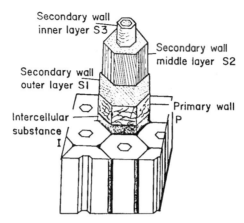

Figure 3 Simplified structure of the cell wall of a softwood tracheid or hardwood fiber.

consist of amorphous pectins and hemicelluloses lacking structural orientation. However, recent studies have shown that the hemicelluloses are partially oriented (48). Immediately below the primary wall is the secondary wall constituting, in fact, nearly all the cell wall. The secondary wall is divided into three layers: S_1, S_2, and S_3. The outer layer of the secondary wall (S_1) is 0.1–0.3 μm thick, with microfibrils oriented with an average angle of 50–70°. The S_2 layer is 1.0–5.0 μm thick and accounts for the major part of the cell wall volume. The microfibrils in this portion of the secondary wall are oriented almost parallel to the fiber axis (10–30°). In the thin S_3 layer (0.1 μm), the microfibrils form a flat helix in the transverse direction (6–90°). The innermost portion of the cell wall consists of the so-called warty layer, containing protoplasmic debris.

The distribution of lignin, cellulose, hemicelluloses, and pectin over the middle lamella and the cell wall in wood fibers has been found to be quite heterogenous.

For both hardwoods and softwoods, the highest concentration of lignin is found in the middle lamella and lowest in the secondary wall. However, because of the thickness of the secondary wall, most of the lignin is located in the secondary wall (Table 2). Most studies also indicate that the lignin in these different parts of the secondary wall is uniquely different.

IV. MICROFIBRILS

In the cell wall, cellulose chains aggregate to form long thin threads called microfibrils. The microfibrils in combination with the other matrix materials provide the necessary rigidity and stress resistance in the plant. According to

Freudenberg (50), the microfibrils in the cell wall act in the same way as reinforcing rods in prestressed concrete. A better descriptive model, suggested by Mark, is furnished by filament-wound, reinforced plastic structures (51), such as pressure vessels.

In wood, the microfibrils are embedded in a matrix of polysaccharides and amorphous lignin. The noncellulosic materials can be removed by a variety of different chemical treatments and the microfibrils can be observed using an electron microscope. An electron micrograph of wood microfibrils is shown in Fig. 2. This micrograph, from a tracheid of Picea jezoensis, illustrates the difference in orientation of the microfibrils in the S_2 and S_3 layers of the plant cell (47).

A long series of chemical, physical, and microscopic investigations has provided some information about the size and shape of the microfibril, the arrangement of the cellulose molecule in the microfibril, and the biogenesis processes involved during formation in the plant cell. In the next section, a review on what is known about the physical structure of the microfibrils will be given.

A. The Dimensions of the Microfibril

The initial studies completed in 1950–1960 on the dimensions of the microfibrils indicated that the diameters varied, depending on the source of cellulose and the position of the microfibril within the cell wall (52, 53). In algae, the microfibril varied from 8.3–38 nm (54); in wood, it varied from 5.0–10.0 nm (55); in ramie, reported values varied from 17–20 nm (56).

Since these initial studies, there have been many improvements in the resolving power of the electron microscope, improved methods of sample preparation for

Table 2 Distribution of Lignin in Birch (White Birch, *Betula papyrifera*) Xylem[a]

Cell	Morphological region[a]	Type of lignin[b]	Tissue volume (%)	Lignin (% of total)	Lignin Concentration (%)
Fiber	S	Sy	73	60	19
	ML	SyGu	5	9	40
	CC	SyGu	2	9	85
Vessel	S	Gu	8	9	27
	ML	Gu	1	2	42
Ray cell	CC	Sy	11	11	27

[a]S, secondary wall; ML, compound middle lamella; CC, cell corner.
[b]Sy, syringyl lignin; SyGu, syringyl-guaiacyl lignin; Gu, guaiacyl lignin.
Source: From Ref. 49.

cellulosic materials, and the development of other physical methods for measuring microfibril dimensions. Many of the more recent electron microscope studies have indicated that the original microfibrils observed are made up of aggregates of even smaller units, which are called elementary fibrils. Studies by X-ray diffraction and electron microscopic observations indicate that the microfibrils are much smaller than was originally believed. However, a considerable amount of controversy exists on the exact width of the microfibril for various types of cellulose. It was proposed by Frey–Wyssling, Mühlethaler, and Muggli (57) in 1966 that there exists a 3.5 nm elementary fibril which they suggested was the true structural unit of higher plant cells. According to these authors and others, a variation in the observed sizes of the microfibrils from various plant sources is due to differences in the amount of aggregation of a common 3.5 nm elementary fibril.

The concept that the 3.5 nm fibril is the ultimate cellulose fibril or that there is a common microfibril for higher plants is still an area of much controversy. Some evidence exists that the microfibril may actually be smaller and have a distribution of sizes.

Smaller fibrils with dimensions around 1.0 nm have been observed by Franke and Ermen (58), Fengel (59), and Hanna and Côté (60). Some have suggested that the smaller microfibrils, the so-called subelementary fibrils, are a fibril formed as an intermediate stage in the development of the final 3.5 nm elementary fibril (60). Other researchers (61) believe that the ultimate size of the fibril depends on the degree of lignification and that the matrix made up of lignin or hemicelluloses restricts the lateral movement of the cellulose fibril and thus prevents crystalline growth from proceeding unrestricted. Recent studies by Harada and Goto (62) with wood and *Valonia* cellulose, Fengel (63) with spruce wood cellulose, and Boylston and Hebert (64) with cotton, ramie, algal, and bacteria cellulose indicated that the microfibril had a distribution of widths that varied depending on the source of cellulose, on the isolation pretreatment, and the degree of delignification (Table 3).

B. Crystalline Areas of the Microfibril

Nägeli (65) in 1958 proposed that cellulose exists in plants as a crystalline micelles. Many years elapsed, however, before the crystalline structure of cellulose was confirmed by X-ray diffraction studies (66–68). The ensuing investigations showed that, with the possible exception of cellulose in the *Halicystis* plant, cellulosic materials from all sources, including bacteria and animals, have the same crystalline structure called cellulose I. When the crystalline structure is destroyed by solution or chemical treatment, the regenerated cellulose exhibits a different X-ray pattern. Based on the X-ray diffraction pattern, Meyer and Misch in 1937 proposed a unit cell for native cellulose (68) (Fig. 4). This proposed unit cell was widely accepted for many years (69, 70).

In 1958, Honjo and Watanabe proposed, based on low temperature electron

Table 3 Size of the Microfibrils from Different Sources of
Cellulose

Source	Electron microscopy (nm)	X-ray line broadening (nm)
Cotton	2.2	4.0
Ramie	2.5	4.4
Bacterial cellulose	5.3	5.5
Algal cellulose	10.7	11.2
Algal cellulose	11.9–14.3	20.0
Wood cellulose	2.02	2.5
Spruce wood	2–3	—

Source: From Refs. 62–64.

diffraction studies (70), a larger unit cell for *Valonia* cellulose with a and b dimensions twice as long as those proposed by Meyer and Misch. Studies by Hebert and Muller (71) using electron diffraction data from cellulose I and from cotton, ramie, bacterial, and *Valonia* cellulose proposed that the type of unit cell may vary, depending on the source of cellulose. The diffraction patterns from ramie and cotton cellulose I could be indexed using the Meyer–Misch cell, but those from *Valonia* and bacterial cellulose require a doubling of the base plane dimensions. The results using neutron diffraction were contradictory for cotton cellulose (72). Besides the two basic unit cells, several modifications of these two types have been proposed and are summarized in a recent review on the structure of cotton cellulose (69).

Several authors in the past decades have attempted to derive the conformation of the cellulose chain and its packing in the crystalline lattice. Historically, Meyer and Misch (68) proposed a conformation that included a two-fold screw axis along the chain. This came to be known as the "straight-chain" conformation. Using X-ray and electron diffraction data and molecular mechanics data, two groups led by Sarko (73) and Blackwell (74, 75) have refined the proposed conformation of *Valonia* cellulose. Both groups used the unit cell proposed by Meyer-Misch. The proposed structure (Fig. 5) contains ribbonlike chains that are arranged in sheets, stabilized by hydrogen bonds in two directions: the OH • 3'—OH • 5 and OH • 2—OH • 6' intrachain hydrogen bonds along the chain and an intermolecular OH•6—OH•3 hydrogen bond roughly perpendicular to the chain axis along the a direction. These hydrogen bonds force the chains into a well-stabilized sheet structure. The nearest O—O distance between the sheets is 3.1–3.2 Å long that, if it is a hydrogen bond, is a very weak one. Originally, Meyer and Misch suggested that the individual chains in the microfibril were antiparallel. More recent evidence has shown that this suggestion may be incorrect. X-ray data collected by Sarko (73) and Blackwell (74, 75) with *Valonia* cellulose support the parallel orientation of the molecules in the microfibril. Much less information is available on the structure of cellulose I of ramie, cotton, and wood. Recent work by French (76) has shown that the current X-ray data for ramie and cotton cellulose cannot be used to distinguish between a parallel and antiparallel structure.

C. Polymorphic Structure of Cellulose

At least four different crystalline forms of cellulose (cellulose I–IV) are recognized on the basis of their X-ray diffraction patterns and infrared spectra (77–81). The unit cell dimensions of each form are shown in Table 4.

The cellulose I form is characteristic of all naturally occurring celluloses, but with some differences. Generally, cellulose I from algal and bacterial sources is more crystalline than the cellulose from cotton, wood or ramie. The proposed structure for the unit cell and conformation of the chain of cellulose I have been discussed earlier (see Figs. 4 and 5).

Cellulose II is obtained from cellulose I by treatment with strong solutions of alkali or by regeneration from solution (e.g., saponification of cellulose esters). Since this form of cellulose is produced by precipitation from solution, it is probably thermodynamically the most stable polymorph. X-ray diffraction studies by Blackwell, Sarko, and others (77–81) favor an antiparallel structure for this form of cellulose. The proposed unit cell contains a larger number of intra and inter sheet hydrogen bonds as compared to cellulose I. These high amounts of hydrogen bonding probably account for the stability of cellulose II.

Cellulose III is produced by the treatment of cellulose I or II with liquid ammonia, followed by complete evaporation of the ammonia. Two types of cellulose III (III_I or III_{II}) are obtained depending on whether the starting material is

Table 4 Unit Cell Dimensions of Cellulose Polymorphs I–IV

Polymorph	a Axis (Å)	b Axis (Å)	c axis (Å)[a]	α(deg)[b]
Cellulose I	7.85	8.17	10.34	96.4
Cellulose II	9.08	7.92	10.34	117.3
Cellulose III	9.9	7.74	10.3	122.0
Cellulose IV	7.9	8.11	10.3	90.0

[a]The c axis of the unit cell is taken as fiber axis.
[b]The convention of an obtuse monoclinic angle is followed. Other angles are 90°.
Source: From Ref. 82.

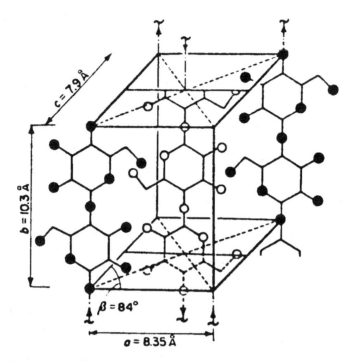

Figure 4 The unit cell of cellulose as proposed by Meyer and Misch (From Refs. 69 and 70.)

cellulose I or cellulose II. Cellulose III_I is assumed to have a parallel-chain structure, whereas III_{II} is believed to have an antiparallel structure. On treatment with boiling water, cellulose III_I is converted into cellulose I and cellulose III_{II} into cellulose II.

Cellulose IV also contains two polymorphs (cellulose IV_I and IV_{II}) that are obtained by heating cellulose III to 280°C. Initial evidence indicates that cellulose IV_I has a parallel chain, whereas cellulose IV_{II} has antiparallel chains. Recent studies by Chidambareswaran and coworkers indicated that cellulose IV_I could be converted into cellulose I and cellulose IV_{II} into cellulose II by heating in ethylenediamine and water (83). More detailed information concerning the polymorphs of cellulose, their X-ray pattern, unit cell, and chain orientation can be found in recent reviews by Sarko (77) and Marchessault and Sundararajan (84).

D. Structure of the Microfibrils

Ideally, any proposed structure for the microfibrils must be consistent with the diffraction and electron microscopy data, explain the results of infrared and magic-

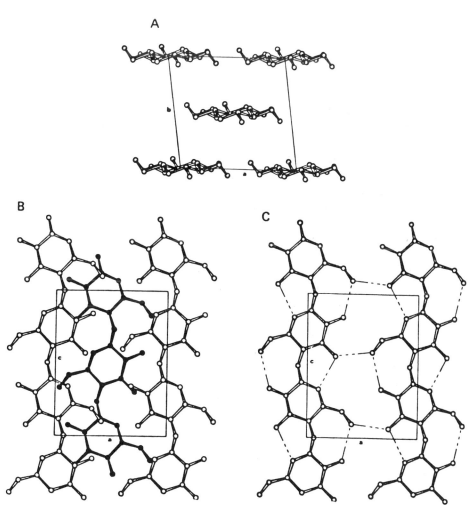

Figure 5 Projection of the proposed parallel two-chain model for cellulose I. (A) unit cell projected perpendicular to the ab plane, that is, along the fiber axis; (B) perpendicular to the ac plane; (C) projection of the (020) plane only showing the hydrogen-bonding network (From Ref. 74.)

angle spinning nuclear magnetic resonance spectroscopy studies, and account for the observed reactivity of the microfibrils. Unfortunately, there is no single model that is completely consistent with these observed results from cellulose in the microfibrils.

Historically, the microfibril was thought to consist of two distinct regions: one area of crystallinity and another area of amorphous cellulose (paracrystalline region). This concept was based on X-ray and accessibility studies using various chemical reagents (Table 5). Originally, it was proposed that cellulose had a bricklike structure in which crystalline micelles (partially ordered region) were surrounded or fringed by amorphous materials (85). The micellar theory, which was subsequently modified to the fringed micellar theory, provided a ready explanation for the diffuse X-ray diffraction pattern, absorption of moisture without modification of the crystalline structure, partial hydrolysis of the microfibrils to rodlike micelles or crystallites of 5.0–10.0 nm in diameter and 50–60 nm in length, commonly known as microcrystalline cellulose, and numerous other observations on partial accessibility or crystallinity of cellulose (86).

According to the fringed micellar theory, the microfibrils are composed of statistically distributed crystalline and amorphous regions formed by the transition of the cellulose chain from an orderly arrangement in the direction of the microfibrils in the crystalline regions to a less orderly orientation in the amorphous area (86).

As more information on the microfibril has accumulated, it has become obvious that the fringed micellar model is not consistent with the experimental observations for cellulose I. Consequently, other models have been proposed that better account for the observed properties of the microfibril. One of the most unusual models was originally proposed by Manley (87), and more recently by Chang (88), Asunmaa (89), Marx-Figini and Schulz (90), and Bittiger and coworkers (91). This model proposes that the cellulose microfibrils consist of a folded-chain structure. According to Manley, the molecular chain of cellulose forms a ribbon 3.5 nm wide by folding in concertina fashion. With the ribbon wound as a tight helix, the straight

Table 5 Percent Noncrystalline and Amorphous Cellulose Content by Various Techniques

Material	Noncrystalline			Accessible		
	Hydrolysis	X-ray	Density	D_2O	Formylation	Thallous ethylate
Cotton	6–8	30–31	40	44	12	0.4
Wood pulp	9-11	30	47	55	—	—
Mercerized cotton	11–14	—	60	—	20	3.3–27
Textile rayon	—	60–63	75	68	39	—
High-tenacity rayon	27–31	60–61	—	86	—	—

Source: From Ref. 85.

segments of the molecular chain become parallel to the helical, fibril axis, as shown in Fig.6. Based on available evidence, the folded-chain concept does not appear to be valid for cellulose I. The extremely small dimensions of the elementary fibril suggest the greater likelihood of an extended-chain conformation. Muggli (92, 93) studied the effect on the molecular weight of sectioning ramie fibers into extremely thin sections, concluding that the majority of the cellulose molecules are very nearly fully extended. Mark (94, 95) analyzed the stress distributions in a circular folded-chain model for cellulose and concluded that the results were not consistent with those observed for native cellulose fiber. Although the extended-chain model cannot be proven, most of the evidence support this type of structure for the microfibril. Probably the strongest evidence is based on the diffraction data that indicates cellulose I has a parallel-chain structure. A folded-chain model, as proposed by Manley, requires an antiparallel chain structure.

The most widely accepted model for the microfibril is the extended-chain model in which the noncrystalline regions are distributed along the microfibril in some type of uniform structure and not as aggregates in alternating regions as suggested in the "fringed-micellar theory."

Several models have been proposed in which a core of crystalline cellulose is surrounded by a paracrystalline region. Frey-Wyssling (96) proposed such a structure in 1954 based on studies with ultrathin sections of ramie fibers. A more recent model has been proposed by Preston (97) in which one central crystalline core is embedded in a paracrystalline cortex of molecular chains that lie parallel to the microfibril length, but otherwise are not stacked in crystalline array.

From an extensive study of the structure and accessibility of cotton cellulose, using X-ray, infrared, microscopy, and chemical techniques, Jeffries (98–100) proposed that the molecules at the surface may account for the observed accessibility and other properties of the cellulose microfibrils. In other words, the so-called amorphous areas are due to surface molecules on the microfibrils. Several models, based on this concept, have been developed that are discussed in the next paragraphs.

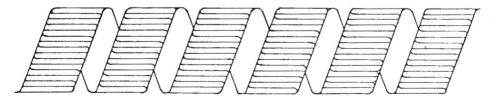

Figure 6   Schematic folded-chain structure of protofibrils as proposed by Manley. (From Ref. 87.)

Frey-Wyssling and Muhlethaler (101–103), using electron microscopy studies and negative staining techniques, proposed the model of an elementary microfibril shown in Fig. 7, in which highly crystalline straight-chain aggregates (36 chains) contain dislocations and chain ends. According to this model, there are no true amorphous areas, but the observed diffuse X-ray diffraction pattern is due to chain-end dislocation.

Rowland and coworkers have proposed another model for the elementary fibrils of cellulose (104–105). This proposed model is similar to Muhlethaler's model, in that there are no true amorphous areas, but only areas of slight disorder. The evidence for this model is based on a series of chemical accessibility studies. The authors concluded that the microfibril consists of a completely crystalline structure in which the irregularities are accounted for by various types of surface imperfections, such as distorted surfaces (A,B) and twisted or strained regions (C) on the crystalline elementary fibrils (see Fig. 8). Other possible models have been suggested by Peterlin and Ingram (106) and others (107).

From this brief outline, it is obvious that no one model can completely account for the observed properties and characteristics of the microfibrils. There is still a considerable amount of controversy and no clear agreement on the structure of the microfibrils. It has even been suggested that the microfibril is an artifact produced by electron microscopy and that its ultimate diameter is the width of the molecule, as has been shown to be true for extended-chain, polyethylene crystals.

V.   BIOGENESIS OF CELLULOSE IN THE CELL WALL

The biogenesis of cellulose in the cell walls involves a series of steps, including formation of the precursor(s), transportation of the precursor(s) to the site of synthesis, and biosynthesis of cellulose in the cell wall. From the previous

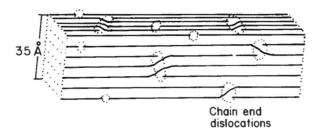

Figure 7   Schematic structural model of elementary fibrils as proposed by Muhlethaler (From Refs. 101–103.)

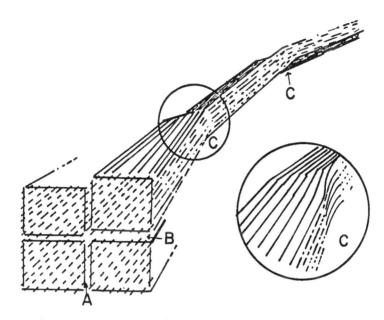

Figure 8 Schematic representation of the elementary fibril as proposed by Rowland showing (A) coalesced surfaces of high order, (B) readily accessible slightly disordered surfaces, and (C) readily accessible surfaces of strain-distorted tilt and twist regions.

discussion about the structure of the cell wall, it is apparent that the biosynthesis process is very complex. A proposed mechanism for the biogenesis of cellulose in the cell wall must account not only for the synthesis but also for the precise arrangement of the cellulose molecules in the microfibril and the different orientations of the microfibril within the plant cell. The large number of events that must be coordinated during synthesis and assembly implies a complex control system. Unfortunately, little is presently known about how such processes are controlled at the molecular level.

Another factor that should be noted is that there are probably at least two different types of cellulose synthesized within the plant. Studies by Marx-Figini (35, 36, 108–111) and others (112) with various types of cellulose in plants have indicated that cellulose from the primary cell wall is different from secondary wall cellulose. For example, cotton cellulose in the primary wall had a degree of polymerization ranging from 3000–6000, whereas the secondary wall contained cellulose with a much higher degree of polymerization of 14,000 and with a very narrow molecular-weight distribution. The same researchers have shown that there are distinct differences in the kinetics of the synthesis of cellulose in higher plants and bacteria. In the secondary wall of the plant, the biosynthesis process occurs by a structure-controlled mechanism in which the molecular weight is not a function of the degree of conversion, the time of synthesis, or the reaction conditions. In contrast to cellulose in plants, the molecular weight of cellulose synthesized in bacteria depends on the polymerization time and generation time of the bacteria (111).

In the next few paragraphs, we will briefly discuss what is known about this process and describe some of the newer theories on how the plant synthesizes the cell wall components. More detailed descriptions of this extremely important process can be found in reviews by Nakaido and Hassid (113), Delmer (114), Karr (115), Franz and Heiniger (116), Villemez (117), and Colvin (118).

Sugar nucleotides are the carbohydrate precursors that form the cell wall polysaccharides. The nucleotides are formed by the combination of purine or pyrimidine bases linked to sugars, which in turn are esterified with phosphoric acid. The two nucleotides found to be involved in cellulose synthesis are uridine 5'-(α-D-glucopyranosyl pyrophosphate) (UDP-D-glucose) and guanosine 5'-(α-D-glucopyranosyl pyrophosphate) (GDP-D-glucose).

The first reports of the synthesis of celluloselike molecules by a cell-free system were given by Glaser (119). He found that a cell-free particulate enzyme preparation from *Acetobacter xylinum* catalyzed incorporation of the D-glucopyranosyl group from UDP-D-glucose into cellulose. Since this initial work, there have been numerous reports of cell-free, particulate enzymes from mung beans (113, 120, 121–123), maturing cotton bolls (124, 125), lupin (126, 127), oat coleoptiles (128), peas (122, 123), string beans (122), corn (122), and squash (122), which use GDP-D-glucose and/or UDP-D-glucose to form celluloselike materials. A more complete list of studies that have been done in this area can be found in a recent review by Franz and Heiniger (116).

It is generally believed that only UDP-D-glucose participates in the biosynthesis of bacterial cellulose, but there is some evidence that both UDP-D-glucose and GDP-D-glucose are involved in the biosynthesis of cellulose in higher plants. However, it should be noted that the involvement of GDP-D-glucose in the synthesis of cellulose in higher plants is still a matter of debate (118, 129, 130). But there is general agreement that the initial precursors for all plant cell wall polysaccharides involve sugar nucleotides.

Hassid (113) has postulated that cellulose is formed in higher plants by the sequence of reactions shown below:

GTP + α-D-glucopyranosyl phosphate pyrophosphorylase
⇌
GDP-D-glucose + PPi
n(GDP-D-glucose) + acceptor glucotransferase →
acceptor -[(1→4)-β-D-glucosyl]$_n$ + n(GDP)
[cellulose]

A considerable amount of study has been done in order to locate the site of polysaccharide synthesis in cellulose. Most of the present evidence, based on autoradiographic studies (117, 131, 132) and chemical studies, indicates that the hemicelluloses and pectic materials are synthesized within the Golgi bodies. In contrast, cellulose synthesis does not appear to occur in the Golgi apparatus. Most evidence indicates that cellulose synthesis occurs outside the cytoplasm at the plasma membrane: cell wall interface; that is, at the site of microfibril deposition. The most convincing evidence for this is derived from autoradiographic studies (131, 133–137), which indicate the synthesis of cellulose only outside the protoplast.

However, the enzymes responsible for cellulose synthesis may be present in the Golgi apparatus or endoplasmic reticulum. There is some evidence that the Golgi apparatus is not only engaged in the synthesis of hemicelluloses, but also may be involved in transferring cellulose synthetase and other cell-surface enzymes from the site of synthesis the polysomes, to the site of action, the plasma membrane (134). This dual function of the Golgi apparatus is especially evident in the lower plants such as the unicellular green alga *Pleurochrysis scherffelii* and others (135). A diagrammatic representation of the synthesis and transportation of plant polysaccharides as proposed by Northcote (136) is shown in Fig. 9.

At the present time, there is still a considerable amount of controversy on whether the sugar nucleotides are the direct glycosyl donors during the formation of cell wall polysaccharides or if the glycosyl unit is transferred to another intermediate, such as a glycolipid (137), complexed with a protein (138, 139), or converted into a disaccharide or oligosaccharide (118, 140, 141) and transported across the plasma membrane. Even though lipid-linked intermediates have been found in cellulose-forming systems, there still is not strong evidence for a lipid or glycoprotein intermediate in cellulose synthesis. The one exception is the alga *Prototheca zopfii* (142), in which both a lipid-linked and a protein intermediate have been identified. Further evidence, albeit less conclusive, is that studies with cotton fiber cell wall have shown some protein is attached to the α-cellulose fraction (143). The importance of these intermediate compounds in the biosynthesis of cell wall polysaccharides remains to be shown.

Once the intermediate compounds reach the site of synthesis, then several events occur: The intermediate polymerizes to form a polymer with a very narrow range of molecular weight in the secondary wall; the cellulosic molecules come together to form a microfibril; and, finally, the microfibrils are oriented and inserted into the cell wall. At the present time, there is still a considerable amount of controversy on how these events occur, in which order they occur, whether they occur simultaneously or concurrently, and what structure or force in the plant controls these processes.

Two general mechanisms have been proposed for the observed properties of the microfibril. One theory, proposed by Colvin, is that the microfibrils are formed and

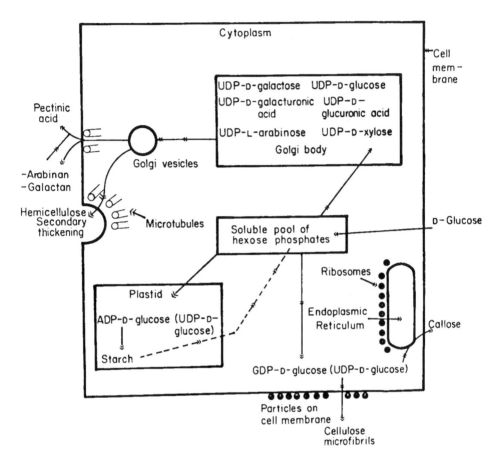

Figure 9 Diagrammatic representation of the synthesis and transportation of various polysaccharides in a growing plant cell (From Ref. 136.)

orientated by physical factors at the cell wall, the so-called "intermediate high-polymer hypothesis" (118, 140, 141–144). Another mechanism that has received wider support is the "granular or template" theory whereby the synthesis and formation of the cell wall are under cellular control and occur in an enzyme complex located at the site of cellulose synthesis.

The intermediate high-polymer hypothesis proposed that polymeric, hydrated intermediates of glucose are extruded from a transient, terminal opening in the bacterial cell. These highly hydrated glucans then associate spontaneously and progressively by hydrogen-bond formation between chains to form an entity that has been called the "nascent microfibril." The nascent microfibril is converted into the final microfibril by loss of water.

The more widely accepted granular or template theory proposes that an enzyme complex is located on the outside of the plasma membrane and attaches glucose residues in a coordinated way to the ends of $(1\rightarrow4)$-β-D-glucans at the tips of mi-crofibrils. This type of complex was originally proposed by Preston (145). A variation of this theory that has been proposed by Brown (146) for bacterial cellulose combines the notion of an active granular, enzyme complex with a multiple spinneret mechanism. An example of this type of granular complex for the synthesis of microfibrils is shown in Fig. 10.

In summary, there is presently no precise information concerning either the control mechanisms that govern cell wall biogenesis or the interactions between cell wall biogenesis processes and general cellular metabolism. Presently, it is not known how many steps are involved in the formation of a polysaccharide from sugar nucleotides, how cellular control is extended beyond the plasma membrane, or how the cell wall is formed from the component polymers. Indeed, the field of cell wall biogenesis provides more questions than answers, and one suspects that most of the major hypotheses about the operation of the biogenesis process have not yet been made (147).

VI. CHEMICAL REACTION OF CELLULOSE

A. Sorption, Swelling, and Dissolving of Cellulose.

The major functional groups on the cellulose molecules are hydroxyl groups that have a strong affinity for polar solvents and solutions which could reach them. The

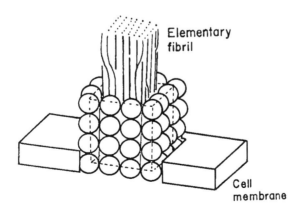

Figure 10 A hypothetical model proposed by Mühlethaler for the biogenesis of cellulosic elementary fibrils. (From Ref. 101.)

interaction of polar compounds can lead to swelling of the cellulose, solubilization of the cellulose if the polar compound is a liquid, or formation of a derivative of the cellulose depending on the structure and reactivity of the polar compound and whether or not it penetrates into the crystalline regions. A good example of swelling of the cellulose involves the reaction of cellulose with water. When cellulose is dry, it absorbs moisture from the air, reaching an equilibrium moisture content that increases with the increasing ambient relative humidity. The adsorbed moisture swells the cellulose, but does not change the crystalline structure, indicating that it enters the accessible rather than the crystalline regions of the microfibril. If adsorption is carried to saturation and the relative humidity is then progressively decreased, the proportion of moisture sorbed decreases, but the new values at any given relative humidity are a little higher than those for the adsorption curve. The explanation for the difference between the adsorption and desorption curve (hysteresis loop) is based on the interconversion of cellulose-water and cellulose-cellulose hydrogen bonds. During swelling, the hydrogen bonds between the cellulose molecules are broken and replaced by hydrogen bonds between cellulose molecules and water. Conversely, desorption and shrinking reform the broken cellulose-cellulose hydrogen bonds with a lag that gives a higher equilibrium moisture content when the cellulose is being dried. This phenomenon accounts for the characteristic hysteresis loop observed on adsorption and desorption of moisture and the corresponding changes in the physical properties of cellulose materials at different relative humidities. A similar phenomenon is observed with other polar liquids, for example, methanol, ethanol, aniline, and benzaldehyde. In general, the higher the polarity of the liquids, the greater the amount of swelling produced; however, in all the above-mentioned examples, swelling is less than that with water. More detailed discussions about the swelling of cellulose with liquids and gases can be found in articles by Browning (148) and Skaar (149).

Interaction with water is an example of intercrystalline swelling or swelling that involves only the accessible portion of the cellulose microfibrils. The most convincing evidence is the fact that the X-ray pattern of native cellulose does not change on wetting.

There are, on the other hand, many liquid solutions of strong acids, strong bases, and a few salts that change the X-ray pattern. This type of swelling is called "intracrystalline swelling" and involves the penetration and swelling of both the accessible and crystalline regions of the microfibril. The most common example of a liquid that can lead to intracrystalline swelling is 14–20% aqueous sodium hydroxide.

Concentrated alkali solution swells the cellulose by forming an alkali cellulose addition compound, but cannot dissolve the macromolecule. The sodium hydroxide addition compound that reacts as sodium cellulosate is highly reactive. It is used as an intermediate for the production of cellulose ether derivatives and cellulose xanthate, discussed later. The latter process, discovered by John Mercer in 1894, involves the treatment of cotton fiber with a solution of 12–18% sodium hydroxide and subsequent regeneration by washing to yield cellulose II. This treatment improves the luster and tensile strength of yarn and cotton fibers.

The interaction of cellulose with highly polar solutions of acids, alkali, salt, and complex-forming reagents is accompanied by increased swelling, which ultimately provides a new crystalline structure accommodating the adsorbed ions or results in the solution of the additional products. Initially, the hydrogen bonding of the crystalline structures is broken with hydroxyl or hydronium ions. The water content of these solutions is very critical because it should be high enough to cause sufficient ionization and low enough to allow the hydroxyl groups of the cellulose to compete for the hydroxyl or hydronium ions. The bulky counter-ion that follows the hydroxyl or hydronium ions into the substrate or the groups can form derivatives like xanthate or a complex like cuprammonium.

Basic reagents that form an additional product and dissolve cellulose include amino and ammonium complexes of copper, cadmium, and other multivalent ions, such as cupriethylenediamine $Cu(H_2N-C_2H_4-NH_2)_2$ $(OH)_2$ and cuprammonium hydroxide $[Cu(NH_3)_4]$ $(OH)_2$. The latter reagent, discovered by Schweizer, is used for the preparation of cuprammonium rayon and characterization of industrial cellulose by viscosity measurement.

Cellulose dissolves in 44% hydrochloric acid, 72% sulfuric acid, and 85% phosphoric acid and causes the homogeneous hydrolysis of cellulose to glucose. Concentrated nitric acid (66%) does not dissolve cellulose, but, like sodium hydroxide, forms an additional compound known as Knecht compound that is an intermediate in the nitration of cellulose. Cellulose also dissolves in some concentrated aqueous solutions of salts, but may require heating, leading to partial degradation of the substrate.

In the last decade, there has been an increased interest in developing organic solvent systems for cellulose. Recently, organic solvent systems containing lithium chloride have been found to be effective with cellulose. These include lithium chloride in dimethylacetamide and lithium chloride in N-methyl-2-pyrrolidinone (150). Other recent examples of organic cellulose-dissolving solvents include the use of amine oxides (151) and of chloral in polar aprotic solvents, such as dimethyl formamide, dimethylacetamide, dimethyl sulfoxide, and N-methyl-2-pyrrolidinone (152). Formaldehyde in dimethyl sulfoxide has been used also to dissolve and characterize cellulose (153–154). More detailed information on cellulose solvent systems can be found in several recent reviews (155–158).

B. Derivatives of Cellulose

Various modifications and derivatives of cellulose are produced in large quantities and used for the industrial production of fiber, film, plastics, explosives, coatings, thickeners, etc. Comprehensive discussion of these products is beyond the scope of this article and can be found in reviews by Spurlin (28), Bolker (38), Ward (155), and others (156–158). Only a brief description of the raw material and the reactions involved in making the major commercial cellulose derivatives will be given.

The production of cellulose derivatives generally requires raw materials having a high content of pure cellulose, technically referred to as α-cellulose. The α-cellulose content is measured by the amount of cellulose that remains undissolved in 18% sodium hydroxide solution, subsequently diluted by water to a lower concentration.

Originally, bleached rags were used for the production of cellulose nitrate, the first derivative produced commercially. As the industry developed, the lack of availability and uniformity of these materials led to the application of a chemical cotton made from linters, which are the short fibers of about 2–3 mm left on the seeds after removal of the long cotton fibers. The shorter fibers are removed from the cotton seeds at different stages in the processing of the seeds and cleaned and purified to give chemical cotton. The latter process involves digestion with dilute sodium hydroxide, bleaching with chlorine, neutralization, and drying.

During World War I, increased demand for cellulose nitrate explosives led to the application of wood pulp in the manufacture of explosives, but, in the United States, chemical cotton remained the choice raw material for the production of various cellulose derivatives until the dissolving pulp industry was developed during World War II because of high-priority demand for chemical cotton.

Dissolving pulp is prepared by removal of the remaining lignin, hemicelluloses, and resin from hardwoods pulped by the sulfite process. The acidic condition of the sulfite process is responsible for removing most of the hemicelluloses. The subsequent purification involves extensive bleaching and extraction with a dilute solution of 1–2% sodium hydroxide at temperatures of 90–170°C and concentrated alkali solution (10–15%) at room temperature. The kraft pulping process can also be used to prepare dissolving pulp from both hardwoods and softwoods, if a prehydrolysis step is used. The purified cellulose obtained by these processes is used as a starting material for a variety of commercial products, including esters and ethers.

A whole series of products can be obtained when the hydroxyl group on cellulose is reacted with another group. The actual chemical and physical properties of the final product are dependent on (1) the type of substituting group (i.e., methyl, ethyl, acetyl, etc.); (2) the degree of substitution (i.e., the relative number of substituted and free hydroxyl groups); and (3) the uniformity of products (i.e., the length of the cellulose molecules and the degree of substitution).

In actual practice, it is very difficult to completely convert all the hydroxyl groups in cellulose into ester or ether groups. This is due, in part, to a steric factor. When one or two of the hydroxyl groups are substituted, steric crowding may inhibit complete reaction. Another factor is "accessibility." Since all the reactions are conducted under heterogeneous conditions, not all the hydroxyl groups may be accessible to reaction, particularly those in the crystalline regions. Finally, not all the accessible hydroxyl groups are equally reactive. The three hydroxyl groups of the D-glucose residues differ in their reactivity to a given reagent, and the relative rates are not necessarily the same for other reagents. As a general rule, most commercial cellulose derivatives are the products of only partial reaction; that is, they contain a certain proportion of unchanged hydroxyl groups. Therefore, the expression "O-methyl cellulose" or "cellulose acetate" by no means characterizes a product; each simply denotes a series of products.

During the formation of any one particular cellulose derivative, the physical and chemical properties are determined largely by the degree of substitution (DS) and the degree of polymerization (DP). The term "degree of substitution" is used to denote the extent of a reaction and is defined as the average number of hydroxyl groups substituted of the three available in the anhydroglucose units. Properties that are most strongly affected by changing the degree of substitution are the solubility, swelling, and plasticity. Derivatives of a low degree of substitution are often more sensitive to water than the original cellulose and may even be dispersible in water. In the case of derivatives having a high degree of substitution with nonpolar substituents, the water solubility is decreased, the sorption of water decreased, and the solubility in organic solvents increased. The plasticity is increased by the substitution of nonpolar groups; and the greater the chain length of the substituent group, the higher the plasticity, because the individual chains are forced further apart.

Another factor that strongly affects the final physical and chemical properties of the cellulose derivative is the average chain length of DP of the cellulose derivative. This is controlled in two ways: by selecting the right type of dissolving pulp and by carefully controlling the reaction condition during formation of the cellulose derivative. During the commercial esterification or etherification of cellulose, two reactions occur: a substitution reaction at the hydroxyl groups and a glycosidic bond cleavage reaction. For materials to be used in spinning fibers, casting films, molding plastics, or spraying lacquers, a reduced viscosity (a lower DP) is required. On the other hand, if the mechanical properties are important in the final product, a balance must be struck between low DP, which will confer ease of fabricability, and the somewhat higher DP necessary for acceptable mechanical properties.

C. Cellulose Esters

Over 100 esters of cellulose have been prepared and described. It is beyond the scope of this chapter to describe in detail the chemical and physical properties of each of them.

The three most important commercial esters are cellulose xanthate, cellulose nitrate, and cellulose acetate. Cellulose xanthate is not a final product, but an intermediate product in the important industrial process for making viscose rayon and cellophane. The manufacturing of these two products involves the steeping of pulp in a 17.5% solution of sodium hydroxide to obtain activated alkali cellulose. The excess of sodium hydroxide is removed and the material aged. The aging process leads to a reduction in molecular weight. After aging, the activated cellulose is treated with carbon disulfide and converted to a partially xanthated derivative that dissolves in alkali to yield a viscose solution. The viscose solution is then forced through an orifice or a slit into an acid bath where the xanthate groups are decomposed and regenerated cellulose is obtained as a filament or sheet. The chemical reactions involved are shown below:

$$Cellulose\text{-}OH + NaOH \rightarrow Cellulose\text{-}OH\text{ - }OH^- \ Na^+$$
$$Cellulose\text{-}OH\text{- }OH^- \ Na^+ + CS_2 \rightarrow Cellulose\text{-}OCS^-_2 \ Na^+ + H_2O$$
$$Cellulose\text{-}OCS_2 \ Na^+ + H^+ \rightarrow Cellulose\text{-}OH + CS_2$$

Another important ester of cellulose is cellulose nitrate. Highly nitrated cellulose (DS 2.4–2.8) is used as an explosive and propellant, whereas less highly nitrated products have been used for the production of films, adhesives, lacquers, and plastics. Commercially, nitrate esters are prepared by reacting cellulose with a mixture of nitric acid, sulfuric acids, and water. The degree of substitution and depolymerization may be controlled by the addition of water and sulfuric acid.

$$Cellulose\text{-}OH + HNO_3 + H_2SO_4 + H_2O \rightarrow$$
$$Cellulose\text{-}O\text{-}NO_2 + H_2O$$

If undegraded nitrates and high nitrogen contents are desired, as, for example, in the determination of molecular weight, nitric acid and phosphorous pentoxide or nitric acid and acetic anhydride (159, 160) may be used.

Cellulose acetate is used widely for the manufacture of films, plastics, fibers, and coatings. The most common industrial process for making cellulose acetate involves a two-stage process. In the first stage, the cellulose is activated by swelling in a mixture of sulfuric and acetic acids. In the second stage, excess acetic acid is removed and acetic anhydride added to begin the reaction:

$$Cellulose\text{-}OH + (CH_3CO)_2O + H_2SO_4 \rightarrow$$
$$Cellulose\text{-}O\text{-}COCH_3 + H_2O + CH_3COOH$$
(Cellulose acetate)

The cellulose acetate can be precipitated directly from solution and redissolved in ethylene dichloride. This material can be forced through an orifice to make cellulose triacetate fibers. If the product is not precipitated and water is added to the solution, hydrolysis occurs and the DS is reduced to approximately 2.0. This product is called "secondary acetate" or cellulose acetate and is used to make fiber or photographic films.

D. Cellulose Ethers

Ether derivatives of cellulose are mainly used as thickeners, dispersants, adhesives, extenders, and films because of their water solubility and gel-forming properties. The introduction of a few ether groups disrupts the regularity and hydrogen bonding of the crystalline structure without seriously affecting the polarity of the molecules, and the product becomes alkali and water-soluble. The polarity, in fact, could be increased by the introduction of ionizing groups, such as a carboxymethyl group. However, as the degree of substitution increases, the material becomes more soluble in organic solvents.

The ether derivatives are generally produced by the reaction of alkali cellulose with the corresponding alkyl halides, such as methyl chloride and carboxymethyl chloride, which gives methyl cellulose (MS) and carboxymethyl cellulose (CMS), two of the best known cellulose-ether derivatives. The highly reactive and accessible alkali cellulose could also react with ethylene oxide and acrylonitrile to give hydroxylethyl cellulose and cyanoethyl cellulose that introduce new functions to the polysaccharide structure as shown below:

$$\text{Cellulose-OH-OH}^-\text{Na}^+ + \text{RCl} \rightarrow \text{Cellulose-OR} + \text{NaCl} + \text{H}_2\text{O}$$
$$\text{Cellulose-OH-OH}^-\text{Na}^+ + \text{CH}_2\text{-CH}_2 \rightarrow \text{Cellulose-CH}_2\text{-CH}_2\text{-OH} + \text{NaOH}$$

$$\text{Cellulose-OH-OH}^-\text{Na}^+ + \text{CH}_2\text{=CH-CN} \rightarrow \text{Cellulose-O-CH}_2\text{-CH}_2\text{-CN} + \text{NaOH}$$

VII. GRAFT AND CROSS-LINKING DERIVATIVES

Other than the ester and ether derivatives, the cross-linking and graft polymerization of cellulose are also of considerable industrial and academic interest. For cross-linking, the substrate is treated with bi- or poly-functional reagents generally derived from formaldehyde in order to bridge the neighboring molecules and reduce their hygroexpansivity. These treatments not only improve the dimensional stability, but also favorably affect the related properties of the fabrics, such as drying and wrinkle resistance.

In graft polymerization, the cellulose molecule is subjected to radiation or specific oxidants to create free radicals that could initiate free radical polymerization of vinyl monomers. The combination with polymers modifies the properties of cellulose.

VIII. DEGRADATION REACTIONS OF CELLULOSE

The fact that cellulose degrades under a variety of circumstances is important from several points of view. For the manufacturer of cellulosic products, degradation is both helpful and undesirable. Thus, a certain amount of degradation, such as that which occurs during the aging of alkali cellulose, is desirable, since it provides an important means of controlling the properties of the final product. However, for the pulp and paper producer the degradation of the cellulose must be kept at a minimum in order to obtain high yields and to retain many of the physical and mechanical properties of the fiber.

There are a variety of different types of degradation: (1) hydrolytic, (2) oxidative, (3) alkaline, (4) thermal, (5) microbiological, and (6) mechanical. Of these only the first five classes, which are perhaps the most important, will be briefly discussed. A more detailed discussion on cellulosic degradation can be found in reviews by Bolker (156), Rydholm (161), and Ward (157).

IX. HYDROLYTIC DEGRADATION

This type of degradation refers to cleavage at the glycoside linkage between C-1 and oxygen by an acid. If the process is carried to completion, the final product will be the monosaccharides. Acid hydrolysis may proceed as a homogeneous process with strong acids, which dissolve the substrate, or as a heterogeneous process with weaker acids.

Under the latter conditions, cellulose maintains its fibrous structure. The heterogeneous reaction occurs in two distinct phases: a rapid initial reaction followed by a much slower reaction. It is believed the easily accessible regions react first and then the less accessible regions react, but more slowly. The hydrolysis, at the initial step, removed 10–12% of the substrate and rapidly reduced the DP to a "leveling off" or limiting valve. Depending on the history of the sample, the leveling off valve usually is approximately 200–300 D-glucose units per chain. Commercially, the process of heterogeneous hydrolysis is used to produce hydrocellulose and gel-forming microcrystalline cellulose (158). There has been a considerable amount of interest in developing heterogeneous acid hydrolysis procedures for converting cellulose to D-glucose. Under optimum conditions, the maximum yield of D-glucose is 50–60%. The major reason that higher yields of glucose are not obtained is due to the acid-catalyzed decomposition of glucose to 2-furaldehyde, 5-hydroxymethyl 2-furaldehyde, and a variety of other products (162–164).

The homogeneous process is a much simpler reaction that proceeds at a uniform rate. Several processes have been developed for industrial saccharification using concentrated hydrochloric or sulfuric acid at low temperature. However, these processes have only been used during wartime or in countries with controlled economics such as the Soviet Union (162–164). The main product from the reaction is D-glucose, which can be converted into a wide range of other products by fermentation and hydrogenation (163). Recent studies have been done using superconcentrated hydrochloric and concentrated hydrofluoric acid for converting various types of cellulosic materials into high yields of D-glucose (164).

The homogeneous process is also used in the laboratory for the analysis of wood pulp. The procedure consists of an initial brief treatment with strong sulfuric acid, followed by treatment with hot dilute acid to complete the hydrolysis (165).

X. OXIDATIVE DEGRADATION

Cellulose is highly susceptible to oxidizing agents. The extent of the resulting degradation depends on the nature of the reagent and on the conditions under which oxidation occurs. The hydroxyl groups and the terminal reducing ends are the sites most susceptible to attack. Most oxidations are random processes and lead to the introduction of carbonyl and carboxyl groups at various positions in the anhydro D-glucose units of cellulose. There are also a variety of secondary reactions that can occur, including chain scission.

"Oxycellulose" is the term used for the product obtained from cellulose by any oxidizing procedure that introduces carbonyl and/or carboxyl groups anywhere in the molecule. Certain oxidizing agents are fairly specific. For example, periodic acid, or sodium metaperiodate, and acidic lead tetraacetate cleave the anhydroglucose units between the C-2 and C-3 positions and form formaldehyde groups at these two positions (dialdehyde cellulose). Nitrogen dioxide—or, more properly, dinitrogen tetroxide—attacks the hydroxy groups at carbon 6 with a fair degree of specificity and converts them into carboxyl groups. However, there is some oxidation at C-2 and C-3 to form ketone groups (166). Hypolodite, chlorine dioxide, and acidified chlorite solutions are mild oxidizing agents that have little or no effect on the hydroxyl groups, but will oxidize the terminal reducing group (or any internal aldehyde group) to carboxyl groups (161, 167). These oxidations are sufficiently specific that they have been used for the quantitative measurement of such reducing groups (167).

Most oxidants are less specific than the aforementioned ones and produce a variety of different types of carbonyl and carboxyl groups in cellulose. A good example is the chlorine and hypochlorite systems that are widely used in bleaching wood pulp. Depending on the pH, these bleaching agents can cause oxidation of the hydroxyl groups to aldehyde, ketone, and carboxyl groups, as well as depolymerization of the polysaccharide (161).

Oxygen can also cause oxidation of cellulose in alkaline solutions. This process is used industrially in the manufacture of viscose rayon. In this process, cellulose is mixed with a concentrated solution of sodium hydroxide. The fiber is shredded and allowed to age. After aging, the cellulose molecular weight has decreased and the fiber contains carboxyl groups.

There are a variety of other reagents that cause the nonspecific oxidation of cellulose. These include ozone, hydrogen peroxide, permanganate, and many others.

XI. ALKALINE DEGRADATION

There are three major types of base degradation of cellulose. One type is an oxidative process that occurs when alkaline solutions of cellulose come in contact with air. This process occurs only at relatively high temperatures and leads to chain scission and the introduction of carboxyl groups. The other two types of degradation are nonoxidative processes.

Probably the most important type of base degradation is the so-called "peeling reactions." This reaction involves the gradual shortening of cellulose chains by a β-elimination mechanism. This reaction normally proceeds from the reducing end of the cellulose molecule. However, if the cellulose has been oxidized at any point in the molecule and a carbonyl or carboxyl group is present, alkaline degradation will proceed at this point with chain scission.

The major steps in the reaction are shown in Fig. 11. The initial step, occurring at the reducing end, involves the extraction of a proton from C-2 and the equilibration of the C-1 epimers and the C-2 ketose 2, the corresponding enediols 3 (Lobry de Bruyn-Alberda van Ekenstein enolization) (168–170). The next step 4 involves the elimination of the C-4 substituent. In the case of cellulose, this would lead to a new reducing end and the formation of a dicarbonyl compound 5 that undergoes a re-arrangement to form isosaccharinic acid 6. The cellulose chain, which was elimi-nated, contains a new reducing end. Therefore, this process will be repeated with progressive shortening of the cellulose chain. The peeling reaction will remove about 50–60 glucose units (171) from each starting point until, finally, another reaction mechanism called the stopping reaction (169) provides the end unit with a configuration stable to alkali (Fig. 11). The stopping reaction leads to the formation of a substituted metasaccharinic acid 11. It follows a very similar mechanism as the peeling reaction. However, elimination occurs at the C-3 position that stabilizes the cellulose chain and prevents further eliminations 7–11.

Another important carbohydrate reaction with alkali involves the alkaline hydrolysis of glycosidic bonds. This reaction leads to the shortening of the polysaccharide chains, which is likely to have some bearing on the strength properties of the final product. This type of alkaline degradation occurs at relatively high temperatures of around 160–180°C (161, 172). It has also been shown that this reaction is independent of the presence of small amounts of oxygen and is, therefore, entirely an alkaline hydrolysis of glycosidic bonds. These same conclusions have been reached in studies of model glycosides (173).

XII. THERMAL DEGRADATION

The thermal degradation of cellulose passes through two types of reactions: a gradual degradation, decomposition, and charring on heating at lower temperature and a rapid volatilization accompanied by the formation of levoglucosan and a

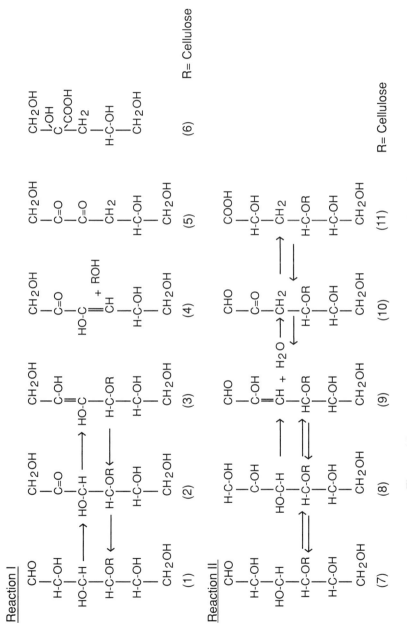

Figure 11   The alkaline degradation reactions of cellulose (168–170).

variety of other organic products at higher temperatures.

At the lower temperatures, ambient to 200°C, it is difficult to draw a line of demarcation between the thermal degradation and the normal aging of cellulose that is accelerated by heating. According to Richter (174), when rag paper is heated at 38°C for about six months, the accelerated aging results in a 19% reduction of the folding strength. Farquhar and coworkers (175) heated raw cotton for 4–24 hr at temperatures of 75–220°C in air and nitrogen. This and other investigations (176–178) have shown that oxygen and water have a profound effect in enhancing the deterioration and thermal degradation of cellulosic materials.

The changes that occur during the low-temperature thermal decomposition of cellulose include reduction in molecular weight; appearance of free radicals; elimination of water; formation of carbonyl, carboxyl, and hydroperoxide groups (especially in air); and evolution of carbon monoxide and carbon dioxide. All these processes occur at a much faster rate in air than in an inert atmosphere such as nitrogen.

In contrast to the relatively slow degradation of cellulose at low temperatures, cellulose heated above 300°C undergoes rapid decomposition. The products of cellulose pyrolysis can be collected in three fractions: a carbonaceous residue (char), a highly viscous syrup that condenses at room temperature (tar), and a mixture of volatile gases. By using a variety of model compounds and the application of thermal and chemical methods of analysis, workers have been able to elucidate some of the chemical pathways involved in the complex pyrolytic transformations (178–180).

The tarry syrup obtained from the pyrolysis of cellulose is produced by a series of inter- and intramolecular transglycosylation reactions accompanied by repolymerization of the resulting monomers. The major product present in this tar is the anhydro sugar, 1,6-anhydro-β-D-glucopyranose, or levoglucosan. Other monomeric substances identified in the tar in lesser amounts are 1,6-anhydro-β-D-glucofuranose (the furanose isomer of levoglucosan), 1,4:3,6-dianhydro-α-D-glucopyranose, 5-hydroxymethyl-2-furaldehyde, and traces of α- and β-D-glucose. In addition, a significant portion of the anhydro sugars have been shown to undergo subsequent transglycosylation and condensation reactions producing a mixture of randomly linked oligosaccharides.

The products of the transglycosylation reactions undergo decomposition through dehydration, fission, and disproportionation reactions to produce volatile products and also undergo condensation and elimination reactions to produce char. A wide variety of volatile products has been identified from the pyrolysis of cellulose, and mechanisms for their formation have been proposed. The extent of these reactions is dependent on the conditions of pyrolysis. The heating of cellulose in the presence of acidic or basic catalysts results in increased charring with a proportionate decrease in tar formation. Pyrolysis of cellulose in the presence of acid catalysts also produces significant changes in the composition of the tar and volatile fractions (180–182).

XIII. ENZYMATIC DEGRADATION

There is a large variety of enzymes present in fungi, bacteria, plants, and animals that can cause the hydrolytic decomposition of cellulose. These enzymes, which are called cellulases, cause the degradation of wood, cotton, and paper and are responsible for losses of millions of dollars each year (163). However, these enzymes also play a very necessary role in nature. Cellulases in organisms that metabolize cellulose play an important role in balancing the carbon cycle. Cellulases and hemicellulases are probably involved, along with related enzymes, in the germination of seeds and in the ripening as well as the rotting of fruits and vegetables. Cellulase, excreted by microflora present in the digestive tracts of herbivores, enables these animals to utilize cellulose as a prime source of energy.

Because of their economic impact and potential, the cellulases of microorganisms have been the most widely investigated. When a fungal spore germinates on wood, there is general dissolution of the primary wall, caused by threadlike structures called hyphae that secrete enzymes as they penetrate the cell wall via bore holes or through pits present in the wood. The cellulases and hemicellulases initially attack the lumen wall and gradually work outward as the polysaccharides are removed from each successive wall layer. Bacteria also produce cellulase enzymes that are capable of degrading cellulose, but, in contrast to the bore holes and inner-fiber damage of fungal growth, bacterial attack appears to involve only the surfaces of fibers.

In recent years, much of the work on cellulases has concentrated on developing an economic enzymatic process for converting waste cellulose into D-glucose (183–184).

The susceptibility of cellulose to enzymatic hydrolysis is determined largely by its accessibility. It is necessary during enzymatic hydrolysis that there be direct physical contact between the cellulose and the enzyme. Since cellulose is an insoluble and structurally complex substrate, this contact can be achieved only by diffusion of the enzymes into the complex structural matrix of the cellulose. Any structural feature that limits the accessibility of cellulose to enzymes will diminish its susceptibility to hydrolysis. Probably the two most important factors that limit the accessibility in most plant materials to enzymatic degradation are the lignin and crystallinity of the cellulose. Other factors that are important are the moisture content, degree of polymerization, and the presence of extraneous constituents in plant materials. More information about enzymatic degradation can be found in reviews by Reese and others (183–185).

REFERENCES

1. R. M. Brown, Jr., Biogenesis of Natural Polymer Systems, in *Proceedings of the 3rd Philip Morris Science Symposium*, Richmond, Va. 1978, pp.51–123.
2. a. E. B. Cowlin and T. K. Kirk, *Biotechnol. & Bioeng.Symp.*, 6:95–123; (1976) 2b. T. E. Timell, *Tappi*, 44:88 (1961).
3. C. F. Cross and E. J. Bevan, *J. Chem. Soc.*, 38:666 (1880).
4. W. G. Van Beckum and G. J. Ritter, *Paper Trade J.*, 105 (18):27 (1937).
5. L. E. Wise, M. Murphy, and A. A. D'Addieco, *Paper Trade J.*, 122 (2):35 (1946)
6. J. W. Green, *Methods Carbohyd. Chem.*, 3:9 (1963).
7. B. L. Browning and L. O. Bublitz, *Tappi*, 36:452 (1953).
8. T. E. Timell, *Advan. Carbohyd. Chem*, 19:247 (1964).
9. T. E. Timell, *Advan. Carbohyd. Chem.*, 20:409 (1965).
10. T. E. Timell, *Methods Carbohvd. Chem.*, 5:100 (1965).
11. H. A. Swenson, *Methods Carbohyd. Chem.*, 3:4 (1963).
12. A. Payen, *Compt. Rend*, 7:1052 (1838).
13. R.Willstatter and L. Zechmeister, *Ber.*, 46:2401 (1913).
14. H. Ost, *Z. Angew Chem.*, 19:993 (1906).
15. W. Crum, *Phil. Mag.*, 30:426 (1847).
16. W. Crum, *Justus Leibigs Ann. Chem.*, 62:233 (1847).
17. G. W. Monier-Williams, J. Chem. Soc., 119:803 (1921).
18. J. C. Irvine and C. W. Soutar, *J. Chem. Soc.*, 117:1489 (1920).
19. J. C. Irvine and E. L. Hirst, *J. Chem. Soc.*, 121:1585 (1922).
20. J. C. Irvine and E. L. Hirst, *J. Chem. Soc.*, 123:518 (1923).
21. K. Freudenberg, E. Plankenhorn, and H. Boppel, *Ber.*, 71:2435 (1938).
22. W. N. Haworth, *Nature*, 116:430 (1925).
23. H. Friese and K. Hess, *Justus Leibigs Ann. Chem.* 456:38 (1927).
24. C. C. Spencer, *Cellulosechemie*, 10:61 (1921).
25. I. Danishefsky, R. L. Whistler, and F. A. Bettelheim, in *The Carbohydrate* (W. Pigman and D. Horton, eds.), Academic Press, New York, 1970, p. 375.
26. J. Dyer, in *Cellulose Technology Research* (A. F. Turbak, ed.), ACS Symposium Series 10, American Chemical Society, Washington, D. C., 1975, p. 181.
27. E. H. Immergut, in *The Chemistry of Wood* (B. L. Browning ed.), Wiley-Interscience, New York, 1963, Vol. 103.
28. P. M. Doty and H. M. Spurlin in *Cellulose and Cellulose Derivatives* (E. Ott, H. M. Spurlin, and M. W. Grafflin, eds.), Wiley-Interscience New York, 1955, Part III, p. 1133.
29. W. J. Alexander and T. E. Muller, *J. Polym. Sci.*, Part C, 36:87 (1971).
30. K. H. Altgelt and L. Segal, *Gel Permeation Chromatography*, Marcel Dekker, New York, 1971, p. 429.
31. S. H. Churms, *Advan. Carbohyd. Chem. and Biochem.*, 25:13 (1970).
32. T. E. Timell, et. al., *Svensk Papperstidnina*, 58:851, 889 (1955); 59:1 (1956); *Tappi*, 40:25 (1957); *Pulp Paper Mag. Can.*, 56 (7):104 (1955).
33. W. G. Harland, *Shirley Inst. Mem.*, 28:167 (1955); *J. Textile Inst.*, 45:T-678 (1954).
34. J. L. Snyder and T. E. Timell, *Svensk Papperstidnina*, 58:851 (1955); *Tappi*, 40:749 (1957); *Ind. Eng. Chem.*, 47:2166 (1955).

35. M. Marx-Figini and E. Penzel, *Makromol. Chem. 87*:507–515 (1965).

36. M. Marx-Figini, *Nature (Lond.), 210*:755 (1966).

37. T. E. Timell, *Wood Sci. and Technol., 1*:45 (1967).

38. H. I. Bolker, *Natural and Svnthetic Polymers*, Marcel Dekker, New York, 1974, p. 37.

39. W. A. Côté, Jr., *Cellular Ultrastructure of Woody Plants*, Syracuse University Press, Syracuse, N.Y. 1965.

40. H. F. J. Wenzl, *The Chemical Technology of Wood*, Academic Press, New York, 1970, p. 92.

41. E. B. Cowling, "Physical and Chemical Constraints in the Hydrolysis of Cellulose and Lignocellulosic Materials," *Biotechnol & Bioeng. Symp., 5*:163–181 (1975).

42. A. B. Wardrop and D. E. Bland, in *Biochemistry of Wood* (K. Kratzl and G. Billek, eds.), Pergamon Press, London, 1959, p. 92.

43. R. Preston, *Annual Rev. Plant Physiol., 30*:55 (1979).

44. P. Albersheim, *Sci. Amer., 232*:80 (1975).

45. J. R. Colvin, *Ultrastructure of the Plant Cell Wall: Biophysical Viewpoint in Plant Carbohydrates II*, Springer-Verlag, Berlin, 1981, Vol. 2, pp. 9-22.

46. J. C. Roland and B. Vian, *Int. Rev. Cytol., 61*:129 (1979).

47. H. Harada, Cellular *Ultrastructure of Woody Plants* (W. A. Côté, Jr. ed.), Syracuse University Press, Syracuse, N.Y., 1965, p. 215.

48. C. Y. Liang, K. H. Bassett, E. A. McGinnes, and R. H. Marchessault, *Tappi, 43*:1017 (1960).

49. B. J. Fergus and D. A. I. Goring, *Holzforschung, 24*:118–124 (1970).

50. K. Freudenberg, *J. Chem. Educ., 9*:1171 (1932).

51. R. Mark, in *Cellular Ultrastructure of Woody Plants* (W. A. Côté, Jr. ed.), Syracuse University Press, Syracuse, N.Y., 1965, p. 493.

52. D. Ohad, D. Danon, and S. J. Hestrin, *J. Cell. Biol., 12*:81 (1962).

53. K. Mühlethalter, *Cellular Ultrastructure of Woody Plants* (W. A. Côté, Jr. ed.) Syracuse University Press, Syracuse, N.Y., 1965, p. 191.

54. R . D. Preston, *Faraday Soc Disc., 11*:165 (1951).

55. A. J. Hodge and A. B. Wardrop, *Nature, 165*:272 (1950).

56. A. Vogel, *Makromal Chem., 111*:111 (1953).

57. A. Frey-Wyssling and K. Mühlethaler, *Ultrastructural Plant Cytology, Elsevier*, Amsterdam, 1965.

58. W. W. Franke and B. Ermen, *Z. Naturforsch, 246:*918 (1969).

59. D. Fengel, *Naturwissenschaften, 61*:31 (1974).

60. R. B. Hanna and W. A. Côté, Jr., *Cytobiologie, 10*(1):102 (1974).

61. R. Marton, P. Rushton, J. Sacco, and K. Sumiya, *Tappi, 55*:1499 (1972).

62. H. Harada and T. Goto, "The Structure of Cellulose Microfibrils in *Valonia*," in *The Biogenesis of Cellulose* (R. M. Brown, Jr., ed.), Plenum Press, New York, 1982, pp. 383–399.

63. V. D. Fengel, *Holzforschung, 32*:2 (1978) 37.

64. E. K. Boylston and J. J. Hebert, *J. Appl. Polym. Sci., 25*:2105 (1980).

65. C. Nägeli, *Die Stärkekorner*, F. Schulthess, Zürich, 1858.

66. R ∩ Herzog and W. Jancke, *Z. Physik*, 3:196 (1920).

67. H. Mark, *Chem. Rev., 26*:169 (1940).

68. K. H. Meyer and L. Misch, *Helv. Chim. Acta.*, *20*:232 (1937).
69. S. G. Shenouda, *Appl. Fiber Sci.*, *7*:275–309(1979).
70. G. Honjo and M. Watanabe, *Nature*, *181*:326(1958).
71. J. J. Hebert and L. L. Muller, *J. Appl. Polym. Sci.*, *18*:3373(1974).
72. M. M. Beg, J. Aslam, Q. H. Khan, N. M. Butt, S. Rolandson, and A. U. Ahmed, *J. Polym. Sci., Polym. Lett. Ed.*, *12*:311–318(1974).
73. A. Sarko and R. Muggli, *Macromolecules*, *7*:486(1974).
74. K. H. Gardner and J. Blackwell, *Biopolym.*, *13*:1975(1974).
75. W. Claffey and J. Blackwell, *Biopolym.*, *15*:1903(1976).
76. A. D. French, *Carbohyd. Res.*, *61*:67–80(1978).
77. A. Sarko, *Tappi*, *61*(2):59–61(1978).
79. J. Blackwell, "The Macromolecular Organization of Cellulose and Chitin," in *Cellulose and Other Natural Polymer Systems*, Plenum Press, New York, 1982, Vol. 20, pp. 403–427.
80. R. E. Hunter and N. E. Dweltz, *J. Appl. Polym. Sci.*, *23*:249–259(1979).
81. R. H. Atalla, Appl. Polym. Symp. *28*:659–669 (1976).
82. R. H. Marchessault and A. Sarko, *Advan. Carbohyd. Chem.*, *22*:421 (1967).
83. P. K. Chidambareswaran, S. Sreenivasan, and N. B. Patil, *J. Appl. Polym. Sci.*, *27*:709(1982).
84. R. H. Marchessault and P. R. Sundarajan, *Advan. Carbodr. Chem. Biochem.*, *33*:387–404(1976); *35*:377–385(1978); *36*:315–327(1979).
85. R. H. Marchessault and J. A. Howsman, *Textile Res. J* .:30–41(1957).
86. J. A. Howsman and W. A. Sisson, in *Cellulose and Cellulose Derivatives* (E. Ott, H. M. Spurlin, and M. W. Grafflin, eds.), Wiley-Interscience, New York, 1954, Part I, p. 231.
87. R. S. J. Manley, J. Polym. Sci. (A-1), 1 (1963) 1875; *Nature*, *204*:1155 (1964).
88. M. M. Y. Chang, *J. Polym. Sci.*, *12*:1349(1974).
89. S. K. Asunmaa, *Tappi*, *49*:319 (1966).
90. M. Marx-Figini and G. V. Schulz, *Biochim. Biophys. Acta*, *112*:81 (1966).
91. H. Bittiger, E. Huseann, and A. Kuppel, *J. Polym. Sci.* Part C, *28*:45 (1969).
92. R. Muggli, *J. Cellulose Chem. Technol.*, *2*:549 (1969).
93. R. Muggli, H. G. Elias, and K. Mühlethaler, *Makromol. Chem.*, *121*:290 (1969).
94. R. E. Mark, *Cell Wall Mechanics of Tracheids*, Yale University Press, New Haven, Conn., 1967, p. 310.
95. R. E. Mark, P. N. Kaloni, R. C. Tang, and P. P. Gillis, *Textile Res. J.*, *39*:203(1969).
96. A. Frey-Wyssling, *Sci.*, *119*:80(1954).
97. R.D. Preston, in *International Review of Cytology* (G. H. Bowine and J. F. Danielli, eds.) 1959 Vol. 8, p.33.
98. R. Jeffries, *J. Appl. Polym. Sci.*, *8*:1213(1964).
99. R. Jeffries, J. G. Roberts, and R. N. Robinson, *Textile Res. J.*, *38*:234(1968).
100. R. Jeffries, D. M. Jones, J. G. Roberts, K. Selby, S. C. Simmens, and J. O. Warwicker, *Cell. Chem. Technol.*, *3*:255(1969).
101. K. Mühlethaler, *J. Polym. Sci.*, Part C, *28*:305 (1969).
102. K. Mühlethaler, *Z. Schweiz, Forstv.*, *30*:53 (1960).
103. A. Frey-Wyssling and K. Mühlethaler, *Makromal. Chem.*, *62*:25(1963).
104. S. P, Rowland and E. J. Roberts, *J. Polym. Sci.*, Part A-1, *10*:867(1972).

105 .   S . P. Rowland, "Selected Aspects of the Structure and Accessibility of Cellulose as They Relate to Hydrolysis in Cellulose as a Chemical and Energy Resource," in *Biotechnology and Bioengineering Symposium*, No. 5 C. R. Wilke, ed., Wiley, New York, 1975, pp. 183–191.

106.   A. Peterlin and P. Ingram, *Textile Res. J.*, *40*:345(1970).

107.   J. Blackwell and F. J. Kolpak, *Macromolecules*, *8*:322(1975).

108.   M. Marx-Figini, *Biochim. Biophys. Acta*, *177*:27–34(1969).

109.   M; Marx-Figini, *J. Polym. Sci.*, Part C, *28*:57–67(1969).

110.   M. Marx-Figini and B. G. Pion, *Biochimica et Biophysica Acta*, *338*:382–393(1974).

111.   M. Marx-Figini, "The Control of Molecular Weight and Molecular-Weight Distribution," in *Biogenesis of Cellulose* (R. M. Brown, Jr., ed.), Plenum Press, New York, 1982, Vol. 13, pp. 243–271.

112.   F. S. Spencer and G. A. MacLachlan, *Plant Physiol.*, *49*:58(1972).

113.   H. Nikaido and W. Z. Hassid, "Biosynthesis of Saccharides from Glycopyranosyl Esters of Nucleoside Pyrophosphates," in *Advances in Carbohydrate Chemistry and Biochemistry*, Academic Press, New York, 1971, Vol. 26, pp. 351–483.

114.   D. P. Delmer, "The Biosynthesis of Cellulose and Other Plant Cell Wall Polysaccharides," in *The Structure, Biosynthesis and Degradation of Wood*, Plenum Press, New York, 1977, Vol. 2, pp. 45–77.

115.   A. L. Karr, "Cell Wall Biogenesis," in *Plant Biochemistry*, 3rd ed. (J. Bonner, ed.), Academic Press, New York, 1976, pp. 405–421.

116.   G. Franz and V. Heiniger, "Biosynthesis and Metabolism of Cellulose and Noncellulosic Cell Wall Glucans," in *Plant Carbohydrates II*, Springer-Verlag, Berlin, 1981, Vol. 5, pp. 47–67.

117.   C. L. Villemez, "The Relation of Plant Enzyme-Catalysed β-(1,4)-Glucan Synthesis to Cellulose Biosynthesis *in vivo*," in *Ann. Proc. Phytochem. Soc.* :183–189, 1974.

118.   J. Ross Colvin, *Tappi*, *12*,(60):59–61, (1977).

119.   L. Glaser, *J. Biol. Chem.*, *232*:627–636, (1958).

120.   W. Z. Hassid, "Biosynthesis of Sugars and Polysaccharides," in *The Carbohydrate* (W. Pigman and D. Horton, eds.), Academic Press, New York, 1970, Vol. IIA, pp. 301–373.

121.   A. D. Elbein, G. A. Barber, and W. Z. Hassid, *J. Amer.Chem. Soc.*, *86*:309 (1964); *J. Biol. Chem.*, *239*:4056 (1964).

122.   T. Liu and W. Z. Hassid, *J. Biol. Chem.*, *245*:1922 (1970).

123.   G. A. Barber and W. Z. Hassid, *Biochim. Biophys. Acta*, *86*:397 (1964).

124.   G. A. Barber and W. Z. Hassid, *Nature*, *207*:295 (1965).

125.   D. P. Delmer C. A. Beasley~and L. Ordin, *Plant Physiol.*, *53*:149 (1974).

126.   D. O. Brummond and A. P. Gibbons, *Biochem. Biophys. Res. Commun.*, 17:156(1964).

127.   D. O. Brummond and A. P. Gibbons, *Biochem. Z*, *342*:308 (1965).

128.   L. Ordin and M. A. Hall, *Plant Physiol.*, *42*:205 (1967); *43*:473 (1968).

129.   D. G. Robinson and H. Quader, *J. Theor. Biol.*, *92*:483–495(1981).

130.   N. C. Carpita, "Cellulose Synthesis in Detached Cotton Fibers," in *Cellulose and Other Natural Polymer Systems* (R. M. Brown, Jr., ed.) Plenum Press, New York, 1982, Vol. 12, pp. 225–242.

131.   D. H. Northcote, *Ann. Proc. Phytochem. Soc.*, *10*:165 (1974).

132.   H. Kauss, *Ann. Proc. Phytochem. Soc.*, *13*:191 (1974).

133. F. B. P. Wooding, *J. Cell. Sci.*, *3*:71 (1968).
134. M. J. Chrispeels, "Biosynthesis, Intracellular Transport, and Secretion of Extracellular Macromolecules," in *Ann. Rev. Plant Physiol.*, *27*:19–38 (1976).
135. G. Shore and G. A. Maclachlan, *J. Cell. Biol.*, *64*:557–571 (1975).
136. D. H. Northcote, in *Plant Cell Organelles* (J. B. Pridham, ed.), Academic Press, New York, 1968, p. 179.
137. A. D. Elbein, "The Role of Lipid-linked Saccharides in the Biosynthesis of Complex Carbohydrates," in *Plant Carbohydrates II* (W. Tanner and F. A. Loewus, eds.), Springer-Verlag, Berlin, 1981, Vol. 8, pp. 166–193.
138. G. Franz, *J. Polym. Sci.*, Part C, *28*:611–621 (1976).
139. C. L. Villemez, *Biochem. Biophy. Res. Comm.*, *40*:636–641 (1970).
140. J. Kjosbakken and J. Ross Colvin, *Can. J. Microbiol.*, *21*(2):111–120 (1975).
141. J. R. Colvin and G. G. Leppard, *Can. J. Microbiol.*, *23*:701 (1977).
142. E. H. Hopp, P. A. Romero, G. R. Daleo, and R. Pont Lezica, *Europ. J. Biochem.*, *84*:561–571 (1978).
143. M. Nowak-Ossorio, E. Gruber, and J. Schurz, *Protoplasma*, *88*:255–263 (1976).
144. G. G. Leppard and J. R. Colvin, *J. Microscopy*, *113*:181–184 (1978).
145. R. D. Preston, "Structure and Mechanical Aspects of Plant Cell Walls with Particular Reference to Synthesis and Growth," in *The Formation of Wood in Forest Trees* (M. H. Zimmermann, ed.), Academic Press, New York, 1964.
146. R. M. Brown, J. H. M. Willison, and C. L. Richardson, *Proc. Natl. Acad. Sci., USA*, *73*:4565–4569 (1976).
147. D. G. Robinson, "The Assembly of Polysaccharide Fibrils," in Plant Carbohvdrates II, Springer-Verlag, Berlin, 1981, pp. 25–28.
148. B. L. Browning, in *The Chemistry of Wood* (B. L. Browning, ed.), Wiley-Interscience, New York, 1963.
149. C. Skaar, *Water in Wood*, Syracuse University Press, Syracuse, N. Y., 1972, Vol. 9, p. 406.
150. A. El-Kafrawy, *J. Appl. Polym. Sci.*, *27*:2435 (1982).
151. A. F. Turbak, 184th American Chemical Society Meeting, Cellulose, Paper and Textile Division, Kansas City, Missouri, Sept. 1982, Paper 35.
152. A. El-Kafrawy and A. F. Turbak, *J. Appl. Polym. Sci.*, *27*:2445 (1982).
153. D. C. Johnson, M. D. Nicholson, and F. C. Haigh, Institute of Paper Chemistry Technical Paper Series #5, April 1975.
154. H. A. Swenson, Institute of Paper Chemistry Technical Paper Series #8, May 1975.
155. K. Ward, Jr. and P. A. Seib, in *The Carbohydrates*, Academic Press, New York, 1970, p. 413.
156. D. M. Jones, *Advan. Carbohyd. Chem.*, *19*:219–244 (1964).
157. S. P. Rowland, "Hydroxyl Reactivity and Availability in Cellulose," in *Modified Cellulosics*, Academic Press, New York, 1978, pp. 148–167.
158. D. F. Durso, "Chemical Modification of Cellulose," in *A Historlcal Review in Modified Cellulosics*, (R. M. Rowell and R. A. Young, eds.), Academic Press, New York, 1978, pp. 24–37.
159. R. L. Mitchell, *Ind. Eng. Chem.*, *45*:2526 (1953).
160. C. F. Bennett and T. E. Timell, *Svensk Papperstidnina*, *58*:281 (1955).
161. S. A. Rydholm, *Pulping Processes*, Wiley-Interscience, New York, 1965.

162. W. Sandermann, *Chemische Holzverwertung*, BLV Verlag, Munchen, 1963.
163. G. J. Hajny, "Biological Utilization of Wood for Production of Chemicals and Food Stuffs," U. S. Dept. of Agriculture, Washington, D. C., Research Paper #FPL 385.
164. I. S. Goldstein, "Chemicals from Cellulose," in *Organic Chemicals from Biomass* (I. S. Goldstein, ed.), CRC Publication, Boca Raton, Fla., 1981, p. 101.
165. J. F. Saeman, W. E. Moore, R. L. Mitchell, and M. A. Millett, *Tappi*, *37*:336 (1954).
166. P. A. McGee, W. F. Fowler, Jr., E. W. Taylor, C. C. Unruh, and W. O. Kenyon, *J. Amer. Chem. Soc.*, *69*:355 (1947).
167. B. Alfredsson, W. Czerwinsky, and O. Samuelson, *Svensk Papperstidina*, *64*:812 (1961).
168. R. L. Whistler and W. M. Corbett, *J. Amer. Chem. Soc.*, *78*:1003 (1956).
169. R. L. Whistler and J. N. BeMiller, *Advan. Carbohyd. Chem*, *13*:289 (1958).
170. G. Machell and G. N. Richards, *J. Chem. Soc.*, 4500 (1957); 1199 (1958); 1924 (1960).
171. O. Samuelson and A. Wennerblom, *Svensk Papperstidnina*, *57*:827 (1954).
172. R. E. Brandon, L. R. Schroeder, and D. C. Johnson, in *Cellulose Technology Research* (A. F. Turbak, ed.), ACS Symposium Series 10, American Chemical Society, Washington, D. C. 1975, p. 125.
173. B. Lindberg, *Svensk Papperstidnina*, *59*:531, 870
174. G. A. Richter, *Ind. Eng. Chem.*, *26*:1154 (1934).
175 R. L . W. Farquhar, D. Pesant, and B. A. McLaren, *Can. Textile J.*, *73*:51 (1956).
176 W. D. Major, *Tappi*, *41*:530 (1958).
177. L. F. McBurney, in *Cellulose and Cellulose Derivatives* (E. Ott, H. M. Spurlin, and M. W. Grafflin, eds.), Wiley-Interscience, New York, 1954, Part I, p. 99.
178. F. Shafizadeh, *Advan. Carbohyd. Chem.*, *23*:419 (1968).
179. F. Shafizadeh, in *Thermal Uses and Properties of Carbohydrates and Lignins* (F. Shafizadeh, K. V. Sarkanen, and D. A. Tillman, eds.), Academic Press, New York, 1976; F. Shafizadeh and P. P. S. Chin, in *Wood Technology: Chemical Aspects* (I. S. Goldstein, ed.) ACS Symposium Series 43, American Chemical Society, Washington, D. C. 1977.
180. F. Shafizadeh, *J. Analyt. and Appl. Pyrolysis*, *3*:283–305 (1982).
181. F. Shafizadeh, T. G. Cochran, and Y. Sakai, *AICHE Symp. Series No. 184*, *75*:24–34 1979.
182. M. J. Antal, Jr., *Advances in Solar Energy*, 1983 (in press).
183. E. T. Reese and M. Mandels, "Enzymatic Degradation," in *Cellulose and Cellulose Derivatives*, Wiley-Interscience, New York, 1971, Part V, p. 1088.
184. D. F. Durso and A. Villarreal, in *Cellulose Technology Research* (A. F. Turbak, ed.), ACS Symposium Series 10, American Chemical Society, Washington, D. C. 1975, p. 125. p. 106.
185. M. Mandels and J. Weber, in *Cellulases and Their Applications*, Advances in Chemistry Series 95 American Chemical Society, Washington, D. C., 1969, p. 391 .

5

Lignins: Occurrence in Woody Tissues, Isolation, Reactions, and Structure

Chen-Loung Chen

North Carolina State University, Raleigh, North Carolina

I. INTRODUCTION

Lignin is one of the integral components of woody tissues in vascular plants, forming about one-quarter of such tissues (1) besides hemi-celluloses and cellulose, and it is the second most abundant organic material next to cellulose in the plant kingdom. Lignins occur in the woody tissues incrusting cellulose fibers in cell walls and filling the space between the cells, i.e., the middle lamella. In addition, lignins are physically and, perhaps, chemically bonded to carbohydrates in the woody tissues imparting mechanical strength to the wood (2). After the senescence and death of the plants, the lignins undergo degradation in nature mostly by soil microorganisms to produce humic substances (3). The latter are probably further decomposed to form carbonaceous substances and even coal occurring in nature (4). Thus, the formation of lignins is not only of primary importance in the physiology of vascular plants, but also plays an important role in the carbon cycle, contributing significantly to the ecological balance of the earth.

In 1838, Payen (5) observed that the treatment of wood with nitric acid resulted in dissolution of part of the wood, leaving a fibrous residue that he named "cellulose." Since the elemental analysis of the wood and Payen's cellulose showed that the former had a higher carbon content than the latter, Payen postulated that woods also contained a carbon-rich component incrusting the cellulose. Thus, Payen is the first to show the composite nature of woods. Later investigations, however, revealed Payen's cellulose consisted of hemi-celluloses and cellulose as it is understood today. The term "lignin" was introduced to describe the dissolved part of wood as characterized by Schulze (6) in 1965; the term is derived from the Latin word for wood, "lignum." In 1868, Erdman (7) isolated catechol and protocatechuic acid from a reaction mixture obtained from the alkali-fusion of wood, indicating that woods contained aromatic constituents in addition to carbo- hydrates, although the nature of lignin was still unknown.

In the 1860s, the development of new pulping processes, in particular the sulfite process introduced first by Tilgman (8) in 1866, engendered considerable interest in the reactions of woods and the formation of pulps during these processes. However, the nature of the major constituents of woods was still not well known at that time, except for the fact that cellulose has the same chemical composition as starch.

In 1874, Tiemann and Harrmann (9) isolated coniferin [4] from the cambium of Norway spruce (*Picea abies*) (see Fig. 1 for structures). The compound underwent hydrolysis under the catalytic influence of emulsin to produce D-glucose and an aglycon, coniferyl alcohol [2], indicating that the compound was a β-D- glucopyranoside of coniferyl alcohol. In the following year, Tiemann (10) eluci- dated the structure of both coniferin and coniferyl alcohol and then suggested (11) that coniferin could be chemically related to the still uncharacterized "aromatic atom complex" in woods, i.e., lignin. Klason (12) expanded this idea in 1897 and postulated that lignin was biogenetically driven from coniferyl alcohol. On the basis of his subsequent investigations, Klason (13–17) proposed that lignin was a macromolecular substance having a structure with basic units comprised of dimers from coniferyl alcohol and coniferyl aldehyde, although he was not able to produce unequivocal experimental evidence for such a proposal.

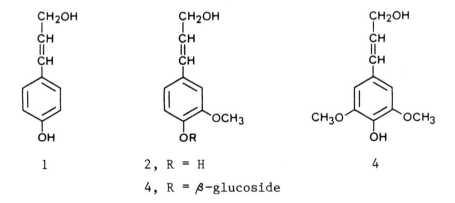

Figure 1   Structures of lignin precursors.

Cousin and Herissey (18) discovered in 1908 that a dimeric dehydrogenative polymerization product was produced when isoeugenol [5] was treated with either ferric chloride or a press-sap of champignon. In 1933, Erdtmann (19) reinvestigated the oxidation of isoeugenol by ferric chloride in order to show that a close relationship exists between phenols of coniferyl alcohol and lignin in terms of biogenesis and chemical structure. He soon found that the dimer, dehydro-diisoeugenol [7], was a phenylcoumaran derivative rather than a biphenyl deriva-tive as originally proposed by Herissey and Doby (20). Since ferric chloride was already known at that time as a one-electron-transferring oxidant, Erdtman proposed that the formation of dehydro-diisoeugenol involved the following steps as shown in Fig. 2: (1) abstraction of a hydrogen atom from the phenolic hydroxyl group of isoeugenol produces a phenoxyl radical $\underline{5a}$ that is resonancing among radicals $\underline{5b}$ and $\underline{5d}$, (2) coupling between radicais $\underline{5b}$ and $\underline{5d}$ with a concomitant aromatization forms a quinonemethide intermediate $\underline{6}$, and (3) a subsequent ring closure by an intramolecular nucleophilic attack of the hydroxyl oxygen atom in ring B on C-α

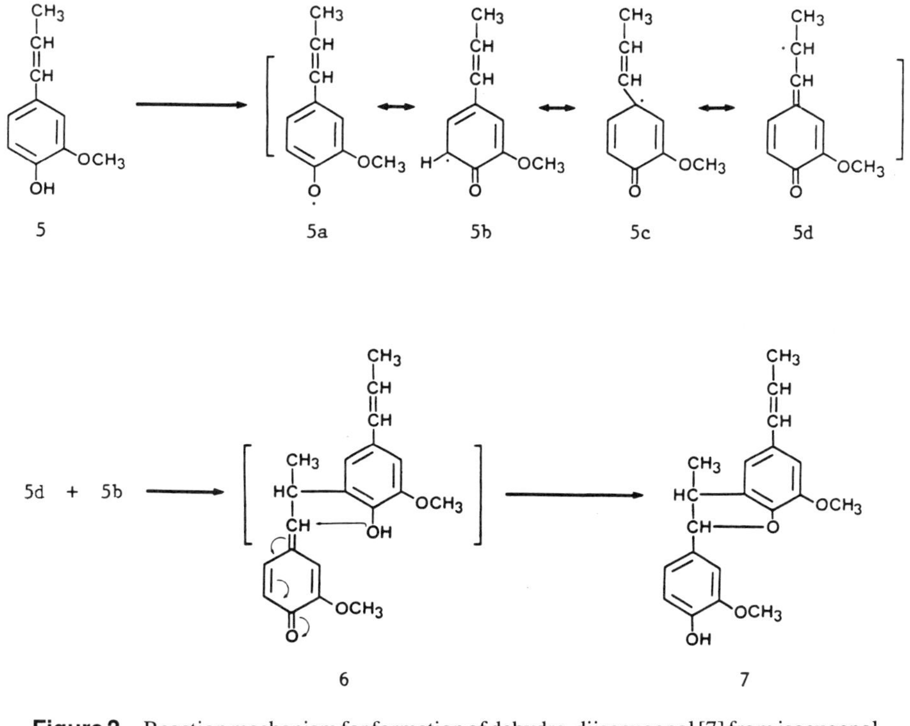

Figure 2 Reaction mechanism for formation of dehydro-diisoeugenol [7] from isoeugenol [5] by dehydrogenative coupling of phenols (From Ref. 7.)

of the quinonemethide moiety in intermediate 6 gives compound 7. A little earlier, Freudenberg (21) proposed a phenylcoumaran structure similar to compound 7 as one of the basic units for spruce lignin on the basis of chemical analysis. All of these led Erdtman to postulate that lignins are produced in vivo from coniferyl alcohol [2] and related compounds, such as 4-hydroxycinnamyl alcohol [1] and sinapyl alcohol [3], by an enzyme-initiated deydrogenative polymerization. This hypothesis has been shown to be correct through comprehensive investigations conducted at several laboratories over the world during the period 1935–1975 (22–25).

The biosynthesis of lignin will be discussed in the following chapter. This chapter, therefore, will be devoted to the nature of lignins, including their distribution in woody tissues, isolation, reactions, and chemical structures.

II. LIGNINS IN PLANT TISSUES

A. Occurrence of Lignins

Lignins occur widely in the woody tissues of vascular plants, but not in the tissues of lower plants such as those of mosses, lichens, fungi, and other microorganisms. The occurrence of lignins in plant tissues can be observed by specific color reactions, such as the Wiesner (26–29), Mäule (30, 31), and Cross-Beavan (32, 33) reactions.

1. Wiesner Reaction

Plant tissues containing lignins give a purple coloration on treatment with phloroglucinol [10] in concentrated hydrochloric acid. This reaction is applicable to all lignins in plant tissues. As shown in Fig. 3, the color is caused by the cationic chromophore [12], the conjugate acid of the extended quinonemethide [11], produced by condensation of 4-O-alkylated 4-hydroxycinnamaldehyde moieties [8] in lignins with phloroglucinol in concentrated hydrochloric acid as shown by Adler and co-workers (26–27) and Pew (28). Recently, Geiger and Fuggerer (29) demonstrated that 4-O-alkylated 4-hydroxycinnamyl alcohol moieties [9] in lignins also condense with phloroglucinol in concentrated hydrochloric acid in the presence of oxygen to produce compound 12, probably by way of oxidation to the corresponding 4-O-alkylated 4-hydroxycinnamaldehyde [8].

2. Mäule Reaction

Lignins containing significant amounts of syringylpropane units produce a rose-red color on successive treatment with aqueous potassium permanganate, hydrochloric acid, and ammonia solutions (30). The reaction is specific to type-15 syringylpropane units in lignins. The reaction is not applicable to 4-hydroxylphenylpropane and guaiacylpropane units of type-13 and -14 in lignins, respectively. Recently, Meshitsuka and Nakano (31) postulated that a chlorinated O-quinone of type-17 is responsible for the color, as shown in Fig. 4. However, the mechanism for the

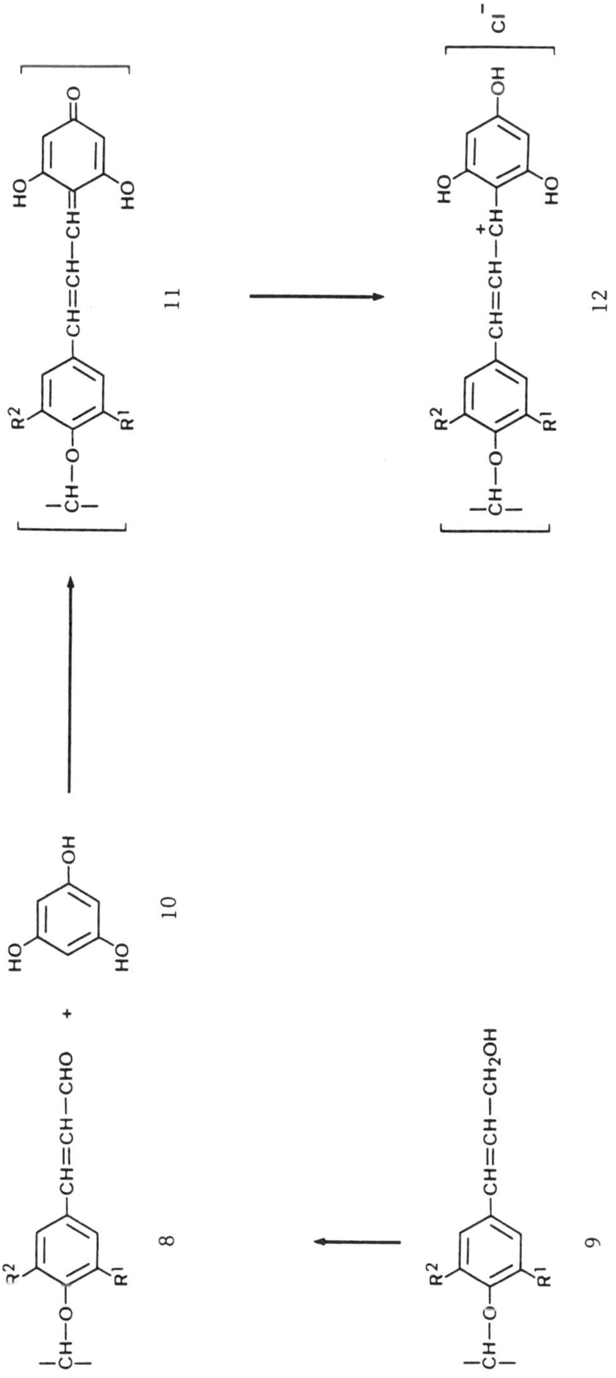

Figure 3 Reaction mechanism for Weisner reaction of lignins. R¹, R² = H or CH₃. (From Refs. 27–29.)

formation of <u>17</u> by way of a chlorinated catechol derivative <u>16</u> is not well established.

3. Cross-Bevan Reaction

Lignins containing syringylpropane units also give a rose-red color on treatment with chlorine-water, then with sodium sulfite solution to destroy excess chlorine (32, 33). The reaction is similar to the Mäule reaction and is also specific to the syringylpropane units in lignins.

B. Classification of Lignins

Until the 1960s, lignins have been classified into three major groups: gymnosperm (softwood), angiosperm (hardwood), and grass lignins according to nitrobenzene oxidation products of woody tissues and/or corresponding lignin preparations in addition to the elemental analysis of the latter.

   The nitrobenzene oxidation of woods, i.e., protolignins, was introduced by Freudenberg and Lautsch (34) for the characterization of the aromatic nature of lignins in woods. This procedure was later intensively used by Creighton and co-

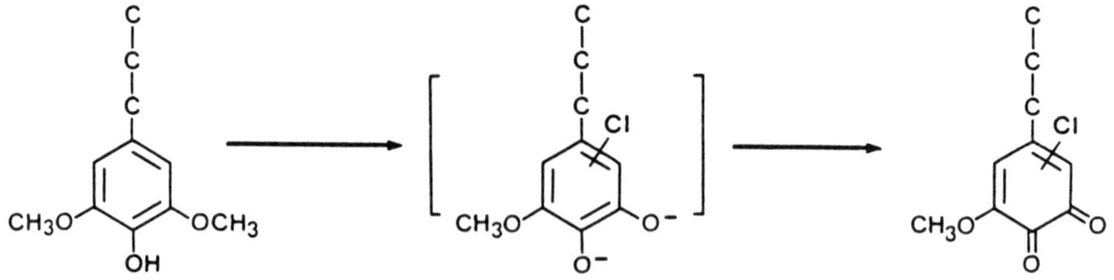

Figure 4   Reaction Mechanism for Mäule reaction of lignins (From Ref. 31.)

workers (35, 36) and others (37) for the systematic classification of lignins. On nitrobenzene oxidation, gymnosperm lignins produced mostly vanillin with negligible amounts of 4-hydroxybenzaldehyde and syringaldehyde, whereas angiosperm lignins yielded mainly vanillin and syringaldehyde, and grass lignins afford all the three phenolic aldehydes in significant amounts. The relative amounts of these aldehyde products in molar ratio produced from a lignin were interpreted as an indication of the relative abundance of uncondensed 4-hydroxyphenylpropane [13], guaiacylpropane [14] and syringylpropane units [15] present in lignin. Thus, the following conclusions have been made on the basis of the nitrobenzene oxidation of lignins: (1) Gymnosperm lignins consist mostly of guaiacylpropane units, (2) angiosperm lignins are mainly composed of guaiacyl- and syringylpropane units, and (3) grass lignins are comprised of 4-hydroxyphenyl-, guaiacyl-, and syringylpropane units, the molar ratio of which depends on the nature of plant species.

At first glance, this classification of lignins seems to be rational because the results are in rather good agreement with those resulting from the taxonomy of plants in biology. The classification is, however, inadequate because it leaves out most of the herbaceous angiosperm (Monocotyledonous Angiospermae) and fern (Pteridophyta) lignins. Moreover, it is incompatible with the chemical characteristics of some lignins. Although genera Tetraclinis, Podicarpus, Welwitschia, Ephedra, and Gentum are members of the conifer (Coniferales; Gymnospermae) family, lignins of these genera were found to be comprised of both guaiacyl- and syringylpropane units (36); hence, the lignins are of the angiosperm type and members of the genera are, therefore, called exceptional conifers. On the basis of these observations, Gibbes (38) proposed in 1958 that lignins should be classified into two major groups: "guaiacyl lignins" and "guaiacyl-syringyl lignins" according to their over-all chemical characteristics rather than in line with the taxonomy of plants in botany. Higuchi and co-workers (39, 40) provided additional support to this proposal as a result of their investigation into the nature of grass (*Greamineae*) lignins. When grass lignins were treated with a cold alkaline solution at a temperature lower than 50°C, the lignins underwent saponification to give a considerable amount of 4-hydroxycinnamic acid and ferulic acid in addition to saponified lignins, i.e., lignin cores. The molar ratio of 4-hydroxycinnamic acid and ferulic acid depended on the nature of grass species. In contrast to the corresponding untreated grass lignins, the lignin cores produced vanillin and syringaldehyde and a negligible amount of 4-hydroxybenzaldehyde on nitrobenzene oxidation. The results indicate clearly that the 4-hydroxycinnamic acid moieties in grass lignins are the major sources for the formation of 4-hydroxybenzaldehyde on nitrobenzene oxidation, and the lignin cores of grass lignins are essentially angiosperm-type lignins, i.e., guaiacyl-syringyl lignins. Japanese researchers have also shown that the 4-hydroxycinnamic acid and ferulic acid peripheral units in grass lignins are mostly bonded to hydroxyl groups of the lignin cores at C-α and C-γ of the side chains in the form of ester linkages (41). Thus, Gibbs's classification of lignins is not only simpler, but also more rational in terms of the chemical characteristics of lignins.

4. Guaiacyl Lignins (Type-G Lignins)

Lignins of this type are comprised of (1) lignins in normal woods of the softwood species (Gymnospermae), except for those in the exceptional conifers (Gymnospermae); (2) lignins in the Cycadales (Gymnospermae) order; and (3) lignins in the Pteridophyta division, except for those in some genera, such as lignins in the Slaginsella genus, which are guaiacyl-syringyl-type lignins (42).

Guaiacyl lignins result in a positive Wiesner reaction, but negative Mäule and Cross-Beavan reactions. On nitrobenzene oxidation, the lignins produce vanillin as their major product and negligible amounts of 4-hydroxybenzaldehyde and syringalehyde. The lignins are polymeric substances consisting mostly of guaiacylpropane units produced in vivo by an enzyme-initiated dehydrogenative polymerization of coniferyl alcohol [2].

5. Guaiacyl-Syringyl Lignins (Type-G–S Lignins)

Lignins of this type include (1) lignins in normal and abnormal woods of the hardwood species (Dicotyledonous Angiospermae); (2) lignins in woody tissues of the Monocotyledonous Angiospermae other than grasses (Gramineae); (3) lignin cores in grass (Gramineae) lignins; (4) lignins in woods of the exceptional conifers (Gymnospermae) genera; (5) lignins in woody tissues of the exceptional Pteriophyta genera.

Guaiacyl-syringyl lignins result in positive Wiesner, Mäule, and Cross-Beavan reactions. On nitrobenzene oxidation, the lignins produce vanillin and syringaldehyde as the major products, in addition to a negligible amount of 4-hydroxybenzaldehyde. The lignins are polymeric substances consisting mostly of guaiacyl-and syringylpropane units produced in vivo by an enzyme-initiated dehydrogenative polymerization of coniferyl and sinapyl alcohols [2 and 3]. The molar ratio of guaiacylpropane units to syringylpropane units in a lignin of this type depends on the nature of the plant species.

6. 4-Hydroxyphenyl-Guaiaacyl-Syringyl Lignins (Type-H-G-S Lignins)

Lignins of this type occur in the woody tissue of grasses (Graminease; Mono-cotyledonous angiospermae). The lignins result in positive Wiesener, Mäule, and Cross-Beavan reactions. On nitrobenzene oxidation, the lignins produce 4-hydroxybenzaldehyde, vanillin, and syringaldehyde as their major products. The lignins are composed of guaiacyl-syringyl-type lignin cores and 4-hydroxycinnamic-acid-type peripheral groups. The peripheral groups consist of 4-hydroxycinnamic acid and ferulic acid units. In general, the former is predominant over the latter. The peripheral groups are bonded to hydroxyl groups in the lignin core at C-α and C-γ of the side chains mostly in the form of esters and to a lesser extent in the form of α-aryl ether linkages. The molar ratio of 4-hydroxycinnamic acid units, guaiacylpropane units including ferulic acid units, and syrinylpropane units in a lignin of this type depends on the plant species (39–41).

7. 4-Hydroxyphenyl-Guaiacyl Lignins (Type-H–G Lignins)

Lignins of this type are found in abnormal woods, e.g., compression wood, of the softwood species (Gymnospermae), except for those catergorized as exceptional conifers. The lignins result in positive Wiesner reactions, but negative Mäule and Cross-Beavan reactions. On nitrobenzene oxidation, the lignins produce 4-hydroxybenzaldehyde and vanillin as their major products and a negligible amount of syringyaldehyde. The lignins are polymeric substances consisting mostly of 4-hydroxyphenyl- and guaiacylpropane units produced in vivo by an enzyme-initiated dehydrogenative polymerization of 4-hydroxycinnamyl and coniferyl alcohols [1 and 2].

The formation of lignins of this type in compression woods is traceable to changes in the physiology of living trees belonging to the softwood species (Gymnospermae) in response to physical stresses and/or injuries against the trees. In order to protect themselves against such stresses and/or injuries, the trees stimulate the biosynthesis of lignin in situ, thereby causing a deficiency in the function of the O-methyl transferase system in the biosynthesis of coniferyl alcohol [2]. This, in turn, results in the biosynthesis of 4-hydroxycinnamyl alcohol [1], in addition to coniferyl alcohol [2], leading to the formation of 4-hydroxyl-guaiacyl-type lignins.

C. Distribution of Lignins in Woody Tissues

The lignin content of woody plants varies considerably depending not only on the nature of the plant but also the woody tissue in the same plant. Generally, normal woods of the soft- and hardwood species (Gymnospermae and Dicotyledonous Angiospermae) in the temperature zone contain 25–33 and 17–25%, respectively, of lignin/oven-dried wood as determined by the Klason method (35–37).

1. Distribution of Lignins in Trees

a. Softwood Species

(1). Normal Wood Tissues.   Except for exceptional conifers, lignins in normal woods of the softwood species (Gymnospermae) are of the guaiacyl type. In the same horizontal cross section of a normal tree trunk, the lignin content was found to be higher in heartwood than in sapwood, i.e., the lignin content decreases from the pith outward in the radial direction for Himalaya cedar (*Cedrus deodora*) and Himalaya cypress (*Cupressus torulosa*) (43) as shown in Table 1. However, this cannot be generalized for all softwood species. The reverse was found to be true for Japanese red pine (*Pinus densifora*) (44) and Insignis pine (*Pinus radiate*) (45). The lignin contents of both sap- and heartwoods were reported to decrease gradually in the vertical direction upward for Himalaya cedar and Himalaya cypress (43) whereas the reverse was found to be true for Japanese red pine (44). The lignin content of earlywood is usually higher than that of laterwood in the same horizontal cross section for a softwood species.

(2). Compression Wood Tissues.   Compression wood develops in the lower parts of leaning tree stems in living trees of the softwood species. The amount of compression wood grows with an increase in the deviation of the tree stem from a vertical direction. As shown in Fig. 5, the part of wood opposite the compression wood zone is called "opposite wood," whereas the part of wood between these two zones is termed "side wood." Except for a narrow area around knots, straight tree stems in the vertical direction are free of compression wood; hence, the stems consist of normal wood tissues if no injury has ever occurred to the stems.

Tissues in compression woods of the softwood species differ considerably from those of corresponding normal woods in terms of morphological, physical, and chemical characteristics; the same is true for the nature of lignins in these tissues

Table 1   Distribution of Lignins in Tree Stem of Some Softwoods and Hardwoods

Tree Species	Height[a]	Sapwood (lignin %/wood)	Heartwood (lignin %/wood) Outer layer	Inner layer
Softwood species (Gymnospermae)				
Himalaya Ceder	1	30.7	32.7	35.6
(*Cedrus deodora*)	7	28.5	31.6	32.1
	11	28.1	30.2	32.1
Himalaya cypress	1	34.9	35.9	35.3
(*Cupressus totulosa*)	6	28.2	31.8	32.6
	10	26.9	29.8	—
Hardwood species (Dicotyledonous angiospermae)				
Sal		29.5	29.3	—
(*Shorea robusta*)				
Siam teak		39.1	34.2	—
(*Tectonia grandis*)				
Sisoo		25.4	30.2	—
(*Dalbergia sisoo*)				

[a]Sample number. Wood samples were collected a distance of every 3 m in height from the bottom of a tree stem.
Source: Ref. 430.

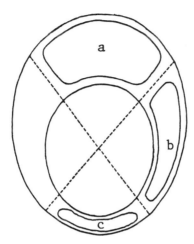

Figure 5   Location of zone in tree stem containing reaction wood. A, compression wood in sapwood of softwood species: (b) side wood (normal wood), (c) compression wood. B, tension wood in sapwood of hardwood species: (a) tension wood, (c) opposition wood.

(46). In contrast, tissues in opposite and side woods are similar to those in corresponding normal woods in the aforementioned characteristics.

Compression woods of the softwood species contain considerably higher amounts of lignin than the corresponding normal woods (47–53). Timell and co-workers (49, 50) found that the lignin contents of pronounced compression woods in balsam fir (*Abies balsamea*) and Alaska larch (Larix *laricina)* were 34% (40.1% vs. 27%) and 43% (38% vs. 27%), respectively, higher than those of the corresponding normal woods. Sakakibara and co-workers (52, 53) obtained similar results for compression woods of Sachalin fir (*Abies sachalinensis* Mast.) and Japanese larch (*Larix letolepis* Gord). Furthermore, compression wood lignins were found to have an appreciably lower methoxyl content than the corresponding normal wood lignins (47, 48, 51–53).

In contrast to the corresponding normal wood lignins, the compression wood lignins produce appreciable amounts of 4-hydroxybenzaldehyde in addition to vanillin, the major product, on nitrobenzene oxidation (48, 51, 53, 54). Bland (48) found that a milled wood lignin (MWL) preparation from the compression wood of Insignis pine (*Pinus radiate)* produced 3.3% 4-hydroxybenzaldehyde and 13.3% vanillin/MWL on nitrobenzene oxidation, with 4-hydroxybenzaldehyde/vanillin (H/V) molar ratio of 0.3 and a total phenolic aldehyde yield of 16.6%/MWL. Latif (51) obtained a similar result from a MWL from the compression wood of Douglas fir (*Pseudotsuga menziesii)* ,with a H/V molar ratio as high as 0.47. Sakakibara and co-workers (53, 54) found a H/V molar ratio of about 0.2 for lignins in the compression woods of Sachalin fir (*A.* sacha*linensis*) and Japanese larch (*L. leptolepis*). Thus, the results of nitrobenzene oxidation indicate that lignins in compression woods of the softwood species contain significant amounts of uncondensed 4-hydroxyphenylpropane units in addition to guaiacylpropane units, as compared to lignins in the corresponding normal woods. Furthermore, Yasuda and Sakakibara (55, 56) found that lignins in the compression wood of Japanese larch (*L. leptolepis*) possessed higher amounts of condensed phenylpropane units than lignins in the normal wood. Consequently, lignins in compression woods are abnormal lignins and are 4-hydroxyphenyl-guaiacyl-type lignins. In addition to higher lignin contents, Timell and co-workers (49, 50) further found that the compression woods of balsam fir (*A. balsamea*) and Alaska larch (*L. laricina*) differed considerably from corresponding normal woods in the composition, quantity, and nature of carbohydrates. Evidently, compression woods in the softwood species are formed in the lower parts of the leaning stems of living trees in response to physical stress placed on the tree stems by changes in the physiology of the trees for the biosynthesis of wood components in order to impart mechanical strength to the tree stems. Moreover, the O-methyltransferase system in the biosynthesis of lignin precursors seems to be affected significantly.

b. Hardwood Species

(1) Normal Wood Tissues.   Lignins in normal woods of the hardwood species (Dicotyledonous Angiospermae) are of the guaiacyl-syringyl type. The lignin content in normal woods of the hardwood species is generally much lower than that in normal woods of the softwood species (Gymnospermae). The lignin content of sapwood was found to be equal to or slightly higher than that of heartwood in the same cross section of a normal tree trunk for white birch (*Betula payyrifera* Marsh.) (57), Sal (*Shorea robusta*), and Siam teak (*Tectonia grandis*) (43). As shown in Table 1, however, the reverse trend was found for Sisoo (*Dalbergia sisso*) (43). No data are available for the distribution of lignin within a tree in the vertical direction for the hardwood species. As in the softwood species, the lignin content of earlywood is generally higher than that of laterwood in the same cross section of a normal tree stem for the hardwood species.

(2) Tension Wood Tissues.   Tension wood develops in the upper part of branches and leaning stems in living trees of the hardwood species. Analogous to compression wood, the part of wood opposite the tension wood zone is called "opposite wood," and the part of wood between these two zones is called "side wood."

In contrast to compression wood of the softwood species, tension woods of the hardwood species contain considerably lower amounts of lignin than the corresponding normal woods (52, 58–60). Wardrop and Dadwell (58) found that the lignin content of pronounced tension woods in mountain grey gum (*Eucalyptus gonloyx*), nitens gum (*Eucalyptus nitens*), and mountain ash (*Eucalyptus regnans*)was 60% (10% vs. 25%), 29% (15% vs. 21%), and 26% (16% vs. 22), respectively, lower than those of the corresponding normal woods. Morohoshi and Sakakibara (52) found that a tension wood of Yachidamo (Japanese ash; *Fraxinus mandschurica* var. japonica) contained 12% less lignin than the corresponding opposite wood. Morohosi and Sakakibara (52) further found that the lignin in the tension wood of Yachidamo (*F. mandschurica*) produced 8.0% vanillin, 12.8% syringaldehyde, and 1.9% syringic acid/Klason lignin on nitrobenzene oxidation, with a total phenolic aldehyde yield of 23.1%/Klason lignin and a syringaldehyde/vanillin (S/V) molar ratio of about 1.6, whereby the syringaldehyde component includes the amounts of syringic acid corrected as syringaldehyde. In contrast, the lignin in, the opposite wood of the same tree section yielded 7.8% vanillin, 9.8% syringaldehyde, and 1.9% syringic acid/Klason lignin on nitrobenzene oxidation, with a total phenolic aldehyde yield of 19.8%/Klason lignin and a S/V molar ratio of about 1.3. Similar results were obtained for tension woods of birch and madrona (61). Thus, the results of nitrobenzene oxidation indicate that lignins in tension woods of the hardwood species are of the guaiacyl-syringyl type and contain more guaiacylpropane units/100 C_9 units, in particular the condensed guaiacylpropane units, than lignins in the corresponding normal and opposition woods. Analogous to the formation of compression woods, tension woods in the hardwood species are formed in the upper parts of branches and leaning stems in response to physical stress placed on the branches and stems by anges in the physiology of the trees for the synthesis of wood components in order to impart mechanical strength to the branches and stems. Here again, the O-methyl-transferase system in the biosynthesis of lignin precursors seems to be significantly affected.

2. Distribution of Lignins in Wood Cells

The distribution of lignins in individual cells in woods of both the soft- and hardwood species can be observed by electronmicroscopy using thin sections of wood with prior removal of carbohydrates in the wood sections by treatment with either 80% hydrofluoric acid or periodate (62) or by fungal degradation (63). Improved procedures used thin sections of woods treated with either 4-(acetoxymercuri)aniline (62) or potassium permanganate (64) without the prior removal of carbohydrates. Although electron micrographs obtained by these methods reveal minute details of the cell wall structure, the methods cannot be used for the quantitative estimation of lignin distribution in wood cells. One of the best methods available for the latter purpose is scanning thin sections of wood by ultraviolet microscopy (65–71). The other method is the determination of the bromine concentration in brominated ultrathin sections of woods by scanning electron microscopy, coupled with energy-dispersive X-ray analysis (SEM-EDXA) (72–74). The ultraviolet microscopic method was first introduced by Lange in 1954 (65) and has been improved by Goring and co-workers (66–71), reducing the thickness of the wood section to 0.5 µm in order to avoid errors caused by nonparallel illumination.

a. Softwood Species   Ritter (75) was one of the first to study the distribution of lignin in softwood tissues. He concluded in 1925 that 75% of lignins in softwoods occurred in the middle lamella, and the remaining 25% were present in the secondary wall. However, Bailey (76) isolated the middle lamella of Douglas fir (*Pseudotsuga menziesii)* and by chemical analysis found that the tissue contained about 71% of the total lignin in the wood. Lange (65) obtained a similar result for the wood tissues of Norway spruce (*Picea excelsa*) by use of ultraviolet microscopy; the weight concentrations.of lignin in the secondary cell wall and the compounded middle lamella were estimated to be approximately 16 and 73%, respectively. Fergus and co-workers (67) determined the distribution of lignin in earlywood and laterwood tracheid tissues of black spruce (*Picea mariana* Mill.) by measuring the ultraviolet light transmitted through ultra thin sections (0.5 μm) of the wood at λ1280 nm. As shown in Table 2, the weight concentrations of lignin were estimated to be 22.5, 49.7, and 84.8% for the secondary cell wall, compounded middle lamella, and cell corner of the middle lamella, in the earlywood tracheids, respectively. In addition, the weight concentrations of lignin in the corresponding layers of the laterwood tracheids were found to be 22.2, 60, and 100%, respectively. Because of its relatively larger tissue volume, the secondary cell wall contains about 72% of the total lignin in the earlywood tracheids and about 82% of the total lignin in the latewood tracheids, leaving only about 28 and 18% of the total lignin in the compounded middle lamella and cell corner of the middle lamella for early and laterwoods, respectively. Similar results were obtained for the corresponding wood tissues of Douglas fir (*P. menziesii)* by Wood and Goring (68) using ultraviolet microscopy, and by Saka and Thomas (73) using the SEM-EDXA technique. Thus, the average concentration of lignin in the middle lamella region determined by modern techniques is in rather good agreement with the value of about 71% obtained for the same tissues by Bailey (76) using the classic method.

Saka and Thomas (74) also determined the distribution of lignin in the tracheid tissue of loblolly pine (*Pinus treda*) by the SEM-EDXA technique. The results were similar to those found for black spruce (*P. mariana*) and Douglas fir (*P. menziesii*). The weight concentration of lignin in the secondary cell wall was found to be about 24% for earlywood. This is in good agreement with the value of 22–24% estimated earlier for the same tissue by Stamm and Sanders (77) on the basis of the density of the wood components.

b. Hardwood Species   Stone (78) suggested in 1955 that in aspen (*Populus tremuloides*), the nature of lignin in different morphological regions of the wood tissues varied in terms of chemical structure on the basis of the following observation: When lignins isolated from the black liquor of the neutral sulfite cook of aspen were degraded by nitrobenzene oxidation, the syringaldehyde/vanillin (S/V) molar ratio increased when the cooking time was increased. A similar observation was made later by Marton (79).

Fifteen years later, Fergus and Goring (70), determined the distribution of lignin

Table 2 Distribution of Lignins in Tracheid of Black Spruce (*Picea mariana* Mill)

Wood	Morphological differentiation[a]	Tissue volume afraction (%)	Lignin , (% of total)	Lignin concentration (g/g)
Earlywood	S	87.4	72.1	0.225
	ML (r, t)	8.7	15.8	0.497
	ML (cc)	3.9	12.1	0.848
Latewood	S	93.7	81.7	0.222
	ML (r, t)	4.1	9.7	0.60
	ML (cc)	2.2	8.6	1.00

[a]S = Secondary wall; ML = middle lammella; r = radial; t = tangential; cc = cell corner.
Source: Ref. 67.

in the xylem tissue of white birch (*Betula papyrifera* Marsh.) by ultraviolet microscopy. As shown in Table 3, the complexity of the lignin distribution in the xylem tissue is compounded not only by the presence of two predominate vertical elements, i.e., fibers and vessels, but also by the fact that hardwood lignins consist mostly of guaiacyl- and syringylpropane units. The weight concentration of lignin in the fiber secondary cell wall was found to be in the range of 16–19%. Because of the relatively larger tissue volume in this region, the fiber secondary cell wall contained about 60% of the total lignin in the xylem tissue. The combined fiber, vessel, and ray cell secondary walls contained over 81% of the total lignin, leaving less than 19% of the total lignin for the primary cell walls and middle lamellae. On the basis of the data given in Table 3, the distributions of lignin in the wood tissues was estimated to be about 77, 11, and 12% based on the Klason lignin content of the wood for the fiber vessel, and ray cell, respectively. The authors (69, 70) further elucidated the nature of lignins in the xylem tissues of white birch on the basis of ultraviolet spectra obtained by scanning ultrathin wood sections (0.5 μm) within the 240–320 nm wavelength range at which lignins of various types are expected to absorb. The lignins in the vessel secondary cell wall and middle lamella gave ultraviolet spectra with a maximum at l 279–280 nm, indicating that the lignins were composed predominantly of guaiacyl-type lignins. The lignins in the middle lamella around fibers and ray cells produced ultraviolet spectra with a maximum of 275–276 nm, indicating that the lignins were mostly of the gualacyl-syringyl type. Finally, the lignins in the fibers and ray parenchyma secondary cell walls afforded ultraviolet spectra with a flat maximum at around 270 nm, suggesting that the lignins were mostly of the syringyl type.

D.   Heterogeneous Nature of Lignins in Wood Tissues

As discussed in the previous sections, lignin is unevenly distributed in the wood

Table 3 Distribution of Lignins in Xylem of White Birch (Betula papyrifera Marsh)

Element	Morphological differentiation[a]	Type of lignin	Tissue volume fraction (%)	Lignin (% of total)	Lignin concentration (g/g)
Fiber	S	syringyl	73.4	59.9	0.16[b]–0.19[c]
	ML(r, t)	gualacyl-syringyl (1:1)	5.2	8.9	0.34–0.40
	ML(cc)	guaiacyl-syringyl (1:1)	2.4	8.8	0.72–0.85
Vessel	S	guaiacyl	8.2	9.4	0.22–0.27
	ML	guaiacyl	0.8	1.5	0.35–0.42
Ray cell	S	syringyl	10.0	11.4	0.22–0.27

[a]S = Secondary wall; ML = middle lammella; r = radial; t = tangential; cc = cell corner.
[b]Calculated using xylem lignin content of 0.199 g/g.
[c]Calculated using xylem lignin content of 0.231 g/g.
Source: Ref. 70.

tissues of the same tree. In addition, lignins in different morphological regions of the same tree differ considerably in terms of chemical characteristics. The latter is particularly obvious in the hardwood species.

1. Softwood Species

Matsukura and Sakakibara (80) fractionated the finely ground wood of Ezomatsu (Japanese sprue; *Picea jezoensis* Carr.) into milled wood lignin (MWL), lignin-carbohydrate complex (LCC), and wood residue (WR) fraction according to the Björkman procedure with yields of about 14, 14, and 68% of the original wood, respectively. These fractions contained about 100, 21, and 14% lignin, respectively, as determined by the Klason procedure. The original wood had a Klason lignin content of 29.3%. Thus, lignins in the MWL, LCC, and WR fractions account for approximately 48, 10, and 32% respectively, of the total lignin in the original wood. The lignins in these fractions were further characterized by nitrobenzene oxidation, potassium permanganate oxidation after O-methylation and acidolysis. The results showed that the lignin in the WR fraction contained more condensed guaiacylpropane units than in the MWL and LCC fractions, and the lignins in the LCC fraction had the lowest degree of condensation.

Recently, Whitting and co-workers (81) isolated wood fractions rich in either the compounded middle lamella or the secondary cell wall tissues from black spruce (*Picea mariana*) wood by the differential sedimentation of the finely ground wood. Whitting and Goring (82) further prepared MWL's from these two wood fractions, as well as from the wood of black spruce. The two wood fractions were individually extracted continuously with dioxane-water (100:4 v/v) for 48 hr at room temperature. After 24 hr of extraction, the lignins extracted represented 39.2% for the lignins in the secondary cell wall fraction and only 10.4% for those in the compounded middle lamella fraction. The amount of lignins extracted from the secondary cell wall fraction was 3.8 times greater than that for the compounded middle lamella fraction. The ratio remained constant thereafter, with extractions carried out continuously for two weeks. As discussed in the previous section, about 77 and 23% of the total lignin in black spruce wood occur in the secondary cell wall and compounded middles lamella, respectively (67). Consequently, the contribution of the lignins in the aforementioned wood tissues to the MWL should be about 30.2 and 2.4%, respectively, of the total lignin in black spruce wood, with a total MWL yield of about 32.6%/ total lignin for the 24 hr of extraction. The yield of MWL prepared from black spruce wood under the same conditions was found to be 32.8%/Klason lignin. Thus, about 93% of the MWL of black spruce, should originate from lignins in the secondary cell wall of the tracheid; the same is perhaps true for the MWL of other softwood species. This together with the results from Ezomatsu (*P. jezoensis*) discussed previously (80) further suggest that in woods of the Picea (spruce) genus, and perhaps in those of other softwood species, lignins in the secondary cell walls contain less aromatic ring-condensed guaiacylpropane units than those in the compounded middle lamellae, although further investigations are required to confirm this postulation.

2. Hardwood Species

Lee and co-workers (83) prepared five sequential MWL fractions from Japanese birch (*Betula maximowicziana* Reg.) in a stepwise fashion according to the Björkman procedure (84). The ball-milling time intervals between two sequential steps were 24, 24, 48, and 72 hr from the initial step. For the 1st–5th MWL preparations, the yields were 3.7, 3.2, 6.0, 6.8, and 8.3% respectively, based on the Klason lignin content of the original wood. The lignin content of these MWL's was found to be about 96.7, 93.0, 91.0, 90.8, and 89.3%, respectively. Thus, the combined total lignin in these MWL's accounts for about 26% of the total lignin in the wood. Elemental analysis of the MWL's showed that their methoxyl contents/C_9 unit increased with an increase in the sequential number. Moreover, on nitrobenzene oxidation, the 1st–5th MWL's gave a syringaldehyde/vanillin (S/V) molar ratio of 1.04, 1.24, 1.35, 1.49, and 1.59, respectively, i.e., the S/V molar ratio also increased with an increase in the sequential number. Thus, the aromatic ring uncondensed syringylpropane units in the MWL's increased with an increase in the sequential number with respect to the aromatic ring uncondensed guaiacylpropane units in the corresponding MWL, indicating that the MWL's, differ considerably in terms of chemical structure. However, further investigations are required to elucidate the chemical characteristics of the MWL's as well as their origin in terms of morphology.

Recently, Lapierre and co-workers (85,86) prepared a MWL from balsam poplar (*Populus trichocarpa*-cv Fridizi Pauley) wood by the Björkman procedure (84) with a yield of about 33%/Klason lignin (18%/wood). The wood residue obtained after removal of the MWL was again ball-milled and then treated with cellulase according to the Pew procedure (87) to obtain a cellulolytic enzyme lignin (CEL) with a yield of about 30%/Klason lignin. Thus, about one-third of the total lignin in the wood remained in the final wood residue. On nitrobenzene oxidation, the MWL produced vanillin and syringaldehyde with a total phenolic aldehyde yield of about 35%/Klason lignin and a S/V molar ratio of about 1.2, whereas the CEL gave a total phenolic aldehyde yield of about 32%/Klason lignin and a S/V molar ratio of about 2, in both cases after correction of the carbohydrate content. In addition, the nitrobenzene oxidation mixtures also contained 4-hydroxybenzoic acid in yields of about 5.4%/MWL and 4.4%/CEL, respectively. Since the amount of syringaldehyde in both oxidation mixtures was almost the same, the MWL should contain substantially more uncondensed guaiacylpropane units than the CEL relative to the uncondensed syringylpropane units. This was shown to be true by subsequent ^{13}C NMR spectroscopic studies of the MWL and CEL (86). Here again, further investigations are required to elucidate the nature of the MWL and CEL in terms of chemical structure, as well as their origin.

III. ISOLATION OF LIGNINS

Lignins in plant tissues are not extractable by organic solvents. The isolation of lignins from plant tissues by any procedure ultimately results in changes in the chemical structure of lignins. Thus, except for technical lignins, it is of primary importance to use a protolignin, i.e., a lignin occurring in extractive-free plant tissues, or a lignin preparation isolated from plant tissues with minimum changes in the chemical structure for characterization of the lignin.

A. Acid Lignins

Before Björkman (84) introduced the milled wood lignin procedure for the isolation of lignins from woody tissues of the soft and hardwood species (*Gymnospermae and Dicotyledonous Angiospermae*) in 1956, lignin preparations used for the structural studies of lignins were prepared from extractive-free woody plant tissues by removal of carbohydrates through acid-catalyzed hydrolysis. For example, the treatment of an extractive-free wood meal with 72% sulfuric acid and superconcentrated hydrochloric acid (about 45%) results in complete removal of carbohydrates and formation of water-insoluble "Klason lignin" (H_2SO_4-lignin) (88) and "Willstätter lignin" (HCl-lignin) (89) in almost quantitative yield, respectively, although hardwoods produce about 3% of acid-soluble (water-soluble) lignins in the Klason procedure. Moreover, it has been shown that lignins in woods (i.e., protolignins) undergo not only hydrolysis but also rather extensive acid-catalyzed condensation during the process of isolation by the acids in these or similar procedures (90–92). Thus, lignin preparations of this type are not appropriate for use in the structural studies of lignins because of the drastic changes that occur in the structure. Nevertheless, modified Klason procedure (93, 94) is used for the determination of lignin content in woody plant tissues. The procedure for the determination of Klason lignin and the acid-soluble lignin contents of woods is described in detail in TAPPI Standard Method T-222 om-83 and TAPPI Useful Method UM-250, respectively.

B. Milled Wood Lignins

Lignins can be isolated from finely ground extractive-free wood meal of the soft- and hardwood species by extraction with neutral solvents. According to Björkman (84), when extractive-free wood meal of a wood species is ground for 48 hr or more in a vibratory ball mill under nitrogen atmosphere either with a nonswelling embedding liquid, such as toluene, or dry, the wood tissues disintegrate, and about 30–50% of lignins in the wood become extractable with aqueous dioxane, usually 90–96% dioxane. The resulting "milled wood lignin" (MWL) or "Björkman lignin" is a cream-colored powder free of ash, and it usually contains a few percent of carbohydrates (84, 95). Almost carbohydrate-free MWL preparations have been prepared by modification of the original purification procedure (96, 97).

A MWL preparation of a wood is chemically similar but not identical to the corresponding lignin occurring in the wood (98). During the grinding of wood meal by a vibratory ball mill, lignins in woods undergo fragmentation by mechanical force, the mechanism of which is not well understood. However, the fragmentation of lignins probably involves the cleavage of aryl ether bonds in the lignins as evidenced by the relatively higher phenolic hydroxyl content of spruce (*Picea excelsa*) and sweetgum (*Liquidambar styraciflua*) MWL´s as compared to the corresponding lignins´occurring in woods, approximately 20–30 vs. less than 10 phenolic hydroxyl groups/100 phenylpropane (C_9) units (98–101). The α-carbonyl content of sweetgum MWL was found to increase with increasing ball-milling time (100), indicating that changes in the structure of lignins by oxidation also occur during the ball-milling of wood meal, in particular during the dry ball-milling, in addition to the fragmentation.

In general, MWL preparations from woods have weight-average molecular weight (Mw) in the range of 11,000–16,000 (84, 87), and relatively higher phenolic hydroxyl and α-carbonyl contents as compared to the corresponding lignins occurring in woods (98–101). Evidently, a MWL from a wood is a part of lignin fragments produced during the ball-milling of the wood meal, which is more susceptible to extraction by organic solvents than the other part remaining in the wood residue under the condition of preparation. For both soft- and hardwood species, these two lignin parts differ considerably in chemical structures and properties, indicating the heterogeneous nature of lignins occurring in woods in terms of chemical structure and distribution in wood tissues as discussed in the previous sections. Thus, a MWL preparation from a wood species is not identical to the lignin occurring in the wood. However, it is the best lignin preparation known so far and has been widely used for the structural studies of lignins.

C. Cellulolytic Enzyme Lignins

Pew (87) showed that most carbohydrates in woods can be removed by a glucosidase treatment of finely ground wood meal by dry grinding of the wood meal in a vibratory ball mill for about 5–8 hr. The resulting wood residue is then purified to give "cellulolytic enzyme lignin" (CEL), which usually consists of lignin and about 12–14% carbohydrates. The characteristics of a CEL preparation are, in general, similar to those of the corresponding MWL preparation obtained from the same wood (100). However, CEL preparations have not been used as widely as MWL preparations in the structural studies of lignins because of the rather higher carbohydrate content and the tedious work involved in their preparation.

IV. REACTIONS OF LIGNINS

Reactions of lignins, in particular degradation reactions, have been extensively investigated since the invention of chemical pulping processes of woods at the end of the 19th century. The major driving force for promoting research has been to clarify the nature of lignins occurring in woods in terms of the chemical structure and ultrastructure of the wood in order to develop economically viable chemically pulping processes and to facilitate the utilization of technical lignins, the major byproducts of the pulping processes.

Classical organic analysis of woods and lignin preparations, as well as model compound experiments, led to the conclusion already by 1940 that lignins in woods are comprised of polymeric substances consisting of phenylpropane skeletons produced in vivo from monomeric precursors, 4-hydroxycinnamyl alcohol [1], coniferyl alcohol [2], and sinapyl alcohol [3], by an enzyme-initiated dehydrogenative polymerization. However, this hypothesis was not accepted unanimously at that time, particularly the aromatic nature of lignins in woods. Some investigators believed as late as the 1950s that the aromatic moieties in lignin preparations were produced from carbohydrates in woods during the process of isolation, and those in technical lignins during the chemical pulping processes. This controversy was finally solved in 1954 by Lange (65) who obtained ultraviolet spectra characteristic of 4-alkylphenols from thin sections of woods by ultraviolet microscopy, scanning the thin wood sections between λ 240–320 nm, and later by other new chemical evidence. In this section, the discussion will focus on the degradative reactions of lignins that led to the elucidation of their structures.

A. Oxidative Degradations

1. Potassium Permanganate Oxidation

As early as 1914, Unger (102) under the direction of Willstätter showed the aromatic nature of lignins in spruce (*Picea excelsa*) wood by the potassium permanganate oxidation of an extractive-free wood meal exhaustedly methylated with ethereal diazomethane. Only veratric acid [19] was isolated from the oxidation mixture in a yield of 4%/Willstätter lignin. In hindsight, this result is significant not only because it shows the aromatic nature of spruce lignin, but also because it indicates the rather low phenolic hydroxyl content of lignin, probably less than 0.1 mole phenolic hydroxyl/phenylpropane (C_9) unit. The latter was overlooked at the time because the nature of lignin was not well understood.

In 1936, Freudenberg and co-workers (103, 104) applied the same procedure with major modifications for the investigation of the nature of aromatic moieties in lignins. A Cuoxam lignin preparation from spruce wood was treated with 70% potassium hydroxide at about 170°C to effect hydrolysis of alkyl aryl ether bonds in the lignin. The phenolic hydroxyl groups thus produced were methylated with dimethylsulfate in a alkaline solution in order to protect the phenolic moieties from decomposition by subsequent oxidation. The methylated lignin was then degraded by oxidation with a potassium permanganate solution at a pH around 7 and at about 90°C (103, 104). Veratric acid [19], isohemipinic acid [21], and dehydrodiveratric acid [26] were isolated from the oxidation mixture in yields of about 14, 4, and 3%/Cuoxam lignin, respectively (see Fig. 6 for structures). In a comparative study (105), a Cuoxam lignin preparation from beech (*Fagus silvatica*) wood was degraded in the same fashion to isolate acids 19, 21, and 3,4,5-trimethoxybenzoic acid [20] in yields of about 2.5, 1, and 4%/Cuoxam lignin, respectively. Since the preparation of a Cuoxam lignin from wood involved alternating the treatment of the extract-free wood meal with boiling 1% sulfuric acid and extraction with Schweizer´s reagent [$Cu(NH_3)_4(OH)_2$], changes in the structure of lignin occurred to some extent during the process of isolating the Cuoxam lignin (106). Parallel investigations were, therefore, conducted on the potassium oxidation of lignin occurring in spruce (*Picea excelsa*) wood (107) and in beech (*Fagus silvatica*) wood (105). Extract-free spruce wood meal was methylated successively with ethereal diazomethane and dimethylsulfate in alkaline solution at room temperature. The fully methylated wood meal (CH_3O content, 31%) was treated with 70% potassium hydroxide solution at 170°C, then remethylated with dimethylsulfate in an alkaline solution. On potassium permanganate oxidation, the remethylated wood meal produced acids 19, 21, and 26 in yields of about 20, 6, and 4%/Klason lignin (28%/wood),

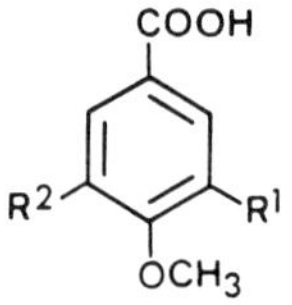

18, $R^1 = R^2 = H$

19, $R^1 = H$; $R^2 = OCH_3$

20, $R^1 = R^2 = OCH_3$

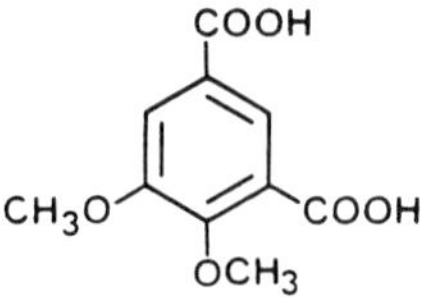

21

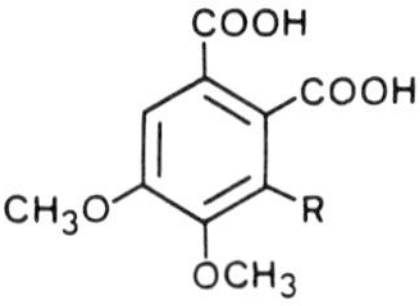

22, R = H

23, $R = OCH_3$

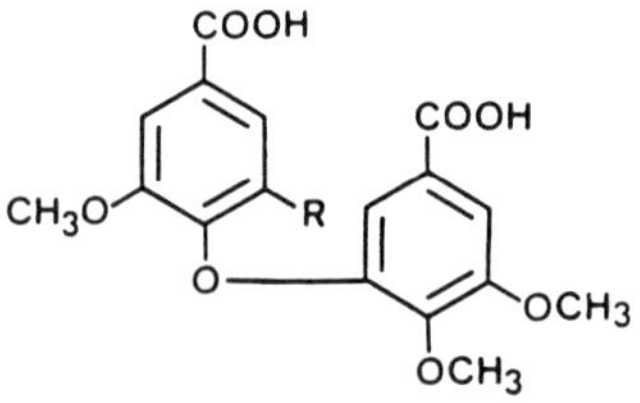

24, R = H

25, $R = OCH_3$

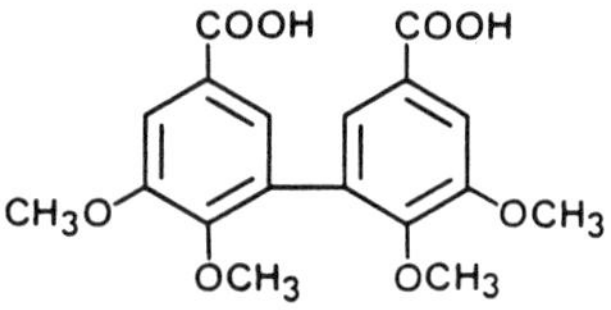

26

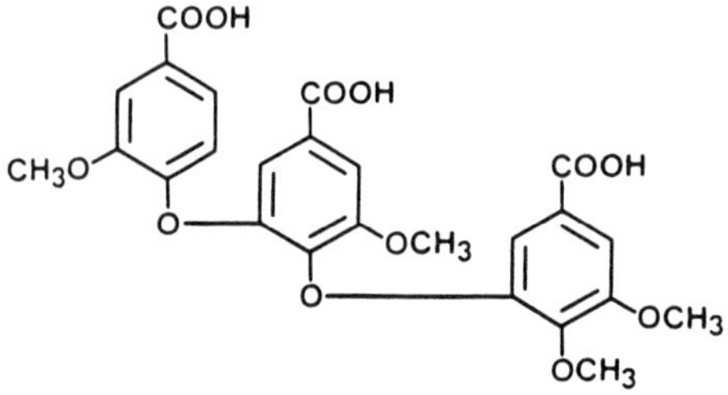

27

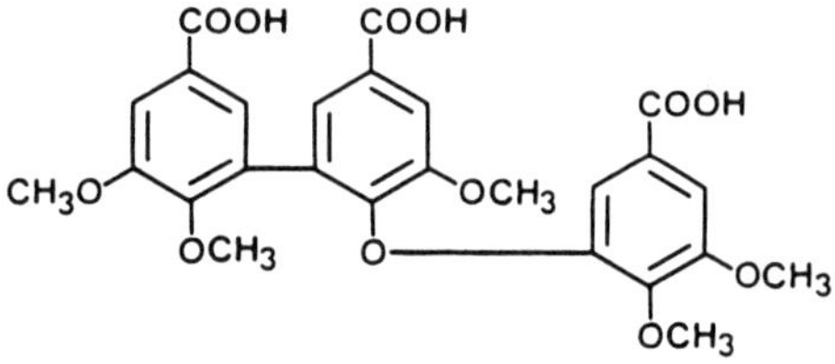

28

Figure 6 Major oxidation products from potassium permanganate oxidation of methylated lignins.

respectively. A small amount of acid 19 was also isolated from the oxidation mixture (107). A fully methylated extract-free beech wood meal was degraded in the same fashion to produce acids 19, 20 and 21 in yields of about 5, 7.5, and 1.5%/ Klason lignin (20.5%/wood), respectively (105).

The aforementioned results revealed that lignins occurring in spruce wood have the following characteristics in terms of chemical structure: (1) The lignins consist mostly of guaiacylpropane units that are bonded to each other predominantly by alkyl aryl ether linkages between carbons in the side chain and C-4 of guaiacyl moieties to form alkyl aryl ether substructures; (2) the lignins also contain appreciable amounts of substructures in which C-5 of guaiacyl moieties are bonded to a carbon atom in the side chain and to C-5 of another guaiacylpropane unit by C—C bonds to form phenylcoumaran substructures similar to dehydrodiisoeugenol [7] and biphenyl substructures, respectively; (3) on the basis of Unger´s result, the lignins contain rather low phenolic guaiacylpropane units, probably less than 10 units per 100 phenylpropane (C_9) units; and finally, (4) alkyl aryl ether substructures are rather susceptible to base-catalyzed. The results also show that beech lignins consist mostly of both guaiacyl- and syringylpropane units and have, in part, characteristics similar to spruce lignins in terms of chemical structure. Thus, the potassium permanganate oxidation of methylated lignins provides not only the first direct evidence for the aromatic nature of lignins, but also for the difference between the nature of softwood and hardwood lignins. The disadvantages of the procedure are as follows: (1) the yields of the degradation products are rather low, and (2) the products do not reveal the chemical structure of side chains. Consequently, the procedure cannot be used directly for the quantitative determination of substructures in lignins. It must be mentioned that phenylcoumaran (β-5') and biphenyl (5-5') substructures are called "condensed units" because a guaiacylpropane unit in the substructures is linked at C-5 of guaiacyl groups to either an alkyl or aryl group of the other C_9 unit by a C—C bond.

In 1959, Freudenberg and co-workers (108–110) reinvestigated the constituents of the oxidation mixture obtained from the potassium permanganate oxidation of fully methylated spruce (*P. excelsa*) wood meal (CH_3O content, 25%) in view of the progress made in the field of organic analysis and lignin chemistry in the intervening years. In addition to acids 19, 20, 21, and 26, anisic acid [18], meta-hemipinic acid [22], 3,4,5-trimethoxyphthalic acid [23], 2,2',3-tri-methoxydiphenylether-4',5-dicarboxylic acid [24], and 11 minor aromatic acids were isolated from the oxidation mixture. Yields of the newly identified acids were less than 0.5%/Klason lignin (28%/wood). Acids 21 and 24 indicate that lignins occurring in spruce wood also contain a condensed substructure of the types 6-alkylguaiacylpropane (β-6') and 5-aroxyguaiacylpropane (4-O-5') units, minor substructures. Acid 22 was isolated earlier by Richtzenhain (111) from the potassium permanganate oxidation of methylated HCl-lignin prepared from Norway spruce (*Picea abies*) wood in a yield of 1.3%/HCl-lignin, in addition to acids 19 and 21. The higher yield of acid 22 obtained from the oxidation mixture of methylated HCl-lignin indicates that the acid originates, in part, from 6-alkylguaiacylpropane substructures produced during the preparation of HCl-lignin by an acid-catalyzed condensation or rearrangement of lignin substructures.

During the period 1967–1973, Miksche and co-workers (98, 112–118) improved the procedure for the potassium permanganate oxidation of methylated lignin, resulting in a substantial increase in the yield of degradation products. The most recent procedure (118, 119) involves the following steps: (1) pretreatment of the lignin preparation or extract-free plant tissue meal with 2N potassium hydroxide solution at 170°C for 2 hr in the presence of cupric oxide; (2) methylation of the resulting base-hydrolyzed and partially oxidized lignin with dimethylsulfate in alkaline solution at pH 12; (3) oxidation of the resulting methylated lignin with a mixture of potassium permanganate and sodium periodate in 0.5N sodium hydroxide solution at 85°C in the presence of t-butanol; and finally, (4) treatment of the resulting oxidation mixture with 5% hydrogen peroxide in alkaline solution at pH 9–11 and at 50°C to effect oxidation of aromatic α-keto-acids possibly present in the oxidation mixture to the corresponding aromatic acids. The aromatic acids in the final oxidation mixture are converted into the corresponding methyl ester by treatment of the mixture with ethereal diazomethane. The constituents of the methyl ester mixture are then analyzed quantitatively by gas chromatography (GC) and gas chromatography-mass spectrometry (GC-MS) after the removal of high-molecular-weight contaminants.

Using the new procedure, Miksche and co-workers (98, 112–118) identified a total of 40 acids from the potassium permanganate oxidation mixture of methylated milled wood (MWL) from wood of Norway spruce (*Picea abies*). These acids included the 19 aromatic acids isolated previously by Freudenberg and co-workers (103–110) and 20 new minor components such as trimeric acids 27 and 28, which were produced in a yield of less than 0.1%. As shown in Table 4, yields of the major components, acids 18, 19, 20, 21, 22, 24, and 26 in the form of the corresponding methyl esters, were determined to be about 0.7, 29.8, 0.5, 5.0, 1.1, 2.1, 6.0, and 2.1%/ MWL, respectively, each representing about a three- to four-fold increase as compared to the earlier results (108–110). In terms of yield in mole% per phenylpropane unit (C_9 unit; average C_9 unit weight for spruce MWL = 185), the yields of these major components were about 0.8, 28.1, 0.4, 3.6, 0.8, 1.0, and 2.9 mole%/C_9 unit, respectively. Thus, the total yield of the major components accounts for approximately 38 mole%/C_9 unit. This value is relatively high as compared to those from degradation products obtained by other methods. It must be noted that one mole of a dimeric degradation product of lignins corresponds to two C_9 units. The total yield of minor components is usually less than 2 mole%/C_9 unit, each less than 0.01%/MWL, which is negligible.

Results from the potassium permanganate oxidation of methylated lignin have the following significance: (1) Oxidation products are produced from C_9 units with free phenolic hydroxyl groups either originally present in lignin or produced as a result of pretreatments such as cupric oxide oxidation or hydrolysis by 70% potassium hydroxide; (2) carboxyl groups in an aromatic ring of an oxidation product are the site of aliphatic substituents bonded to the aromatic ring by C—C

Table 4 Yield of Major Aromatic Acids from Woods and Milled Wood Lignins (MWL's) of Spruce and Birch on Potassium Permanganate-Sodium Periodate Oxidation

Wood species	Specimen	Yield, mole % per C_9 unit[a]									
		18	19	20	21	22	23	24	25	26	Total
Norway spruce	MWL	0.8	28.1	0.4	3.6	0.8	+	1.0	+	2.9	37.6
(*Picea abies*)											
White spruce	wood	0.8	28.6	0.5	4.2	0.8	+	1.2	+	3.4	39.5
(*Picea glauca*).	MWL	0.9	29.2	0.5	4.0	0.7	+	1.1	+	3.2	39.6
White birch	wood	+	10.9	20.4	1.2	0.3	0.3	0.4	1.1	0.9	35.5
(*Betula Payrifera*)	MWL	+	11.2	20.2	1.3	0.3	0.2	0.2	1.2	1.0	35.6

[a]+ = trace amount; - = not detected.
Source: Refs. 98 and 119.

bonds; (3) a dimeric or oligomeric product reveals the type of linkage between any two aromatic moieties in phenylpropane units of a substructure giving rise to the product; (4) the yield of an oxidation product in mole%/C_9 unit is the minimum quantity of a phenylpropane structure with the aromatic moiety having a particular substitution pattern in lignin, giving rise to the product; (5) the approximate quantity of a particular phenylpropane structure can be estimated from the yield of a corresponding oxidation product in mole%/C_9 unit, provided that the oxidation efficiency of the C_{alkyl}—C_{alkyl} bonds is known; and finally, (6) the approximate quantity of uncondensed phenylpropane units can be estimated from the total yield of monomeric aromatic monocarboxylic acids in mole%/C_9 unit; the quantity of condensed phenylpropane units in mole%/C_9 unit is then approximately equal to (100-x), where x is the approximate quantity of uncondensed units. The latter is also called the degree of condensation when it is expressed in mole%/C_9 unit.

The average oxidation efficiency has been determined to be in the range of 0.6–0.45/COOH for the formation of aromatic carboxylic groups from the corresponding C_{alkyl}—C_{alkyl} bonds on the potassium permanganate oxidation of methylated spruce lignin (98, 117, 118) The mean oxidation efficiency of C_{alkyl}—C_{alkyl} bonds in spruce lignins was estimated to be 0.5/COOH for monocarboxylic acid products and 0.6/COOH for dicarboxylic acid products. This follows that for an aromatic monocarboxylic acid product, a conversion factor of approximately $(1/0.5) = 2$ is required to estimate the approximate quantity of a phenylpropane structure, giving rise to the acid on the basis of yield in mole%/C_9 unit, i.e., 2 x (yield of the product in mole%/C_9 unit). For aromatic dicarboxylic acid products, the conversion factor is approximately $2 \times (1/0.6) = 3.33$. Thus, on the basis of the data given in Table 4, approximately 1.5, 56, 1, 12, 2.5, 7, and 19 phenylpropane units/100 C_9 units are

involved in moieties of the type 29, 30, 31, 32, 33, 34,and 35, respectively (see Fig.
7 for structures) in the MWL from wood of Norway spruce (*Picea abies*). The
substitution pattern of aromatic moieties in most of the phenylpropane units in the
spruce MWL is thus characterized by potassium permanganate oxidation. In
addition, the total uncondensed phenylpropane units is approximately 60 units/100
C_9 units. Thus, the degree of condensation is about 0.4/C_9 unit.

2. Nitrobenzene Oxidation

The nitrobenzene oxidation of lignins was introduced in 1939 by Freudenberg and
co-workers (34, 120, 121) in order to obtain further evidence for the aromatic nature
of lignins. In addition, the conversion of technical lignins such as lignosulfonates
into vanillin was regarded at that time in the pulping industry as a potential

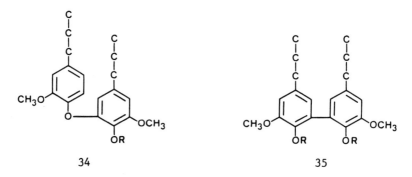

Figure 7 Major substructures of lignins determined by potassium permanganate oxidation of methylated lignins. R = H or alkyl.

application of technical lignins as chemical feedstocks for chemical and related industries.

Vanillin [37] was obtained in a yield of about 25%/Klason lignin by heating wood meal of Norway spruce (*Picea abies*) with a 2*M* sodium hydroxide solution in the presence of nitrobenzene at 160°C for 3 hr. When wood residue from the oxidation was treated again in the same fashion, an additional 2% of vanillin was obtained besides about 1% guaiacol, about 8% vanillic acid [39] and less than 1% 5-carboxyvanillin [40] (see Fig. 8 for structures). A lignosulfonate from sulfite pulping of the same wood species produced vanillin in a yield of about 20%/lignin on nitrobenzene oxidation. The procedure was an application of an earlier invention by Bischler (122) for the conversion of isoeugenol [5] into vanillin [37] in about 90 mole% yield.

On nitrobenzene oxidation, softwoods (Gymnospermae) produce mostly vanillin [37] (34, 37, 121, 123), whereas hardwoods (Dicotyledonous Angiospermae) afford a mixture of vanillin [37] and syringaldehyde [38] (35, 36), and grasses (Monocotyledonous Angiospermae) a mixture of 4-hydroxybenzaldehyde [35],

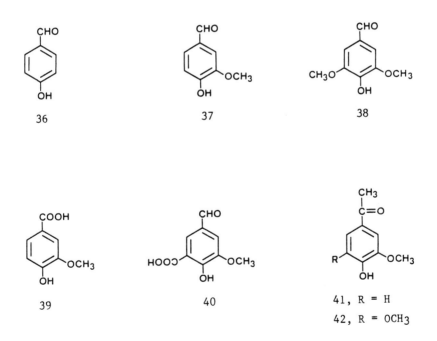

Figure 8 Major oxidation products from nitrobenzene and cupric oxide oxidations of lignins.

vanillin [37] and syringaldehyde [27] (39, 40, 124). The molar ratio of the aldehydes produced depends on the nature of plant species. Thus, the procedure has been adapted for the taxonomy of plants as discussed in Sec. II. B.

Wacek and Kratzl (125–129) investigated the nitrobenzene oxidation of lignin compounds during the years 1942–1948. The results indicated that vanillin was produced in rather good yield from phenolic guaiacylpropane derivatives containing a hydroxyl, carbonyl, or sulfonyl group at C-α or a double bond in the side chain on nitrobenzene oxidation in alkaline solution. In contrast, phenyl analogues of these compounds gave only benzoic acid with a yield in the 40–90% range. The rather high yield of benzoic acid seems to indicate that the resulting benzaldehyde undergoes further oxidation by nitrobenzene under the reaction conditions, in addition to the Cannizaro reaction. In contrast, 4-hydroxybenzaldehyde [36] and its derivatives, vanillin [37] and syringaldehyde [38], did not undergo both the further oxidation by nitrobenzene and Cannizaro reaction under the same reaction conditions. Evidently, as shown in Fig. 9, the aldehydes 36–38 are resonancing between phenolate form a and enolate form b in alkaline solution caused by the -M effect of the carbonyl group at C-4. This would result in the resonance stabilization of the molecules and prevent the attack of nucleophiles on the carbonyl group. The nucleophilic attack of a hydroxide anion on a tertiary aldehyde group is the initial step of the Canizzaro reaction.

For phenolic guaiacyl-type model compounds, similar results were later obtained by Leopold (130, 131) and Pew (132). The results from the studies on the nitrobenzene oxidation of lignin model compounds strongly indicate that the phenolic aldehydes 36–38 are derived from the oxidative degradation of the corresponding uncondensed 4-hydroxyphenylpropane units and their 4-O-alkyl ethers in lignins under the reaction conditions. Moreover, aromatic acids related to phenolic aldehydes 36–38 are produced only in negligible amounts from either lignin model compounds or lignins on nitrobenzene oxidation. As discussed previously, benzaldehyde and 4-O-alkylated vanillin readily undergo the Canizzaro reaction under the conditions of nitrobenzene oxidation to produce benzoic acid, benzyl alcohol, and their 4-O-alkylated derivatives, respectively. These findings indicate that the Canizzaro reaction does not play an important role during the nitrobenzene oxidation of lignins. It follows that the base-catalyzed cleavage of α- and β-aryl ether bonds producing the corresponding phenolate anions precedes an oxidative cleavage of C_α—C_β bonds, resulting in the formation of the corresponding 4-formylphenolate anions. The phenolate anions are not susceptible to nucleophilic attack on the carbonyl group by a hydroxide anion under the conditions of nitrobenzene oxidation because of resonance stabilization. However, the reaction mechanism for the nitrobenzene oxidation of lignins is not well understood. Although nitrobenzene in an alkaline solution has been usually regarded as two-electron-transfer oxidant (133), recent investigation into the reaction mechanism for the nitrobenzene oxidation of lignins and related compounds indicates that

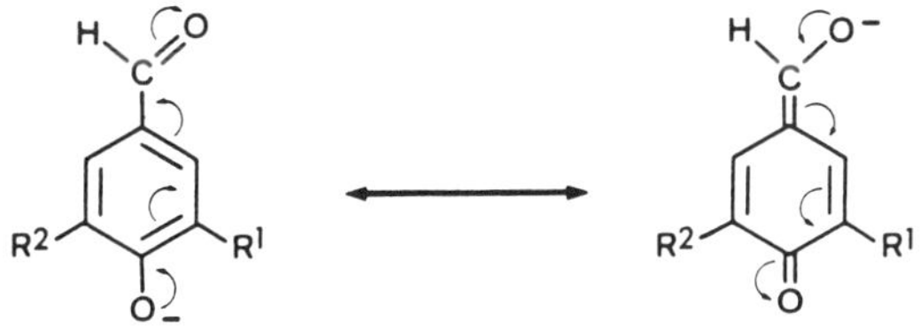

Figure 9   Resonance stablization of 4-formylphenolate anions. R^1, $R^2 = H$ or OCH_3.

nitrobenzene in an alkaline solution acts as an one-electron-transfer oxidant (134).

The optimum reaction conditions for the nitrobenzene oxidation of lignins have been described by Leopold (130) as a result of model compound experiments. The highest yields of phenolic aldehydes were obtained when lignin-containing plant tissues and lignin preparations were oxidized in a $2M$ sodium hydroxide solution with nitrobenzene at 170–180°C for 2–3 hr. These results were confirmed by later investigations (135). In general, the nitrobenzene oxidation of lignins is conducted in the laboratory according to the procedure of Leopold (130). However, extensive modifications have been made in the qualitative and quantitative determination of the oxidation products because of the progress made in the field of instrumental analysis during the past three decades. The constituents of the oxidation mixture are qualitatively determined by means of high-performance liquid chromatography (HPLC), gas chromatography (GC), and gas chromatography-mass spectrometry (GC-MS), (119).

The results obtained from the nitrobenzene oxidation of a lignin indicate both the type of the lignin and the minimum quantity of uncondensed phenylpropane units present in the lignin. As shown in Table 5, birch wood (*Betula papyrifera*) produces vanillin [37] and syringaldehyde [38] in 14.5 and 36.2 mole%/C_9 unit, respectively, with a total phenolic aldehyde yield of 50.7 mole% and a syringaldehyde/vanillin molar ratio (S/V molar ratio) of 2.5. The results show that birch lignin is of the guaiacylsyringyl type (G-S type) and contains at least about 15 and 36 uncondensed guaiacyl- and syringylpropane units/100 C_9 units, respectively, representing a total of about 51 uncondensed phenylpropane units/100 C_9 units (119). Since the oxidation efficiencies of lignin substructures differ considerably from one another or are undetermined, the results cannot be interpreted to mean that birch lignin necessarily contains a total of about 49 condensed phenylpropane units/100 C_9 units. In addition, the S/V molar ratio does not imply that uncondensed phenylpropane units in the lignin consist of guaiacyl- and syringylpropane units in a molar ratio corresponding to the S/V molar ratio. Thus, the S/V and H/V/S molar ratios (H = 4-

Table 5   Yield of Aromatic Aldehydes from Woods, Bamboo, Milled Wood, and Bamboo Lignins (MWL's and MBL) of Some Plant Species on Nltrobenzene Oxidation

Plant species	Specimen	Yield, mole % per C_9 unita			
		p-Hydroxy benzaldehyde	Vanillin	Syring-aldehyde	Total aldehyde
White spruce	wood	+	33.4	+	33.4
(*Picea glauca*)	MWL	+	33.9	+	33.9
White birch	wood	-	14.5	36.2	50.7
(*Betula papyrifera*)	MWL	-	14.1	34.2	49.5
Zhong-Yang Mu	wood	-	26.7	10.4	37.1
(*Bischofia polycarpa*)	MWL	-	26.8	11.8	38.6
Bamboo	MBL	7.9	19.0	25.7	25.7
(*Phyllostachys pubescens*)					

a+ = trace amount; - = not detected.
Source: Refs. 119 and 124.

hydroxybenzaldehyde) are not absolute measures of the amounts of uncondensed phenylpropane units in lignins. However, the results do provide a relative measure of the extent of condensation in the aromatic moieties of lignins. Generally, the S/V molar ratio of a lignin divided by 3 gives an approximate value for the syringylproapne/guaiacylpropane molar ratio of the lignin (136).

3. Metal-Oxide Oxidation

The successful conversion of softwood lignins into vanillin by nitrobenzene oxidation stimulated the search for an economically viable process for the manufacture of vanillin from technical lignins. The nitrobenzene process for the manufacture of vanillin from technical lignins is not cost-efficient because relatively expensive nitrobenzene undergoes reduction and is not recoverable.

In 1942, Pearl (137) found that vanillin [37] could be obtained in a yield of about 20%/lignin when hemlock sulfite liquor solid (lignosulfonate content about 40%) was heated with a 4*M* sodium hydroxide solution in the presence of cupric sulfate hydrate in an autoclave at 160°C for several hours (see Fig. 8 for structures). At the initial stage of the reaction, cupric sulfate was converted into cupric oxide that was the actual oxidant in the reaction. The formation of cupric oxide *in situ*, however, has two disadvantages: (1) The reaction must be carried out in a very dilute solution because cupric sulfate hydrate has a rather low solubility in water, and (2) the reaction mixture contains a considerable amount of sodium sulfate. Consequently, cupric sulfate hydrate was later replaced by cupric hydroxide when the latter became commercially available (138). Cupric hydroxide reverts to cupric oxide with the loss of one mole of water upon heating.

During investigation into the cupric oxide oxidation of a *Torulopsis utilis*-fermented spent liquor from the sulfite pulping of white spruce (*Picea glauca*), Pearl and Beyer (138) found further that the best yield of vanillin was obtained when a mixture of lignosulfonate, cupric oxide, and 1.5–2M sodium hydroxide in an approximate ratio of 1:6:2.9 (w/w/w) was heated in an autoclave at 170°C for 5 hr. The yield of vanillin was about 21.5%/lignin, whereas that of the total ether-extractable oxidation products was about 56%/lignin. In terms of economics, the cupric oxide process for the manufacture of vanillin has two advantages over the nitrobenzene process: (1) Cupric oxide is recoverable by the oxidation of cuprous oxide produced during the reaction, and (2) the absence of byproducts derived from reduction of nitrobenzene.

The optimum reaction conditions for the cupric oxide oxidation of lignins have been demonstrated to be the heating of a mixture consisting of 0.2–0.3 g of the lignin sample or 1 g of wood meal, and 1.2 g of cupric hydroxide in 10 ml of a 2M sodium hydroxide solution in a small stainless autoclave at 170–180°C for 2-3 hr (138, 139). As shown in Table 6, the cupric oxide oxidation of a lignin gives, in general, a result comparable to that obtained from the nitrobenzene oxidation of the same lignin under optimum conditions. However, on cupric oxidation, softwood or hardwood lignins produce an appreciably lower amount of vanillin [37] or vanillin [37] and syringaldehyde [38] and a correspondingly higher amount of acetoguaiacone [41] or acetoguaiacone [41] and acetosyringone [42] than with the nitrobenzene oxidation of corresponding lignins, respectively. Pepper and co-workers (140) showed that compounds 41 and 42 in an alkaline solution are stable toward oxidation by cupric oxide, but susceptible to oxidation by nitrobenzene. Thus, the higher yields of compounds 41 and 42 are gained in part at the expense of the yield of compounds 37 and 38, respectively. Nevertheless, the reaction mechanism for the cupric oxide oxidation of lignin is not well established, although cupric oxide is known to be a one-electron-transfer oxidant.

Investigations were also conducted on the oxidation of lignosulfonates in an alkaline solution by silver oxide (141–148), mercuric oxide (142, 143, 149) and oxides of vanadium, chromium, and nickel (150, 151). These oxides have higher oxidation potential than cupric oxide; hence, they are stronger oxidants than cupric oxide.

The silver oxidation of lignin was originally discovered by Pearl (141) in 1942 for the quantitative conversion of vanillin into vanillic acid. When one mole of vanillin in 400 ml of a 2.5M sodium hydroxide solution was treated with 0.5 mole of silver oxide at 55°C, vanillic acid was produced in quantitative yield. The reaction is exothermic, and thus it required careful control of the reaction temperature.

The silver oxide oxidation of lignosulfonates is usually conducted in a 6M sodium hydroxide solution with a rather large excess of alkali at reflux temperature

Table 6 Oxidation Products and Their Yields from Woods of Norway Spruce and Aspen on Nitrobenzene and Cupric Oxide Oxidations

Wood species	Oxidation products	Yield (%/Klason lignin)	
		Nitrobenzene	Cupric oxide
Norway spruce			
(*Picea abies*)	guaiacol	—	1.4
	vanillin [37]	27.5	15.9
	acetoguaiacone [41]	—	3.9
	total yield	27.5	21.5
Aspen			
(*Populus tremuloides*)	4-hydroxybenzaldehyde [36]	-	1.8
	vanillin [37]	12.4	7.8
	syringaldehyde [38]	30.0	20.0
	acetoguaiacone [41]	-	2.0
	acetosyringone [42]	-	5.3
	total yield	42.4	36.9

Source: Refs. 124 and 135.

(105–110°C) for a short period, less than 10 min. In large-scale experiments, the reaction temperature is difficult to control because of the exothermic nature of the reaction. On silver oxide oxidation, lignosulfonates produce oxidation products similar to those produced on cupric oxide oxidation of corresponding lignins.

It is evident that the use of metal oxides to convert technical lignins into useful chemical feedstocks is not an economically viable process because of the rather high costs involved and the complex nature of the reaction mixture. However, the cupric oxidation of lignin is useful as a method for the characterization of lignins.

4. Oxidation of Lignins in Alkaline Solution by Oxygen

As early as 1922, Fischer and co-workers (152–154) observed that technical lignins including HCl-lignins were susceptible to oxidation by pressurized oxygen (55 atm) at about 200°C. Phenolic substances thus produced were further decomposed under the reaction conditions to afford formic acid, acetic acid, oxalic acid, and benzenepentacarboxylic acid as the major oxidation products (154). Even at room temperature, oxidation occurred when the reaction was prolonged, as indicated by a decrease in the methoxyl content of the lignin (155).

About two decades later, Lautsch and co-workers (156) showed that vanillin could be obtained in a yield of about 10%/lignin by the oxidation of either the wood meal of Norway spruce (*Picea abies*) or spent liquor from the sulfite pulping of the same wood species with pressurized oxygen at moderate reaction conditions. The optimum reaction conditions were found to be a reaction temperature in the 105–140°C range, oxygen pressure of less than 20 atm, and alkaline concentration of less than $2M$ with a lignin (C_9)/alkali molar ratio of less than 0.25. At a reaction temperature above 140°C, the vanillin produced underwent further degradation, while at a reaction temperature below 105°C, the base-catalyzed hydrolytic cleavage of aryl ether bonds is not sufficiently effective.

Vanillin [37] was produced from the wood meal of Norway spruce (*Picea abies*) by the following procedure: a mixture consisting of 40 g of air-dried wood meal (Klason lignin content, 25.5%) and 400 ml of 10% potassium hydroxide solution (the total amount of KOH, 40 g) was placed in a 1-L autoclave, into which oxygen with a pressure of 10 atm was introduced. The autoclave was heated at 120°C for 2 hr. From the reaction mixture, about 1 g of purified vanillin was obtained, corresponding to a yield of about 9.5%/Klason lignin (156).

A two-step process was used for the preparation of vanillin from a spruce sulfite spent liquor: 400 ml of the sulfite spent liquor containing 8.7 g of lignosulfonates and 14.7 g of sodium hydroxide was preheated in an autoclave at 135°C; then the mixture in the autoclave was subjected to a post-oxidation of the lignin with pressurized oxygen at 125°C for 2–2.5 hr. In the postoxidation step, the pressure of oxygen was controlled in such manner that three atoms of oxygen were involved in the oxidation of one phenylpropane (C_9) unit. The yield of purified vanillin was 7.7%/lignin. The yield of vanillin was somewhat lower when air was used instead of oxygen (156). Processes based on these findings are now employed in the chemical industry for the commercial production of valine from softwood lignosulfonates. Although the yield of vanillin is substantially lower in these oxygen processes than nitrobenzene and cupric oxide processes, the oxygen processes have the following three major advantages over the latter two processes: (1) the considerably lower alkali consistency, (2) the absence of byproducts derived from reduction of the oxidant, and (3) the low cost of oxygen. Thus, in terms of overall economics, the oxygen processes for the manufacture of vanillin from softwood lignosulfonates are more favorable than the other two processes.

It must be mentioned that the degradation of lignins by oxygen in an alkaline solution are the important basis for both the soda-oxygen pulping and soda-oxygen bleaching of pulps in nonchlorine bleaching processes recently applied widely in the pulp industry.

B. Hydrolysis

1. Ethanolysis and Acidolysis

Klason (157) observed in 1893 that part of the lignin occurring in wood was degraded and dissolved into the solvent when wood meal was refluxed in ethanol in the presence of a few % of mineral acid. No attempt was, however, made to investigate the constituents of the degradation mixture, probably because of the complex nature of the reaction mixture. This procedure is known as the ethanolysis of lignins.

In 1939, Hibbert and co-workers (158) isolated 2-ethoxy-1-(4-hydroxy-3-methoxyphenyl)-1-propanone [43] in a yield of about 3%/Klason lignin from a lignin degradation mixture obtained by refluxing spruce wood meal (Klason lignin, 28.6%) in absolute ethanol containing 2% of hydrogen chloride for 48 hr. Subsequent investigations into the constituents of the ethanolysis mixture resulted in the further isolation of 1-(4-hydroxy-3-methoxyphenyl)-2-oxopropane [44] (159), 1-ethoxy-1-(4-hydroxy-3-methoxyphenyl)-2-oxopropane [45] (160), and 1-(4-hydroxy-3-methoxyphenyl)-1,2-dioxopropane [46] (161, 162), each in a yield of less than 2%/Klason lignin, in addition to a trace amount of vanillin (see Fig. 10 for structures). The total yield of compounds 43 through 46 was less than 10%/Klason lignin. On ethanolysis, maple wood meal (Klason lignin, 23%) produced compounds 43 through 46 and their syringyl analogues with a somewhat higher total yield (159, 163, 164'). Compounds of these types are called "Hibbert ketones." Hibbert and co-workers (165) postulated that ketones were produced from lignins by way of either 3-(4-hydroxy-3-methoxyphenyl)-2-oxopropanol [47] or 3-(4-hydroxy-3,5-dimethoxyphenyl)-2-oxopropanol [48]. Although the yields of Hibbert ketones were rather low, the isolation of these compounds from the hydrolysates of ethanolysis is of primary importance in lignin chemistry in terms of the structure of lignins, since it provides the first direct, chemical evidence for the postulation that lignins are composed of phenylpropane (C_6-C_3 or C_9) units.

Two decades later, Adler and co-workers (166) found that the refluxing of wood meal with $0.2M$ hydrogen chloride in dioxane-water (9:1, v/v) for 4 hr resulted in the hydrolysis of lignins occurring in the wood with the formation of ether-extractable phenols to a considerable extent, in addition to high-molecular-mass lignin products. This procedure has been termed "acidolysis" in order to differentiate it from ethanolysis, although the phenolic fraction from acidolysis also contains appreciable amounts of Hibbert ketones as the corresponding fraction from an ethanolysis mixture of the same wood meal.

As shown in Fig. 11, subsequent studies on the acidolysis of lignins and related compounds revealed that Hibbert ketones 43 through 48 originate from arylglycerol-β-aryl ether (β-O-4') substructures of the types 49 and 50 (167–169). On acidolysis, β-O-(2-methoxyphenyl)guaiacylglycerol [51] undergoes acid-catalyzed dehydration to give β-guaiacoxyconiferyl alcohol intermediate [53] by way of the corresponding benzylium ion intermediate 52. The intermediate 53 produces 3-(4-hydroxy-3-methoxyphenyl)-2-oxopropanol [47] and guaiacol [54] on the acid-catalyzed hydrolysis of enol ether. Presumably, the intermediate 53 also undergoes an allylic rearrangement to yield the α-hydroxy-β-guaiacoxyeugenol intermediate [55], which gives 1-hydroxy-1-(4-hydroxy-3-methoxyphenyl)-2-oxopropane [56] and

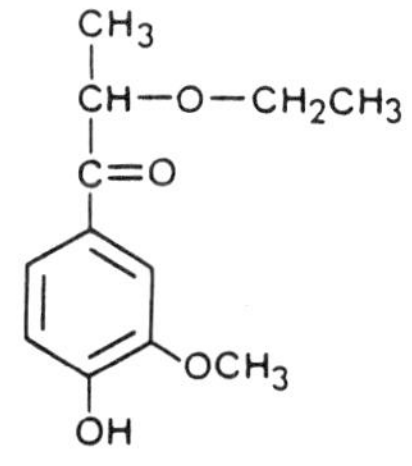

43

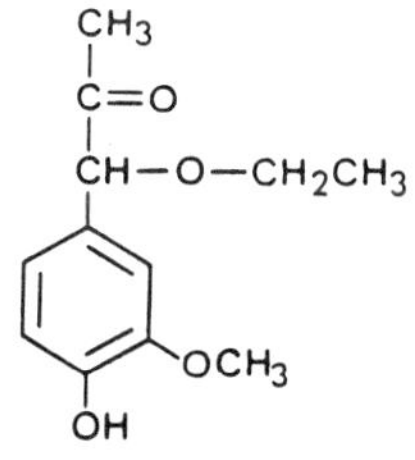

44

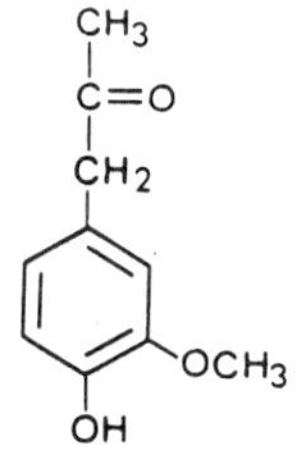

45

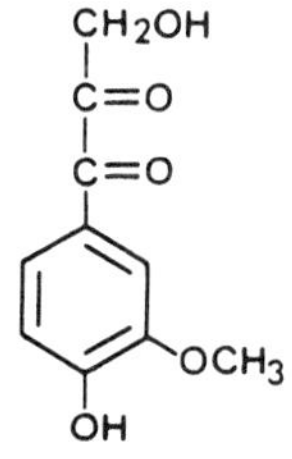

46

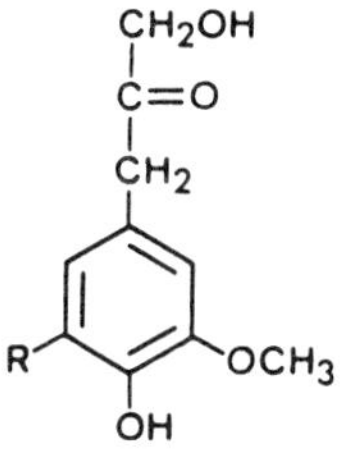

47, R = H

48, R = OCH_3

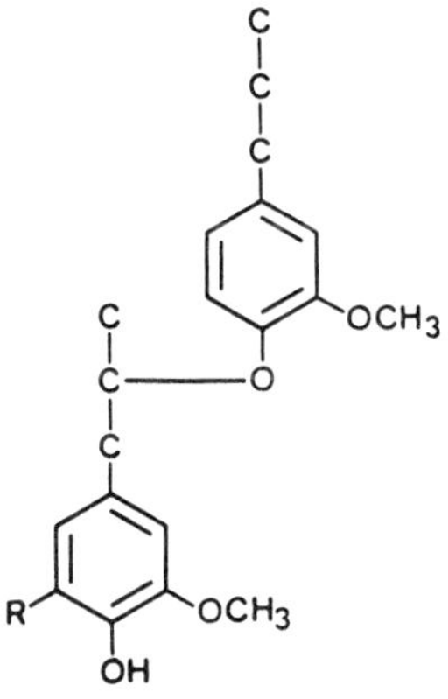

49, R = H

50, R = OCH_3

Figure 10 Major degradation products from ethanolysis of lignins.

guaiacol [54] on acid-catalyzed hydrolysis. Compound 56 also is produced in part from compound 47, probably through keto-enol tautomerism to the β-hydroxyconiferyl alcohol intermediate [58] followed by an allylic rearrangement. Compound 56 interconverts with 2-hydroxyl-(4-hydroxy-3-methoxyphenyl)-propanone [57] through keto-enol tautomerism under the reaction conditions; at the equilibrium state, compounds 56 and 57 exist in a molar ratio of 0.25 (168). Hibbert and co-workers (165) earlier postulated that compounds 44 and 46 are probably produced from 56 by dehydration to dimeric ether intermediate 59, followed by the cleavage of the ether bond with a concomitant hydride shift involving the hydrogen atom at C-α. In addition, intermediate 52 also undergoes a reverse Prins reaction to give a β-guaiacoxy-4-vinylguaiacol intermediate 60 with the elimination of formaldehyde. The intermediate 60 produces homovanillin [61] and guaiacol [54] on

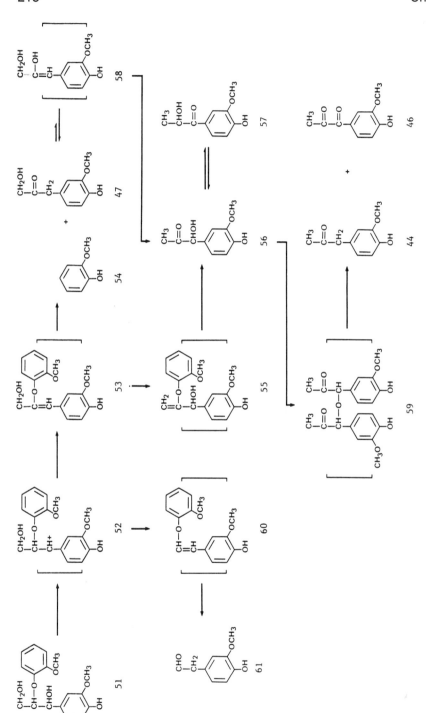

Figure 11 Reaction mechanism for acidolysis of β-O-(2-methoxyphenyl) guaiacylglycerol [51] (From Refs. 167–169.)

acid-catalyzed hydrolysis. The major products of the acidolysis of compound 51 were compounds 47, 54, and a mixture of compounds 56 and 57 with yields of 53, 74, and 18 mole%, respectively. Yields of other products were insignificant.

In general, reactions involved in acidolysis are similar to those in ethanolysis, expect for the formation of compounds 56 and 57 instead of compounds 43 and 45 in the former. Evidently, compounds 43 and 45 are secondary products formed by substitution of the benzylic hydroxyl group in compounds 56 and 57, respectively, with ethoxyl group during ethanolysis. Thus, the advantage of acidolysis over ethanolysis is that the former does not produce O-ethyl derivatives as secondary products and a shorter reaction time.

Lundquist and co-workers (170–173) investigated the contents of an acidolysis mixture from milled wood lignin (MWL) of Norway spruce (*Picea abies*). The acidolysis mixture was fractionated by gel permeation chromatography to obtain a low-molecular-mass fraction in a yield of approximately 17%/MWL. Dimeric degradation products 62 through 68 were isolated from the low-molecular-mass faction in addition to monomeric degradation products 45 through 48 (see Figs. 10–13 for structures). Compound 47 was the major constituent of the fraction with a yield of about 5%/MWL. It is evident from the model compound experiment discussed previously (169) that the monomeric products were produced from either guaiacylglycerol-β-aryl ether type-49 substructures on acidolysis, except for compound 48 that was a degradation product of the syringylglycerol-β-aryl ether type-50 substructure, a minor unit in softwood lignins. Model compound studies (173–176) established the fact that reaction mechanisms for the acidolysis of phenylcoumaran (β-5') and 1,2-bis-guaiacylpropane-type (β-1') substructures [73 and 74] are essentially similar to those for the acidolysis of β-O-4' substructures, except for deviation in the major pathways caused by the presence of an aryl group instead of an aroxyl group at C-β. Thus, dimeric products 62 and 63 are produced from a β-5' substructure. The presence of an α-ketol group in the side chain of the compounds can be further interpreted as evidence of the presence of a third guaiacylpropane unit bonded to C-β' in the side chain of the β-5' substructures in the form of β-O-4' linkage. Consequently, spruce MWL must contain a trimeric type-75 substructure. The other dimeric products 64 through 67 are produced from β-1' type-74 substructures (176). Compounds 66 and 67 are presumably produced from the β-1' substructures via intermediates, produced by the allylic rearrangement of the benzylium ion intermediate leading to the formation of the major product 65 by benzopinacol rearrangements, respectively. Recent studies on the structure of spruce MWL by ^{13}C MNR spectroscopy (177) show that the lignin does not contain β-1' substructures; if anything, the quantity of the substructure must be so insignificant that it cannot be detected even by a high-resolution ^{13}C MNR spectrometer. Conceivably, the β-1' substructures could be parts of secondary byproducts formed during the acidolysis of spruce MWL, which in turn undergoes subsequent acidolysis to produce compounds 64 through 67. The formation of secondary byproducts

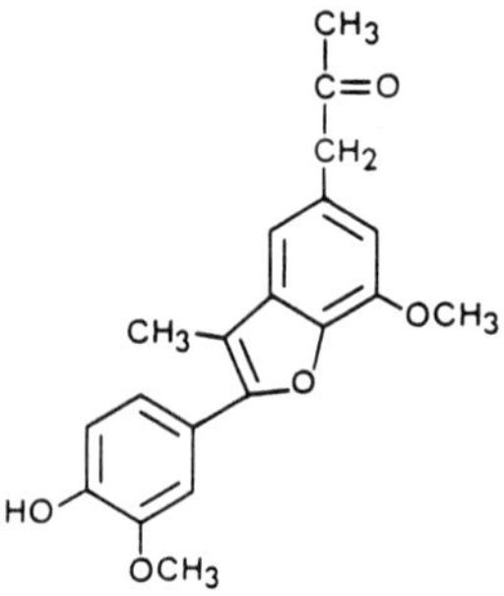

Figure 12   Major degradation products from acidolysis of lignins.

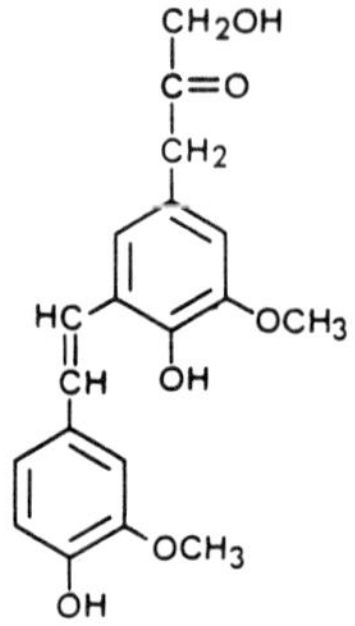

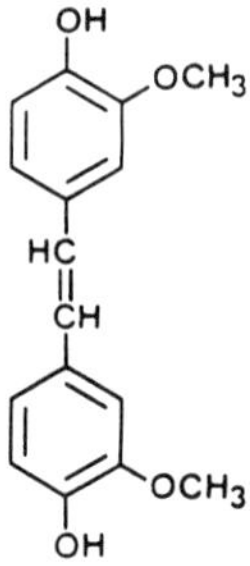

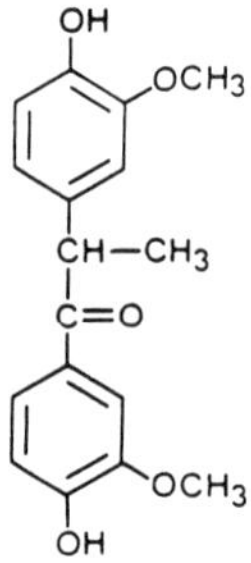

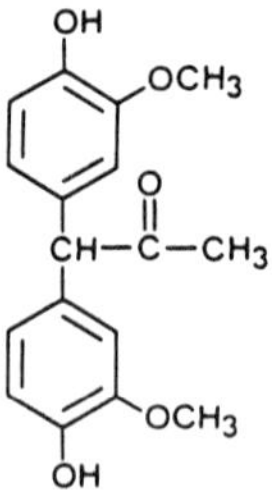

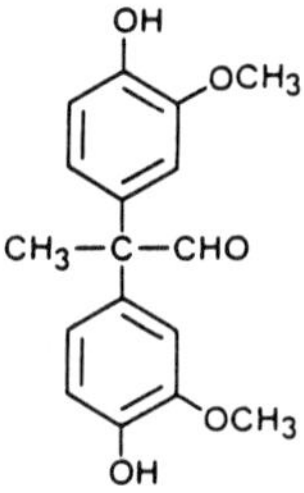

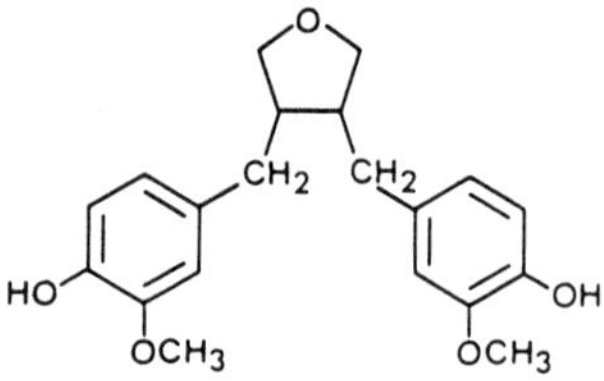

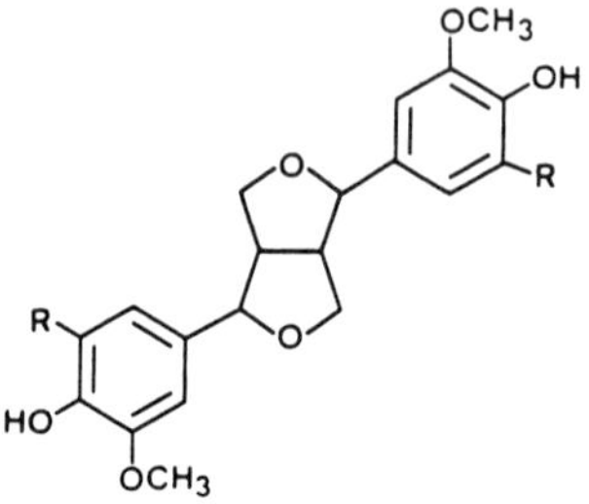

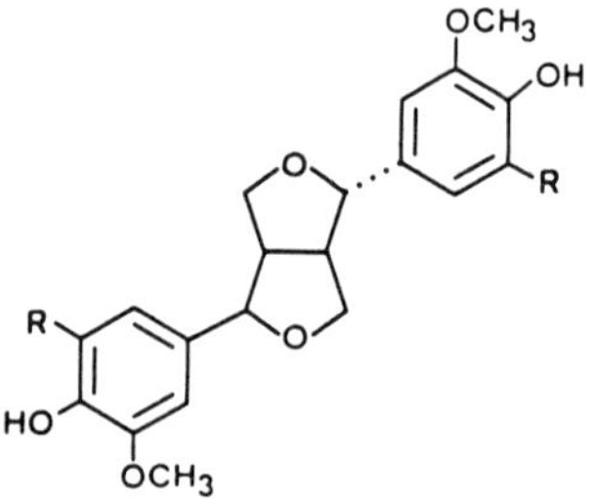

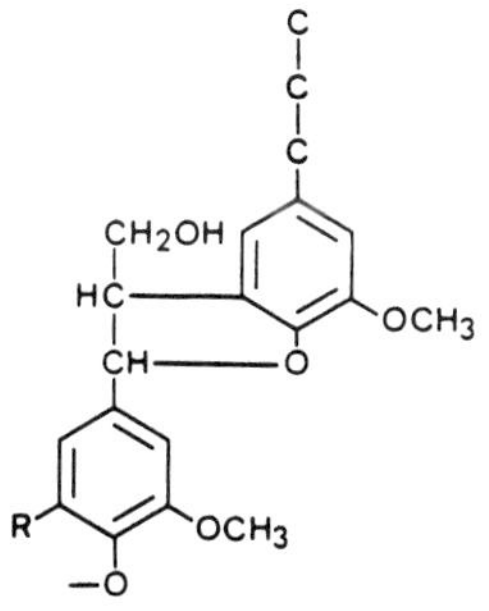

Figure 13 Major substructures of lignins determined by acidolysis of lignins. R = H or OCH$_3$.

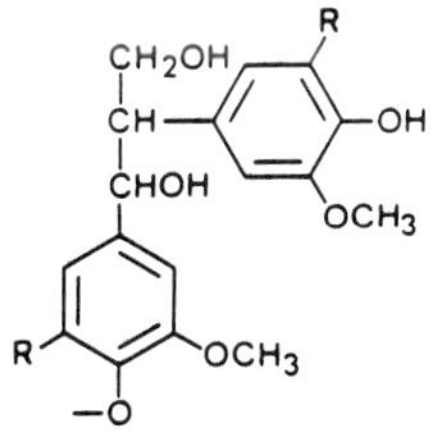

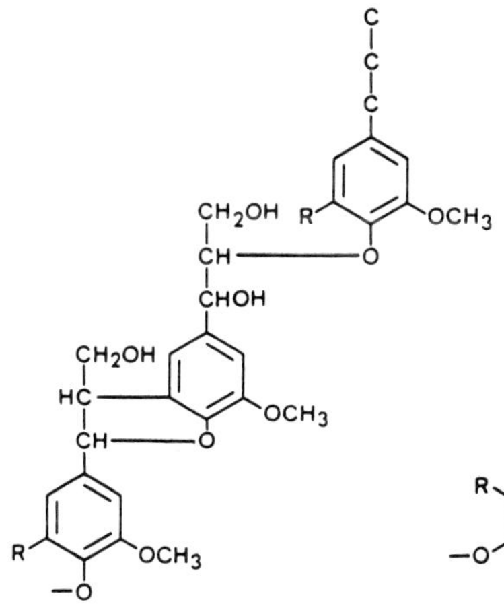

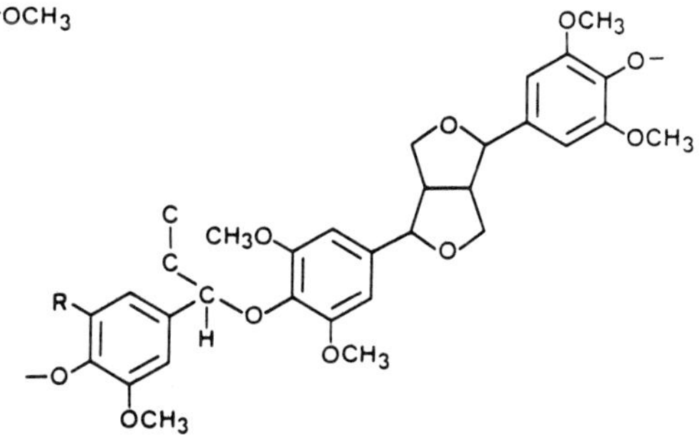

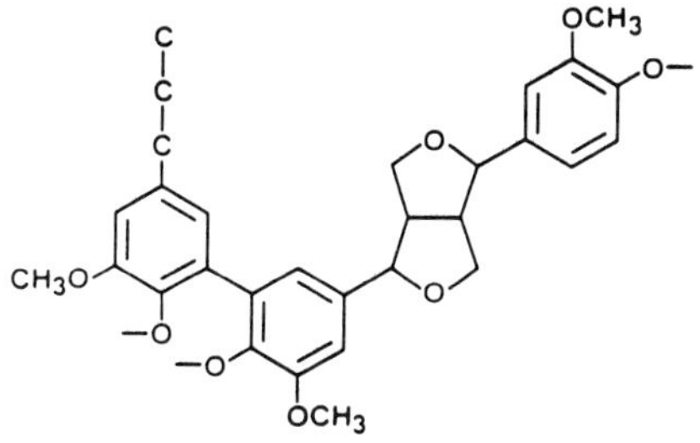

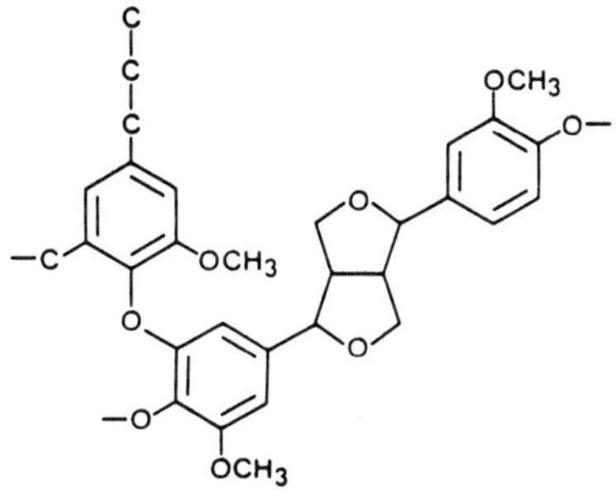

containing a β-1' substructure has been found during mild hydrolysis (178) and the hydrolysis of lignins in dioxane-water at 160–180°C (179, 180), which will be discussed in the next sections. The origin of compound 68 is not known.

On acidolysis, MWL from birch (*Betula verrucosa*) gave a low-molecular-mass fraction in a yield of about 30%/MWL. The constituents identified in this fraction were the aforementioned acidolysis products from spruce MWL, their syringyl analogues with one or two syringyl groups, D, L-syringaresinol [70],and D, L-epi-syringaresinol [72] (181) (see Figs. 12 and 13 for structures). The major products were compound 47 and its syringyl analogue, compound 48, isolated in yields of about 3 and 5%/MWL, respectively. Since birch MWL consists mostly of guaiacyl- and syrinylpropane units in a molar ratio of approximately 1:1 (110), the relatively higher yield of compound 48 as compared to the yield of compound 47 indicates that more uncondensed syringylpropane units are involved in β-O-4' substructures than uncondensed guaiacylpropane units. D, L-pinoresinol [69] and D,L-syringaresinol [70] are rather stable under the conditions of acidolysis, except for the partial conversion of these compounds into D, L-epipinoresinol [71] and D,L-episyringaresinol [72], respectively (169, 182). Thus, it was postulated that compounds 71 and 72 are produced from syringaresinol-type (syringyl-type β-β') substructures, oxygen atoms at C-4 and C-4 ' of those are bonded to either C-α or C-β of the side chains of two other arylpropane units in the form of α—O—4' or β—O—4' bonds such as a type-76 substructure.

As discussed above, it is evident that birch MWL (guaiacyl-syringyl lignin) is more susceptible to acidolysis than spruce MWL (guaiacyl lignin). This is attributable to the fact that birch lignin contains less condensed substructures than spruce MWL. Although D, L-syringaresinol [70] and D, L-episyringaresinol [72] were produced in appreciable amounts on the acidolysis of birch MWL (181), D, L-pinoresinol [69] and D, L-epipinoresinol [71] were not detected in the acidolysis mixture from either spruce or birch MWL, even in a trace amount. The results imply that the pinoresinol (guaiacyl-type β-β') substructures are not present in these MWL's, or they are present in the MWL's as a part of condensed substructures in which one of the guaiacylpropane units in the pinoresinol moiety is bonded to another guaiacylpropane unit at C-5 to form either a biphenyl (5-5') or a diaryl ether (5-O-4') moiety, as in the type-77 and -78 substructures. Nimz and co-workers (183) showed that spruce MWL contains a rather low content of guaiacyl-type β-β' substructures by ^{13}C NMR spectroscopy; less than 10 guaiacylpropane units/100 C_9 units are involved. Thus, the guaiacyl-type β-β' substructures are minor substructures in both spruce and birch MWL's and present in MWL's as condensed substructures that are rather stable in terms of acidolysis. In contrast, syringyl-type β-β' substructures are susceptible to acidolysis, resulting in the formation of D, L-syringaresinol [70] and D, L-episyringaresinol [71].

2. Mild Hydrolysis

It has been known for several decades that a substantial part of wood is hydrolyzed to produce a water-soluble hydrolysate when wood meal—in particular, hardwood meal—is heated with water at a temperature between 100–200°C for a long period of reaction time (184, 185). The process involves the acid-catalyzed hydrolysis of wood components, since acetic acid is produced by the hydrolysis of O-acetyl groups in hemicellulose present in the wood at the initial stage of the process. Thus, the yield of the hydrolysate depends on the wood species, as well as reaction conditions. Until quite recently, no attempt had been made to investigate comprehensively the constituents of hydrolysate, particularly components derived from the degradation of lignins. Kratzl and Silbernagel (186), however, observed that the condensation of lignin degradation products occurs rapidly above 130°C during the process.

In view of the observation made by Kratzl and Silbernagel, Nimz (187–189) introduced in 1965 the percolation of wood meal with water or 2% acetic acid solution at 100°C with a flow rate of solvent approximately 2 L/day for several weeks to bring about the hydrolysis of lignin in wood, in addition to other wood components. Since effluent was removed from the system immediately after passing through the reactor, secondary condensations, rearrangements, and further degradation of the resulting lignin degradation products should be negligible. The procedure is, therefore, referred to as the "mild hydrolysis of lignin."

When preextracted beech (*Fagus silvatica*) wood meal (Klason lignin, 22.2%) was percolated with water at 100°C for 65 days, about 41% of the wood meal was hydrolyzed and passed into water. The remaining wood meal contained about 18% of lignin as determined by the Klason procedure. Thus, approximately 50% of the lignin originally present in the wood meal was hydrolyzed during the process. The use of a 2% acetic solution as a solvent instead of water did not result in an increase of the amount of lignin hydrolyzed under the same reaction conditions (189). Since then, water has been used as the solvent for the mild hydrolysis of lignins. From the phenolic low-molecular-mass fraction of the hydrolysate, five constituents of the fraction were isolated: syringylglycerol [80], 1,2-bis-guaiacyl-1,3-propanediol [81] 2-syringyl-1-guaiacyl-1,3-propanediol [82], 1,2-bis-syringyl-1,3-propanediol [83], and D,L-syringresinol [70] (187–191) (see Fig. 14 for structures). Compounds 70, 80, and 83 were the major constituents of the low-molecular-mass fraction, and they were isolated in yields of approximately 1, 2.5, and 1.5%/Klason lignin, respectively. Compounds 81 and 82 were the minor components, each isolated in a yield of less than 0.1%/Klason lignin. Compounds 80 through 83 consisted of D, L-*threo* and D, L-*erythro* diasteroisomers that were isolated.

As compared to beech, lignins in spruce (*Picea excelsa*) wood are less susceptible to mild hydrolysis. This is attributable to the fact that softwoods contain, in general, less hemicelluloses such as xylan having O-acetyl groups than hardwoods.

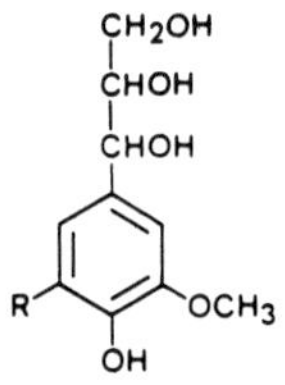

Figure 14 Major degradation products from mild hydrolysis of lignins.

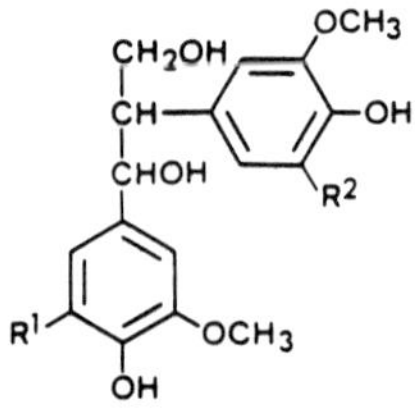

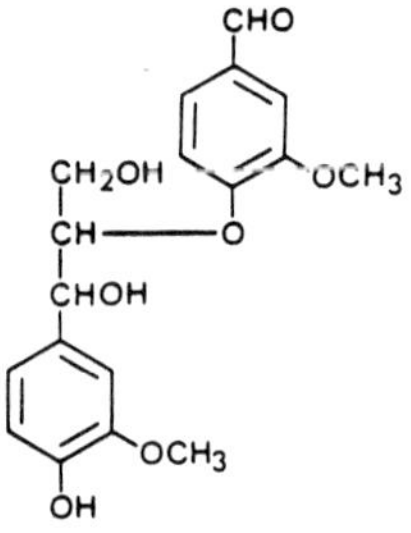

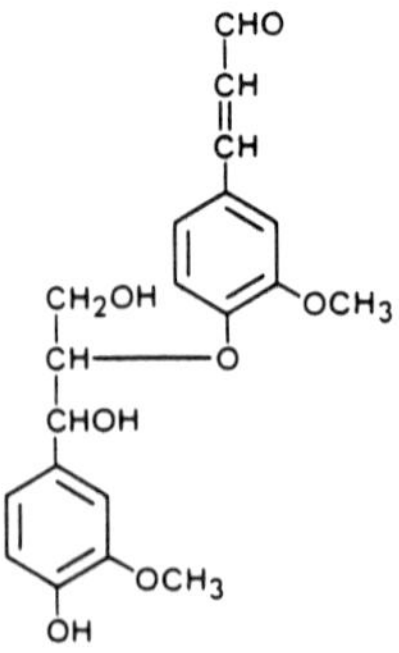

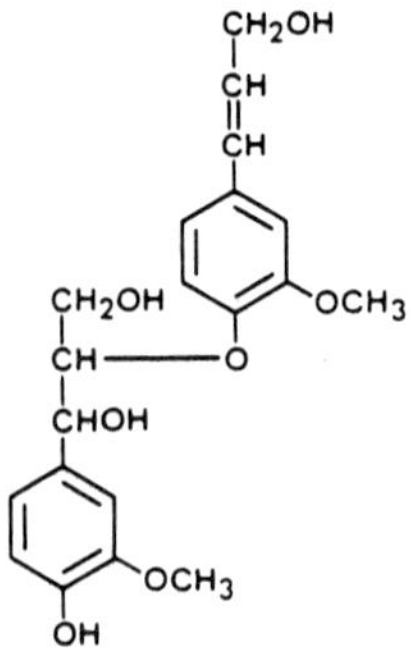

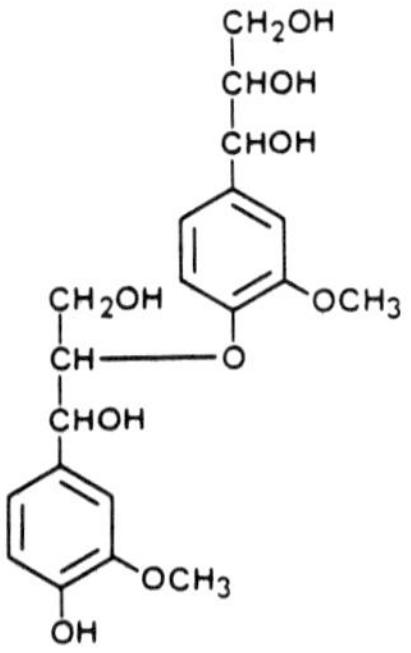

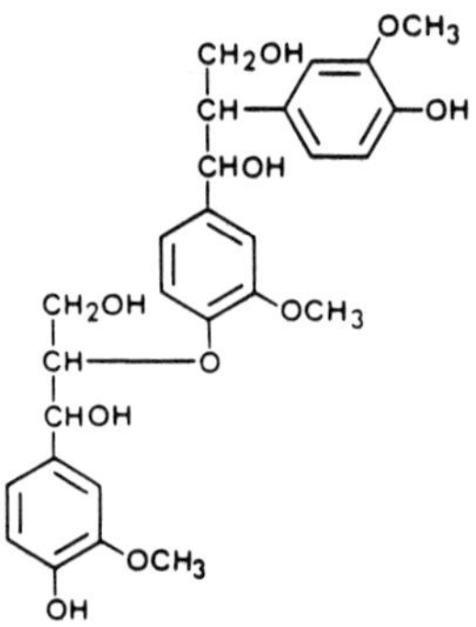

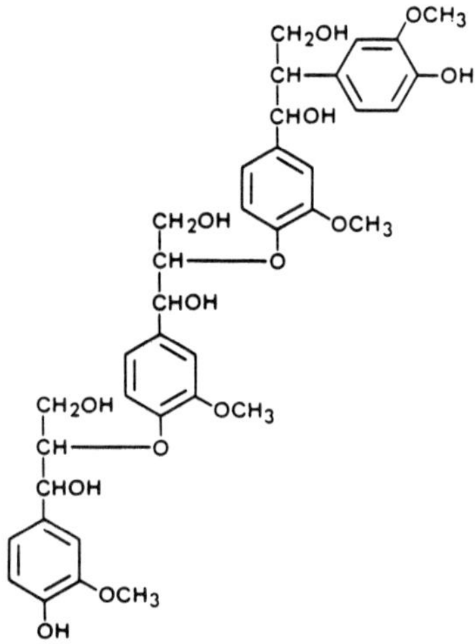

On mild hydrolysis with water at 100°C for 10 days, only about 10% of the lignin originally present in preextracted spruce wood meal (Klason lignin, 28.2%) was hydrolyzed. A prolonged percolation under the same reaction conditions did not increase significantly the amount of lignin hydrolyzed (178, 192). From the phenolic low-molecular-mass fraction of the hydrolysate, eight lignin degradation products were isolated. The products include guaiacylglycerol [79], 1,2-bis-guaiacyl-1,3-propanediol [81], guaiacylglycerol-β-vanillin ether [84], guaiacylglycerol-β-coniferyl aldehyde ether [85], guaiacylglycerol-β-coniferyl alcohol ether [86], guaiacylglycerol-β-guaiacylglycerol [87], two trimeric diasteroisomers with structure 88 and one tetrameric product with structure 89 (178, 191–194) (see Fig. 14 for structures). Compounds 81, 88, and 89 were the major constituents of the fraction, each isolated with a yield of about 0.2–0.3%/Klason lignin. The other compounds were minor components, each with a yield of less than 0.05%/Klason lignin. Moreover, compounds 79, 81, and 84 through 86 consisted of a D, L-*threo* and D, L-*erythro* diasteroisomer on the basis of ^{1}H NMR spectral data, although the sterochemistry of the compounds was not studied further. It was observed, however, that the compounds did not have optical activity.

Sano (180) demonstrated that on heating at 180°C for 20 min, β-O-(2-methoxyphenyl)guaiacylglycerol [51] in dioxane-water (1:1, v/v) undergoes acidolysis at pH 4 to produce most of the acidolysis products (see Fig. 11) In addition, as shown in Fig. 15, a homolytic cleavage of the β—O—4' bond in the

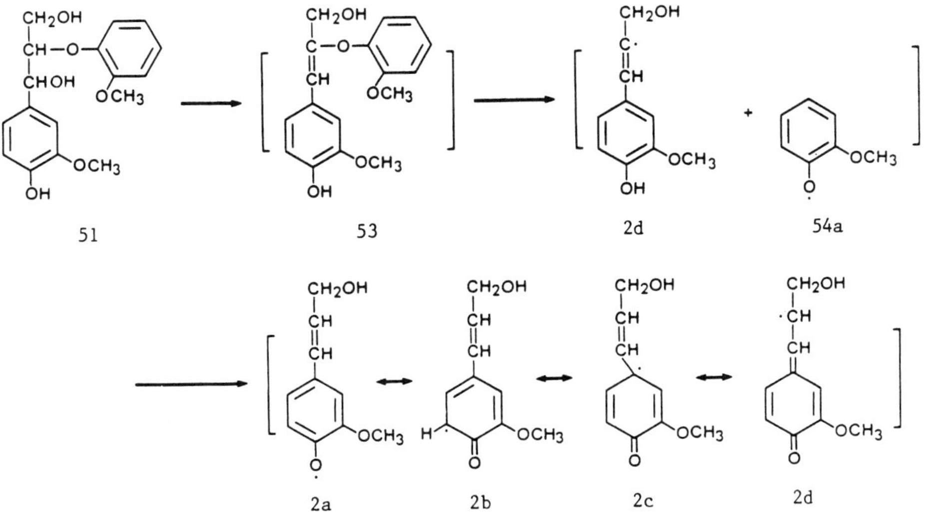

Figure 15   Possible mechanism for homolytic cleavage of β-aryl ether bonds in β-O-4' substructures in lignins under acid-catalyzed hydrolysis of lignins. (From Ref. 180.)

compound occurs simultaneously, resulting in the formation of the quinone-methide-type $\underline{2d}$ radical, a mesomeric form of the phenoxyl radical $\underline{2a}$ produced by the removal of one hydrogen atom from the phenolic hydroxyl group of coniferyl alcohol [2]. The $\underline{2d}$ radical, in turn, initiates polymerization as well as radical chain reactions to yield secondary reaction products such as dimeric and oligomeric dehydrogenative polymerization products of coniferyl alcohol, including compound $\underline{88}$ and related compounds, in particular compounds containing a β-β and/or β-1' substructure. Thus, some of the products, compounds $\underline{70}$ and $\underline{81}$ through $\underline{89}$, isolated from the hydrolysates of beech and spruce woods on mild hydrolysis could be secondary reaction products. In spite of this shortcoming, the compounds isolated from the hydrolysates of beech and spruce woods offer qualitative evidence for the presence of syringyl-type β-β' type-76 substructures and β-1' substructures of types $\underline{81}$–$\underline{83}$ in beech lignin, and for the presence of β-O-4' substructures of types $\underline{84}$–$\underline{87}$ and β-1' type-81 substructures in spruce lignin. Compounds $\underline{85}$ and $\underline{86}$ were isolated earlier in vitro from the dimeric fraction of the reaction mixture from dehydrogenative polymerization of coniferyl alcohol by Freudenberg and co-workers (195, 196). Moreover, except for compound $\underline{84}$, compounds $\underline{70}$ and $\underline{81}$ through $\underline{89}$ have an intact side chain with respect to arylglycerol, 4-O-alkylconiferyl aldehyde, 4-O-alkylconiferyl alcohol, or aryl-1,3-propanediol moiety. Thus, the results provide the first direct evidence that lignins are indeed composed mostly of arylpropane units derived from an enzyme-induced dehydrogenative polymerization of p-hydroxycinnamyl alcohol derivatives, as demonstrated by Freudenberg (21) during in vitro experiments with coniferyl alcohol. The rather low yield of low-molecular-mass lignin degradation products from woods on mild hydrolysis led Nimz (192) to postulate that most of the products are originally bonded to C-α of arylpropane units in the periphery of lignin macromolecules in the form of terminal α-aryl ether groups. These types of linkages are noncyclic benzyl aryl ether bonds and are prone to acid-catalyzed hydrolysis under the reaction conditions.

3. Hydrolysis in Dioxane-Water At High Temperature

Sakakibara and Nakayama (179, 197, 198) observed in 1961 that about 40–60% of lignins in preextracted wood meals of Ezomatsu (Japanese spruce; *Picea Jezoensis* Carr.) and Buna (Japanese beech; *Fagus crenata* Blum) are hydrolyzed and passed into the solvent when the wood meals are heated in dioxane-water (1:1, v/v) and adjusted to pH 4 with acetic acid at 180°C for 120 min in an autoclave. In addition, about 10–25% of carbohydrates in the wood meals are also degraded. In general, the amounts of wood compounds hydrolyzed increase with increasing reaction temperature, and the optimum reaction temperature has been determined to be 180°C. Moreover, hardwood lignins are more susceptible to hydrolysis than softwood lignins, probably because hardwood lignins contain less condensed phenylpropane units than softwood lignins (198). In contrast to the mild hydrolysis of lignins (see the preceding section), the content of O-acetyl groups in hemicelluloses present in woods does not play a significant role in hydrolysis since the solvent used in the reaction is adjusted to pH 4 with acetic acid.

On hydrolysis in dioxane-water at 180°C wood meal of Yachidamo (Japanese ash; *Fraxinus mandshurica* var. japonica) yielded a hydrolysate, from which the following compounds were isolated: syringylglycerol-β-syringylglycerol ether [90], 2-(4-hydroxy-3,5-methoxyphenyl)-3-hydroxymethyl-5-(β-formylvinyl)-7-methoxycoumaran [91], 2-(4-hydroxy-3,5-methoxyphenyl)-3-hydroxymethy-4-(4-hydroxy-3,5-dimethoxybenzoyl)-tetrahydrofuran [92], guaiacylglycerol-β-syringaresinol ether [93], and syringyl-glycerol-β-syringaresinol ether [94] (see Fig. 16 for structures) in addition to compounds 70, 72, 79, 80, and 83 (199–203). The latter four compounds were isolated previously from the hydrolysate obtained from beech wood on mild hydrolysis (187–191) (see Fig. 14 for structures). The isolation of compound 90 is particularly interesting in connection with the finding of Fergus and Goring (70) that lignins in the fibers and ray parenchyma secondary cell walls of white birch (*Betula papyrifrea* Marsh.), a hardwood species, consist mostly of syringylpropane units. Because both ortho positions to phenolic hydroxyl group in a sinapyl alcohol [3] are not available for the dehydrogenative coupling of phenols, it is conceivable that β-O-4' linkage is the most predominate bonding involving syringylpropane units in hardwood lignins.

The following compounds were isolated from the hydrolysate obtained from the wood meal of Ezomatsh (*P. Jezoensis*) on hydrolysis in dioxane-water (204–206): α'-hydroxy-β-guaiacyl-α', β'-dihydrodehydro-diconiferyl alcohol [95] in addition to compounds 2, 79, 81, 87, and 88. The latter four compounds are also isolated from the hydrolysate of spruce wood obtained on mild hydrolysis by Nimz and co-workers (178, 191–194). In general, the hydrolysate from softwoods contains a considerably less total amount of low-molecular-mass compounds (MW < 1000) than that from hardwoods. Evidently, this is attributable to the fact that syringylpropane units in a hardwood lignin consist of less condensed units than guaiacylpropane units in the same lignin. Consequently, hardwood lignins contain less condensed C_9 units than softwood lignins.

As discussed in the preceding section, Sano (180) showed that on dioxane-water hydrolysis, β-O-(2-methoxyphenyl) guaiacylglycerol [51] yields some acidolysis products in addition to compounds 2, 69, 81, 84, 95 and similar trimeric products. The formation of the aforementioned dimeric and trimeric products under the reaction conditions is possible only when compound 51 undergoes a concomitant dehydration to compound 53, followed by a homolytic cleavage of the β—O—4' bond during the reaction to give free radical 2d as shown in Fig. 15. This implies that the characterization of lignins by acid-catalyzed hydrolysis would eventually lead to incorrect conclusions because of the formation of secondary byproducts, acidolysis through allylic and benzopinacol rearrangements, mild and dioxane-water hydroly-ses through homolytic cleavages of aryl ether bonds in β-O-4'-type substructures to produce radical species and subsequent radical reactions.

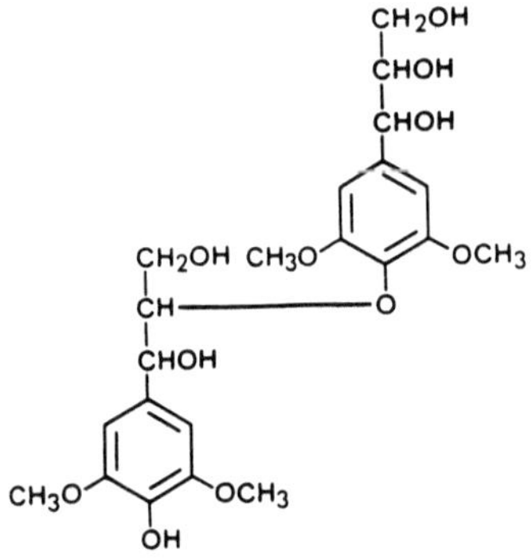

Figure 16 Major degradation products from dioxane-water hydrolysis of lignins at high temperature (170–180°C).

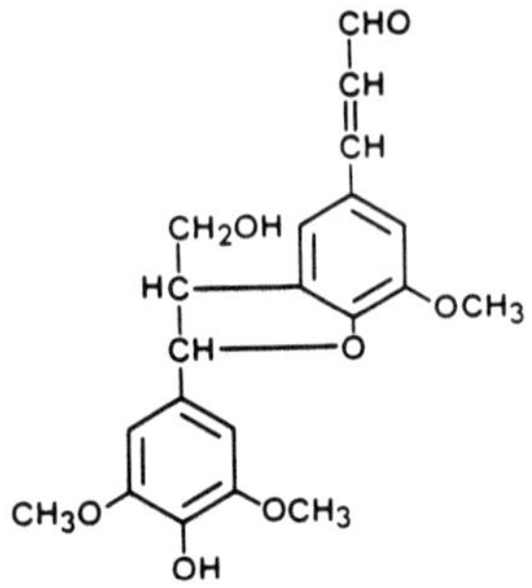

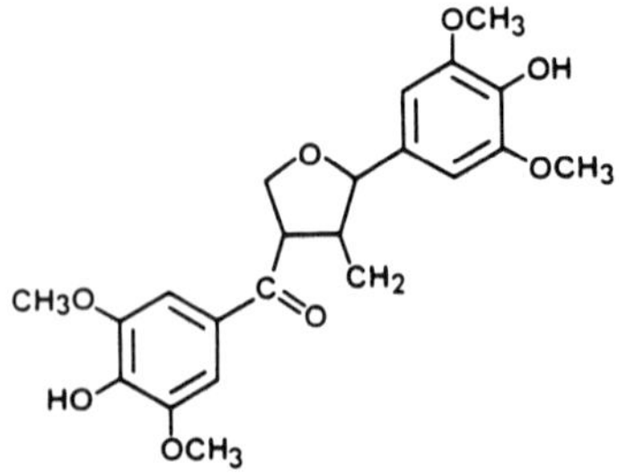

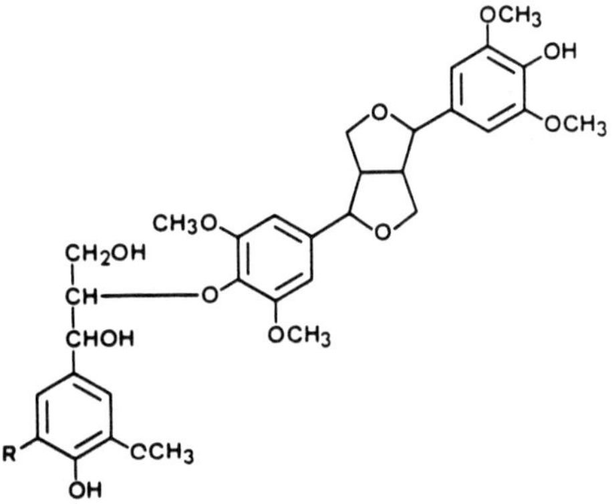

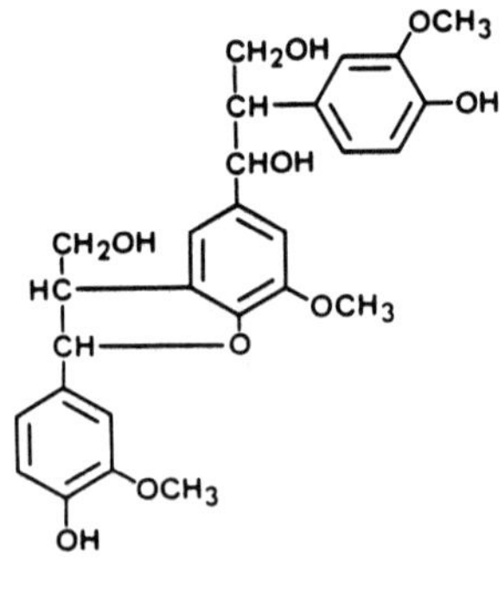

4. Thioacetolysis

After many discouraging results in searching for an effective method for the chemical characterization of lignins, Nimz and co-workers (207–209) introduced in 1969 the hydrolysis of lignins with thioacetic acid in the presence of a catalytic amount of boron trifluoride at room temperature. The resulting lignin thioacetate was treated successively with dilute alkaline solution and Raney-nickel to obtain low-molecular-mass reductive degradation products of the lignin. This procedure is called "thioacetolysis."

As shown in Fig. 17, when a lignin is treated with thioacetic acid in the presence of boron trifluoride, C-α hydroxyl and aroxyl groups in some of the phenylpropane (C_9) units are substituted by thioacetyl groups. These reactions proceed via formation of the corresponding benzylium cations by elimination of water or the corresponding phenols under the catalytic influence of boron trifluoride as a Lewis acid, and the subsequent nucleophilic attack of a sulfur atom in thioacetic acid on C-α. In addition, the resonance effect of the electron lone pairs of oxygen atom at C-4 of aryl groups promotes the elimination of water or phenols as nucleofugals. In the case of β-O-4' type-49 or-50 substructures the reaction results in the formation of thioacetate [96]. The thioacetate is then hydrolyzed by treatment with a dilute sodium hydroxide solution at 60°C to the corresponding benzylthioxide anion 97 under the reaction conditions because the pKa of the thiols is in the range of 8–9. Moreover, the thioxide anions are rather strong nucleophiles, stronger than alkoxide anions. Consequently, an intramolecular nucleophilic substitution of the substituent at the neighboring C-β by the benzylthioxide anion facilitates the cleavage of the β-aryl ether bond in intermediate 97 via a trans elimination with neighboring-group participation under rather mild reaction conditions. As a result, the aroxyl substituent at C-β is easily released as a corresponding phenolate anion with the formation of a thirane intermediate 98 that undergoes polymerization to give thioether intermediate 99. On heating in the presence of Raney-nickel, side chains of the thioester-reduced low-molecular-mass degradation products 100 or 101 form the substructures 49 and 50, respectively.

Since β-aryl ether (β—O—4') bonds are the most abundant linkages between C_9 units in lignins, in particular hardwood lignins, the yield of low-molecular-mass degradation products by this method is relatively high. In the case of lignin in beech wood (*Fagus silvatica*), the yield of low-molecular-mass fraction is approximately 90%/Klason lignin by this procedure (210). The fraction (MW < 1000) includes monomeric, dimeric, trimeric, and tetrameric degradation products. From the dimeric fraction, 20 compounds are isolated and identified (208, 209). The compounds (see Fig. 18 for structures) include 1,2-bis-guaiacylethane [102], 5-(2-guaiacylethyl)guaiacylpropane [104], 2,3-di-(syringylmethyl)butane [106], 2-guaiacylmethyl-3-guaiacylpentane [107] 2,2'-di-hydroxy-3,3-dimethoxy-5,5'-dipropylbiphenyl [109], 3,4-di(syringylmethyl)tetrahydrofuran [110], 5-(4-propylguaiacoxy)guaiacylpropane [111], and 2-hydroxy-1,3-dimethoxy-6,7-dimethyl-

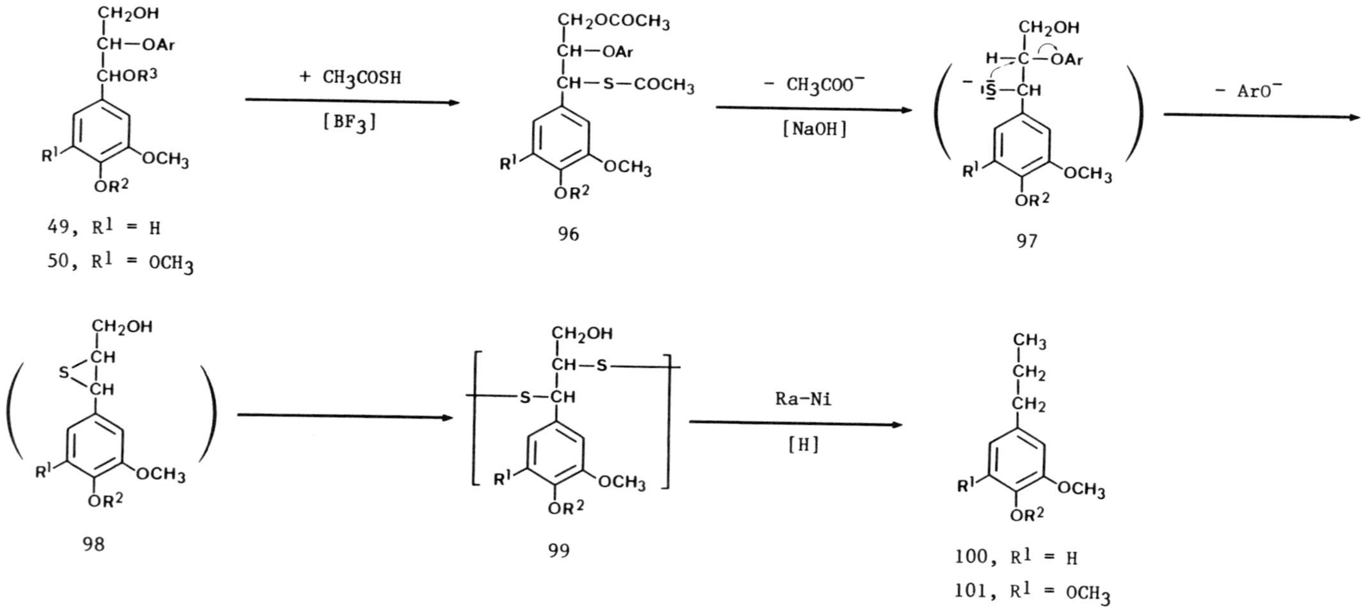

Figure 17 Reaction mechanism for thioacetolysis of lignins. R^2, R^3 = H, aryl, or alkyl. (From Res. 208–210.)

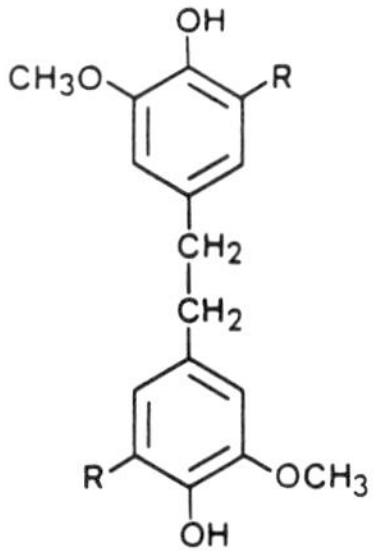

102, R = H
103, R = OCH_3

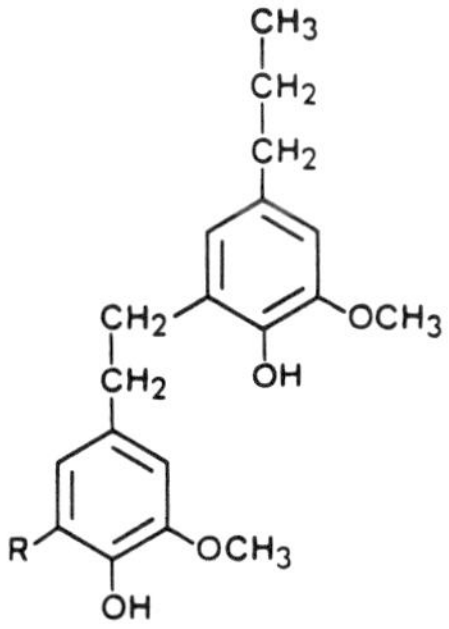

104, R = H
105, R = OCH_3

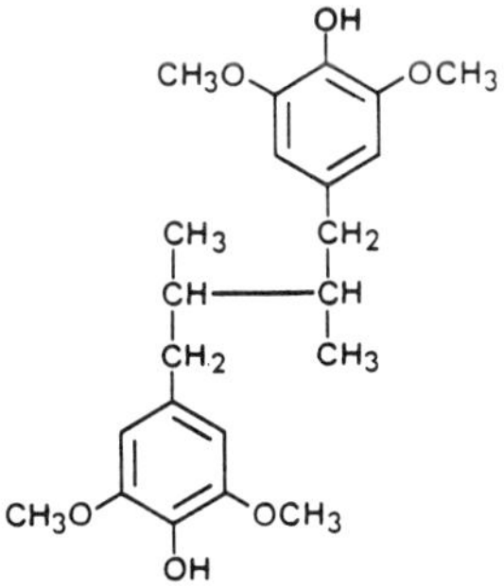

106

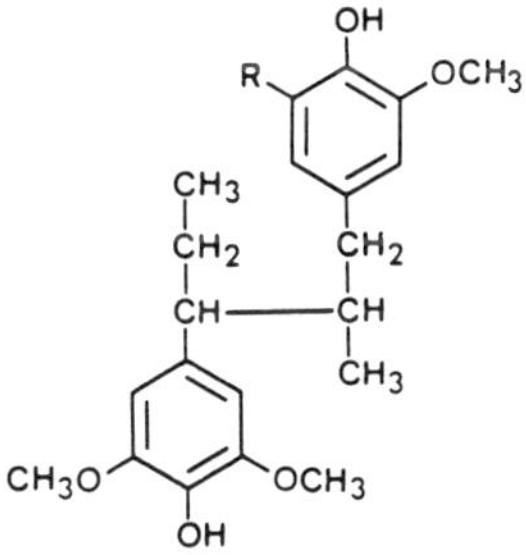

107, R = H
108, R = OCH_3

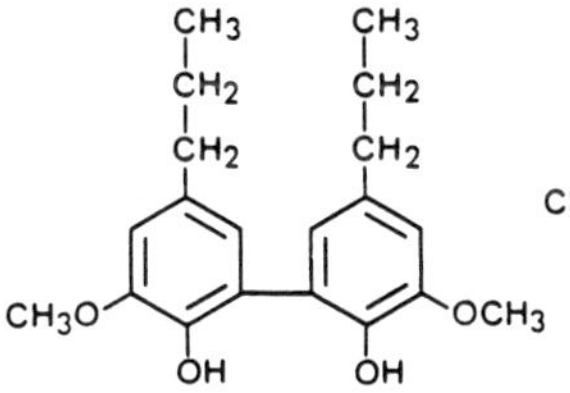

109

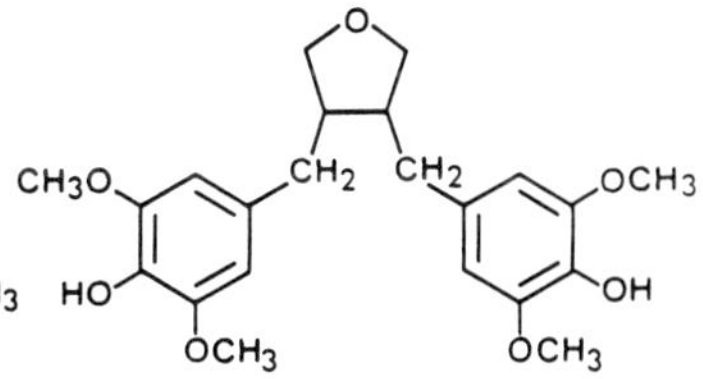

110

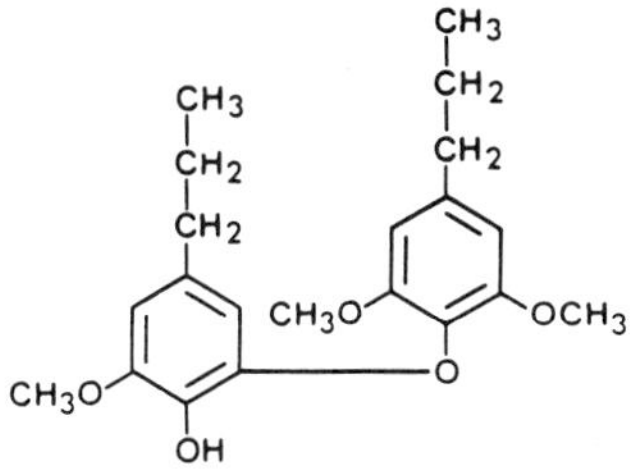

111

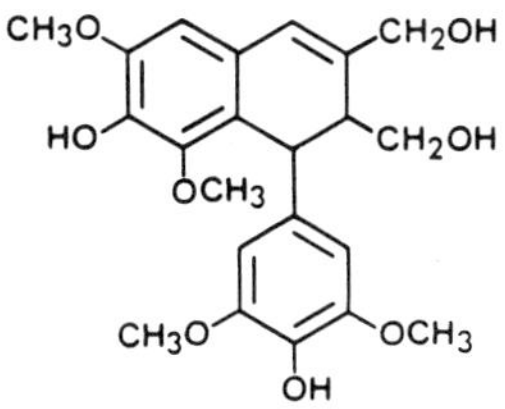

112

Figure 18 Major dimeric degradation products from thioacetolysis of lignins.

8-syringyl-7,8-dihydronaphthalene [112], in addition to compounds <u>103</u>, <u>105,</u> and <u>108</u>, syringyl analogues of compounds <u>102</u>, <u>104</u>, and <u>107</u>, with a yield for each of less than 0.8%/Klason lignin. Although the yield is rather low (each less than 0.3%/Klason lignin), the isolation of compounds <u>107</u> and <u>108</u> is interesting in terms of the chemical structure of lignins. The degradation products indicate that a, β-sub-structures, such as in an intermediate of type 115 (see Fig. 19 for structures), are also present in lignins, in addition to other substructures such as β-O-4', β-1, β-5', 5-5', β-β', 5-O-4', and tetralin substructures. As shown in Fig. 19, the formation of the α, β-substructure <u>115</u> requires not only dehydrogenation of immediate lignin precursors such as coniferyl alcohol [2] or sinapyl alcohol [3], but also the recombination among any two of the resulting radical species to a quinonemethide intermediate such as <u>113</u>, followed by the radical addition of a radical species of the type 2d derived from the dehydrogenation of a lignin precursor on C-α of the quinonemethide to a quinonemethide-radical intermediate <u>114</u>. An internal nucleophilic addition of the hydroxyl group at C-γ of C_9 unit A on C-α of the quinonemethide moiety in C_9 unit C in the intermediate <u>114</u> would lead to the formation of a oligomeric phenyloxane radical species of the type 115 containing an α, β'-substructure. Alternatively, the nucleophilic addition of a mole of water or a lignin precursor or an oligolignol on C-α of the quinonemethide moiety would lead to a lignin moiety containing an α, β'-substructure. A reaction of this type could be one of the propagation reactions during the process of lignification. Chen and Connors (210) demonstrated that a dehydrogenation-radical substitution mechanism occurs as a propagation reaction, in addition to the dehydrogenation-radical coupling mechanism in the dehydrogenative polymerization of phenols by one-electron-transfer oxidant.

5. Thioacidolysis

Recently, Lapierre and co-workers (212–217) have improved the efficiency of acidolysis by modifying the procedure in view of the rather low yield of oligomeric degradation products, in particular that of monomeric products, and the occurrence of secondary reaction products in acidolysis. In the modified procedure, a wood tissue or lignin sample is treated in dioxane-ethanethiol in the presence of boron trifluoride etherate or fluorboric acid etherate, acting as a Lewis acid, in an autoclave at 100°C for 4 hr under a nitrogen or argon atmosphere instead of in $0.2M$ hydrogen chloride in dioxane-water (9:1,v/v) at a refluxing temperature.

As shown in Fig. 20, an α-aryl ether bond in β-O-4' type-43 substructures undergoes substitution of aroxyl group at C-α a by an ethanethioxyl group under catalytic influence of boron trifluoride etherate by way of the corresponding conjugate acid intermediate <u>116</u>. This results in the formation of 1-S-ethyl-2-O-aryl(guaiacylpropane-1-thiol-2,3-diol) (118) and a phenolic fragment corresponding to the substituent at C-α in 49. The β-aroxyl group also undergoes thioacidolysis in the same fashion described above via corresponding intermediates <u>119</u> and <u>120</u> to produce the major product, 1,2,3-S-triethyl(guaiacylpropane-1,2,3-trithiol) [121].

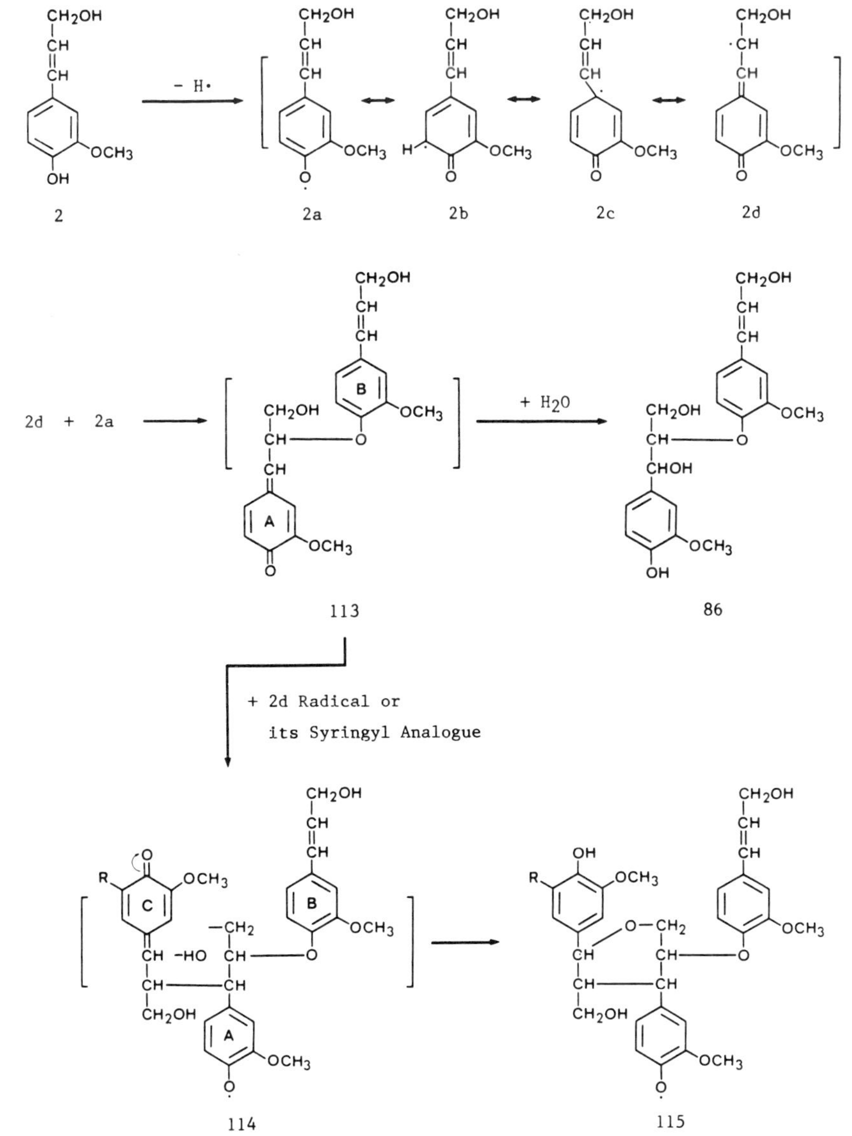

Figure 19 Possible reaction mechanism for formation of β-O-4' and a-β' substructures during in vivo lignificatlon process. R = H or OCH$_3$.

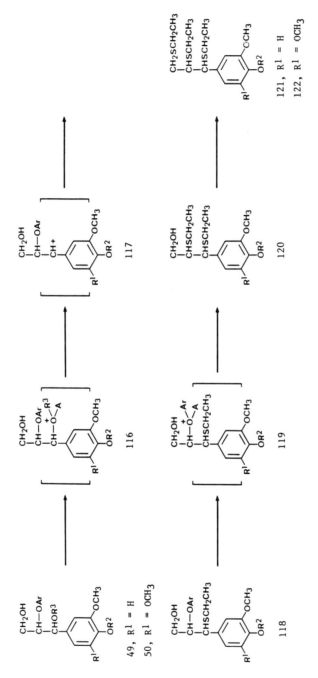

Figure 20 Reaction mechanism for thioacidolysis of lignins, R^2, $R^3 =$ H, aryl, or alkyl; A = Lewis acid. (From Ref. 212.)

In general, the α-aryl ether bond is more susceptible to thioacidolysis than the β-aryl ether bond because of the +M effect of the oxygen atom at C-4. This indicates that the substitution of an α-aroxyl group by an ethanethioxyl group probably follows a SN1cA reaction mechanism in which the reaction proceeds via the corresponding benzylium ion intermediate 117. The subsequent displacement of β-aroxyl and γ-hydroxyl groups by ethanethioxyl groups leads to the formation of the major products, compounds 121. Whether intermediate 120 and compound 121 are produced by a SN1cA or SN2cA reaction mechanism is not known. From syringyl (substructure 50) and 4-hydroxyphenyl analogues of substructure 49, 1,2,3-S-triethyl(syringylpropane-1,2,3-trithiol) [122] and 1,2,3-S-triethyl(4-hydroxyphenylpropane-1,2,3-trithiol) [123] are obtained as the major products, respectively.

In the laboratory, the thioacidolysis of lignins is conducted according to the procedure of Lapierre and co-workers (213–215). The reaction mixture is adjusted to pH 3, and the resulting solution is extracted with methylene dichloride. Compounds 121 through 123 and related degradation products in the resulting solution are quantitatively determined by gas chromatography (GC) and gas chromatography-mass spectrometry (GC-MS) after sillylation.

The results from the thioacidolysis of lignins unambiguously show that lignins occurring in plant tissue are composed mostly of phenylpropane units in the form of arylglycerol ether structures. In addition, the total yield of thioacidolysis products derived from uncondensed guaiacylpropane units, compound 121 and related compounds, indicate the minimum quantity of uncondensed guaiacylpropane units present in the lignin being characterized. Similarly, the total amount of compound 122 and related compounds or that of their 4-hydroxyphenyl analogues show the minimum amount of uncondensed syringyl- or 4-hydroxyphenylpropane units present in the lignin, respectively. Thus, the significance of the results from thioacidolysis of lignins is similar to those from nitrobenzene oxidation of the corresponding lignins. As shown in Table 7 (217), however, results of nitrobenzene oxidation is better than those of thioacidolysis in terms of yields of degradation products as expressed in both mole%/C_9 unit and moles/g, although the former does not reveal the arylglycerol ether structures of lignins as the latter does.

V. STRUCTURE OF LIGNINS

Lignins occurring in woody tissues of vascular plants are produced in vivo by an enzyme-initiated dehydrogenative polymerization of three immediate precursors, 4-hydroxycinnamyl alcohol [1], coniferyl alcohol [2] and synapyl alcohol [3] (see Fig. 1 for structures). The molar ratio of lignin precursors contributing to formation of lignins depends not only plant species but also morphological regions of the same plant. For example, most of the lignins in softwood species (Gemnospermea) consists mainly of guaiacylpropane units derived from the in vivo dehydrogenative

Table 7 Total Yield of Uncondensed Monomeric Products from Woods and Their Lignin
Preparations of Some Wood Species on Thioacidolysis and Nitrobenzene Oxidation

Wood species	Specimen	Total yield (μmoles/Klason Lignin or MWL[a])	
		Thioacidolysis	Nitrobenzene oxidation
Norway spruce	wood	1006	—
(*Picea abies*)	MWL	—	—
White spruce	wood	—	1798
(*Picea glauca*)	MWL	—	1826
Birch	wood	1866	—
(*Betula verrucosa*)	MWL	—	—
White birch	wood	—	2418
(*Betula papyrifera*)	MWL	—	2360
Poplar	MWL	1024	1660
(*Populus euramericana*)	EL[b]	1243	1770

[a]MWL = Milled wood lignin.
[b]EL = Enzymatically isolated lignin prepared from ball-milled poplar wood meal residue after removal of MWL.
Source: Refs. 119 and 217.

polymerization of coniferyl alcohol [2]. In contrast, lignins in hardwood species (Dicotyledonous Angiospermae) are composed mostly of guaiacyl- and syringylpropane units derived from dehydrogenative polymerization of coniferyl alcohol [2] and sinapyl alcohol [3] in different molar ratio, the magnitude of which depends on the nature of plant species. Similarly, lignins in grasses (Monocotyledonous Angiospermae) are comprised of guaiacyl-syringyl-type lignin cores and 4-hydroxycinnamic acid-type peripheral groups bonded to the lignin cores mostly in the form of cinnamic acid ester linkage at C-α and C-γ.

Unlike monomeric units in other naturally occurring biopolymers, the monomeric phenylpropane units in lignins thus produced are linked to each other not buy a single intermonomeric linkage but by several different carbon-to-carbon and ether linkages, most of which are not readily degradable under mild reaction conditions without undergoing extensive changes in the original structures such as extensive degradation of side chains and oxidative cleavage of aromatic moieties. Therefore, characterization of lignins in terms of chemical structure by chemical degradation is not only difficult but also tedious. However, these difficulties are alleviated by introduction of ^{13}C NMR spectrum of a lignin preparation which provides information about the nature of all carbons in the lighin (218).

In this section, the structure of lignins will be discussed on the basis of chemical evidence such as the results of elemental analysis, functional group analysis, chemical degradations as discussed in Sec. IV, and a combination of these methods with ^{13}C NMR spectroscopy (218, 219).

A. Chemical Composition

1. Elemental Composition

Lignin preparations isolated from the corresponding plant tissues usually contain some carbohydrates. These carbohydrates are present in lignin preparations either as a part of the lignin-carbohydrate-complex (LCC) or simply as contaminants. Before characterizing a lignin preparation, a carbohydrate analysis must, therefore, be conducted. In case the carbohydrate content of the sample is rather high, i.e., more than 5%/lignin, then the sample must be purified again in order to remove the carbohydrate contaminants as much as possible. The data from the elemental analysis of a lignin preparation are usually corrected for the carbohydrate content on the basis of the carbohydrate analysis of the sample.

Table 8 shows the carbohydrate content, elemental composition, and C_9 unit formula of some representative milled wood and bamboo lignins (MWL and MBL). It can be observed that the methoxyl content of softwood (spruce) MWL is smaller than 1 mole/C_9 unit, whereas those of hardwood MWL's and MBL are larger than 1 mole/C_9 unit. Evidently, the data indicate that spruce lignin is a guaiacyl-type lignin, whereas hardwood and bamboo lignins are guaiacyl-syringyl-type lignins. It must be mentioned that bamboo lignin, one of the grass lignins, consists of a guaiacyl-syringyl-type lignin core and 4-hydroxycinnamic acid-type peripheral units bonded to the lignin core by ester linkages at C-α and C-γ of the lignin core.

2. Functional Group Analysis

In addition to 4-hydroxylphenyl, guaiacyl and syringyl groups, their 4-O-alkyl and 4-O-aryl analogues and methoxyl groups as discussed in the previous sections, lignins also contain aliphatic and aromatic hydroxyl groups, and carbonyl groups including conjugate carbonyl groups such as cinnamaldehyde-type and benzoyl-type carbonyls.

Table 9 shows the total hydroxyl, phenolic, and aliphatic contents of milled wood and bamboo lignins (MWL's and MBL) from some plant species. In general, the phenolic hydroxyl content of MWL's from the wood of soft- and hardwood species is in the range of 0.2–0.3 mole/C_9 unit. However, lignins in the wood of these plant species have a phenolic hydroxyl content of approximately 0.1 mole/C_9 unit (99, 101). The higher phenolic hydroxyl contents observed in the MWL's are probably caused by the cleavage of α-aryl ether bonds during the process of isolating MWL's. The total hydroxyl content of lignins is determined by acetylation and the subsequent determination of O-acetyl content in the resulting acetylated lignin by saponification according to the Kuhn-Roth procedure (220). The phenolic hydroxyl

Table 8 Carbohydrate Content, Elemental Composition, and C_9 Unit Formula of Milled Wood and Bamboo Lignins (MWL's and MBL) from Some Plant Species

Lignin preparations	Carbohydrate content (% lignin)	Elemental composition[a]				C_9 unit formula
		C%	H%	O%	OCH%	
Milled wood lignins						
White spruce (*Picea glauca*)	2.2	64.30	6.21	29.48	15.72	$C_9H_{7.66}O_2(H_2O)_{0.48}(OCH_3)_{0.94}$
White birch (*Betula papyrifera*)	4.8	61.88	6.32	33.44	22.43	$C_9H_{6.87}O_2(H_2O)_{0.86}(OCH_3)_{1.52}$
Sweetgum (*Liqidambar stryaciflua*)	3.6	58.74	5.67	35.59	22.16	$C_9H_{5.00}O_2(H_2O)_{1.28}(OCH_3)_{1.55}$
Zhong-Yang Mu (*Bischofia polycarpa*)	6.2	61.36	5.96	32.68	17.75	$C_9H_{6.49}O_2(H_2O)_{0.92}(OCH_3)_{1.13}$
Milled bamboo lignin						
Bamboo (*Phyllostachys pubescens*)	5.0	61.66	5.53	32.81	19.42	$C_9H_{5.52}O_2(H_2O)_{0.84}(OCH_3)_{1.25}$

[a](1) Corrected for carbohydrate content, (2) based on average of two elemental analyses, and (3) 0% by difference.
Source: From Refs. 101, 119, and 125.

Table 9 Total, Phenolic, and Aliphatic Hydroxyl Contents of Milled Wood and Bamboo Lignins (MWL's and MBL) from Some Plant Species

Lignin preparations	Total -OH	Phenolic -OH	Aliphatic -OH
	(moles/C_9 unit)		
White spruce (*Picea glauca*)	1.46	0.28	1.18
Zhong-Yang Mu (*Bischofia polycarpa*)	1.33	0.22	1.11
Bamboo (*Phyllostachys pubescens*)	1.49	0.36	1.13

Source: From Refs. 125 and 219.

content of lignins is estimated by ultraviolet spectrophotometry according to the Goldschmid procedure (221). However, the procedure usually yields considerably lower phenolic hydroxyl content for alkaline technical lignins such as soda lignins and kraft lignins due to the presence of catechol-type structures in the lignins. Alternatively, the phenolic hydroxyl content of lignins is determined by the ammonolysis of the corresponding acetylated lignins according to the Månsson procedure (222). The aliphatic hydroxyl content of a lignin is then calculated by subtracting the phenolic hydroxyl content from the total hydroxyl content. The aliphatic and phenolic hydroxyl contents of a lignin can also be estimated from the ^{1}H NMR spectrum of the corresponding acetylated lignin in case the total O-acetyl content of the acetylated lignin is known (218). Aliphatic hydroxyl groups of lignins include primary and secondary hydroxyl groups, mostly present in C-γ and C-α, respectively. The latter consist of phenolic and nonphenolic benzyl alcohol-type hydroxyl groups (phenolic and nonphenolic α-hydroxyl groups). As shown in Fig. 19, α-hydroxyl groups are produced by the nucleophilic addition of a mole of water on C-α of a quinonemethide intermediate such as intermediate 113 to β-O-4'-type substructure 86 during the dehydrogenative polymerization of lignin precursors in the in vivo lignification process. As shown in Table 8, therefore, lignins have an oxygen content of more than 2 atoms/C_9 unit, excluding the oxygen atom in methoxyl groups since lignin precursors, 4-hydroxycinnamyl alcohol [1], coniferyl alcohol [2], and sinapyl alcohol [3] (see Fig. 1 for structures), have two oxygen atoms/molecules, excluding the oxygen atom in the methoxyl groups. Consequently, the oxygen content of lignins in excess of the 2/C_9 unit, excluding the oxygen atom in the methoxyl groups, is usually expressed in terms of H_2O that is called "added water." Softwood lignins usually contain the added water in the range of 0.3–0.5/C_9 unit and hardwood lignins in the range of 0.5–0.8/C_9 unit. Theoretically, the value for added water of a lignin should approximately correspond to the total content of α-hydroxyl groups. In practice, however, this is not the case, probably because α-hydroxyl groups undergo oxidation to the corresponding α-carbonyl groups and other reactions during the in vivo lignification process and/or during the process of isolation. The value for added water in sweetgum MWL (*Liquidambar styraciflua*) (see Table 8) is considerably high, approximately 1.3 moles/C_9 unit. Since lignins are hygroscopic, the high added water value for sweetgum MWL is caused by insufficient drying before conducting elemental analysis and/or absorption of moisture during the analysis.

   Lignins contain carbonyl groups including conjugated carbonyl groups of the 4-alkoxycinnamaldehyde type [8] (see Fig. 3 for structure) and 4-hydroxy, and 4-alkoxyphenyl-3-oxopropne types (phenolic and nonphenolic α-carbonyl groups). The total carbonyl content of Norway spruce (*Picea abies*) MWL was determined to be approximately 0.2 mole/C_9 unit by reaction with hydroxylamine hydrochloride (223, 224). The presence of 4-alkoxycinnamaldehyde-type structure 8 in lignins can be qualitatively shown by the Weisner reaction as discussed in Sect. II.A.1 (26–28). Among the methods for the determination of the total carbonyl content in lignins, the hydroxylamine hydrochloride method is the best one because hydroxylamine hydrochloride reacts with almost all types of carbonyl groups including diketones and o- and p-quinones under rather mild reaction conditions to give the corresponding oxaimes quantitatively with the liberation of one mole of hydrochloric acid per one carbonyl group. The contents of 4-alkoxycinnamaldehyde and 4-hydroxy and 4-alkoxyphenyl-3-oxopropane types in spruce MWL were determined to be approximately 0.03, 0.01, and 0.06 mole/C_9 unit by sodiumborohydride reduction-ultraviolet spectrophotometry, with a total conjugated carbonyl content of approximately 0.1 mole/C_9 unit (224). The difference between the total and conjugated carbonyl contents, approximately 0.1 mole/C_9 unit, is assumed to be the quantity of phenyl-2-oxopropane-type carbonyl groups (β-carbonyls) present in the MWL. However, the presence of β-carbonyls in spruce MWL is not well established. No comprehensive investigations have been conducted on the total carbonyl contents, types of carbonyl groups and their content in lignins from other plant species.

3. Major Lignin Substructures

The results of degradation reactions and the functional group analysis of lignins showed that lignins are comprised of the following major substructures, as shown in Fig. 21, besides numerous minor substructures: arylglycerol-β-aryl ether (β-O-4', A); glycerol-2-aryl ether (displaced β-O-4', B); noncyclic benzyl aryl ether (noncyclic α-O-4', C); phenylcoumaran (cyclic β-5' and cyclic α-O-4', D); C-2/C-6 condensed arylpropane (α-2'/6' and β-2'/6', E); biphenyl (5-5', F); diphenyl ether (4-O-5', G); 1,2-bis-arylpropane (β-1', H); β-β' linked (β-β', I); arylglycerol-α-4-hydroxycinnamic acid ester (α-HC ester, J); and arylglycerol-γ-4-hydroxycinnamic acid ester (γ-HC ester, K) substructures. There are more than 10 minor substructures

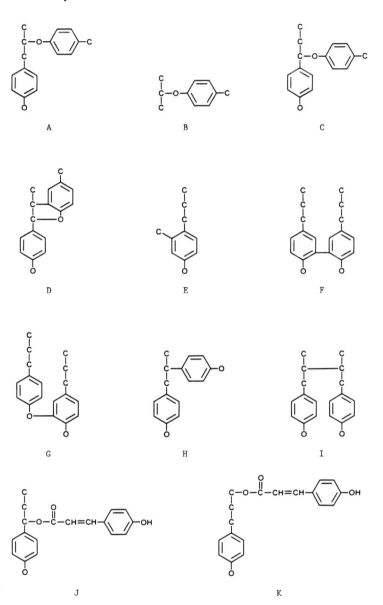

Figure 21   Major substructures in lignins. (Fom Refs. 24 and 98.)

that are present in lignins in an insignificant amount, usually in a trace amount, and are not included in Fig. 21.

A substructure of type A (β-O-4') is the major substructure in lignins. The frequency of a β-O-4' substructure in guaiacyl lignins (softwood lignins) is approximately in the range of 45–50/100 C_9 units. In guaiacyl-syringyl lignins (softwood lignins and lignin cores of grass lignins), approximately 75–85% of syringylpropane units are involved in the β-O-4' substructure, of which approximately 80–90% are associated with nonphenolic β-O-4' substructures. Since the molar ratio of syringylpropane to guaiacylpropane of guaiacyl-syringyl lignins depends on the nature of plant species, the quantity of β-O-4' structures in the lignins varies considerably among hardwood and grass species. Consequently, the proportion of the aforementioned substructures in the lignins also differs considerably among plant species belonging to angiospermae. Only grass lignins contain substructures of the ester-types J and K (α- and γ-HC esters). The proportion of these structures in grass lignins is in the range of 5–25 mole%/C_9 unit, depending on the nature of grass species. In general, the lignins in soft- and hardwood species do not contain α- and γ-HC ester-type substructures. However, lignins in woods belonging to genus Poplus (poplar) contain a 4-hydroxybenzoic acid ester group bonded to C-α of the arylpropane units.

B. Structures

1. Spruce Lignin

One of the important results from the characterization of lignins by potassium permanganate is that the difference between the total yields of aromatic acid products in mole%/C_9 with and without cupric oxide pretreatment corresponds to the number of phenolic hydroxyl groups/100 C_9 units liberated during the cupric oxide pretreatment. The latter, in turn, is approximately equal to the total number of α- and β-aryl ether bonds, including cyclic α-aryl ether bond/100 C_9 units (98). Thus, Erickson and co-workers (98) was able to characterize the structure of milled wood lignin (MWL) from Norway spruce (*Picea abies*) in terms of its substructures and types of intermonomeric bonds and their frequency by potassium permanganate oxidation of the MWL and other chemical methods. The results are as follows (see Fig. 21 for structures):

1.   The frequency of an arylglycerol-β-aryl ether (β—O—4') bond in substructure A was estimated to be approximately 48/100 C_9 units by subtracting the frequencies of the displaced β—O—4' bond in B and the noncyclic and cyclic α—O—4' bond in C and D from the total phenolic hydroxyl groups/100 C_9 units liberated during the cupric oxide pretreatment in the potassium permanganate oxidation of the MWL (98).

2.   The frequency of a glyceraldehyde-2-aryl ether (displaced β—O—4') bond in substructure B was estimated to be approximately 2/100 C_9 units by acidolysis

(225) and ^{1}H NMR spectroscopy (226).

3.   The frequency of a noncyclic benzyl aryl ether (noncyclic α—O—4') bond was estimated to be approximately 6–8/100 C_9 units through the selective acid-catalyzed hydrolysis of spruce MWL under mild reaction conditions (227, 228). Freudenberg and co-workers (229) obtained a similar value for the bond by another method.

4.   The frequency of a β-aryl (β-5') bond in substructure D is estimated to be approximately 9–12/100 C_9 units on the basis of the yield of isohemipinic acid [21] produced on potassium permanganate oxidation of the MWL (98).

5.   The frequency of a cyclic benzyl aryl ether (cyclic α-O-4') bond in substructure D is the same as that of the β-5' bond in D.

6.   The total frequency of α- and β-aryl (α-6' and β-6') bonds in C-6-condensed arylpropane substructures E was estimated to be a total of 2.5–3/100 C_9 units on the basis of the yield of meta-hemipinic acid [22] produced on potassium permanganate oxidation of the MWL (98).

7.   The frequency of a biphenyl (5-5') bond in substructure F was estimated to be approximately in the range of 9.5-11/100 C_9 units on the basis of the yield of dehydrodiveratric acid [26] produced on potassium permanganate oxidation of the MWL (98). A recent study on the structure of spruce MWL by ^{13}C NMR spectroscopy.indicates that the frequency of the 5—5' bond in the lignin approximately 12–13/100 C_9 units (230).

8.   The frequency of a diphenyl ether (4—O—5') bond in substructure G was estimated to be approximately in the range of 3–4/100 C_9 units on the basis of the yield of 2,2',3-trimethoxydiphenyl-ether-4'5-dicarboxylic acid [24] produced on potassium permanganate oxidation of the MWL (98).

9.   The frequency of a β-aryl (β-1') bond in 1,2-bis-arylpropane of the type H was estimated to be approximately 7/100 C_9 units by acidolysis of the MWL [171].

10.   The frequency of a β-β' bond in a β-β' linked substructure of type I was estimated to be approximately 2/100 C_9 units on the basis of the total yield of pinoresinol [69] derivatives obtained by acidolysis of spruce MWL (171). Structural studies of spruce MWL by ^{13}C NMR spectroscopy also showed that the lignin contains a rather low content of pinoresinol substructures (183, 231, 132).

Table 10 summarizes the types of substructures and intermonomeric bond types and their frequency in spruce MWL. Figure 22 shows a structural scheme of spruce lignin that is comprised of only the prominent substructures and bond types discussed above (24).

2. Hardwood Lignins

Larsson and Miksche (116) determined the molar ratio of syringylpropane to guaiacylpropane units in milled wood lignin (MWL) from birch (Betula verrucosa) to be approximately 1:1 on the basis of the yields of aromatic acid produced on potassium permanganate oxidation of the MWL. In addition, the bond types and

Table 10 Frequency of Major Intermonomeric Bond Types in MIlled Wood Lignin (MWL) from Spruce (*Picea abies*)

Intermonomeric bondtype[a]	Frequency per 100 C_9 units
β—O—4', bond in substructure A	48
Displaced β—O—4', bond in substructure B	2
Noncyclic α—O—4', bond in substructure C	6–8
β—5' bond in substructure D	9–12
Cyclic α—O—4' bond in substructure D	9–12
α—6' and β—6' bonds in substructure E	2.5–3
5—5' bond in substructure F	9.5–11
4—O—5' bond in substructure G	3.5–4
β—1' bond in substructure H	7
β—β' bond in substructure I	2

[a]See Fig. 21 for the Structure of substructures.
Source: Refs. 24 and 98.

Table 11 Frequency of Major Intermonomeric Bond Types in Milled Wood Lignin (MWL) from Birch (*Betula verrucosa*)

Intermonomeric bond type[a]	Frequency per 100 C_9 units		
	Guaiacyl-type	Syringyl-type	Total
β—O—4' bond in substructure A	22–28	34–39	58–65
Displaced β—O—4' bond in substructure B			2
Noncyclic α—O—4' bond in substructure C			6–8
β—5' Bond in substructure D			6
Cyclic α—O—4' bond in substructure D			6
α—6' and β—6' bonds in substructure E[b]	1–1.5	0.5–1	1.5–2.5
5—5' bond in substructure F	4.5		4.5
4—O—5' bond in substructure G	1	5.5	6.5
β—1' bond in substructure H			7
β—β' bond in substructure I			3

[a]See Fig. 21 for the structure of substructures.
[b]In the case of syringyl-type α—2' and β—2' bonds instead of α—6' and β—6' bonds, respectively.
Source: Ref. 115.

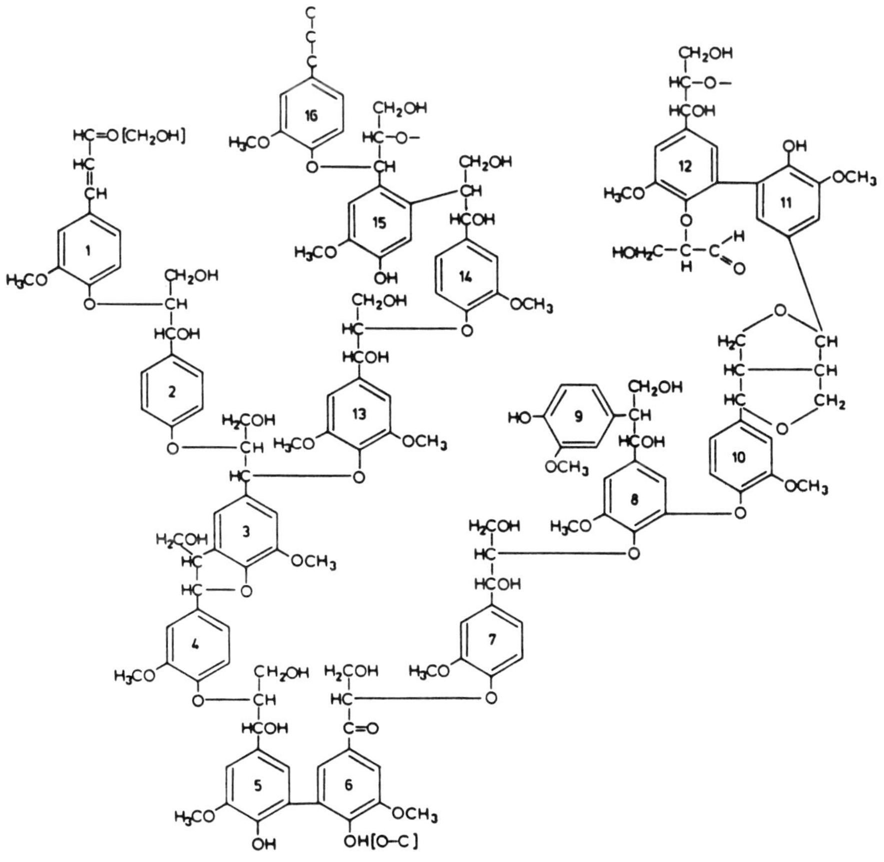

Figure 22 Structural scheme of spruce lignin. (From Ref. 24.)

their frequency per $100\,C_9$ unit in the birch MWL were estimated in the same fashion as described for the estimation of those in spruce MWL (98). The results are given in Table 11. Evidently, about 70–80% of syringylpropane units in birch lignin are involved in β-O-4' substructures of type A. Recently, Robert and co-workers (233, 234) concluded that approximately 80–90% of the syringylpropane units involved in β-O-4' substructures are nonphenolic on the basis of ^{13}C NMR spectral data of birch MWL.

Nimz (235) presented major substructures and calculated the proportion of major bond types from the relative yields of monomeric and dimeric degradation products isolated from the thioacetolysis mixture obtained from the wood meal of beech

Table 12 Proportion of Major Intermonomeric Bond Types in Beech (*Fagus silvastica*) Lignin

Intermonomeric bond type[a]	Proportion (in %)
β—O—4' and noncyclic α—O—4' bonds in substructes A and C	65
β—5' bond in substructure Db	6
α—6' in substructure E of syringyl-type tetralin structure (112)[c]	0.5
5—5 bond in substructure F	2.3
4—O—5' bond in substructure G	1.5
β—1' bond in substructure H	15
β—β' bond in substructure I of Syrinaresinol- and pinoresinol types	5
β—β' bond in substructure I of dibenzyltetrahydrofuran type	2
α—β' bond in substructure of type-107	2.5

[a]See Figs. 18 and 21 for the structure of substructures.
[b]Substructure D also contains 6% of the cylcic α—O—4' bond.
[c]Tetralin-type substructure also contains 0.5% of β—β' bond.
Source: Ref. 235.

(*Fagus silvatica*). Table 12 summarizes the major bond types and their proportion in beech lignin. Figure 23 shows a constitution scheme of beech lignin. Considering the difference in the molar ratio of syringylpropane to guaiacylpropane units between beech and birch milled wood lignins, approximately 1:1 vs. 1:2, the proportion of major bond types in beech lignin is rather similar to that of birch lignin.

C.   Characterization of Lignins by ^{13}C NMR Spectroscopy

Zhong-Yang Mu (*Bischofia polycarpa*; Euphorbiaceae) is a hardwood timber species widely distributed in southeastern China south of the Yangtze river (236). Recently, approximately 6600-year-old logs of Zhong-Yang Mu were recovered from the river bed of the Yangtze river (237). Although the surface of the logs is covered by a layer of carbon, the inner layer of the logs is rather well preserved. Since the physical properties of the intact ancient wood are similar in magnitude to those of the corresponding part of recent wood, the chemical composition of wood components in Zhong-Yang Mu has drawn considerable interest. The characterization of milled wood lignin (MWL) from recent wood of Zhong-Yang Mu is a part of comparative studies on the chemical composition of wood components in recent

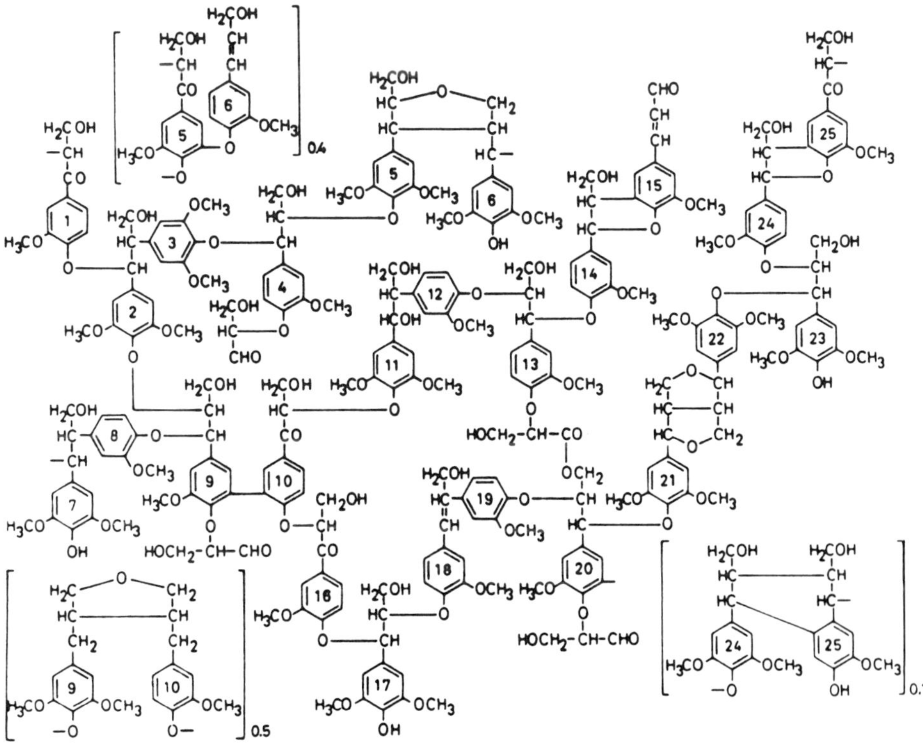

Figure 23 Constitutional scheme of beech lignin. (From Ref. 235.)

and ancient woods of Zhong-Yang Mu (B. *polycarpa*) (219).

Fig. 24 shows the quantitatative ^{13}C NMR spectrum of the MWL from the recent wood of Zhong-Yang Mu by an inverse gated decoupling (IGD) sequence (238). The quantitative nature of the spectrum is evidenced by the fact that signal 29 (aromatic methoxyl carbon) in the spectrum integrates about 1.16 carbons/phenyl group, as shown in Table 13. This corresponds to a methoxyl content of approximately 1.16 moles/C_9 unit. The elemental analysis of the MWL gives a methoxyl content of 1.13 moles/C_9 unit. Thus, the methoxyl content of the MWL estimated from the IGD spectrum has a deviation of about 3% less than the value obtained from elemental analysis. This represents within a 5% deviation from the latter value, the limit of error allowed for the ^{13}C NMR spectroscopic estimation. Fig. 25 shows CH, CH_2, and CH_3 ^{13}C NMR subspectra of the MWL, edited by the distortionless enhancement by polarization transfer (DEPT) sequence (234, 239–241), whereas Fig. 26 depicts the quarternary C ^{13}C NMR subspectrum of the MWL obtained by

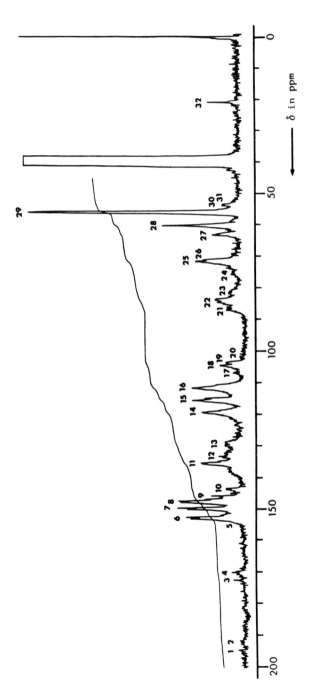

Figure 24 Quantitative ^{13}C NMR spectrum of milled wood lignin (MWL) from recent wood of Zhong-Yang Mu (*Bischofia polycarpa*) by inverse gated decoupling sequence. Solvent: DMSO-d_6 (From Ref. 219.)

Table 13 Integral of Spectral Regions in ^{13}C NMR Spectrum of Milled Wood Lignin (MWL) from Recent Wood of Zhong-Yang Mu (*Bischofia polycarpa*)

Spectral region	Chemical shift Range (δ in ppm)	Integral[a]	Number of carbons per phenyl group
Aromatic quarternary[b]	156–128	33.0	3.61
Aromatic tertiary C	128–103	23.0	2.51
Syringyl C—2/C—6	108–103	3.3	0.36
Oxygenated aliphatic C[c]	90–57.5	28.6	3.13
Aromatic methoxyl C	57.5–54.0	10.6	1.16
Aliphatic C (C—β of β—β' and β—5')	54.5–53.0	0.8	0.09

[a]The MWL contains approximately 0.03/phenyl group of cinnamaldehyde-type and 0.03/phenyl group of coniferyl alcohol-type substructures. Therefore, the aromatic region of the spectrum integrates a total of 6.12 carbons/phenyl group, and,factor for one carbon is $56/6.12 = 9.15$.
[b]Include 0.12 vinyl carbons/phenyl groups.
[c]Consist of carbons in lignin side chains and carbohydrates.
Source: Ref. 219.

subtracting the DEPT-edited CH, CH$_2$, and CH$_3$ subspectra from the IGD spectrum. In Table 13, the IGD spectra of the MWL is divided into several spectral regions on the basis of the DEPT-edited subspectra of the MWL. Table 14 shows the assignment of signals in the IGD spectrum of the MWL. The signals are assigned on the basis of published data (218, 230–234, 242).

As shown in Table 8, the MWL from the recent wood of Zhong-Yang Mu contained a carbohydrate content of approximately 6.2%. Indeed, the IGD spectrum of the MWL (see Fig. 24) reveals that the MWL contains a substantial amount of carbohydrates, as evidenced by the presence signals 4, 20, 24, and 32 at δ 169.6, 101.6, 76–73, and 20.9 ppm corresponding to O-acetyl C=O, C-1, and C-2 C-4 of xylan, and O-acetyl CH$_3$, respectively . Thus, the ^{13}C NMR spectral data further support the results of the carbohydrate analysis of the MWL. The IGD spectrum of the MWL further shows that the lignin is of the guaiacyl-syringyl type. The presence of guaiacylpropane structures is evidenced by signals 14, 15, and 16 at δ 119.6, 115.8, and 112.2 ppm corresponding to C-6, C-5, and C-2 of the guaiacyl group, respectively. In contrast to the rather weak signal 9 at δ 145.6 ppm corresponding to C-4 of the guaiacyl group, signals 7 and 8 at δ 149.9–149.4 and 147.7–147.2 ppm

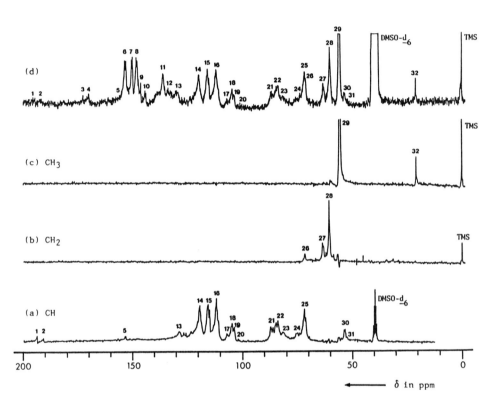

Figure 25 DEPT-edited ^{13}C NMR subspectra of milled wood lignin (MWL) from recent wood of Zhong-Yang Mu (*Bischofia Polycarpa*): (a) CH subspectrum, (b) CH$_2$ subspectrum, (c) CH$_3$ subspectrum, and (d) spectrum by inverse decoupling sequence. (From Ref. 219.)

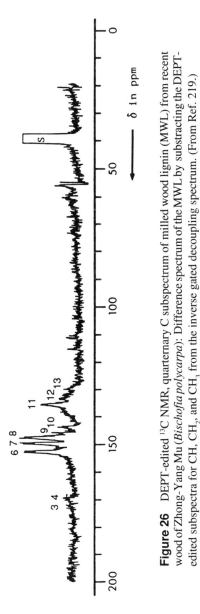

Figure 26 DEPT-edited ^{13}C NMR, quaternary C subspectrum of milled wood lignin (MWL) from recent wood of Zhong-Yang Mu (*Bischofia polycarpa*): Difference spectrum of the MWL by substracting the DEPT-edited subspectra for CH, CH$_2$, and CH$_3$ from the inverse gated decoupling spectrum. (From Ref. 219.)

Table 14 Assignment of Signals in ^{13}C NMR Spectrum of Milled Wood Lignin (MWL) from Recent Wood of Zhong-Yang Mu (*Bischofia polycarpa*)

Signal	Chemical shift (δ in ppm)	Intensity	Assignment[a]
1	194.0	vw	C=O in Ar-CH-CH-CHO and Ar-CO-C-
2	191.6	vw	-C=O in Ar-CHO
3	172.0	vw	-C=O in -COOH of aliphatic acids
4	169.6	w	-C=O in O-acetyl groups of xylan
5	152.9	vw	C-α in Ar-CH=CH-CHO
6	152.9–152.3	s	C-3/C-3' in etherified biphenyl (5-5') C-3/C-5 in etherified S
7	149.4–149.2	s	C-3 in etherified G
8	147.7–147.2	s	C-3 in nonetherified G C-4 in etherified G C-3/C-5 in nonetherified S
9	145.6	w	C-4 in nonetherified G
10	143.2	m	C-4' in phenylcoumaran (β-5') C-4/C-4' in etherified 5-5'
11	135.6	s	C-1 in etherified G C-4 in etherified S
12	135–131	m	C-1 in nonetherified G and S
13	130–129	m	C-α/C-β in Ar-CH=CH-CH$_2$OH C-β in Ar-CH=CH-CHO
14	119.0	s	C-6 in etherified and nonetherified G
15	115.8	s	C-5 in etherified and nonetherified G
16	112.2	s	C-2 in etherified and nonetherified G
17	106.5	w	C-2/C-6 in etherified and nonetherified S with α-C=O
18	104.7	m	C-2/C-6 in etherified and nonetherified 5
19	103.7	m	C-2/C-6 in etherified and nonetherified S and S β-β'
20	101.6	vw	C-1 in xylan
21	87.0	m	C-β in S β—O—4' (*erythro*)
22	84.6	s	C-β in G β—O—4' (*erythro*)
23	81.2	m	Unknown
24	76–73	w	C-2/C-3/C-4 in xylan
25	72.2	s	C-α in G β—O—4' (*erythro*)
26	71.6	w	C-α G and S β—O—4' (both *threo*) C-γ G and S β—β'
27	62.9	s	C-γ in G and S β-5' and β-1' C-γ in G and S β—O—4' with α-C=O C-5 in xylan

Table 14 (continued)

Signal	Chemical shift (δ in ppm)	Intensity	Assignment[a]
28	60.2	s	C-γ in G and S β—O—4'
29	55.8	vs	C in methoxyl group of Ar-OCH$_3$
30	53.8	w	C-β/C-β' in G and S β—β'
31	53.4	vw	C-β in G and S β-5'
32	20.4	s	C in CH$_3$ of O-acetyl groups of xylan

[a]vs = Very strong; s = strong; m = moderate; w = weak; vw = very weak.
[b]G = Guaiacyl-type; S = sryigyl-type.
Source: From Ref. 219.

are very strong. Since signals 7 and 8 correspond to C-3 and C-4 of the 4-O-alkylated guaiacyl group (242), the guaiacyl moieties must be present in the lignins predominantly in the form of 4-O-alkyl ethers. Moreover, the 4-O-alkylated guaiacyl moieties seem to be present in the lignin in the form of β-O-4' substructures because of the presence of relatively strong signals 22, 25, and 28 at δ 84.6, 71.8, and 60.2 ppm. These signals correspond to C-β, C-α, and C-γ of the β-O-4' substructure, respectively (243). The presence of syringylpropane structures is evidenced by signals 18 and 19 at δ 104.9 and 103.7 ppm, both corresponding to C-2/C-6 of the syringyl group, in addition to signal 6 at δ152.6–152.3 ppm corresponding to C-3/C-5 of the 4-O-alkylated syringyl group (242). The presence of signal 21 at d 87.0 ppm in addition to the signals 25 and 26 indicates further that 4-O-alkylated syringylpropane units are present in the lignins mostly in the form of β-O-4' substructures. Signals 30 and 31 at δ 53.8 and 53.4 ppm indicate the presence of β-β' and β-5' substructures in the lignin. The aromatic region of the IGD spectrum is chosen as the standard to analyze the spectra quantitatively, since the region does not contain the signals of carbohydrate contaminants.The DEPT-edited CH subspectrum (Fig. 25a) of the MWL indicate the presence of cinnamaldehyde- and cinnamyl alcohol-type structures as evidenced by signals 1, 5, and 13 at δ 194.0, 152.9, and 129.7–129.4 ppm, respectively. From the integral, the quantity of cinnamaldehyde- and cinnamyl alcohol- type structures, e.g., coniferyl aldehyde- and alcohol-type structures, in the lignin is estimated to be each on the order of approximately 0.03 mole/phenyl group. Consequently, the integral of the aromatic region of the IGD spectrum should correspond to 6.12 carbons/phenyl group. Since one uncondensed syringylpropane unit always has two C-2/C-6 carbons, an approximate mole% of the syringylpropane unit/C$_9$ unit in the MWL can be estimated from the total number of carbons in the syringyl C-2/C-6 spectral region in Table 13, i.e., 0.18 mole%/C$_9$ unit [= 0.36/2] for the MWL. Because p-hydroxyphenylpropane units in the MWL are negligible as indicated by the result of nitrobenzene oxidation as shown in Table 5, the MWL consists mostly of guaiacylpropane and syringylpropane units in an approximate molar ratio of 0.82:0.18. The methoxyl content of the MWL would then be 1.18 moles/C_9 unit. The methoxyl content of the MWL thus estimated is somewhat higher than the elemental analytical value of 1.13 moles/C_9 unit as shown in Table 8 and the value of 1.16 moles/C_9 unit estimated from the aromatic methoxyl spectral region in Table 13. The excess value probably is due to the presence of diphenyl ether (4—O—5') substructures. The chemical shift of C-2 and C-6 in 5-aroxy-guaiacylpropane is in the range of δ 110–103 ppm. If we consider the possible experimental errors, in particular those from the ^{13}C NMR experiments, the frequency of the 4—O—5' bond can be estimated to be approximately 0.03–0.04/C_9 unit [= 1.18 - (1.13 +1.16)/2] for the MWL. On the basis of potassium permanganate oxidation, the frequency of the 4—O—5' bond has been estimated to be approximately 0.03–0.04/C_9 unit for the spruce MWL (98) and 0.06-0.07/C_9 unit for birch MWL (115). Thus, the value for frequency of 4—O—5 ' bond in the MWL estimated here is rather reasonable. It follows that the methoxyl content of the MWL is about 1.14 moles/C_9 unit . The MWL then consists of guaiacylpropane and syringylpropane units in an approximate molar ratio of 0.86:0.14 [=1:0.16]. From the results of nitrobenzene oxidation as given in Table 5, a guaiacylpropane-syringylpropane molar ratio of approximately 1:0.15 [= 11. 8/ (26.8 x 3)] is obtained (136), which is in good agreement with the value estimated from the ^{13}C NMR spectra data. If we assume that the lignin does not contain aromatic ring-condensed units, then the number of aromatic tertiary carbons must be 2.86 carbons/C_9 unit [= 0.86 x 3 + 0.14 x 2]. From the aromatic tertiary carbon spectral region in Table 13, a value of 2.51 carbons/C_9 unit is obtained. The degree of condensation for aromatic moieties in the MWL is then about 0.35/C_9 unit [= 2.86-2.51]. The MWL from the recent wood of Zhong-Yang Mu (*Bischofia polycarpa*) is, therefore, a rare guaiacyl-syringyl lignin (hardwood lignin), in which guaiacylpropane units are exceedingly predominant over syringylpropane units.

As demonstrated above, ^{13}C NMR spectroscopy is one of the most effective tools for the characterization of lignins when it is used imaginatively, in particular in combination with the chemical degradation of lignins.

REFERENCES

1. K. Freudenberg, *Sci.*, *148*:595 (1965).
2. K. Freudenberg, *Cellulosechemis*, *12*:263 (1931).
3. W. Flaig, in *The Use of Isotopes in Soil Organic Matter Studies*, Pergamon Press, New York, 1966, p. 103.
4. F. J. Stevenson and J. H. A. Butler, in *Organic Geochemistry*, (G. Englinton and M. T. J. Murphy, eds.), Springer-Verlag, Berlin, 1969, p. 534.
5. A. Payen, *Compt. Rend. Acad. Sci.* (Paris), *7*:1052 (1838).
6. E. Sjöström, *Wood Chemistry*, Academic Press, New York, 1981.
7. E. Erdman, *Ann. Chem. Pharm.*, *Suppl. V*:223 (1868).
8. B. Tilgman, British Patent 2924 (1866).
9. F. Tieman, *Ber. Dtsch. Chem. Ges.*, *7*:608 (1874).
10. F. Tieman, *Ber. Dtsch. Chem. Ges.*, *8*:1127 (1875).
11. F. Tieman and B. Mandelshon, *Ber. Dtsch. Chem. Ges.*, *8*:1136 (1875).
12. P. Klason, *Svensk Kem. Tidskr.*, :135 (1897).
13. P. Klason, *Ber. Dtsch. Chem. Ges.*, *53*:706 (1920).
14. P. Klason, *Ber. Dtsch. Chem. Ges.*, *53*:1864 (1920).
15. P. Klason, *Ber. Dtsch. Chem. Ges.*, *56*:300 (1920).
16. P. Klason, *Ber. Dtsch. Chem. Ges.*, *62*:635 (1920).
17. P. Klason, *Ber. Dtsch. Chem. Ges.*, *62*:2523 (1920).
18. H. Cousin and H. Herissey, *Compt. Rend. C. R. Acad. Sci.* (Paris), *147*:247 (1908).
19. H. Erdman, *Biochem. Z.*, *258*:172 (1933).
20. H. Herissey and G. Dolby, *J. Pharm. Chim.* (Paris), *30*:289 (1909).
21. K. Freudenberg and W. Durr, in *Handbuch der Pflanzenalyse*, Vol, 3 (G. Klein, ed.), Springer-Verlag, Wien, Austria, 1933, p. 125.
22. K. Freudenberg and A. C. Niesh, *The Constitution and Biosynthesis of Lignin*, Springer-Verlag, Berlin.
23. K. V. Sarkanen and C. H. Ludwig, *Lignins—Occurrence, Formation, Structure and Reactions*, Wiley-Interscience, New York, 1971.
24. E. Adler, *Wood Sci. Technol.*, *11*:169 (1977).
25. A. Sakakibara, in *The Structure, Biosynthesis and Degradation of Wood* (F. A. Loewus and V. C. Runeckles, eds.), Plenum Press, New York, 1977, p. 117.
26. E. Adler, K. J. Bjorkquist, and S. Haggroth, *Acta Chem. Scand.*, *2*:9 (1948).
27. E. Adler, *Svensk Papperstidn.*, *54*:445 (1951).
28. J. C. Pew, *J. Amer. Chem. Soc.*, *73*:1678 (1951).
29. H. Geiger and H. Fuggerer, *Z. Naturforsch.*, *34b*:1471 (1979).
30. G. H. N. Towers and R. D. Gibbs, *Nature* (London), *172*:25 (1952).
31. G. Meshitsuka and J. Nakano, *Mokuzai Gakkaishi*, *24*:563 (1978).
32. F. C. Crocker, *Ind. Eng. Chem.* :625 (1921).
33. D.E. Bland, *Holzforschung*, *20*:12 (1966).
34. K. Freudenberg and W. Lautsch, *Naturwiss.*, *27*:227 (1939).
35. R. H. J. Creighton, R. D. Gibbs, and H. Hibbert, *J. Amer. Chem. Soc.*, *66*:32 (1944).
36. R. H. J. Creighton and H. Hibbert, *J. Amer. Chem. Soc.*, *66*:37 (1944).
37. B. Leopold and I. L. Malmstrom, *Acta Chem. Scand.*, *6*:49 (1952).

38. R. D. Gibbs, in *The Physiology of Forest Trees* (K. V. Thimann, ed.), Ronald Press, New York, 1958, p. 269.

39. I Kwamura and T. Higuchi, in *Chim. Biochem. Lignine, Cellulose et Hemicellulose*, Grenoble, France, p. 469.

40. T. Higuchi, Y. Ito, and Kawamura, *Photochem.*, 6:875 (1967).

41. M. Shimada, T. Fukuzuka, and T. Higuchi, *TAPPI*, 54:72 (1971).

42. M Erikson and G. M. Miksche, *Proceedings of the 167th ACS National Meeting*, Abstract CELL 22, Los Angeles, Calif., April 1974.

43. D. Narayanmurti and N.R. Das, *Holz Roh-u. Werkstoff*, 13:52 (1955).

44. K. Hatta, *J. Japan For. Soc.*, 32:257 (1950).

45. E. Anderson and W. W. Pigman, *Sci.*, 105:601 (1951).

46. W. A. Côté, Jr. and A. C. Day, in *Cellular Ultrastructure of Woody Plants*, (W. A. Côté, Jr., ed.), Syracuse Univer. Press, Syracuse, N. Y.,1965, p. 391.

47. E. Hagglund and S. Ljungren, *Papierfabrikant*, 31:35 (1953).

48. D. E. Bland, *Holzforschung*, 15:102 (1961).

49. W. A. Côté, Jr., N. P. Kutscha, B. W. Simon, and T. E. Timell, *TAPPI*, 50:350 (1967).

50. T. E. Timell, in *Cellular Ultrastructure of Woody Plants* (W. A. Côté, Jr., ed.), Syracuse Univer. Press, Syracuse, N. Y.,1965, p. 127.

51. A. M. Latif, Ph. D. Thesis, Univer. of Washington, Seattle, Wash., 1968.

52. N. Morohoshi and A. Sakakibara, *Mokuzai Gakkaishi*, 17:393 (1971).

53. S. Yasuda and A. Sakakibara, *Mokuzai Gakkaishi*, 21:363 (1975).

54. N. Morohoshi and A. Sakakibara, *Mokuzai Gakkaishi*, 17:400 (1971).

55. S. Yasuda and A. Sakakibara, *Mokuzai Gakkaishi*, 21:370 (1975).

56. S. Yasuda and A. Sakakibara, *Mokuzai Gakkaishi*, 21:639 (1975).

57. G. J. Ritter and L. C. Fleck, *Ind. Eng. Chem.*, 14:1050 (1922).

58. A. B. Wardrop and H. E. Dadwell, *Austr. Sci. J. Res.*, B1:3 (1948).

59. G. Jayme and M. Harder-Steinhauser, *Papier*, 4:104 (1950).

60. L. P. Clermont and F. Bender, *Pulp and Paper Mag. Can.*, 59:139 (1958).

61. K. V. Starkanen and H. L. Hergett, in *Lignins—Occurrence, Formation, Structure and Reactions* (K. V. Starkanen and C. H. Ludwig, eds.) Wiley-Interscience. New York, 1971, p. 73.

62. I. B. Sacks, I. T. Clark, and J. C. Pew, *J. Polym. Sci.*, Part C, No. 2,:213 (1963).

63. H. Meier, *Holz Roh-u. Werstoff*, 13:323 (1955).

64. A. B. Wardrop, in *Cellular Ultrastructure of Woody Plants* (W. A. Côté, Jr., ed.), Syracuse Univer. Press, Syracuse, N. Y., 1965, p. 61.

65. P. W. Lange, *Svensk Paperstidn.*, 57:525 (1954).

66. J. A. N. Scott, A. R. Procter, B. J. Fergus, and D. A. I. Goring, *Wood Sci. Technol.*, 3:73 (1969).

67. B. J. Fergus, A. R. Procter, J. A. N. Scott, and D. A. I. Goring, *Wood Sci. Technol.*, 3:117 (1969).

68. J. R. Wood and D. A. I. Goring, *Pulp and Paper Mag. Can.*, 72:61 (1971).

69. B. J. Fergus and D. A. I. Goring, *Holzforschung*, 24:113 (1970).

70. B. J. Fergus and D. A. I. Goring, *Holzforschung*, 24:118 (1970).

71. Y. Musha and D. A. I. Goring, *Wood Sci. Tech.*, 9:45 (1975).

72. S. Saka, R. J. Thomas, and J. S. Gratzl, *TAPPI*, 61:73 (1978).

73. S. Saka, R. J. Thomas, *Wood Sci. Technol.*, 16:1 (1982).

74. S. Saka, R. J. Thomas, *Wood Sci. Technol.*, *16*:167 (1982).
75. G. Ritter, *Ind. Eng. Chem.*, *17*:1194 (1925).
76 A. J. Bailey, *Ind. Eng. Chem. Anal. Ed.*, *8*:52 (1936).
77 A. J. Stamm and H. T. Sanders, *TAPPI*, *49*:397 (1966).
78. J. E. Stone, *TAPPI*, *38*:610 (1955).
79. D. E. Marton, *TAPPI*, *42*:301 (1951).
80. M. Matsukura and A. Sakakibara, *Mokuzai Gakkaishi*, *15*:35 (1969).
81. P. Whitting, B. D. Favis, F. G. T. St.-Germain, and D. A. I. Goring, *J. Wood Chem. and Technol.*, *1*:29 (1981).
82. P. Whitting and D. A. I. Goring, *Svensk Papperstidn.*, *84*:R120 (1981).
83. Z.-Z. Lee, G. Meshitsuka, N. C. Cho, and J. Nakano, *Mokuzai Gakkaishi*, *27*:671 (1981).
84. A. Björkman, *Svensk Papperstidn.*, *59*:477 (1956).
85. C. Lapierre and B. Monties, *Proceedings of 1981 International Symposium on Wood and Pulping Chemistry*, Vol. V, Stockholm, Sweden, June 1981, p. 35.
86. C. Lapierre, J. Y. Lallemand, and B. Monties, *Holzforschung*, *36*:275 (1982).
87. J. C. Pew, *TAPPI*, *40*:553 (1957).
88. P. Klason, *Cellulosechemie*, *4*:81 (1923).
89. R. Willstätter and L. Zechmeister, *Ber. Dtsch. Chem. Ges.*, *46*:2401 (1913).
90. P. Odincovs and Z. Kreicberga, *Voprosy Leosokhim. i. Khim Drevesing Trudy Inst. Lesokhz. Problem, Akad. Nauk. Latv. U.S.S.R.*, *6*:51 (1951).
91. K. Freudenberg and W. Durr, *Ber. Dtsch. Chem. Ges.*, *62*:1814 (1920).
92. H. Richtzenhain, *Acta Chem. Scand.*, *4*:589 (1950).
93. T. Ploetz, *Cellulosechemie*, *18*:49 (1940).
94. K. Freudenberg and T. Ploetz, *Ber. Dtsch. Chem. Ges.*, *73*:754 (1940).
95. A. Björkman and B. Person, *Svensk Papperstidn.*, *60*:553 (1957).
96. K. Freudenberg, in *The Constitution and Biosynthesis of Lignin* (K. Freudenberg and A. C. Niesch, eds.), Springer-Verlag, Berlin, 1968, p. 64.
97. K. Lindquist, B. Ohlson, and R. Simonson, *Svensk Papperstidn.*, *80*:143 (1977).
98. M. Erickson, S. Larsson, and G. E. Miksche, *Acta Chem. Scand.*, *27*:903 (1973).
99. J.-M. Yang and D. A. I. Goring, *Can. J. Chem.*, *58*:2411 (1980).
100. H.-M. Chang, E. B. Cowling, W. Brown, E. Adler, and G. M. Miksche, *Holzforschung*, *29*:153 (1975).
101. M. H. Winston, C.-L. Chen, J. S. Gratzl, and I. S. Goldstein, *Holzforschung*, *40* (suppl.): 45 (1986).
102. E. Unger, Ph. D. Dissertation, Universitat Munhen, Germany, 1914.
103. K. Freudenberg, A. Janson, E. Knopf, and A. Haag, *Ber. Dtsch. Chem. Ges.*, *69*:1415 (1936).
104. K. Freudenberg, M. Meister, and E. Flickinger, *Ber. Dtsch. Chem. Ges.*, *70*:500 (1937).
105. K. Freudenberg, and H. Fr. Muller, *Ber. Dtsch. Chem. Ges.*, *71*:1821 (1938).
106. K. Freudenberg, and E. Plankenhorn, *Ber. Dtsch. Chem. Ges.*, *75*:857 (1942).
107. K. Freudenberg, K. Enger, E. Flickinger, A. Sobek, and F. Klink, *Ber. Dtsch. Chem. Ges.*, *71*:1810 (1938).
108. K. Freudenberg and C.-L. Chen, *Chem. Ber.*, *93*:2533 (1960).
109. K. Freudenberg, C.-L. Chen, and G. Cardinale, *Chem. Ber.*, *95*:2814 (1962).
110. K. Freudenberg and C.-L. Chen, *Chem. Ber.*, *100*:3683 (1967).
111. H. Richtzenhain, *Acta Chem. Scand.*, *4*:589 (1950).

112. S. Larsson and G. E. Miksche, *Acta Chem. Scand.*, *21*:1970 (1967).

113. S. Larsson and G. E. Miksche, *Acta Chem. Scand.*, *23*:917 (1969).

114. S. Larsson and G. E. Miksche, *Acta Chem. Scand.*, *23*:3339 (1969).

115. S. Larsson and G. E. Miksche, *Acta Chem. Scand.*, *25*:647 (1971).

116. S. Larsson and G. E. Miksche, *Acta Chem. Scand.*, *25*:673 (1971).

117. S. Larsson and G. E. Miksche, *Acta Chem. Scand.*, *26*:2031 (1972).

118. M. Erikson, S. Larsson, and G. E. Miksche, *Acta Chem. Scand.*, *27*:127 (1973).

119. C.-L. Chen, "Characterization of Lignin by Oxidative Degradation: Use of Gas Chromatography-Mass Spectometry Technique," in *Methods in Enzymology* (W. A. Wood and S. T. Kellogg, eds.), Academic Press, Orlando, Florida, 1987, Vol. 161B, p. 110.

120. K. Freudenberg, *Angew. Chem.*, *52*:362 (1939).

121. K. Freudenberg, W. Lautsch, and K. Engler, *Ber. Dtsch.Chem. Ges.*, *73*:167 (1940).

122 F. Ullmann, *Enz. d. tech. Chem.*, *2*, Aufl. VIII:817 (1931).

123. R. H. J. Creighton, J. L. McCaththy, and H. Hibbert, *J. Amer. Chem. Soc.*, *63*:312 (1941).

124. B. Leopold, *Acta Chem. Scand.*, *6*:38 (1952).

125. D.-S. Tai, C.-L. Chen, and J. S. Gratzl, *J. Wood Chem. and Technol.*, *10*:75 (1990).

126. A. V. Waeck and K. Kratzl, *Ber. Dtsch. Chem. Ges.*, *76*:891 (1943).

127. A. V. Waeck and K. Kratzl, *Ber. Dtsch. Chem. Ges.*, *77*:516 (1944).

128. K. Kratzl, *Ber. Dtsch. Chem. Ges.*, *77*:717 (1944).

129. K. Kratzl and I. Khautz, *Mh. Chem.*, *78*:376 (1948).

130. B. Leopold, *Acta Chem. Scand.*, *4*:1523 (1950).

131. B. Leopold and I.-L. Malmstrom., *Acta Chem. Scand.*, *5*:936 (1951).

132. J. C. Pew, *J. Amer. Chem. Soc.*, *77*:2831 (1955).

133. H.-M. Chang and G. G. Allan, in *Lignins—Occurrence, Formation, Structure and Reactions* (K. V. Sarkanen and C. H. Ludwig, eds.), Wiley-Interscience, New York, 1971, pp. 435–436.

134. T. P. Schultz and C. M. Templeton, *Holzforschung*, *40*:93 (1986).

135. K. R. Kavanagh and J. M. Pepper, *Can. J. Chem.*, *33*:24 (1955).

136. K. V. Sarkanen and H. L. Hergett, in *Lignins—Occurrence, Formation, Structure and Reactions* (K. V. Sarkanen and C. H. Ludwig, eds.), Wiley-Interscience, New York, 1971, p. 69.

137. I. A. Pearl, *J. Amer. Chem. Soc.*, *68*:1429 (1942).

138. I. A. Pearl and D. L. Beyer, *TAPPI*, *33*:544 (1950).

139. K. R. Kavanagh and J. M. Pepper, *Can. J. Chem.*, *33*:24 (1955).

140. J. M. Pepper, B. W. Casselman, and J. C. Karapally, *Can. J. Chem.*, *45*:3009 (1967).

141. I. A. Pearl, *J. Amer. Chem. Soc.*, *68*:429 (1946).

142. I. A. Pearl, *J. Amer. Chem. Soc.*, *70*:2008 (1948).

143. I. A. Pearl, *J. Amer. Chem. Soc.*, *71*:2196 (1949).

144. I. A. Pearl, *J. Amer. Chem. Soc.*, *72*:1427 (1950).

145. I. A. Pearl and D. L. Beyer, *TAPPI*, *33*:508 (1950).

146. I. A. Pearl and D. L. Beyer, *TAPPI*, *42*:800 (1959).

147. I. A. Pearl and D. L. Beyer, *For. Prod. J.*, *11*:442 (1961).

148. I. A. Pearl and D. L. Beyer, and D. Whitney, *TAPPI*, *44*:479 (1961).

149. R. A. Davis, E. T. Reaville, Q. P. Peniston, and J. L. McCarthy, *J. Amer. Chem. Soc.*, *77*:2495 (1955).

150. L. A. Pershina and V. P. Vasileva, *Izy. Tomsk. Polytekh. Inst.*, *136*:33 (1965).
151. L. A. Pershina, V. L. Kusina, and V. P. Vasileva, *Izy. Tomsk. Polytekh. Inst.*, *136*:36 (1965).
152. F. Fischer, H. Schrader, and W. Triebes, *Gesammelte Abh. Kenntnis Kohle*, *5*:211 (1922).
153. F. Fischer, H. Schrader, and W. Triebes, *Gesammelte Abh. Kenntnis Kohle*, *5*:315 (1922).
154. F. Fischer, H. Schrader, and W. Triebes, *Gesammelte Abh. Kenntnis Kohle*, *6*:1 (1923).
155. H. Schrader, *Gesammelte Abh. Kenntnis Kohle*, *5*:276 (1922).
156. W. Lautsch, E. Plankenhorn, and F. Klink, *Angew. Chem.*, *53*:450 (1940).
157. P. Klason, *tekn. Tidskr. at Kemi och Metallurgi*, *23*:11 (1893).
158. A. B. Cramer, M. J. Hunter, and H. Hibbert, *J. Amer. Chem. Soc.*, *61*:509 (1939).
159. M. Kulka and H. Hibbert, *J. Amer. Chem. Soc.*, *65*:1180 (1943).
160. E. West, A. S. MacInnes, and H. Hibbert, *J. Amer. Chem. Soc.*, *65*:1187 (1943).
161. L. Brickman, W. L. Hawkins, and H. Hibbert, *J. Amer. Chem. Soc.*, *62*:2149 (1940).
162. M. Kulka, W. L. Hawkins, and H. Hibbert, *J. Amer. Chem. Soc.*, *63*:2371 (1941).
163. A. B. Cramer, M. J. Hunter, and H. Hibbert, *J. Amer. Chem. Soc.*, *61*:516 (1939).
164. M. Kulka, F. Fischer, and H. Hibbert, *J. Amer. Chem. Soc.*, *66*:39 (1944).
165. L. Mitchell and H. Hibbert, *J. Amer. Chem. Soc.*, *66*:602 (1944).
166. J. M. Pepper, P. E. T. Baylis, and E. Adler, *Can. J. Chem.*, *37*:1241 (1959).
167. E. Adler, K. Lundquist and G. E. Miksche, *Adv. Chem. Ser.*, *59*:22 (1966).
168. K. Lundquist and K. Hedlund, *Acta Chem. Scand.*, *21*:1750 (1967).
169. K. Lundquist and R. Lundgren, *Acta Chem. Scand.*, *26*:2005 (1972).
170. K. Lundquist, *Acta Chem. Scand.*, *18*:1316 (1964).
171. K. Lundquist, *Acta Chem. Scand.*, *24*:889 (1970).
172. K. Lundquist and L. Ericsson, *Acta Chem. Scand.*, *24*:3681 (1970).
173. K. Lundquist and K. Helund, *Acta Chem. Scand.*, *18*:1316 (1964).
174. E. Adler, S. Ellmer, and K. Lundquist *Acta Chem. Scand.*, *13*:2149 (1972).
175. E. Adler and K. Lundquist, *Acta Chem. Scand.*, *17*:13 (1963).
176. K. Lundquist and G. E. Miksche, *Tetrahedron Lett.*, 2131 (1965).
177. D. Robert and C.-L. Chen, unpublished results (1987).
178. H. Nimz, *Chem. Ber.*, *99*:2638 (1966).
179. A. Sakakibara and N. Nakayama, *Mokuzai Gakkaishi*, *7*:13 (1962).
180. Y. Sano, *Mokuzai Gakkaishi*, *21*:508 (1975).
181. K. Lundquist, *Acta Chem. Scand.*, *27*:2597 (1973).
182. K. Weiges, *Chem. Ber.* , *94*:2522 (1961).
183. H. Nimz, I. Mogharob, and H. D. Ludeman, *Markromol. Chem.*, *175*:2563 (1974).
184. F. E. Brauns, *The Chemistry of Lignin*, Academic Press, New York, 1952.
185. F. E. Brauns and D. A. Brauns, *The Chemistry of Lignin*, Supplemental Volume, Academic Press, New York, 1952.
186. K. Kratzl and H. Silbernagel, *Mh. Chem.*, *83*:1022 (1953).
187. H. Nimz, *Chem. Ber.*, *98*:538 (1965).
188. H. Nimz and H. Gaber, *Chem. Ber.*, *98*:538 (1965).
189. H. Nimz, *Chem. Ber.*, *98*:538 (1965).
190. H. Nimz, *Chem. Ber.*, *98*:3160 (1965).
191. H. Nimz, *Chem. Ber.*, *99*:469 (1966).
192. H. Nimz, *Chem. Ber.*, *20*:105 (1966).
193. H. Nimz, *Chem. Ber.*, *100*:181 (1967).

194. H. Nimz, *Chem. Ber.*, *100*:2633 (1967).
195. K Freudenberg and H. Schluter, *Chem. Ber.*, *88*:617 (1955).
196. K Freudenberg and B. Lehmann, *Chem. Ber.*, *93*:1354 (1960).
197. A. Sakakibara and N. Nakayama, *Mokuzai Gakkaishi*, 8:153 (1962).
198. A. Sakakibara and N. Nakayama, *Mokuzai Gakkaishi*, 8:157 (1962).
199. S. Omori and A. Sakakibara, *Mokuzai Gakkaishi*, *17*:464 (1971).
200. S. Omori and A. Sakakibara, *Mokuzai Gakkaishi*, *18*:355 (1972).
201. S. Omori and A. Sakakibara, *Mokuzai Gakkaishi*, *18*:577 (1972).
202. S. Omori and A. Sakakibara, *Mokuzai Gakkaishi*, *20*:388 (1972).
203. S. Omori and A. Sakakibara, *Mokuzai Gakkaishi*, *21*:170 (1973).
204. Y. Sano and A. Sakakibara, *Mokuzai Gakkaishi*, *16*:76 (1970).
205. Y. Sano and A. Sakakibara, *Mokuzai Gakkaishi*, *16*:121 (1970).
206. Y. Sano and A. Sakakibara, *Mokuzai Gakkaishi*, *21*:461 (1975).
207. K. Lundquist, *Appl. Polym. Symp.*, *28*:2393 (1976).
208. H. Nimz, *Chem. Ber.*, *102*:799 (1969).
209. H. Nimz, K. Das, and N. Minemura, *Chem. Ber.*, *104*:1871 (1971).
210. H. Nimz, K. Das, *Chem. Ber.*, *104*:2359 (1974).
211. C.-L. Chen and W. J. Connors, *J. Org. Chem.*, *39*:3877 (1974).
212. C. Lapierre, B. Monties, and C. Rolando, *J. Wood Chem. and Technol.*, *5*:277 (1985).
213. C. Lapierre, B. Monties, and C. Rolando, *Holzfroschung*, *40*:47 (1986).
214. C. Lapierre, B. Monties, and C. Rolando, *Holzfroschung*, *40*:113 (1986).
215. C. Lapierre and C. Rolando, *Holzfroschung*, *42*:1 (1988).
216. C. Lapierre, B. Monties, and C. Rolando, *Holzfroschung*, *42*: (1986).
217. C. Lapierre, "Heterogeneite des Lignies de Peuplier: Mise en Evidence Systematique," These de Doctorat d´Etat, Universite Paris-Sud, Paris, France, 1986.
218. C.-L. Chen and D. Robert, "Characterization of Lignin by ^{1}H and ^{13}C NMR Spectroscopy, in *Methods in Enzymology* (W. A. Wood and S. T. Kellog, eds.), Academic Press, Orlando, Florida, 1987, Vol. 161B, pp. 137.
219. D. Pan, D. Tai, C.-L. Chen, and D. Robert, *Holzfroschung*, *44*:7 (1990).
220. R. Kuhn and H. Roth, *Ber Dtsch. Chem. Ges.*, *66*:1274 (1933).
221. O. Goldscmid, *Anal. Chem.*, *26*:1421 (1954).
222. P. Månsson, *Holzfroschung*, *37*:143 (1983).
223. J. Gierer and S. Soderberg, *Acta Chem. Scand.*, *13*:127 (1959).
224. E. Adler and J. Marton, *Acta Chem. Scand.*, *13*:175 (1959).
225. L. Brendtson, K. Helund, L. Hemra, and K. Lundquist, *Acta Chem. Scand.*, *B28*:333 (1974).
226. K. Lundquist and T. Olsson, *Acta Chem. Scand.*, *B31*:788 (1977).
227. E. Adler, H.-D. Becker, T. Ishihara, and A. Stamivik, *Holzfroschung*, *20*:3 (1966).
228. E. Adler, G. E. Miksche, and B. Johansson, *Holzfroschung*, *22*:171 (1968).
229. K. Freudenberg, J. M. Harkin, and H.-K. Werner, *Chem. Ber.*, *97*:909 (1964).
230. M. Drumond, M. Aoyama, C.-L. Chen, and D. Robert, *Wood Chem. and Technol.*, *9*:421 (1989).
231. D. Robert and G. Brunow, *Holzfroschung*, *38*:85 (1984).
232. D. Robert and C.-L. Chen, *Holzfroschung*, *43*:323 (1989).
233. D. Robert and D. Gagnaire, Ekman Days, *Proc. 1st Int. Symp. Wood and Pulping Chem.*, *1*:86 (1981).

234. M. Bardet, M.-F. Foray, and D. Robert, *Makromol. Chem.*, *186*:1495 (1985).
235. H. Nimz, *TAPPI*, *56*(5):124 (1973).
236. G. Krüsman, "Handbuch der Laugeholze," Band I Verlag Paul Parey, Berlin und Hamburg, Germany, 1976.
237. J. Zhang, W. Wang, Y. Wi, J. Song, and Y. Tang, *Namlin Ke Ji* (*Journal of Science Technology*, Nanjing Technological College of Forest Products), 1979(3), 1 (1978).
238. D. Robert, M. Bardet, G. Gellerstedt, and E. L. Lindfors, *J. Wood Chem. and Technol.*, *4*:239 (1984).
239. M. R. Bendall, D. T. Pegg, D. M. Doddrell, and W. E. Hull, "DEPT Bruker Information," Bruker Analytische Messtechnik, Karlsruhe, Germany, 1982.
240. D. M. Doddrell, D. T. Pegg, and M. R. Bendall, *J. Mag. Reson.*, *48*:323 (1982).
241. R. Benn and G. Gunther, *Angew. Chem. Int. Ed. Engl.*, *22*:350 (1983).
242. H. Y. Hassi, M. Aoyama, D. Tai, C.-L. Chen, and J. S. Gratzl, *J. Wood Chem. and Technol.*, 7:555 (1987).
243. M. Bardet, D. Gagnaire, R. Nardin, D. Robert, and M. Vincedon, *Holzfroschung*, *40*, Suppl:17 (1986).

6

Lignin Biosynthesis

Ronald Sederoff and Hou-min Chang

North Carolina State University, Raleigh, North Carolina

I. INTRODUCTION

A. Lignin, a Major Component of the World's Biomass

Lignin, in combination with cellulose and hemicelluloses in wood, constitutes the most abundant organic material on the surface of the earth. The total amount of wood in the world's forests has been estimated at about 1.5 billion metric tons (1, 2). Wood components play an important role in the carbon cycle and the global ecology through effects on soils, forest ecology, and climate.

About one-fourth of the woody plant biomass is lignin (3). The biosynthesis of this component therefore requires a major fraction of the biosynthetic resources of living plants. Lignin is highly reduced, with a carbon content 50% higher than that of the polysaccharides. It is therefore relatively rich in potential energy; however, this energy is not readily utilized in its natural state (3). Lignin is degraded slowly and persists as organic matter in soils and water for long periods of time.

B. The Function of Lignin

Hemicelluloses and lignin form the embedding matrix of both primary and secondary cell walls, reinforcing the cellulose microfibrils and imparting rigidity to the wall. In the middle lamella between cell walls, lignin functions as a binding agent, giving woody stems compressive strength and bending stiffness. Lignification is necessary for water transport in conductive xylem to limit lateral diffusion, thus facilitating longitudinal transport. Lignified cell walls are resistant to microbial attack because the penetration of destructive enzymes secreted by invading organisms is reduced (4-6).

C. Chemistry of Lignin

Woody stems of arborescent gymnosperms and angiosperms contain 15–36% of lignin as their dry weight (2, 5). The main features of lignin structure are well known (2,7,8). Higher-plant lignin is found predominantly in two structural types: guaiacyl lignin and guaiacyl-syringyl lignin. Guaiacyl lignin is the major component of gymnosperm (softwood) lignin and is derived primarily from coniferyl alcohol. Angiosperms contain a characteristic hardwood lignin that is produced by the copolymerization of coniferyl and sinapyl alcohols (guaiacyl-syringyl lignin). Normal lignin contains small amounts of a third lignin component, para-hydroxyphenyl lignin, derived from para-coumaryl alcohol. In the grasses (*Graminaceae*) and in compression wood of gymnosperms, this third type of phenylpropane unit, para-hydroxyphenylpropane, is more prevalent than in normal wood (2).

D. Evolution of Lignin Biosynthesis

Lignin is presumed to be a basic adaptation of land plants and a major step in land plant evolution (3, 9). In higher plants, lignin is needed to prevent the compression of vascular tissue that would collapse the cell walls, and to support the weight of the plant. Early vascular plants of the late Silurian (Cooksonia) are only a few centimeters high and contain lignified tissue. Although the lignified tissue was confined to the xylem, it is not believed to have provided mechanical support. Later, lignification enabled the plants to attain larger size by withstanding forces of bending and compression.

Lignin is synthesized in the cell walls of all true vascular plants, ferns, and club mosses (10). Recently, ligninlike material has been found in the green alga *Coleochaete* (11). The material in the algae resembles lignin because it gives a positive reaction for syringyl groups and shows autofluorescence with ultraviolet light. The ligninlike material in the algae was similar to that found in a species of hornwort *Anthoceros*, an early land plant related to mosses. Well-organized algal plants have been found in sediments 2.3 billion years old (9), suggesting an early age for the origin of ligninlike compounds. This result further suggests that lignin functioned first as an antimicrobial agent and only later did it take on a role in water transport and mechanical support in the evolution of land plants.

E. Lignin Biosynthesis in Secondary Metabolism

Lignin shares common biosynthetic pathways with a wide variety of secondary metabolites such as flavonoids, suberin, coumarins, stilbenes, and lignans. All of these compounds are derived from phenylalanine, which is the precursor for all the phenylpropanoid (C_6C_3) biosynthetic pathways (Fig. 1). Phenylalanine is derived from sucrose through the synthesis of dehydroshikimic acid. Lignin is derived from phenylalanine through the formation of cinnamic acids. These cinnamic acids are activated by forming cinnamoyl-CoA thioesters that are reduced to cinnamyl alcohols that are polymerized into lignin.

Phenylalanine itself is converted into two acid derivatives: phenylpyruvic acid and cinnamic acid. Phenylpyruvic acid and para-hydroxyphenyl acetic acid have been shown to be growth regulators in algae (12, 13) and phenyl acetate has been identified as a natural auxin in higher plants (14). Cinnamic acid and its derivatives are precursors for large groups of phenylpropanoid derivatives, particularly, coumarins, benzoic acids, cinnamoyl esters, and cinnamoyl amides.

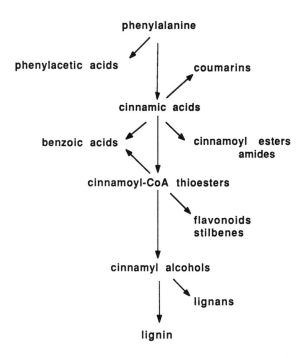

Figure 1   Metabolic pathways related to the synthesis of lignin precursors. (Modified after Ref. 23.)

Cinnamoyl-CoA thioesters and malonyl-CoA condense to form chalcone (15). Chalcone is a common precursor for all flavonoids, which include many plant pigments such as anthocyanins. The anthocyanins are common plant pigments that are responsible, in part, for flower color. Cinnamoyl-CoA thioesters also give rise to stilbenes. Pinosylvin is a stilbene found in pines that acts as a phytoalexin in seedlings and as a toxin and pest repellent in older trees (16).

The metabolic pathways that branch from intermediates common to the lignin pathway could affect the biosynthesis of lignin. However, it is not known how the competing metabolic pathways are regulated, nor is it known to what extent major pathways are tissue-specific in any single plant system. For example, many extractives found in heartwood are derived from the phenylpropanoid pathway (17). Most evidence supports the view that these extractives are synthesized in the cambial zone and transported centripetally, perhaps by ray parenchymal cells, suggesting that the site and timing of the synthesis of extractives may be different from that of the biosynthesis of most lignin in xylem. More work is needed to determine the developmental specificity of these related biosynthetic pathways.

F. Developmental Regulation of Lignin Biosynthesis

In differentiating xylem, the lignification of cell walls is initiated in the cell corners at the completion of cell enlargement (18). This event signals the onset of secondary cell wall thickening and stiffening. Lignification first proceeds through the middle lamella moving away from the corners, then into the primary cell wall (18). Secondary wall lignification occurs after the cellulose microfibrils have been laid down. In the tracheids of xylem, lignification is the last function of the cell and is followed by cell death.

Lignin content and location can vary due to mechanical stress and during stage of development. If excessive mechanical stress is applied to conifer stems, compression wood is formed on the side receiving compressive stress. The quality and distribution of lignin in compression wood is different from normal wood. Lignin content is increased and the lignin contains an increased proportion of linkages derived from para-coumaryl alcohol compared to normal wood (2). In compression wood of larch the concentration of lignin is higher in the cell wall but no lignin is present in the cell corners (18). During the juvenile growth period of conifers, the wood formed has a higher content of lignin and hemicellulose (19).

II. MONOLIGNOL BIOSYNTHESIS

The major features of the pathway for the biosynthesis of lignin precursors are well established. Studies using radioactive tracers and enzyme activity provide strong evidence for the formation of lignin from carbohydrates via shikimic acid, then to phenylalanine, to cinnamic acids, to cinnamyl alcohols (monolignols), and finally to lignin. All the proposed enzymes of the pathway for lignin biosynthesis have been identified and at least partially purified and characterized. In addition, plausible mechanisms for the storage and transport of monomers and the subsequent poly-merization of lignin from the monomers have been proposed (Fig. 2).

The first segment of the lignin biosynthetic pathway (Fig. 2) that is not shared with general cellular metabolism is the biosynthesis of ferulic acid (f) and sinapic (h) acids from phenylalanine (b) through cinnamic acid (c). After phenylalanine is deaminated to form cinnamic acid, two hydroxylation steps follow to form caffeic acid (e). Subsequently, caffeic acid is methylated to form ferulic (f) acid. In angiosperms, ferulic acid is further hydroxylated to form 5-hydroxyferulic acid (g), then methylated to form sinapic acid (h). Ferulic acid is the precursor for coniferyl alcohol (r) and guaiacyl lignin, whereas sinapic acid is the precursor for sinapyl alcohol (s) that, together with coniferyl alcohol, copolymerizes to form guaiacyl-syringyl lignin.

A. The Phenylalanine Cinnamic Acid Pathway

Phenylalanine is converted by nonoxidative deamination to cinnamic acid in one step by phenylalanine ammonia-lyase (PAL) (Fig. 2, step 1A) (20). The enzyme has been studied extensively (21). It is a tetrameric enzyme and the expression of PAL is highly regulated. Levels of PAL expression are affected by wounding, infection, hormones, light, and development.

An analogous activity for the deamination of tyrosine to para-coumaric acid (tyrosine ammonia-lyase; TAL) (Fig. 2, step 1B) was detected by Neish (22), but this activity, found mainly in grasses, may be another activity of PAL (23) because it never occurs independently of PAL. DNA clones for PAL mRNA sequences have been selected from cDNA libraries of parsley (24) and bean (25, 26). Therefore, the complete nucleotide and protein sequences for many of these PAL genes are known.

The conversion of cinnamic acid to para-coumaric acid occurs by the addition of a hydroxyl group at the C4 position (Fig. 2, step 2). Cinnamate-4-hydrolase (C4H) catalyzes this step (27). The hydrolase is a hemoprotein that is part of a multienzyme system tightly bound to membranes and linked with cytochrome P-450. The second hydroxylation is catalyzed by para-coumarate-3-hydrolase (C3H, step 3), adding a hydroxyl to the C3 position, producing caffeic acid (e). C3H is a copper-containing enzyme bound to chloroplast lamellae and has been purified to homogeneity from spinach beet (28). It is a monomeric protein with a molecular weight of 40 kD. The inhibition of C3H in mung bean tissues did not block the formation of caffeic acid, suggesting that another enzyme may be present with monooxygenase activity (23). It is interesting to note that lignin levels are reduced in monterey pines grown in copper-deficient soils (29), possibly due to the reduced activity of C3H.

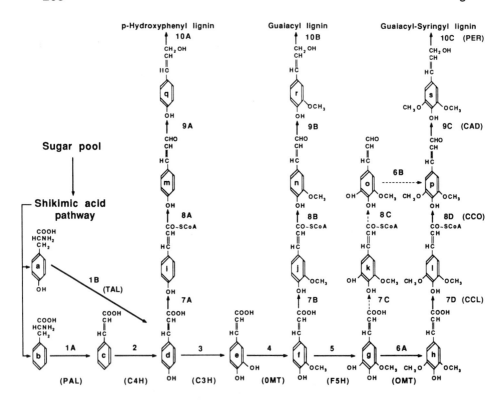

Figure 2 General pathway of lignin biiosynthesis. The pathway is described, including
some abbreviations and designations for compounds and enzymes. Each step in the pathway
is designated by a number, and when different sustrates are converted, the steps are
designated 1A, 1B, etc. The compounds used as substrates are designated by lower case
letters inside the rings of the compounds (a through s). The enzymes are abbreviated with a
three-letter code, e.g., OMT for O-methyl transferase.

Steps in the pathway: Step 1A. Nonoxidative deamination of phenylalanine (b) to
cinnamic acid (c) by phenylaline ammonia-lyase (PAL). Step 1B. Nonoxidative deamination
of tyrosine (a) to para-coumaric acid by (d) tyrosine ammonia-lyase. Step 2. Hydroxylation
of cinnamic acid (c) to para-coumaric acid by (d) by cinnamate 4-hydrolase C4H). Step 3.
Hydroxylation of para-coumaric acid by (d) to caffeic acid (e) by para-coumarate 3-hydrolase
(C3H). Step 4. Tranmethylation of caffeic acid (e) to ferulic acid (f) by O-methyl transferase
(OMT). Step 5. Hydroxylation of ferulic acid (f) acid to 5-hydroxyl ferulic acid (g) by ferulate
5-hydrolase (F5H). Step 6A. Transmethylation of 5-hydroxyl-ferulic acid (g) to sinapic acid
(h) by O-methyl transferase (OMT). Step 6B. Transmethylation of 5-hydroxyconiferaldehyde
(o) to sinapaldehyde (p) by O-methyl transferase (OMT). Step 7. Activation of cinnamic
acids to cinnamoyl-Coenzyme A, by cinnamyl CoA ligase (hydroxcinnamate: CoA ligase)
(CCL). A, para-coumaric acid (d) to para-coumaroyl-CoA (i). B, Ferulic acid (f) to feruloyl-
CoA (j). C, 5-hydroxyferulic acid (g) to 5-hydroxyferuloyl-CoA (k). D, sinapic acid (h) to

B. Biosynthesis of Ferulic and Sinapic Acids

The conversion of caffeic acid (e) to ferulic acid (f) is catalyzed by an O-methyl-transferase (OMT) (Fig. 2, step 4), often called catechol O-methyltransferase. In angiosperms, ferulate-5-hydrolase (F5H) (Fig. 2, step 5) then converts ferulic acid (f) to 5-hydroxyferulate (g), which may again be methylated to sinapic acid (h). The O-methyl-transferases from both gymnosperms and angiosperms have been puri-fied and show important differences in substrate specificity. Both can use caffeic acid as a substrate-producing ferulic acid. However, only angiosperm OMT can utilize 5-hydroxyferulate as a substrate to give sinapic acid (30, 31). It has been argued that the lack of the angiosperm "difunctional" OMT is at least partially responsible for the difference in the types of lignin in gymnosperms and angio-sperms.

Ferulate-5-hydrolase (F5H), the enzyme that catalyzes the conversion of ferulic acid (f) to 5-hydroxyferulic acid (g), has been detected in angiosperms and has been partially purified from poplar (32). It is also associated with a microsomal fraction and was characterized as a cytochrome-P-450-dependent mixed function monooxygenase that is distinct from the cinnamate-4-hydrolase (C4H).

C. Reduction of Cinnamic Acids to Cinnamyl Alcohols

The reduction of cinnamic acids to cinnamyl alcohols proceeds through three enzymatic steps (Fig. 2, steps 7,8,9). The first step is the formation of cinnamoyl coenzyme A thioesters (step 7), followed by the reduction of the thioester to the cinnamyl aldehydes (step 8) and then to the cinnamyl alcohols (step 9). Cinnamoyl-CoA esters occupy a central role in the metabolism of plant phenolic compounds (33). The general name for the enzymes is hydroxycinnamate: CoA ligases (CCL) (Fig. 2, steps 7A, 7B, 7C, 7D). Alternatively, the enzymes may be more specifically defined such as para-coumarate: CoA ligase (step 7A). The enzymes depend strictly on ATP as a substrate, and by analogy with the activation of fatty acids, the reaction

---

sinapoyl-CoA (1). Step 8. Reduction of cinnamoyl-CoAs to cinnamyl aldehydes by cinnamoyl-CoA reductase (hydroxycinnamoly-CoA:NADPH oxidoreductase) (CCO). A, para-coumaroyl-CoA (i) to para-coumaraldehyde (m). B, feruloyl-CoA (j) to coniferaldehyde (n). C, 5-hydroxyferuloyl-CoA (k) to 5-hydroxyconiferaldehyde (o). D, sinapoyl-CoA (1) to sinapaldehyde (p). Step 9. Reduction of cinnamyl aldehydes to cinnamyl alcohols by cinnamyl alcohol dehydrogenase (CAD). A, para-coumaraldehyde (m) to para-coumaryl alcohol (q). B, coniferaldehyde (n) to alcohol (r). Sinapaldehyde (p) to synapyl alcohol (s). Step 10. The polymerization of cinnamyl alcohols into lignin is a complex process that may involve serveral enzymes or enzyme systems. Monolignols may be glucosylated and deglucosylated for storage and/or transport to the lignifying zone. Peroxidases (PER) are thought to be involved in monolignols for polymerization. See the text for more details.

appears to proceed through an intermediate acyl adenylate, which reacts with CoA to form the thioester. Most enzyme preparations have an estimated molecular weight of about 55 kD (23). The enzymes may show specificity for particular hydroxylated substrates. The biosynthesis of many secondary products proceeds from thioesters of cinnamic acids (23) (Fig. 1).

In some hardwoods, sinapate-CoA ligase activity (step 7D) has not been detected. This result has suggested the presence of an alternate pathway (Fig. 2) through the formation of a thioester of S-hydroxyferulate (k), which could be reduced to the aldehyde (o) and then methylated to sinapaldehyde (p) (steps 7C, 8C, 6B). Alternatively, the timing of the appearance of syringyl lignin lags behind that of guaiacyl lignin, suggesting that the appearance of the enzymes involved in the synthesis of sinapyl alcohol may be delayed and therefore not detectable until later in the period of lignin biosynthesis.

The first reductive step in the biosynthesis of cinnamyl alcohols is catalyzed by cinnamoyl-CoA oxidoreductase (CCO) (Fig. 2, step 8), an enzyme that has been purified extensively from soybean and spruce (34, 35). These enzymes are single polypeptides of about 38 kD in molecular weight. Considerable specificity was observed in that sinapoyl-CoA was reduced only by angiosperm enzymes, whereas enzymes from both gymnosperms and angiosperms could reduce feruloyl-CoA.

The final step in the synthesis of cinnamyl alcohols is catalyzed by cinnamyl alcohol dehydrogenase (CAD) (Fig. 2, step 9). CAD has been studied in a number of plants, including Populus (36), bean (*Phaseolus vulgaris*) (37), Japanese black pine (38), and loblolly pine (39). CAD is a dimeric zinc requiring enzyme-utilizing NADPH to reduce cinnamyl aldehydes to alcohols. Molecular weights vary significantly. For example, the subunit size for the enzyme isolated from Populus is 38 kD compared to 44 kD in loblolly pine. Isozymes have been observed that differ in substrate specificity. In loblolly pine (39), genetic analysis has shown that a single gene codes for the major CAD. This enzyme has relative specificity for coniferyl alcohol. The gene for CAD is highly polymorphic; five different electrophoretic alleles have been identified.

In bean, a CAD has been isolated and characterized that is stimulated in suspension cells by a fungal elicitor, prepared from fragments of fungal cell walls (37). This CAD also is encoded by a single gene, but this enzyme may be different from the activity involved in xylogenesis. This CAD has a 65 kD subunit size, which is significantly larger than the size of the other enzymes that have been studied. cDNA clones of this elicitor-response CAD gene have been isolated. The fungal elicitor stimulates transcription of this gene, leading to a rapid, transient induction of CAD mRNA. The response precedes that of the other enzymes, suggesting a role for CAD in the activation of other enzymes in the lignin pathway or in the disease response.

It is sometimes difficult to be certain of the identity of specific clones on the basis of indirect evidence. Walter (40) has found that the presumptive CAD clone shows a strong sequence similarity to the sequence of malic enzyme from monocots.

III. DEHYDROGENATIVE POLYMERIZATION OF LIGNIN

A. Transport of Monolignols to the Lignifying Cell Wall

Monolignols are not abundant in their free form in lignifying tissues, but are found as the corresponding 4-O-β-D-glucosides, where the phenolic hydroxyl is masked by the sugar moiety (23). Enzymes catalyzing the synthesis of such glucosides have been isolated and characterized (41–44). The enzymes require UDP-glucose and show specificity for coniferyl or syringyl alcohol. β-glucosidases have been found attached to cell walls of spruce seedlings during the onset of lignin biosynthesis (45), suggesting that the cinnamyl alcohol glucosides represent the metabolic forms in which cinnamyl alcohols are excreted from the cytoplasm into the lignifying zone (46, 47).

Beech bark contains significant amounts of the Z isomers as well as E isomers of coniferyl and sinapyl alcohols. Although more work is needed, the formation of the Z isomer may be produced at the level of the hydroxycinnamyl alcohols (48). These results suggest that the lignification process may involve specific isomers of the monolignols and that more attention should be paid to the isomeric specificity of lignin biosynthetic enzymes in other plants and tissues.

B. Role of Peroxidase in Lignification

It is generally accepted that peroxidase (PER) (Fig. 2, step 10) is involved in the polymerization process that converts cinnamyl alcohols to lignin (2, 49). The staining of cross sections of a variety of plant stems shows high peroxidase activity associated with cells undergoing lignification (49, 50). Specific peroxidases, associated with cell walls and lignification have been purified and characterized in a variety of plants, the best characterized peroxidase being from tobacco (51). The moderately anionic peroxidases are cell-wall-associated, highly expressed in wound tissue, and have high activity for the polymerization of cinnamyl alcohols in vitro (52).

Two anionic peroxidases have been purified and a sequence determined from the N-terminus. A cDNA clone was obtained and sequenced for this enzyme. mRNA coding for this cDNA was found to be abundant in stem tissue, but not in leaf or root tissue (51). Experiments that increased the levels of these anionic peroxidases in tobacco resulted in plants that showed severe wilting associated with flowering (53). The wilting was due to a dramatic loss of turgor, but the mechanism of this effect is not known.

Peroxidases are thought to be involved in the generation of hydrogen peroxide and in the oxidative polymerization of hydroxylated cinnamyl alcohols. Different peroxidases may mediate these two mechanisms (50, 54). It is now generally accepted (55) that hydrogen peroxide can be generated by two different pathways at the expense of NADH and involving the activity of peroxidases (56). Furthermore, NAD and malate can be used instead of NADH, suggesting a role for cell-wall-bound malate dehydrogenase in the polymerization process (55).

Several workers have suggested that the polymerization of lignin involves the formation of superoxide radicals (2, 57). It is argued that peroxidase is too large to penetrate cell walls directly, but that superoxide radicals would be sufficiently stable and diffusible to move into the lignifying zone. However, peroxidase is present in abundance in the lignifying zone, and more work is needed to test this proposal.

C. Role of Quinone Methides in Polymerization

Lignin is a polymeric natural product derived from an enzyme-initiated dehydrogenative polymerization of the three specific monolignol precursors: coniferyl alcohol, sinapyl alcohol, and para-coumaryl alcohol. Gymnosperm lignin is formed by the polymerization of coniferyl alcohol, and angiosperm lignin is formed by the copolymerization of coniferyl alcohol and sinapyl alcohol. Para-hydroxyphenyl propane units derived from para-coumaryl alcohol are usually present in small amounts in both gymnosperm and angiosperm lignins. The polymerization of the monolignol precursors to lignin is a complex process, leading to a wide variety of linkages in lignin structures. In gymnosperms, the polymerization of lignin is less complex because it essentially occurs with only one monolignol precursor, coniferyl alcohol. Therefore, the following discussion will use coniferyl alcohol as an example.

Coniferyl alcohol is dehydrogenated by peroxidase/hydrogen peroxide to yield phenoxy radicals (58). These radicals are found in several mesomeric (resonant) forms and are very stable (Fig. 3). As a first step in lignin formation, two radicals couple nonenzymatically, but nonrandomly, to form dilignols. Although 10 different linkages are possible as a result of the 5 mesomeric structures, only 3 dilignols are found: β-5, β-O-4, and β-β coupling products (59). All three involve at least one β radical, β-5 being the predominant linkage. The formation of these three dilignols is shown in Fig. 4.

For each coupling involving a β radical, a quinone methide intermediate is formed that is stabilized and rearomatized by either the inter- or intramolecular nucleophilic addition of any groups having an acidic proton. These groups include organic acids, lignols, water, and other alcohols. As shown in Fig. 4, the quinone methide intermediates in the β-5 and β-β coupling products are stabilized by the intramolecular addition of phenolic and alcoholic hydroxyl groups, respectively. The quinone methide intermediate in the β-O-4 coupling product, on the other hand,

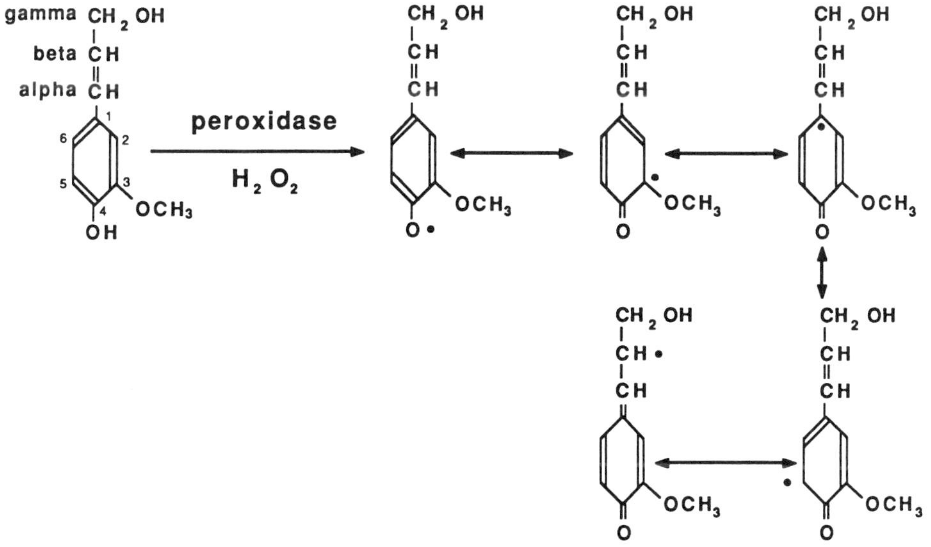

Figure 3. Mesomeric forms of the phenoxy radical formed by the dehydrogenation of coniferyl alcohol.

is stabilized by intermolecular nucleophilic addition. This has three important consequences in the lignification process: (1) the formation of an α-O-4 linkage in lignin when a lignol is added; (2) formation of lignin-carbohydrate linkages when the acidic or alcoholic groups in carbohydrates are added to quinone methide; and (3) formation of α-ester groups that are found in abundance in lignin in grasses (60) and in trace amounts in both hardwood and softwood lignins (61, 62). The possibility of the addition of a monolignol radical to quinone methide has been proposed to explain the presence of a small amount of α-β linkage in lignin (see the chapter by C.L. Chen in this volume).

D. Repetitive Dehydrogenation and Coupling

As shown in Fig. 4, the three dilignols formed as a result of the coupling of monolignol radicals have at least one phenolic hydroxyl group that is readily dehydrogenated by the peroxidase/hydrogen peroxide system to form dilignol radicals. The dilignol radicals can couple with each other to form tetralignols or with a monolignol to form a trilignol. These trilignols and tetralignols all have at least one phenolic hydroxyl group for further dehydrogenation. The repetitive dehydrogenation and coupling of lignol radicals constitute the main mechanism of lignification.

Dilignol radicals and oligolignol radicals differ from the monolignol radicals in

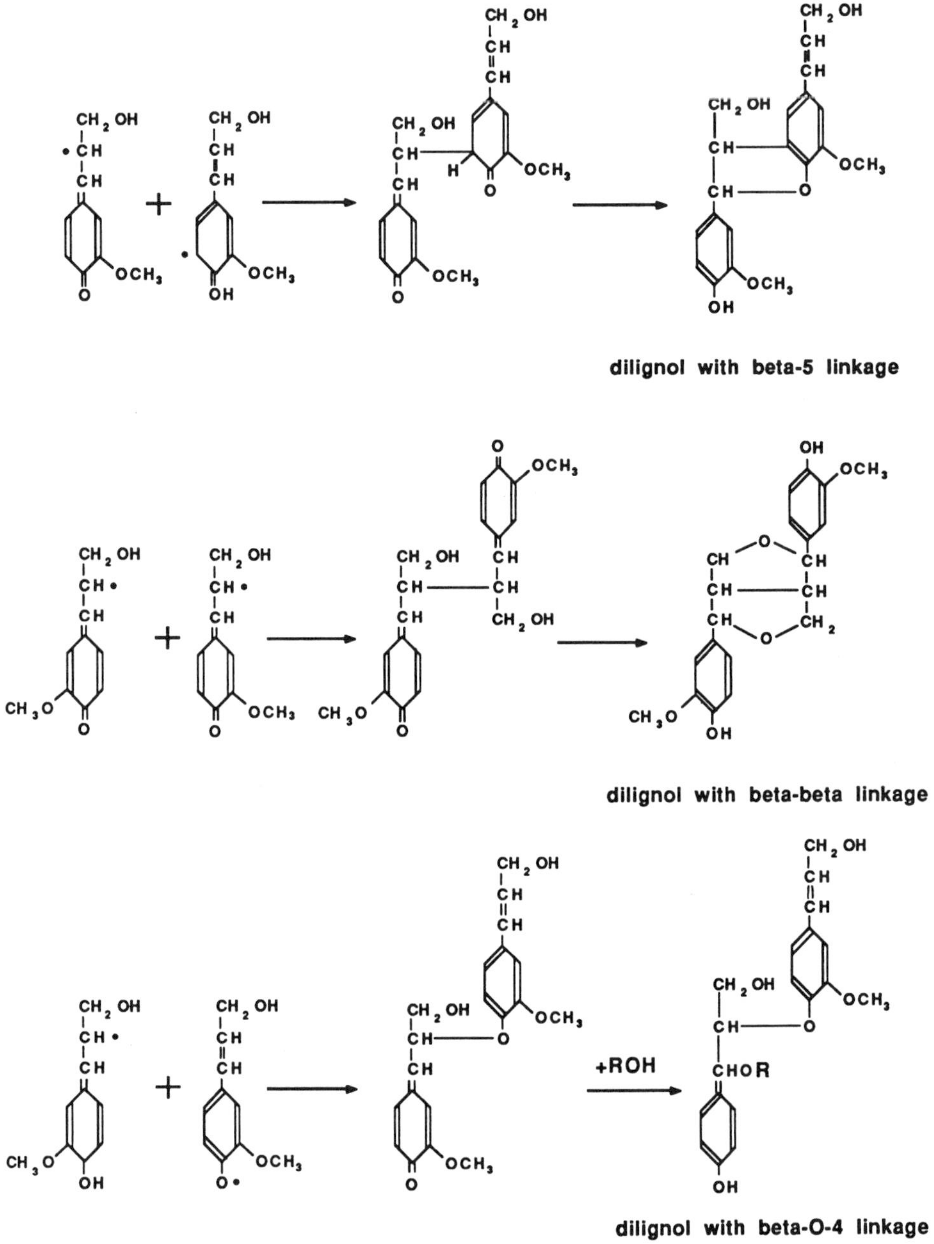

Figure 4   Formation of three dilignols from coniferyl alcohol.

that they do not have a double bond conjugated with the aromatic ring that has a free phenolic hydroxyl group. Hence, dilignols and oligolignols lack the β radical form and consequently further coupling between dilignols or oligolignols cannot involve β linkages; only 5-5, 4-0-5, 1-5, and 1-0-4 linkages prevail. In contrast, coupling between a monolignol radical and a dilignol or oligolignol radical always involves one β radical that is supplied from the monolignol (59), predominantly forming β-O-4 linkages.

E. Determination of Lignin Structure

The behavior of the lignol radicals has significant implications for the structure of lignin. Monolignol radicals are different from the other radicals; coupling is not random and operates according to well-defined rules. Consequently, the structure of lignin is influenced by the rate of release of monolignols into the lignifying zone.

Studies on the dehydrogenative polymerization of coniferyl alcohol in vitro (7) demonstrated that the structure of the in vitro lignin differed significantly if coniferyl alcohol was added all at once ("zulauf" method), or if it was added gradually ("zutropf" method) [see also (2) for a discussion of these methods]. In the plant, if monolignols were released at a slow rate, monolignols would be continuously added to a pre-existing polymer and the β-O-4 linkage would predominate. Alternatively, if the monolignols were released at once, monolignols would be rapidly converted into dilignols and oligolignols that would result in a preponderance of β-5, 5-5, 5-0-4 types of linkages.

Therefore, the structure of lignin is determined by the types of monolignols produced and the rate that they are released into the lignifying zone. Analysis of the types of linkages found in natural lignin shows that lignin is formed predominantly, but not exclusively by a gradual supply of monolignol to the lignifying zone. Consequently, the structure of lignin might vary dramatically if the rate of formation of monolignols or their transportation changed during the lignification process. For example, the lignin in the middle lamella is formed very early in the process of cell wall biosynthesis when growth is vigorous and the relative concentration of monolignols could be high, compared to the structure of lignin in the cell wall formed later during the terminal differentiation of the tracheids.

IV. REGULATION OF LIGNIN BIOSYNTHESIS

A. Genetic Regulation

Lignin synthesis is developmentally regulated and cell-specific. During xylogenesis, cambial fusiform initials differentiate to produce lignin. The enzymes of the lignin biosynthetic pathway have greatly increased activity. The properties of the lignin that is produced depend on the amount, rate, and kind of substrates (monolignols) that are produced during the time of lignification. Lignin biosynthesis does not appear to be dependent on or concurrent with the synthesis of other major pathways of phenylpropanoid metabolism, such as flavonoids or extractives in coniferous species. Therefore, it is interesting to inquire whether lignin biosynthesis could be genetically modified within limits that would not interfere with key biological processes of growth and development, but could make sufficient differences in wood quality to improve yield, reduce waste, and improve the nature and quality of the wood-derived products.

To explore the prospects of the genetic engineering of lignin, it is important to know the detailed mechanisms of the biosynthetic pathway of lignin biosynthesis and the relationship of this pathway to other major metabolic and developmental processes in the tree. The ability to modify a process by genetic engineering is strongly related to how well the genetic regulatory mechanisms are understood; otherwise, proposed modifications are not likely to work. However, very little is known about the genetic and developmental regulation of the lignin biosynthetic pathway.

Some information is available on the ability to modify wood properties by traditional selective breeding. Wood characteristics, particularly specific gravity, are responsive to improvement (63–65). Heritability for lignin content in loblolly pine is at least moderate (66) and strong in slash pine and quaking aspen (67, 90) indicating that selection could be used to modify lignin content.

B. Mutations in Lignin Biosynthesis in Grasses

In maize, a mutation called brown midrib has been characterized that has lowered lignin content (68). The lignin is reduced from 0.88% dry matter to 0.05% in the silage of the brown midrib 3 mutation and digestability is significantly increased (69). Similar mutations have been found in sorghum and (70) and pearl millet (71). Two lines of evidence indicate that O-methyl transferase is involved in this mutation in maize. Thioacidolysis revealed the presence of 5-hydroxyguaiacyl monomers in the brown rib mutant that were absent in wild-type maize (72), suggesting that the methylation of 5-hydroxyferulic acid is reduced in the mutant. Direct measurement of the catechol O-methyl transferase (OMT) activity has been shown to be reduced in brown midrib maize seedlings compared to normal maize (91). These results further support the view that a mutant enzyme may be responsible for the low levels of lignin in the brown rib mutant.

C. Genetic Regulation of PAL

The extent to which the pathway of lignin biosynthesis may be regulated by differential gene expression in development or environmental stimulation is indicated by the regulation of a key enzyme, PAL, that initiates the lignin biosynthetic pathway from phenylalanine (Fig. 2, step 1A). The genetic and developmental regulation of PAL is best characterized in bean. In bean, PAL is encoded by a family of three genes, each of which exists in polymorphic forms based on minor structural differences (25). The distinct PAL isoforms show marked differences in developmental specificity and response to environmental stimuli (26).

All three PAL genes are expressed at high levels in roots, and all show induction by mechanical wounding. Fungal infection only activates PAL1 and PAL3, whereas illumination activates PAL1 and PAL2 but not PAL3. PAL1 and PAL2 are expressed in shoots but only PAL1 is expressed in leaves. PAL2 is expressed at very high levels in petals, whereas PAL1 is weakly expressed and PAL3 is not expressed at all.

The complexity of the PAL gene family demonstrates that there are alternate regulatory modes for phenylpropanoid biosynthesis during growth and development. We do not know if a specific subset of genes is always expressed in lignifying tissues or if different genes are alternatively expressed for the synthesis of lignin under different developmental or environmental circumstances.

Studies of cells grown in a liquid suspension culture suggest that the levels of expression of PAL can be modified, because cell lines with increased levels of PAL have been obtained from tobacco, carrot, and sycamore maple (21, 74, 75). More research is needed to demonstrate if the level of PAL or amount of lignin could be modified in a whole plant or specific tissue.

D. Effects of Chemical Inhibitors

Lignin biosynthesis is intimately involved with the biosynthesis of other cell wall components in the formation of wood. It is important to know to what extent the biosynthesis of the components is coupled such that a modification of one component could affect the others. Evidence that bears on this question comes from several sources where the synthesis of lignin has been modified or inhibited either by chemical inhibition, genetic mutation, or from a disease state. These studies provide interesting information on the role of lignin in plant growth and development, as well as information on the degree to which lignin biosynthesis is coupled to the biosynthesis of other major components of the primary and secondary cell wall.

AOPP (2-aminooxy-3-phenylpropionic) acid is an amino-oxy analogue of L-phenylalanine and, therefore, is a powerful inhibitor of the action of PAL in a specific and competitive manner (76, 89). When AOPP is fed to mung bean seedlings, the result is inhibition of growth, wilting, and death. The synthesis of lignin and anthocyanin is inhibited with dramatic effects on the structure of the plants. In treated seedlings, the xylem vessels of the hypocotyls and roots collapse. Cellulose microfibrils are separated from one another in the unlignified secondary wall and lie disorganized in the lumen of the mature xylem cells (77). However, many aspects of the differentiation of the cell wall appear unaffected by the reduced level of lignification. The basic organization of the secondary wall in AOPP-treated tissue appears undisturbed until at maturity the secondary wall detaches and fragments. The cells programmed to become xylem vessels continue to express their developmental program. The lack of lignification appears to have no feedback effect on the regulation of other features of the secondary wall (77).

Similarly, the walls of phloem fibers are lignified in untreated mungbean seedling tissues and completely unlignified in AOPP-treated tissues. These fibers were not collapsed or disorganized in any apparent way. No deleterious effects were observed on the structure or organization of any of the cells within or without the endodermis of the root or the hypocotyl. AOPP-treated seedlings appear more susceptible to fungal infection, supporting the view that normal lignification plays an important role in disease resistance (6, 77).

The specific inhibition of CAD is also known to reduce lignification in poplar tissues (73). Two organic compounds OHPAS [N-(O-hydroxyphenyl)-sulfinamoyl-t-butyl acetate] and NH2PAS [N-(O-aminophenyl)-sulfinamoyl-t-butyl acetate] function as inhibitors in vivo. The treatment of poplar tissues with these inhibitors reduced the synthesis of lignin by 45%.

Glyphosate [N-(phosphonomethyl)glycine] is known to inhibit the synthesis of aromatic amino acids (78). Lignin biosynthesis in asparagus spears has been shown to be inhibited by glyphosate, presumably through the inhibition of the synthesis of phenylalanine that is the precursor for the biosynthesis of cinnamyl alcohols (79). If freshly harvested asparagus spears are treated with glyphosate, there is decreased toughening and lignification on storage. These results provide further support for the hypothesis that the amount of lignification in wood biosynthesis also depends on the level of the available precursor.

E. Modification of Lignin in Diseases

An interesting modification of lignin in a woody plant has been studied in apple (*Malus x domestica* var. Lord Lambourne). In this variety, a disease called rubbery wood disease was first identified in 1944 (80, 81). The disease was originally thought to be caused by a virus and was subsequently thought to be due to infection by a mycoplasma (82). This view has been challenged. The disease is now thought to be due to a xylem-limited bacteria similar to rickettsia (83). The disease symptoms include excessive flexibility in the branches, the soft "cheesy" texture of the wood, and weeping habit of the branches (84). In severe cases, young trees cannot support themselves; they grow along the ground and have shortened internodes (85).

Analysis of the wood produced by the affected parts of the tree shows that it is rich in cellulose, but poor in lignin (85). The lignin itself has a lower methoxyl content than the lignin found in uninfected wood (86). It is not known how mycoplasmal infection causes a decrease in the level and reduction in the methoxyl content of the lignin. One possibility is that the infection reduces the level of a precursor such as phenylalanine for lignin biosynthesis.

A similar disease has been observed in citrus. It is graft transmissible and shows unusual flexibility and willowing habit (87). The disease-producing agent has been presumed to be a virus-like organism.

F. Potential Modification of Lignin by Genetic Engineering.

There are many approaches to the potential modification of lignin. The inhibition of precursor synthesis might be possible if antisense constructs could be made that would inhibit the biosynthesis of any of the enzymes for the lignin biosynthetic pathway. CAD, peroxidase, Co-A reductase, and enzymes involved in transport are potential candidates for this strategy. To test this strategy, the appropriate specific genes and demonstration of antisense inhibition are needed. Lignin might be modified if the level of the precursor could be increased. If one of the enzymes, such as PAL, regulated the level of the precursor, increasing the number of copies of PAL or modifying its regulation could increase the precursor level.

Modification of the kind of precursors polymerized into lignin has been considered as an approach to modifying lignin (88). Hardwoods use sinapyl alcohol as a precursor, and as a result, the lignin that is produced is more readily hydrolyzed in paper processing. If it were possible to move the genes that produce sinapyl alcohol into conifers, and to have them properly expressed, then such modified lignin could be produced. What is not clear is the number of enzymes that would have to be transferred into conifers from hardwoods.

Much more information is needed about lignin biosynthesis before we can evaluate these strategies. We do not have essential information about the regulation of gene expression and cell physiology to know if any of these strategies will work, nor do we have the isolated genes or the technology to transfer genes into conifers to test the hypotheses upon which these strategies are based. However, progress in this field is rapid and it may not be long before such experiments are possible.

REFERENCES

1. J. Gammie, "World Timber to the Year 2000," *The Economist*, Intelligence Unit Special Rep. 98, Economist Newspaper Ltd., London, England, 1986.
2. T. Higuchi, "Biosynthesis of Lignin," in *Biosynthesis and Biodegredation of Wood Components* (T. Higuchi, ed.), Academic Press, New York, 1985, pp. 141–160.
3. A. Brown, "Review of Lignin in Biomass," *J. Appl. Biochem.*, 7:371–387 (1985).
4. Y. Asada, T. Ohguchi, and I. Matsumoto, "Lignin Formation in Fungus Infected Plants," *Rev. Plant Protein Res.*, 8:104–113 (1975).
5. H. Grisebach, "Lignins," in *The Biochemistry of Plants*, Vol. 7 (P. K. Stumpf and E. E. Conn, eds.) Academic Press, New York, 1981, pp. 457–478.
6. C. P. Vance, T. K. Kirk, and R. T. Sherwood, "Lignification as a Mechanism of Disease Resistance," *Ann. Rev. Phytopathol.*, 18:259–288 (1980).
7. K. Freudenberg and A. C. Neish, in *Constitution and Biosynthesis of Lignin* (A. Kleinzeller, G. F. Springer, and H. G. Wittmann, eds.), Springer-Verlag, New York, 1968, pp. 78–122.
8. A. Sakakibara, "Chemical Structure of Lignin Related Mainly to Degradation Products," *Recent Advances in Lignin Biodegradation Research* (T. Higuchi, H-M. Chang, and T. K. Kirk, eds.), Proc. 2nd Int. Seminar UNI Pub. Co., Tokyo, Japan, 1983, pp. 12–33.
9. E. S. Barghoorn, "Evolution of Cambium in Geologic Time," in *The Formation of Wood in Forest Trees* (M. H. Zimmermann, ed.), Academic Press, New York, 1964, pp. 3–17.
10. I. Kawamura and T. Higuchi, "Comparative Studies of Milled Wood Lignins from Different Taxonomical Origins by Infrared Spectroscopy," in *Symposium International sur la Chimie et al Biochimie de al Lignine, de la Cellulose et des Hemicelluloses*, Grenoble, Universite Faculte des Sciences, Reunies de Chambery, 1964, pp. 439–456.
11. C. F. Delwiche, L. E. Graham, and N. Thomson, "Lignin-like Compounds and Sporopollenin in Coleochaete, an Algal Model for Land Plant Ancestry," *Sci.*, 245:399–401 (1989).
12. L. Fries, "Growth Regulating Effects of Phenyl Acetic-acid and P Hydroxyphenl Acetic-acid on Fungus Spiralis Phaeophyceae Fucales in Axenic Culture," Inst. Physiol. Bot., Uppsala, Sweden, 1977, Vol. 16, No. 4, pp. 451–456.
13. L. Fries and S. Aberg, "Morphogenetic Effects of Phenyl Acetic-acid and P Hydroxyphenyl Acetic-acid on the Green Alga Entermorpha-compressa in Axenic Culture, "Inst. Physiol. Bot., Uppsala, Sweden, 1978, Vol. 88, No. 5, pp. 383–388.
14. F. Wightman and D. L. Lighty, "Identification of Phenyl Acetic-acid as a Natural Auxin in the Shoots of Higher Plants.," Biol. Dept., Carleton Univ., Ottawa, Canada, 1982; see also *Physiol. Plant*, 55(1):17–24 (1982).
15. H. Grisebach, "Biosynthesis of Flavonoids," in *Biosynthesis and Biodegradation of Wood Components* (T. Higuchi, ed.), Academic Press, New York, 1985, pp. 291–324.
16. H. Kindl, Biosynthesis of Stilbenes. In *Biosynthesis and Biodegradation of Wood Components*(T. Higuchi, ed.), Academic Press, New York, 1985, pp. 349–377.

17.   W. E. Hillis, "Occurence of Extractives in Wood Tissue," in *Biosynthesis and Bio-degradation of Wood Components* (T. Higuchi, ed.), Academic Press, New York, 1985, pp. 209–228.

18.   S. Saka and D. A. I. Goring, "Localization of Lignins of Wood Cell Walls," in *Biosynthesis and Biodegradation of Wood Components* (T. Higuchi, ed.), Academic Press, New York, 1985, pp. 51–62.

19.   A. J. Panshin and C. de Zeeuw, *Wood Technology: Structure, Identification, Properties, and Uses of Commercial Woods of the United States and Canada*, 4th ed., McGraw-Hill, New York, 1980.

20.   J. Koukol and E. E. Conn, "Purification and Properties of the Phenylaline Deaminase of Hordeum vulgare," *J. Biol. Chem.*, *236*:2692–2698 (1961).

21.   K. R. Hanson and E. A. Havir, "Phenylalanine Ammonia-lyase," in *Biochemistry of Plants: A Comprehensive Treatise* (P. K. Stumpf and E. E. Conn, eds.), Academic Press, New York, 1981, Vol. 7, pp. 577–625.

22.   A. C, Neish, "Formation of m- and p-Coumaric Acids by Enzymatic Deamination of the Corresponding Isomers of Tyrosine," *Phytochem.*, *1*:1–24 (1961).

23.   G. G. Gross, "Biosynthesis and Metabolism of Phenolic Acids and Monolignols," in *Biosynthesis and Biodegredation of Wood Components*, (T. Higuchi, ed.) Academic Press, New York, 1985, pp. 229–271.

24.   D. Kuhn, J. Chapell, A. M. Boudet, and K. Hahlbrock, "Induction of Phenylalanine Ammonia-lyase and 4-Coumarate: CoA Ligase mRNAs in Cultured Plant Cells by UV Light or Fungal Elicitor," *Proc. Natl. Acad. Sci. USA*, *81*:1102–1106 (1984).

25.   C. L. Cramer, K. Edwards, M. Dron, X. Liang, S. L. Dildine, G. P. Bolwell, R. A. Dixon, C. J. Lamb, and W. Schuch, "Phenylalanine Ammonia-lyase Gene Organization and Structure," *Plant Mol. Biol.*, *12*:367–383 (1989).

26.   X. Liang, M. Dron, C. L. Cramer, R. A. Dixon, and C. J. Lamb, "Differential Regulation of Phenylalanine Ammonia-lyase Genes During Plant Development and by Environmental Cues," *J. Biol. Chem.*, *264*(24):14, 486–14, 492 (1989).

27.   R. Pfandler, D. Scheel, H. Sandermann, and H. Grisebach, "Stereospecificity of Plant Microsomal Cinnamic Acid 4-Hydroxylase," *Arch. Biochem. Biophys.*, *178*:315–316 (1977).

28.   P. T. F. Vaughan, R. Eason, J. Y. Paton, and G. A. Ritchie, "Molecular Weight and Amino Acid Composition of Purified Spinach Beet Phenolase," *Phytochem.*, *14*:2383–2386 (1975).

29.   G. Downes and N. D. Turvey, "Reduced Lignification in Pinus radiata D. Don," *Aust. For. Res.*, *16*:371–377 (1986).

30.   M. Shimada, H. Fushiki, and T. Higuchi, "Mechanism of Biochemical Formation of the Methoxyl Groups in Softwood and Hardwood Lignins," *Mokuzai Gakkaishi*, *19*:13–21 (1973).

31.   H. Kuroda, M. Shimada, and T. Higuchi, "Characterization of a Lignin-Specific O-methyltransferase in Aspen Wood," *Photochem.*, *20*:2635–2639 (1981).

32.   C. Grand, "Ferulic Acid 5-hydroxylase: A New Cytochrome P-450-dependent Enzyme form Higher Plant Microsomes Involved in Lignin Synthesis," *FEBS Lett.*, *169*:7–11 (1984).

33.   M. H. Zenk, "Recent Work on Cinnamoyl-CoA Derivatives," in *Recent Advances in Phytochemistry Series*, (T. Swain, J. B. Harborne, and C. F. van Sumere, eds.),

Plenum Press, New York, 1979, Vol. 12, pp. 139–176.

34. H. Wengenmayer, J. Ebel, and H. Grisebach, "Enzymic Synthesis of Lignin Precursors. Purification and Properties of a Cinnamoyl-CoA: NADPH Reductase from Cell Suspension Cultures of Soybean (Glycine max.)," *Eur. J. Biochem.*, *65*:529–539 (1976).

35. T. Luderitz and H. Griesbach, "Enzymic Synthesis of Lignin Precursors Comparison of Cinnamoyl-CoA Reductase and Cinnamyl Alcohol: NADP+ Dehydrogenase form Spruce (Picea abies L.) and Soybean (Glycine max L.)," *Eur. J. Biochem.*, *119*:115–124 (1981).

36. F. Sarni, C. Grand, and A. M. Boudet, "Purification and Properties of Cinnamoyal-CoA Reductase and Cinnamyl Alcohol Dehydrogenase from Poplar Stems (Populus X euramericana)," *Eur. J. Biochem.*, *139*:259–265 (1984).

37. M. H. Walter, J. Grima-Pettenati, C. Grand, A. M. Boudet, and C. J. Lamb, "Cinnamyl-alcohol Dehydrogenase, a Molecular Marker Specific for Lignin Synthesis: cDNA Cloning and mRNA Induction by Fungal Elicitor," *Proc. Natl. Acad. Sci. USA*, *85*:5546–5550 (1988).

38. H. Kutsuki, M. Shimada, and T. Higuchi, "Regulatory Role of Cinnamyl Alcohol Dehydrogenase in the Formation of Guaiacyl and Syringyl Lignins," *Phytochem.*, *21*(1):19–23 (1982).

39. D. M. O'Malley, R. Sanozky-Dawes, T. Presnell, and R. R. Sederoff, "Characterization and Genetic Control of Cinnamyl Alcohol Dehydrogenase in Loblolly Pine," *South. For. Tree Improv. Conf.*, *20*:441 (1989).

40. M.H. Walter, J. Grima-Pettenati, C. Grand, A. M. Boudet, and C. J. Lamb, "Extensive Sequence Similarity of the Bean CAD4 Clone to a Maize Malic Enzyme," *Plant Mol. Biol.*, *15*:525–526(1990).

41. R. K. Ibrahim and H. Grisebach, "Purification and Properties of UDP-glucose: coniferyl Alcohol Glucosyltransferase from Suspension Cultures of Paul's Scarlet Rose," *Arch. Biochem. Biophys.*, *176*:700–708 (1976).

42. R. K. Ibrahim "Glucosylation of Lignin Precursors by Uridine Diphosphate Glucose:coniferyl Alcohol Glucosyltransferase in Higher Plants," *Z. Planzenphysiol.*, 85:253–262 (1977).

43. G. Schmid and H. Grisebach, "Enzymic Synthesis of Lignin Precursors—Purification and Properties of UDP Glucose: coniferyl-alcohol Glucosyltransferase and Cambial Sap of Spruce (Picea abies L.)," *Eur. J. Biochem.*, *123*:363–370 (1982).

44. G. Schmid, D. K. Hammer, A. Ritterbusch, and H. Grisebach, "Appearance and Immunohistochemical Localization of UDP-glucose: coniferyl Alcohol Glucosyltransferase in Spruce (Picea abies L., Karst.) Seedlings," Pla*nta*, *156*:207–212 (1982).

45. S. Marcinowski and H. Grisebach "Enzymology of Lignification Cell-wall-bound B-Glucosidase for Coniferin from Spruce (Picea abies) Seedlings," *Eur. J. Biochem.*, *87*:37–44 (1978).

46. G. G. Gross, "Recent Advances in the Chemistry and Biochemistry of Lignin," in *Biochemistry of Plant Phenolics* (T. Swain, J. B. Harborne, and C. F. Van Sumere, eds.), *Rec. Adv. Phytochem.*, *12*:177–220 (1979).

47. G. G. Gross, "The Biochemistry of Lignification," *Adv. Bot. Res.*, *8*(2):25–63 (1980).

48. N. G. Lewis, P. Dublesten, T. L. Eberhardt, E. Yamamoto, and G. H. N. Towers, "The E/Z Isomerization Step in the Biosynthesis of Z-coniferyl Alcohol in Fagus

grandifolia," *Phytochem.*, *26*:2729–2734 (1987).

49.   J. M. Harkin and J. R. Obst, "Lignification in Trees: Indication of Exclusive Peroxidase Participation,"*Science 180*:296–298(1973).

50.   R. Goldberg, T. Le, and A. M. Catesson, "Localization and Properties of Cell Wall Enzyme Activities Related to the Final Stages of Lignin Biosynthesis," *J. Exp. Bot.*, *36*(164):503–510. (1985).

51.   L. M. Lagrimini, W. Burkhart, M. Moyer, and S. Rothstein, "Molecular Cloning of Complementary DNA Encoding the Lignin Forming Peroxidase from Tobacco: Molecular Analysis and Tissue-Specific Expression," *Proc. Natl. Acad. Sci.*, *84*:7542–7546 (1987).

52.   M. Mader, A. Nessel, and M. Bopp, "On the Physiological Significance of the Isoenzyme Groups of Peroxidase from Tobacco Demonstrated by Biochemical Properties. II. pH Optima, Michaelis-Constants, Maximal Oxidation Rates," *Z. Planzenphysiol.*, *82*, 247–260 (1977).

53.   L. M. Lagrimini, S. Bradford, and S. Rothstein, "Peroxidase-Induced Wilting in Transgenic Tobacco Plants," The Plant Cell, 2:7–18 (1990).

54.   M. Mader, J. Ungemach, and P. Schloss, "The Role of Peroxidase Isoenzyme Groups of Nicotiana tabacum in Hydrogen Peroxide Formation," *Planta, 147*:467-470 (1980).

55.   R. Goldberg, M. Liberman, C. Mathieu, M. Pierron, and A. M. Catesson, "Development of Epidermal Cell Wall Peroxidases Along the Mung Bean Hypocotyl: Possible Involement in the Cell Wall Stiffening Process," *J. Exp. Bot.*, *38*(193):1378–1390 (1987).

56.   G. G. Gross, C. Janse, and E. F. Elstner, "Involvement of Malate, Monophenols, and the Superoxide Radical in Hydrogen Peroxide Formation by Isolated Cell Walls from horseradish (Amoracia lapathifolia Gilib.)," *Planta*, *136*:271–276 (1977).

57.   U. Westermark, "Calcium Promoted Phenolic Coupling by Superoxide Radical—a Possible Lignification Reaction in Wood," *Wood Sci. Technol.*, *16*:71–78 (1982).

58.   K. Freudenberg, "Lignin: Its Constitution and Formation from p-Hydroxycinnamyl Alcohols," *Sci.*, *148*:595–600 (1965).

59.   K. V. Sarkanen, "Precursors and Their Polymerization," in *Lignins: Occurrence, Formation, Structure and Reactions* (K. V. Sarkanen and C. H. Ludwig, eds.) Wiley-Interscience, New York, 1971, pp. 95–163.

60.   T. Higuchi, Y. Ito, and I. Kawamura, "p-Hydroxyphenyl-propane Component of Grass Lignin and the Role of Tyrosine Ammonia-lyase in Its Formation," *Phytochem.*, 6:875–881 (1967).

61.   D. C. C. Smith, "Ester Groups in Lignin," *Nature*, *176*:267–268 (1955).

62.   D. C. C. Smith, "Contribution of Residues Containing Carbonyl to the Ultra-Violet Absorption of Lignins," *Nature*, *176*:927–928 (1955).

63.   J. P. van Buijtenen, "Heritability Estimates in Wood Density of Loblolly Pine," *TAPPI*, *45*:602–605 (1962).

64.   B. J. Zobel and J. Talbert, *Applied Tree Improvement*, Wiley, London, 1983.

65.   B. J. Zobel and J. P. van Buijtenen, *Wood Variation: Its Causes and Control*, Springer-Verlag, New York, 1989.

66.   J. P. van Buijtenen, personal communication, (1987).

67.   D. W. Einspahr, J. P. van Buijtenen, and J. R. Peckman, "Natural Variation and Heritability in Triploid Aspen,"*Silvae Genet.*, *12*:51–58(1963).

68. L. D. Muller, R. F. Barnes, L. F. Bauman, and V. F. Colenbrander, "Variations in Lignin and Other Structural Components of Brown Midrib Mutants of Maize," *Crop Sci.*, *11*:413–415 (1971).

69. R. F. Weller and R. H. Phipps, "The Effect of the Brown Midrib Mutation on the in vivo Digestibility of Maize Silage," *6th Silage Conference*, 1981, pp. 73–74.

70. T. S. Bittinger, R. P. Cantrell, and J. D. Axtell, "Allelism Tests of the Brown Midrib Mutants of Sorghum," *J. Heredity*, *72*:147–148 (1981).

71. J. H. Cherney, J. D. Axtell, M. M. Hassen, and K. S. Anliker, "Forage Quality Characterzation of a Chemically Induced Brown Rib Mutant in Pearl Millet," *Crop Sci.*, *28*:783–787 (1988).

72. C. Lapierre, M. T. Tollier, and B. Monties, "Occurrence of Additional Monomeric Units in the Lignins form Internodes of a Brown-Midrib Mutant of Maize bm3," *Compte Rendu de l'Academie des Sciences III*, *307*:723–728 (1988).

73. C. Grand, F. Sarni, and A. M. Boudet, "Inhibition of Cinnamyl-alcohol Dehydroge-nase Activity and Lignin Synthesis in Poplar (Populus x euramericana, Dode) Tissues by Two Organic Compounds," *Planta*, *163*:232–237 (1985).

74. J. E. Palmer and J. Widholm "Characterization of Carrot and Tobacco Cell Cultures Resistant to p-Fluorophenylalanine," *Plant Physiol.*, *56*:233–238 (1975).

75. R. W. E. Gather Cote and H. E. Street, "Isolation, Stability and Biochemistry of a p-Fluorophenylalanine-resistant Cell Line of Acer Pseudoplatanus L," *New Phytol.*, *77*:29–41 (1976).

76. N. Amrhein and K.-H. Goedeke, "Alpha Aminooxy-beta-phenyl Propionic-acid, a Potent Inhibitor of L Phenylalanine Ammonia-lyase EC-4.3.1.5 in vitro and in vivo, *Plant Sci. Lett.*, *8*(4):313–317 (1977).

77. C. C. Smart and N. Amrhein, "The Influence of Lignification on the Development of Vascular Tissue in Vigna radiata L.," *Protoplasma*, *124*:87–95 (1985).

78. R. A. Jensen, "The Shikimate/Arogenate Pathway: Link Between Carbohydrate Metabolism and Secondary Metabolism," *Physiol. Plant.*, *66*:164–168 (1986).

79. M. E. Saltviet, Jr., "Postharvest Glyphosate Application Reduces Toughening, Fiber Content, and Lignification of Stored Asparagus Spears," *J. Amer. Soc. Hort. Sci.*, *113*:569–572 (1988).

80. M. B. Crane, "The Mystery of Lord Lambourne," *The Grower*, *22*:10–12 (1944).

81. T. Wallace, T. Swarbrick, and L. Ogilvie, "Some New Troubles in Apples with Special Reference to the Variety Lord Lambourne," *The Grower*, *22*:12–13 (1944).

82. A. B. Beakbane, M. D. Mishra, A. F. Posnette, and C. H. W. Slater, "Mycoplasma-like Organism Associated with Chat Fruit and Rubbery Wood Disease of Apple (Malus domestica Borkh.), Compared with Those in Strawberry with Green Petal Disease," J. Gen. Mic*robiol.*, *66*:52–62 (1971).

83. P. Friedland, "Virus and Viruslike Diseases of Pome Fruit Simulating Noninfective Disorders," Washington State Univ., Pullman, Washington, 1989, Special Publication 0003.

84. I. W. Prentice, "Experiments on Rubbery Wood Disease of Apple Trees," Ann. Rept., East Malling Res. Sta., Kent, England, 1949, pp. 122–125.

85. E. B. Cowling, "The Engineering Mechanics of Pathogenesis," in *Plant Diseases* (J. G. Horsfall and E. B. Cowling, eds.), Academic Press, New York, 3, pp. 311–312.

86.   E. Sondheimer and W. G. Simpson, "Lignin Abnormalities of Rubbery Apple Wood,"
       Can. J. Biochem. Physiol., *40*:841:846 (1962).

87.   Y. S. Ahlawat and V. V. Chenulu, "Rubbery Wood—A Hitherto Unknown Disease
       of Citrus," *Current Sci.*, *54*:580–581 (1985).

88.   R. C. Bugos, V. Chiang and W. H. Campbell, "Cloning and Expression of Lignin by
       Specific O–Methly Transferase". Plant Physiology, vol 93 Supplement, abstract 83,
       (1990) p. 15.

89.   N. Amrhein, G. Frank, G. Lemm, and H.–B. Luhmann, "Inhibition of Lignin
       Formation by L–Alpha–Aminooxy–Beta–Phenylproprionic Acid – an Inhibitor of
       Phenylalanine Ammonialyase". *Eur. J. Cell Biol.,* *29*:139–144 (1983).

90.   D. W. Einspahr, R. E. Goddard, and H. S. Gardner, "Slash Pine Wood and Fiber
       Property Heritability Study", *Silvae Genet.,* *13:*103–109 (1964).

91.   C. Grand, P. Parmentier, A. Boudet, and A. M. Boudet, "Comparison of Lignin and
       of Enzymes Involved in Lignification in Normal and Brown Midrib (bm3) Mutant
       Corn Seedlings," *Physiologie Vegetale, 23*:905–911 (1985).

7

Hemicelluloses

Roy L. Whistler

Purdue University, West Lafayette, Indiana

Chyi-Cheng Chen

Hoffmann-La Roche Inc., Nutley, New Jersey

I. INTRODUCTION

Hemicelluloses constitute 20–30% of wood. They are found predominantly in the primary and secondary cell walls. A smaller amount also occurs in the middle lamella. Crude preparations of wood gum were obtained by Thomsen (1) and by Wheeler and Tollens (2) through extraction of wood with alkaline solution, followed by ethanol precipitation. In 1891, Schulze (3) found that polysaccharides extracted from plants with dilute alkaline solution were hydrolyzed more readily than cellulose. Schulze considered them to be related chemically and structurally to cellulose and, therefore, designated them as hemicelluloses. It was soon learned that these polysaccharides are a distinct group of cell wall components.

Hemicelluloses are generally classified according to the types of sugar residues present. Thus, xylan is a polymer of D-xylosyl residues and mannan a polymer of D-mannosyl residues. Heteroglycans containing two to four, or rarely five or six, different sugar types are frequently encountered, whereas homoglycans are not common. The sugar units most often found in hemicelluloses are D-xylose, D-mannose, D-glucose, D-galactose, L-arabinose, 4-O-methyl-D-glucuronic acid, D-galacturonic acid, and D-glucuronic acid. Less common sugar units are L-rhamnose, L-galactose, L-fucose, and O-methylated neutral sugars.

Hemicelluloses from gymnosperms (softwoods) and angiosperms (hardwoods) are not the same. In hardwoods, the predominant hemicellulose is a partially acetylated (4-O-methylglucurono) xylan with a small proportion of glucomannan.

In softwoods, the major hemicellulose is partially acetylated galactoglucomannans. A smaller amount of an arabino-(4-O-methylglucurono)xylan is also present. Among the softwoods, the larches are unique in that their major hemicellulose is an arabinogalactan. The qualitative and quantitative differences in hemicelluloses (4) of softwoods and hardwoods are shown in Table 1. There have been a number of reviews on hemicelluloses (5–14).

II. ISOLATION

A. Extraction

The most common method of isolation is alkaline extraction. Extractives such as resins, terpenes, and phenolic compounds can be removed almost completely by Soxhlet extraction with azeotropic benzene-ethanol (15) or with acetone. Most of the lignin is then removed from the extractive-free wood using a mixture of acetic acid and sodium chlorite at 70–75 °C (16). The extractive-free and delignified carbohydrate residue is called holocellulose. Hemicelluloses are extracted either directly from extractive-free wood or more frequently from holocellulose. Although some hemicelluloses are water-soluble after isolation, they may not be water-extractable from wood prior to delignification. The arabinogalactan of larches is one of the few major polysaccharides that can be isolated in high yield by extracting the lignified wood with water.

Extraction of hemicelluloses from holocellulose rather than from wood results in a more complete removal of hemicelluloses with less contamination from lignin. Nevertheless, the extractive method has some disadvantages because some soluble hemicelluloses may be lost during the holocellulose preparation (17). Delignification

Table 1 Major Hemicelluloses of Softwoods and Hardwoods

Hemicellulose type	Softwood[a]	Hardwood[a]
O-Acetyl-galactoglucomannan	16	—
Glucomannan	—	3
Arabino-(4-O-methylglucurono)xylan	10	—
O-Acetyl-(4-0-methylglucurono)xylan	—	26

[a]Five common species were used to obtain the average values, and all values are in percent of extractive-free wood.
Source: From Ref. 4.

by acid-chlorite may oxidize reducing-end residues of hemicelluloses to aldonic acids (18) and cause the slight depolymerization and oxidation of 2,3-glycol groups (19). The addition of manganese salts increases the rate of delignification, but retards the rate of depolymerization of polysaccharides (20). Most of the residual lignin in holocellulose is alkali-soluble and thus is concentrated in the alkaline solution during hemicellulose extraction. The presence of lignin in crude hemicellulose hinders the fractionation procedure (21,22). Much of this lignin can be removed by treating the crude hemicellulose with buffered sodium chlorite (23).

Various alkaline concentrations from 2–24% have been used for hemicellulose extraction. Alkaline extraction can result in many changes in the polysaccharides even under oxygen-free conditions (24). A significant proportion of the hexuronic acid residues in 4-O-methylglucuronoxylans are cleaved during alkali treatment (25). Alkaline degradation occurs at the reducing ends of the polysaccharide and, among other things, gives rise to saccharinic acids by way of a β-elimination reaction. This leads to the exposure of new reducing groups and such a "peeling" reaction proceeds in a stepwise manner unless interrupted by a "stopping" reaction. In a (1→4)-linked polysaccharide, the action of alkali on a reducing sugar unit bearing a branch at C-3 has been thought to cause elimination of the side chain and the reducing sugar unit, which rearranges to an alkali-stable metasaccharinic acid reside. However, it now appears that the alkaline degradation of (1→4)-linked polysaccharides does not necessarily occur at such branching points (26).

Extraction of hemicelluloses with dilute alkali under nitrogen lowers oxidative and alkaline degradation. The addition of iodide ion increases the stability of polysaccharides during extraction of wood pulp with alkali (27). Alkaline degradation can also be minimized by reducing the aldehydic end groups with borohydride (26). Hemicellulose extraction may be performed by stepwise increase in alkali concentration. Successive treatments with alkali of increasing concentration lower exposure of the more soluble hemicelluloses to a high alkali concentration.

Acetyl groups that are commonly a part of hemicellulose are lost during alkaline extraction. O-acetyl groups are effective in preventing the development of lateral order in O-acetyl glucomannans (28). Thus, the loss of the acetyl groups during alkaline extraction may cause such polysaccharides to become insoluble in water (29). Various solvent systems have been tried in efforts to isolate intact polysaccharides. In most instances, solvents are single solvents such as methyl sulfoxide, but sometimes more than one solvent is employed. Extraction with methyl sulfoxide and then with water removes about half of the xylans of birch wood with their acetyl groups still intact (30).

By taking advantage of differences in alkaline solubilities, it is possible to effect a crude separation. A sodium hydroxide solution is better than a potassium hydroxide solution as a solvent for polysaccharides. Therefore, extraction with potassium hydroxide solution effects a greater separation between xylans and the less soluble glucomannans. The addition of borates to form a complex increases the extractability of certain polysaccharides. For example, a 24% potassium hydroxide solution extracts (4-O-methylglucurono) xylan and galactoglucomannan from holocellulose of both gymnosperm (17) and angiosperm (31) woods, but does not remove the glucomannan. Glucomannans can be extracted subsequently with a 17.5% sodium hydroxide solution containing 4% borate. Since hemicelluloses are generally a mixture of polysaccharides, further purification of the fractions from alkaline extraction is necessary in order to obtain hemicellulose components representing particular structural types.

B. Purification

1. Fractional Precipitation

Hemicelluloses obtained from alkaline extraction may be separated into three fractions. Acidification of the extract with 50% aqueous acetic acid to pH 4.5–5.0 (32) gives a water-insoluble fraction, termed hemicellulose A. The addition of three volumes of ethanol to the supernatant gives another precipitate, hemicellulose B. The carbohydrate left in the solution is termed hemicellulose C. A, B, and C fractions obtained in replicate determinations from the same sample are not very consistent. One separation method involving dialysis and treatment with an ion-exchange resin provides more consistent results (33). Other specific fractionation procedures are described as follows.

a. With Ethanol.   Fractional precipitation with ethanol has been used extensively for the purification of hemicelluloses (34–36). Pure polysaccharides are frequently obtained by the gradual addition of ethanol to their dilute aqueous solution. Fractionation at or near neutral pH is common. However, separation at pH 2–4, at which the ionization of carboxyl groups is repressed, is useful for acidic hemicellulose. Since acid hydrolysis can occur at low pH, these separations should be conducted rapidly and preferably in the cold. Even then, the acid-labile L-arabinofuranoside linkages may be cleaved (37).

b. With Complexing Agents.   Some polysaccharides form insoluble complex. Linear and branched polysaccharides may be differentiated by iodine precipitation. Usually, only the linear polysaccharides are precipitated by iodine, thus leaving the branched polysaccharides in the solution (38–41).

An alkaline solution of copper(II) salts can be added to a hemicellulose solution to precipitate those containing significant amounts of D-mannosyl or D-xylosyl units. Copper is removed from the insoluble polysaccharide-copper complex by an acidified alcohol solution or by chelating agents. A Fehling solution precipitates hemicellulosic complexes enriched in xylose and impoverished in arabinose, galactose, and glucose (42). Copper (II) acetate can also be used to separate neutral polysaccharides from acidic polysaccharides (43).

Barium hydroxide precipitates polysaccharides containing $(1 \rightarrow 4)$-D-mannosyl units and provides a method for the separation of mannans from xylans. Galactoglucomannan and glucomannan are completely precipitated from an aqueous solution at barium hydroxide concentrations of $0.003M$. 4-O-Methylglucuronoxylans from hardwoods are not precipitated by barium hydroxide until the concentration reaches $0.15M$. Arabino-(4-O-methylglucurono)xylans and arabinogalactans from softwoods are not precipitated by barium hydroxide (44).

Acidic hemicelluloses are separated from neutral hemicelluloses by precipitation with quaternary ammonium salts (45,46), such as cetyltrimethylammonium bromide or cetyltrimethylammonium chloride. The quaternary ammonium complexes formed in a neutral or slightly alkaline solution can be fractionated by varying the salt concentration or by varying the pH. With proper manipulation, separation may be effected not only between acidic and neutral polysaccharides, but even among acidic polysaccharides. Neutral polysaccharides convertable to borate complexes can subsequently be precipitated as quaternary ammonium salts (47). Under strongly alkaline conditions, even neutral polysaccharides can be precipitated directly as their quaternary ammonium salts (48).

Fractional precipitation with complexing agents requires careful manipulation. The solubility of some polysaccharides is decreased by the addition of the complexing agent to a certain concentration, but beyond which the solubility of the complex may increase or decrease depending on the type of complexing agent. Therefore, before fractionation, the required concentration of the complexing agents should be determined by the development of a preliminary solubility curve (37).

2. Chromatography

Fractional precipitation gives good results only if the difference in the solubility of the hemicelluloses or their derivatives is large. Chromatographic techniques can purify polysaccharides having a small difference in solubility or having samples available only in small quantity.

Purification can be accomplished by various chromatographic techniques, such as paper chromatography and column chromatography. Among column packings used are alumina, carbon, cellulose, DEAE-cellulose, and synthetic ion-exchange resins. Evaluation of a number of these was made by Blake and Richards (37).

Polysaccharides of different molecular weight can be separated by gel permeation chromatography (GPC) (49). Column materials used have generally been Sephadex (50) or polyacrylamide gels (51, 52). The introduction of high-performance liquid chromatography (HPLC) has considerably reduced the time of separation. A mixture of mono-, oligo-, and polysaccharides from a wood extract has been analyzed in 20 min by GPC over a starch gel (53).

After a polysaccharide is isolated, its homogeneity in chemical composition and molecular weight may be determined. The homogeneity of isolated hemicelluloses can be determined by ultracentrifugation (54) or by free-boundary electrophoresis (55,56). Isolates are more likely to redissolve if they have been kept wet, or were lyophilized, or dried by solvent exchange (57). Hemicelluloses obtained after precipitation and drying through solvent exchange are usually completely amorphous.

III. CHARACTERIZATION OF HEMICELLULOSE STRUCTURE

Once a polysaccharide is shown to be homogeneous, its structure can be determined. This proceeds through the establishment of monosaccharide composition, anomeric configuration, sequences of glycosyl units, kind of branching, if any, and their distribution. Molecular weight and molecular-weight distribution can also be established.

A. Monosaccharide Composition

Determination of the kind and quantity of glycosyl units present in a hemicellulose is usually the first step. Hemicelluloses are commonly hydrolyzed by mineral acids, formic acid (58), or trifluoroacetic acid (59,60). Trifluoroacetic acid is advantageous in that it can be removed by evaporation, hence, neutralization is not necessary.

Uronic acid units are not readily hydrolyzed and are linked usually to another sugar as an aldobiouronic acid. One of the common aldobiouronic acids is 2-O-(4-O-methyl–D-glucopyranosyluronic acid)-D-xylose. The stability of the glycosidic bond of aldobiouronic acids to acid hydrolysis is so great that extensive destruction of the uronic acid and of other sugar units occurs on attempts to completely hydrolyze the polysaccharide. Thus, when uronic acid units are present, they are best reduced to glycosyl units before hydrolysis of the polysaccharide. This can be done by treating the polysaccharide with a water-soluble carbodiimide and then with sodium borohydride (61). The qualitative and quantitative determination of the monosaccharides obtained from acid hydrolysis are often made by GC, sometimes, in combination with mass spectrometry (MS) (62) or, increasingly, HPLC (53,63,64).

The absolute configuration of the constituent monosaccharides and their linkages are often assigned from their optical rotation or by enzymic methods. Nevertheless, the former method requires a substantial amount of material and the latter method specific enzymes that are not always available. Recently, GC has been used to identify enantiomeric sugars.

Two approaches are used for the separation of enantiomeric carbohydrates by GC. One is to derivatize the carbohydrates with achiral agents and separate the derivatives on a chiral stationary phase. Trifluoroacetyl and methyl trifluoroacetyl glycoside derivatives of enantiomeric sugars can be separated in a capillary column coated with XE-60/L-valine-S-α-phennyethylamide (65,66). Alternatively, the enantiomeric sugars are derivatized with chiral agents and separated as their diastereomeric derivatives. Separation of the enantiomers of arabinose, ribose, fucose, and lyxose as their (-)-methyloxime pertrifluoroacetates can be achieved on a OV-225 wall-coated capillary column (67). The configurations of arabinose and galactose from larch arabinogalactan are thus confirmed as L-arabinose and D-galactose by gas chromatographic separation of their acetylated (+)-2-octyl glycosides on a SP-1000 wall-coated glass-capillary column (68). The absolute configurations of rhamnose, mannose, galactose, and glucose obtained from methanolysis of a lipopolysaccharide have been identified as their trimethylsilylated (-)-2-butyl glycosides on a SE-30 wall-coated glass-capillary column (69).

B. Linkage and Sequence

Once the monosaccharide composition of a hemicellulose is known, the nature of the linkages between units is determined by procedures involving methylation, periodate oxidation, fragmentation analysis, and nuclear magnetic resonance. Determining the sequence of monosaccharide units is commonly done by partial hydrolysis of the polysaccharide with subsequent characterization of the oligosaccharide fragments.

1. Methylation

Methylation analysis is a classical procedure but still valuable in the structural characterization of polysaccharides (70,71). Acid hydrolysis or methanolysis of a completely methylated polysaccharide followed by characterization of the resulting partial methylated monosaccharides provides information on the polysaccharide structure. Since permethylated polysaccharides are often not soluble in an aqueous solution, they are usually subjected to a preliminary depolymerization in methanolic hydrogen chloride, 90% formic acid, or 72% sulfuric acid. The latter two reagents produce less loss of methyl ether groups than does hydrogen chloride (72). Any methylated sugar having only a single unsubstituted hydroxyl group must be from a nonreducing end unit in the polysaccharide and must have been a terminal glycosidic unit. Methylated sugars with more than one free hydroxyl group must have been linked within the chain structure or served as the reducing end unit. A methylated sugar with three hydroxyls free must be from a branch point in the polysaccharide. Structural work on the oligosaccharide derivatives obtained by partial acid hydrolysis of fully methylated polysaccharides also provides information on the positions at which the linkage occurred in the original molecule.

Two classical methylation methods involve treating polysaccharide with methyl sulfate and potassium hydroxide (Haworth) (73) or with methyl iodide and silver oxide (Purdie) (74,75). However, because of the difficulty of methylating the polysaccharide to completion in an aqueous system, it has been common to first acetylate the polysaccharide (76) and to conduct several methylations, finally using an organic solvent such as dimethylformamide (77) or tetrahydrofuran (78).

Methylation of polysaccharides is now usually achieved by the Hakomori procedure (79). The polysaccharide is treated with sodium methylsufinymethide (dimsyl sodium) (80), followed by reaction with methyl iodide. The Hakomori procedure yields, in many instances, complete methylation in one step. Dimsyl potassium has been used in place of dimsyl sodium for Hakomori methylations (81). The preparation using dimsyl potassium is easier and the methylated products are significantly more free from impurities. Some polysaccharides that show strong resistance to permethylation in methyl sulfoxide by the Hakomori method may be completely methylated in a relative short time in a mixture of methyl sulfoxide and 1,1,3,3-tetramethylurea (1:1 ratio). Resistance to permethylation is rationalized as due to a three-dimensional structure in which the reactivity of hydroxyl groups are restricted by inter- and intramolecular hydrogen bonds. 1,1,3,3-Tetramethylurea causes relaxation of the hydrogen bonding and facilitates methylation (82).

Polysaccharides that are not soluble in methyl sulfoxide may be methylated by one Haworth methylation and the methyl sulfoxide soluble product is then fully methylated by a single Hakomori methylation. O-Acetyl groups present in the polysaccharide are saponified during methylation. The locations of O-acetyl groups in the glycosyl units are determined by reacting the free hydroxyls with methyl vinyl ether under an acid catalyst (83) to make an acetal that on methylation and acid hydrolysis yields a mixture of sugars and methylated sugars in which the methoxyl groups mark the positions of the original O-acetyl groups. An example of the method is its use to locate O-acetyl groups in a lipopolysaccharide (84).

Methylation analysis of a polysaccharide containing uronic acid units is facilitated by reducing the carboxyl groups of the original polysaccharides or the fully methylated polysaccharides to hydroxyl groups. Reduction is accomplished by treating an aqueous solution of the polysaccharide with a water-soluble carbodiimide and then with sodium borohydride (61). When sodium borodeuteride is used, the glycosyl units derived from the uronic acid units are readily distinguished in their mass spectra from the nondeuterated derivatives (85,86) because of the two deuterium atoms introduced at C-6. The carboxyl groups of methylated polysaccharides may be reduced also with lithium aluminum hydride. To lessen the alkaline degradation of polysaccharides containing 4-O-substituted uronic units, it is better to reduce the polysaccharide as a first step.

The site of attachment of uronic acid units in birch xylan is confirmed by a base-catalyzed degradation of the methylated polysaccharide. After treating the methylated xylan with sodium methylsulfinylmethanide in methyl sulfoxide for 1 hr, the uronic acid residues are completely removed. The degraded material is then ethylated, followed by acid hydrolysis. The presence of 2-O-ethyl-3-O-methylxylose in the hydrolyzate indicates that uronic acid residues were connected to the O-2 positions of the xylopyranosyl chain (87).

2. Fragmentation

Hemicelluloses or their derivatives may be subjected to acetolysis or partially hydrolysis either by acids or by enzymes to give a mixture of monosaccharides and oligosaccharides. Glycosidic chain linkages can be established through these carbohydrate fragments. Rates of monosaccharide production also provide information on the location of single sugar units as branches and the nature of their ring forms.

a. Acid Hydrolysis. The acid lability of glycosidic bonds varies depending on the monosaccharide composition, ring form, and configuration of glycosidic linkages. Glycosidic linkages of 6-deoxyhexoses are hydrolyzed about five times faster than those of corresponding hexoses. Furanoside linkages are more labile than pyranoside linkages. The glycosidic linkage between the D-xylopyranosyl residues of xylan has been determined through the examination of oligosaccharide fragments obtained from partial acid hydrolysis of the polysaccharide (88). Thus, from xylan a homologous series of oligosaccharides is obtained from xylobiose to xyloheptaose, in which the β-D-xylopyranosyl residues are linked by, $(1\rightarrow4)$ bonds .

The resistance of the glycosidic linkage of the uronic acid unit in the polysaccharide chain can be advantageous. Acid hydrolysis of the polysaccharide leads to cleavage of the usual glycosidic bonds and results in the accumulation of aldobiouronic disaccharide that can be separated easily, and its glycosidic linkage determined after reduction of the carboxyl group to produce the corresponding neutral disaccharide that can be characterized readily (89).

Uronic acid residues may be introduced into certain polysaccharides by oxidation of the primary hydroxyl groups with oxygen in the presence of platinum (90). Hydrolysis of an oxidized larch arabinogalactan yields a hydrolysate containing 6-O-(L-arabinofuranosyluronic acid)-D-galactose, thus showing the location of the L-arabinofuranosyl residues in the polysaccharide (91). Similar treatment of an arabinoxylan yields 3-O-(L-arabinofuranosyluronic acid)-D-xylose, revealing that the terminal L-arabinofuranosyl units are attached directly to the xylan chain (92). Terminal D-galactopyranosyl groups may be transformed into D-galactosyluronic acid groups by oxidation, first with galactose oxidase and then with hypoiodite (93).

It is also possible to intentionally introduce acid-labile linkages into polysaccharides. As an example, the α-D-glucopyranosyl units attached by $(1\rightarrow6)$ linkages to the main chain of a fungal glucomannan can be converted into 3,6-anhydro derivatives by successive tritylation, acetylation, detritylation, p-toluenesulfonylation and treatment with sodium methoxide. The 3,6-anhydro-glycosyl unit, which is acid-sensitive, can be selectively removed from the polysaccharide chain by mild acid hydrolysis (94).

b. Acetolysis. Partial acetolysis of polysaccharides can provide mixtures of fully acetylated monosaccharides and oligosaccharides. Acetolysis involves treatment of the polysaccharide or acetylated polysaccharide in either acetic anhydride or a mixture of acetic anhydride and acetic acid containing 3–5% sulfuric acid. It is complementary to acid hydrolysis, as the relative rates of cleavage of the glycosides in the two reactions are sometimes reversed. Whereas the $(1\rightarrow6)$ linkage in polysaccharides is somewhat resistant to acid hydrolysis, it is the least stable to acetolysis (95,96). The depolymerization of polysaccharides by hydrolysis and acetolysis can sometimes be of value in structural analysis. As a method for linkage determination for those polysaccharides that are insoluble in aqueous solution, acetolysis may have an experimental advantage over direct partial acid hydrolysis since the soluble polysaccharide acetate may be more uniformly depolymerized.

c. Enzymic Hydrolysis. Enzymic hydrolyses are used as a supplement to chemical depolymerizations. Biochemical methods can provide valuable information about the fine structures of polysaccharides if the specificity and mode of action of the enzymes are known. A comprehensive review on hemicellulases has been written by Dikker and Richards (97).

Hydrolysis of a glucomannan from Jack pine (*Pinus banksiana*) by a crude hemicellulase produces mannobiose and mannotriose in 25% yield, indicating the possibility of long chains of (1→4)-β-D-mannopyranosyl units. Other hydrolysis products are (1→4)-β-D-linked disaccharides containing D-glucosyl and D-mannosyl units. These disaccharide residues appear to position randomly in a linear polymer chain (98).

A 4-O-methylglucuronoxylan from white birch was hydrolyzed with a pectinase in a semipermeable membrane. The oligosaccharide in the hydrolysate was identified as O-(4-O-methyl-β-D-glucosyluronic acid)-(1→2)-O-D-xylopyranosyl-(1→4)-O-β-D-xylopyranosyl-(1→4)-D-xylose (99). The L -arabino-(4-O-methyl-D-glucurono)-D-xylan from redwood (*Sequoia sempervirens*) is hydrolyzed by an endoxylanase to give a dialyzable oligosaccharide mixture in 80% yield. The mode of action and the nature of the dialysis-accumulated products suggest that the L-arabinosyl and 4-O-methyl-D-glucuronic units are irregularly distributed on the main chain (100).

3. Periodate Oxidation

Periodate oxidation is useful in structural characterization since it indicates the number of α-glycol group present. For each glycol unit oxidized, one mole of periodate is reduced to iodate and the carbon chain cleaved with the formation of two carbonyl groups. Oxidation of α,β,τ,-hydroxyls requires two molecular portions of periodate and results in a double cleavage of the carbon chain with the formation of aldehydic groups on the α- and τ-carbons and the liberation of one mole of formic acid from the β-carbon. In addition to glycol units, oxidation occurs in α-hydroxyaldehydes, α-hydroxyketones and α-hydroxyhemiacetals if present. Formaldehyde is produced from the oxidation of primary hydroxyls, and formic acid from an aldehyde or carbinol unit between two other hydroxyl-bearing carbons (101).

For proper structural evaluation, it is essential that the oxidation of glycol units be complete but that no over-oxidation occurs. Oxidations are conducted in a dilute solution of sodium metaperiodate in the dark at low temperature and at pH 3–3.5 (101). The use of oxygen-free water and the addition of propyl alcohol, as a radical scavenger, lessen radical-induced depolymerization during oxidation (102). Properly conducted, oxidation is quantitative and reproducible. Comparison of the amount of periodate consumed and the amount of formic acid and formaldehyde produced provides information on the molecular structure, nature of the end groups, and points of linkage between constituent monosaccharides.

Problems may arise with nonspecific oxidation (103,104) and with the production of formyl ester or intermolecular hemiacetal that may slow or prevent the theoretical consumption of periodate. In a $(1\rightarrow4)$-β-D-xylan, after 0.6 mole of periodate per residue is oxidized, the initial second-order rate constant for the oxidation is lower by 2% due to the formation of temporary intermolecular hemiacetals (102).

Lead tetraacetate also cleaves glycols similarly to periodates. However, lead tetraacetate is decomposed by water and must be used in nonaqueous solvents, usually glacial acetic acid. Lead tetraacetate may also be used in methyl sulfoxide (105), which is a good solvent for many polysaccharides. Any formic acid produced in the reaction is further oxidized to carbon dioxide by lead tetraacetate, in contrast to periodate.

4. Smith Degradation

Additional structural information may be obtained by hydrogenation of the periodate oxidation products and subsequent mild hydrolysis, a sequence of reactions called Smith degradation (106). Normally, sugar units of a polysaccharide that are cleaved by periodate and reduced are acid-sensitive, being acetals. When a sugar unit not cleavaged is joined to a cleaved unit, it appears as a glycoside that is relatively stable to acid. Because of the marked difference in stability between true acetals and glycosides, controlled hydrolysis using a mild acid furnishes simple polyalcohols, glycoaldehyde, and glycosides of mon-, di, and oligo-saccharides. Structural analysis of these hydrolytic products provides information on the fine structure of the parent polysaccharide.

The polyaldehyde formed from periodate oxidation can be reduced with hydrogen and platinum, but sodium borohydride is more convenient. The use of a concentrated solution of borohydride also lessens the competing β-elimination reaction (107).

5. β-Elimination

In the presence of alkali, a hydroxyl or alkoxyl group in the β-position to an electron-withdrawing group, commonly a carbonyl or carboxylic ester is eliminated. The use of the β-elimination reaction has helped in solving sequence problems in certain polysaccharides.

When a polysaccharide containing uronic acid is methylated, the carboxyl groups are esterified and are effective electron sinks. The alkaline treatment of the methylated polysaccharide results in β-elimination and rapid cleavage of the chain if it is attached to C-4 of the uronic acid unit and slower cleavage occurs if attachment is at C-3. Selective β-elimination reactions can provide useful information on unit sequences in acidic polysaccharide (108). Applications of β-elimination to the structural determination of polysaccharides have been reviewed by Lindberg and co-workers (109)

6. Chromium Trioxide Degradation

Chromium trioxide preferentially oxidizes fully acetylated gly osides having equatorial aglycones (110). A peracetylated aldopyranoside, in which the aglycone occupies an equatorial position in the chair form, is oxidized by chromium trioxide to yield a 5-aldulosonate. An anomer with an axial aglycone is oxidized slowly (111). Fully acetylated furanosides are easily oxidized to 4-aldulosonates, irrespective of their anomeric configuration.

These reactions have been used to determine the anomeric configuration of sugar residues in oligo- and polysaccharides (112). In such an analysis, a completely acetylated sample is treated with chromium trioxide in acetic acid. Glycosides with equatorial glycosidic linkages are oxidized, while the surviving residue contains axial linkages. An exception is 4-O-methyl-α-D-glucose, derived from 4-O-methyl-α-D-glucuronic acid that is attached to a xylan chain through a (1→2) linkage. The glucoside is oxidized readily by chromium trioxide, presumably due to the presence of the methoxyl group (113). Less exact information on the anomeric nature of glycosidic linkage in a polysaccharide can also be obtained by following the optical rotatory changes that occur during acid hydrolysis.

7. NMR Spectroscopy

Proton NMR spectroscopy is useful for the quantitative estimation of specific functional groups such as O-methyl and O-acetyl substituents and for determining the anomeric character of glycosidic linkages in polysaccharides (114). When used on mixtures of Western red cedar hemicelluloses, however, no discrete peaks were obtained in the region associated with anomeric protons (115). ^{13}C-NMR has been used effectively to determine the composition, anomeric form, ring structure, linkage, conformation, and dynamic properties of polysaccharides (116,117). In many cases, it can give the same information as the more tedious chemical methods.

Native and modified polysaccharides are not always amenable to NMR analysis because of such intrinsic properties as insolubility, gel formation, or high viscosity. These problems are partly avoided by using a variety of techniques in sample preparations. Thus, to avoid high viscosity, a galactomannan has been partially acid-hydrolyzed and shown to give higher-quality spectra bearing close resemblances to native glucomannans (118). Sometimes, polysaccharide can be examined in solid state by ^{13}C-NMR measured with magic angle spinning and cross polarization (MAS-CP). The use of the MAS-CP ^{13}C-NMR method is exemplified with the spectra of xanthan gum. The solid state NMR technique provides more detailed information than the corresponding solution experiments in an equivalent 3 hr period (119).

The chemical shifts of the monosaccharide residues in a polysaccharide in ^{13}C-NMR are similar to those of free monosaccharides, except for substituent effects. Substituent effects cause an increase in the chemical shift of the carbons directly involved in linkages and smaller changes in the chemical shift of the neighboring β-carbons. Thus, the comparison of the chemical shifts between monosaccharide and polysaccharides may help in the determination of the position and configuration of linkages.

^{13}C-NMR is used to determine the ratio of mannosyl to galactosyl units in guar gum, locust bean gum, and fenugreek gum (118). The results compare favorably with oxidation results (120) or with parallel GC/alditol acetate determinations (121). Xylan isolated from red seaweed (*Nemalion vermiculare*) has been subjected to structural analysis. Methylation analysis shows that the polysaccharide consists of (1→3)- and (1→4)-linked β-D-xylopyranosyl units in the ratio of about 1:4 (122). The NMR results enable correction of the previous conclusion (122) concerning the proportion of the two types of linkage, which is now shown to be 1:6. In addition, the ^{13}C-NMR spectrum shows that all of the, D-xylopyranosyl units in the polysaccharide are pyranoid, whereas methylation analysis alone cannot distinguish 4-O-substituted pyranosyl units from 5-O-substituted furanosyl units (123).

8. Gas Liquid Chromatography-Mass Spectrometry

Gas chromatography in combination with mass spectrometry (GC-MS) provides another powerful tool for the structural elucidation of polysaccharides. The quantitative and qualitative analysis of hydrolyzates from polysaccharides or from methylation analysis has been simplified by the introduction of a variety of analytical techniques, based on GC and MS analysis of their volatile derivatives. Early procedures rely mainly on the formation of trimethylsilyl derivatives (124); but the formation of multiple products from each aldose makes the separation and quantitative work difficult. These difficulties are alleviated by the use of alditol acetates and aldononitrile acetates derivatives (62,125,126). Both derivatives can be readily prepared and are used extensively.

Alditol acetates are analyzed by GC on packed columns (127) or glass capillary columns (128,129). Presently, flexible fused-silica wall-coated open-tubular (WCOT) columns are available for the separation of alditol acetates. A fused-silica WCOT column with a polar phase (FFAP) provides a more efficient and rapid separation of alditol acetates under isothermal operation at 205°C (130). After reduction, acetylation, and hydrolysis of the acidic polysaccharide isolated from the sap of the lac trees (*Rhus vernicifera*), the hydrolyzate is treated with sodium borodeuteride and then applied to the FFAP column to determine its monosaccharide composition. The acetates of rhamnitol, arabinitol, 4-O-methyl-D-glucitol-6-d, and D-glucitol are readily identified by GC-MS analysis.

A systematic examination by electron-impact MS of a series of synthetic, fully methylated xylobioses, xylotrioses, and a xylotetraose provides information for the linkage analysis, especially the establishment of branch points, of permethylated xylans (131). Partially methylated monosaccharides obtained by the acid hydrolysis of the methylated crude hemicellulose from a rice endosperm cell wall are analyzed in the form of their alditol acetates by GC-MS analysis on a Silar-10C glass capillary column (132).

C. Molecular Weight

Hemicelluloses generally occur in a homologous series with a normal distribution of molecular weight. Hemicelluloses from hardwoods have average degrees of polymerization of 150–200. Osmometry, viscometry, sedimentation equilibrium, sedimentation diffusion, light-scattering measurement, GPC are used to determine the molecular weight and to provide information on the size, shape, and molecular-weight distribution of the polysaccharides (133,134).

The degree of polymerization of neutral polysaccharides may be determined by the method devised by Yamaguchi and co-workers (135–137) and improved by Tanaka (138). The reducing-end residue of the polysaccharide is reduced to an alditol residue and then the polysaccharide is hydrolyzed. The alditol and the relatively large amounts of monosaccharides in the hydrolysate are separated on a Dowex column in hydroxyl form and determined as their alditol acetates by GC. The degree of polymerization is calculated from the ratio of the monosaccharides to the alditol.

IV. HEMICELLULOSE FINE STRUCTURES

A. Softwood Hemicelluloses

Because of the higher lignin content in softwoods species, the wood is generally delignified before hemicellulose is extracted. Although there are many ways to isolate and separate hemicelluloses from softwoods, a scheme proposed by Timell (139) and modified by Dutton and co-workers (140), based on the fractionation of Ponderosa pine hemicelluloses, is representative (Fig. 1). The lignin-contaminated hemicellulose (4–7% lignin) is first treated with buffered sodium chlorite to remove residual lignin, which otherwise would hinder the fractionation. Crude galactoglucomannan is separated from the crude xylan by barium hydroxide precipitation and further purified by fractional precipitation with barium hydroxide. The crude xylan is contaminated with galactoglucomannan and is purified by barium hydroxide precipitation and ethanol titration.

1. O-Acetyl-Galactoglucomannan

Galactoglucomannan was first identified in 1956 (141) as a constituent of the alkaline extract of sulfite pulp from Western hemlock and was later found in many softwood species. Galactoglucomannan represents about 16% of the wood and is the major hemicellulose component. Notable exceptions are incense cedar where xylan predominates (142) and some species of larches where arabinogalactan is the major hemicellulose component.

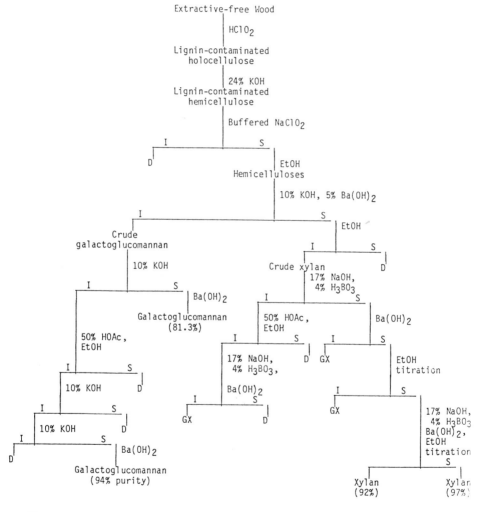

Figure 1 Isolation and separation of softwood hemicelluloses (ponderosa pine). I, insoluble fraction; S, soluble fraction; D, fraction that may be discarded or needs further purification; GX, mixture of galactoglucomannan and xylan. Purity is defined as the percentage of hexose or pentose in total neutral sugar for galactoglucomannan and xylan, respectively.

Galactoglucomannans can be roughly divided into two fractions having different contents of galactosyl units. The fraction that has a low content of galactosyl units has a ratio of galactosyl to glucosyl to mannosyl units of 0.1:1:4, whereas in the galactosyl-rich fraction, the corresponding ratio is 1:1:3. The fraction with a low galactosyl content is often referred to as glucomannan. Galactoglucomannans have a main chain consisting of (1→4)-linked β-D-glucopyranosyl units and β-D-mannopyranosyl units. The D-glucosyl units and D-mannosyl units in the main chain are in random arrangement (9). The isolation of cellobiose from the galactoglucomannans of Western hemlock (143), Amabilis fir (144), and European larch (145) indicates the contiguity of D-glucosyl units in the polysaccharide; however, no cellobiose has been isolated by the fragmentation of the galactoglucomannan from loblolly pine, which must contain only isolated D-glucosyl units. The α-D-galactopyranosyl units are linked as a single-unit side chain to both D-glucosyl units and D-mannosyl units of the main chain by (1→6) bonds. About half of the D-mannosyl units are substituted by O-acetyl groups equally distributed between C-2 and C-3 (146). A representative structure of galactoglucomannans is shown in Fig. 2.

Although galactoglucomannans are insoluble in water in the native state, they often show water solubility after isolation. Water solubility increases with increasing D-galactose content. The presence of a galactosyl side chain presumably prevents the association of the linear chain segments and, thus, increases water solubility. Galactoglucomannans are soluble also in methyl sulfoxide, a solvent particularly useful as an extractant when the chemical modification of the native polysaccharide is to be avoided.

A galactoglucomannan isolated from Scotch pine (*Pinus sylvestris*) consists of a linear chain of at least 60 D-glucosyl and D-mannosyl units, with approximately 3% of the hexose units substituted at C-6 with an α-D-galactosyl residue (147). The polysaccharide is composed of D-mannosyl, D-glucosyl, and D-galactosyl units in a molar proportion of 72:20:8. The O-acetyl groups are irregularly distributed along the backbone of the glucomannan (148). About half of the D-mannosyl units are substituted with O-acetyl groups equally distributed between C-2 and C-3 (146). The weight-average molecular weight is M_w 23,300 (148).

A galactoglucomannan isolated from Monterey pine (*Pinus radiata*) has a mo-

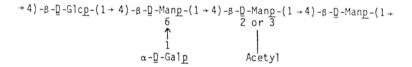

Figure 2   Representative structure of O-acetyl-galactoglucomannan.

lecular weight of 9.3 x 10 and is composed of 45 D-glucosyl and D-mannosyl units, with D-mannose as the nonreducing end group. D-galactosyl units are attached to the chain, probably through (1→6) linkages, in the ratio of one galactosyl unit to every 75 sugar residues of the chain (149).

The galactomannans isolated from black spruce (*Picea mariana*) and Sitka spruce (*Picea sitchensis*) have similar structures, namely (1→4)-linked mannan backbone with a nonreducing end group consisting of D-mannosyl units and a few single-unit D-galactosyl side chains (150).

2. Arabino (4-O-Methylglucurono) Xylan

Another major hemicellulose in softwood is arabinoglucuronoxylan, representing 7–15% of the wood. The hemicellulose is composed of about 200 (1→4)-linked β -D-xylopyranosyl units that are partially substituted mainly at C-2 by 4-O-methyl-α-D-glucuronic acid groups, on the average, one acid residue per five or six xylosyl units (151). Although 4-O-methyl-D-glucuronic acid is most often linked at C-2 of a D-xylopyranosyl unit, C-3 linkages are observed also (152). In addition, it contains α-D-arabinofuranosyl units linked predominantly at the C-3 position of the main xylan chain, with an average of 1.3 residues per 10 xylopyranosyl units. Because of their furanosidic structure, the arabinosyl side chains are readily removed under mild acid conditions. Often, the L-arabinosyl groups occur as single-unit side chains, but they may consist of side chains several sugar residues in length. Such branches may be terminated with a 4-O-methyl-D-glucopyranosyluronic acid group or with a D-xylopyranosyl group. A representative structure of arabino(4-O-methylglucurono)xylan is shown in Fig. 3.

Arabino(4-O-methylglucurono)xylan can be obtained in fairly good yields by direct extraction with alkali of the fully lignified wood, although some species are more suitable in this respect than others. Arabino(4-O-methylglucurono)xylan is isolated in a yield of 2.7% from Norway spruce (*Picea abies*) by extraction with 24% aqueous potassium hydroxide containing 1% potassium borohydride and separated from the mixture as a barium complex. The polysaccharide contains one 4-O-methyl-α-D-glucuronic acid per 5.9 D-xylosyl units and one L-arabinosyl unit

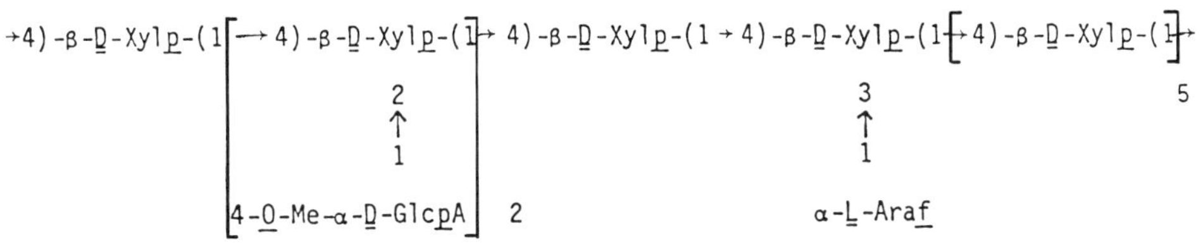

Figure 3   Representative structure of arabino-(4-O-methylglucurono)xylan.

per 7.4 D-xylosyl units. The average macromolecule consists of 128 D-xylosyl units (153).

A hemicellulose of slash pine (*Pinus elliottii*) has been shown by methylation analysis to have a (1→4)-β-D-xylan chain with many branch points, and 4-O-methyl-D-glucopyranosyluronic acid, L-arabinofuranosyl, and D-xylopyranosyl units as nonreducing end groups (154).

An O-(4-O-Me-α-D-GlcAp)-(1-2)-O-β-D-Xylp-(1-4)-O-(4-O-Me-α-D-GlcAp)-O-(1-2)-D-Xyl is isolated from a partial hydrolyzate of the hemicellulose precipitated from a spruce neutral sulphite liquor. This polymer has the 4-O-methyl-D-glucuronic acid residues located on adjacent D-xylosyl units in the main xylan chain (155). This finding is confirmed also in larch xylan (156).

After digesting a xylan from red wood (*Sequoia sempervirens*) with a β-D-endoxylanase (157), the ratio of uronic acid to xylose in the residue is half that of the original xylan. This observation suggests that the uronic acid residues are localized in regions of the polysaccharide rather than being randomly distributed (158). A similar distribution is indicated for birch xylan using a method based on the selective cleavage of the methylated polysaccharide (159).

The xylose-based hemicelluloses in both softwoods and hardwoods are often simply called xylan. Hardwoods contain 10–35% of xylans and softwoods 7–15%. The softwood xylans are structurally similar to hardwood xylans, with the exception of a few differences. Hardwood xylans do not contain α-L-arabinofuranosyl units, whereas softwood xylans do not contain acetyl groups. Softwood xylans are more acidic than hardwood xylans because of their greater content of 4-O-methyl-D-glucuronic acid side chains. One acid group is present in every 5–6 D-xylosyl units in the softwoods, compared to one in every 10 D-xylosyl units in hardwoods (9,10). The distribution of uronic acid side chains is different for hardwood and softwood xylans. In softwood xylans, a large portion of the uronic acid residues is located in adjacent xylose residues (156).

3. Arabinolgalactan

Arabinogalactans are present as a minor constituent in softwood species, with the exception of the genus Larix, in which arabinogalactans constitute up to 35% of the wood (160). Because arabinogalactans have numerous branches, they are the only major wood polysaccharides that can be extracted in high yield from untreated wood with water. Usually, defatted wood is extracted with hot water, followed by alcohol precipitation, to produce a relatively pure, structurally unaltered polysaccharide. Soluble arabinogalactans can be precipitated with cetyltrimethylammonium hydroxide as the borate complex (161). The gum can be fractionated by gel filtration with little chemical transformation (162,163).

Arabinogalactans from larch wood have a backbone of (1→3) -linked β-D-galactopyranosyl units, each of which bears a side chain at the C-6 position. The majority of such side chains consist of (1→6) -linked β-D-galactopyranosyl units, with an average of two such residues per side chain. Another common side chain is composed of single L-arabinofuranosyl units or of 3-O-β-L-arabinopyranosyl-L-arabinofuranosyl units. Small amounts of D-glucuronic acid residues are also present in arabinogalactans from European larch (*L. decidua*) and tamarack (*L. lyallii*). The ratio of galactosyl to arabinosyl units is about 6:1 for most species. Arabinogalactans from larch wood are composed of two fractions having molecular weights of 100,000 and 16,000 (164). They can be separated by electrophoresis or fractional precipitation with a mixture of quaternary ammonium salt and boric acid (161). The structure of arabinogalactan from larch is shown in Fig. 4.

Arabinogalactans from Western larch (*L. occidentalis*) contain two components of molecular weight 78,000 (20%) and 18,000 (80%). Smith degradation of the two polysaccharides yields a single product of molecular weight 2200. Prolonged acid treatment does not cause further degradation. Thus, the arabinogalactans seem to consist of (1→3)-linked blocks of about 12 β-D-galactosyl residues separated at regular intervals by sugar units vulnerable to periodate. Three such regions exist in the component of molecular weight 18,000, and a correspondingly larger number in the component of molecular weight 78,000 (165).

Arabinogalactan is found in a small amount in other softwood species, such as hemlock, black spruce, parana pine, mugo pine, Douglas fir, incense cedar, and juniper (166–168). The arabinogalactans from other softwood species are similar to larch arabinogalactans. One difference seems to be that, in the larch arabinogalactans, only two-thirds of the L-arabinosyl units occur as terminal L-arabinofuranosyl residues, the remainder being present as 3-O-β-L-arabinopyranosyl-L-arabinofuranosyl units. In other softwood arabinogalactans, a larger proportion of the L-arabinose occurs as L-arabinofuranosyl end groups.

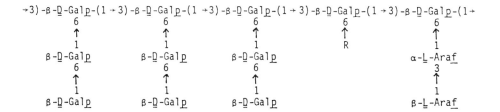

Figure 4 Representative structure of larch arabinogalactan.
R: b-D-galactopyranosyl or, less frequently, a-L-arabinofuranosyl, or a b-D-glucopyranosyluronic acid residue

4. Minor Constituents

The existence of arabinans in various softwoods has been reported (167). An impure arabinan, consisting of 75% of L-arabinosyl units, is isolated from Scotch pine (*Pinus sylvestris*). Its composition is consistent with a main chain of (1→5)-linked α-L-arabinofuranosyl units, many of which are branched at C-3 and a few at both C-2 and C-3 (169). Another arabinan is isolated from maritine pine (*Pinus pinaster*) (170).

B. Hardwood Hemicelluloses

1. O-Acetyl-(4-O-Methylglucurono)xylan

Hardwoods contain 10–35% xylan. The basic skeleton of hardwood xylans is a linear or singly branched backbone of about 200 (1→4)-linked β-D-xylopyranosyl units (171). The common side chains are 4-O-methyl-α-D-glucopyranosyluronic acid residues that are attached by α linkages to C-2 and to a lesser extent to the C-3 position of the D-xylopyranosyl units. On the average, every tenth D-xylopyranosyl unit is substituted by a 4-O-methyl-D-glucuronic acid residue.

An additional structural feature of hardwood xylans is the presence of acetyl groups (172,173). The amount of acetyl groups in hardwood xylans ranges from 8–17% corresponding to approximately 3.5–7 acetyl groups per 10 D-xylopyranosyl units. The acetyl groups are linked to D-xylopyranosyl units of the main chain predominately at C-2 positions, although a small proportion occur at the C-3 positions (174,175). A general formula for an O-acety-(4-O-methylglucurono)xylan is shown in Fig. 5.

Besides arabinogalactan, O-acetyl-(4-O-methylglucurono)xylan is another major hemicellulose that can be isolated in reasonable yield by direct extraction from the lignified wood. Extraction with aqueous potassium hydroxide generally removes 70–80% of the total xylan. For quantitative isolation, the wood is first delignified and xylan is then isolated from the holocellulose. The xylan obtained is similar to native xylan, except for its lack of acetyl groups.

An alkali-soluble xylan containing both 4-O-methyl-D-glucopyranosyluronic and D-galactopyranosyluronic acids has been isolated from birch (*Betula verrucosa*). The D-galacturonic acid moieties are integral parts of the xylan structure and are linked to nonreducing, terminal D-xylosyl units in the highly branched polysaccharide (176). Work on birch xylan has revealed an interesting structural feature, as shown in Fig. 6. The unit next to the reducing xylopyranosyl end group is D-galacturonic acid, which is linked to a L-rhamnosyl unit through the C-2 position. The rhamnosyl unit, in turn, is connected through its C-3 position to the xylan chain (177). O-acetyl groups in birch (*Betula verrucosa*) xylan occur at a degree of substitution of 0.62 and are attached to C-2 or C-3 in a ratio of about 1. Substitution on both C-2 and C-3 also occurs (172).

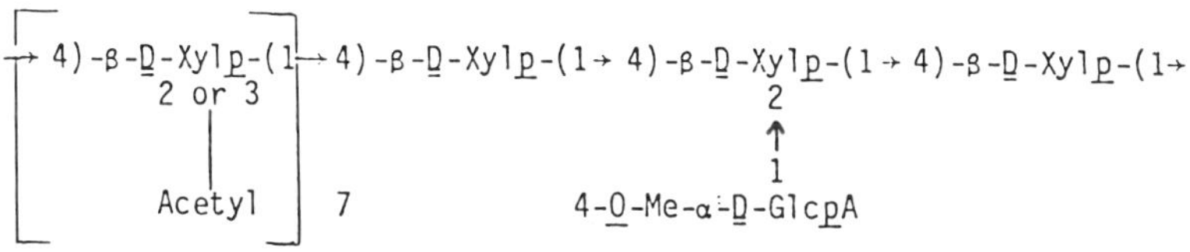

$$\left[\begin{array}{c} \rightarrow 4)\text{-}\beta\text{-}\underline{D}\text{-Xyl}\underline{p}\text{-}(1 \\ \underset{|}{2\ or\ 3} \\ | \\ \text{Acetyl} \end{array}\right] 7 \quad \begin{array}{c} \rightarrow 4)\text{-}\beta\text{-}\underline{D}\text{-Xyl}\underline{p}\text{-}(1\rightarrow 4)\text{-}\beta\text{-}\underline{D}\text{-Xyl}\underline{p}\text{-}(1\rightarrow 4)\text{-}\beta\text{-}\underline{D}\text{-Xyl}\underline{p}\text{-}(1\rightarrow \\ 2 \\ \uparrow \\ 1 \\ 4\text{-}\underline{O}\text{-Me-}\alpha\text{-}\underline{D}\text{-Glc}\underline{p}A \end{array}$$

Figure 5   Representative structure of O-acetyl-(4-O-methylglucurono)xylan.

4-O-methyl-D-glucuronic acid residues, linked to the xylan chain by (1→2) linkages, have been found to retard the alkaline peeling reaction (178,179). In alkaline media, the reducing xylose end group of birch xylan is easily isomerized to a xylulose group and removed by β-elimination, which exposes a reducing galacturonic acid end group. The presence of (1→2)-linked L-rhamnosyl and D-galacturonosyl units also has a retarding effect on the peeling reaction. This retardation may account, in part, for the fact that hardwood, having a high xylan content, gives a high yield of pulp in alkaline pulping. The analysis of partial hydrolyzed xylan indicates that the 4-O-methyl-D-glucuronic acid units linked to the C-2 position are randomly distributed along the birch xylan backbone (159,180).

Hemicelluloses of trembling aspen (*Populus tremuloides*) (181,182) have O-acetyl-(4-O-methyl-D-glucurono)xylan as a major component and a minor amount of glucomannan. The 4-O-methylglucuronoxylan of D.P. 212 in aspen wood contains one acid side chain per nine D-xylosyl units (183,184) and has 2–3 branches per molecules with branch points located at C-3 on the xylan chain (181).

4-O-methylglucuronoxylan, accounting for 86% of the alkali-extractable xylan from basswood (*Tilia americana*), is composed of a backbone of 180 xylosyl units, with every ninth D-xylosyl unit carrying, on the average, a terminal (1→2)-linked side unit of 4-O-methyl-α-D-glucuronic acid (185).

A O-4-methylglucuronoxylan isolated from white willow (*Salix alba*) consists approximately of 120 β-D-xylosyl units. On the average, every eleventh D-xylosyl unit carries one 4-O-methyl-D-glucuronic acid or D-glucuronic acid, attached to the backbone through the C-2 of a xylosyl unit as a simple terminal branching unit (186).

$$\rightarrow\beta\text{-}\underline{D}\text{-Xyl}\underline{p}\text{-}(1\rightarrow 4)\text{-}\beta\text{-}\underline{D}\text{-Xyl}\underline{p}\text{-}(1\rightarrow 3)\text{-}\alpha\text{-}\underline{L}\text{-Rha}\underline{p}\text{-}(1\rightarrow 2)\text{-}\alpha\text{-}\underline{D}\text{-Gal}\underline{p}A\text{-}(1\rightarrow 4)\text{-}\underline{D}\text{-Xyl}$$

Figure 6   The structure of the reducing end group in birch xylan.

2. Glucomannan

Hardwoods contain 2–5% of glucomannans. Hardwood glucomannans are pure copolymers that upon hydrolysis yield D-glucose and D-mannose in the approximate ratio of 1:2, respectively, with the exception of the genus Betula where the ratio is 1:1. Glucomannans consist of linear chains of (1→4)-linked β-D-glucopyranosyl and β-D-mannopyranosyl units with a random arrangement of the two sugar units. No D-galactose is obtained from hardwood glucomannans. The weight-average D.P. of birch and aspen glucomannans are 120 and 108, respectively. A general formula for a glucomannan is shown in Fig. 7.

To isolate glucomannan, holocellulose is extracted with aqueous potassium hydroxide to remove most of the 4-O-methyl-D-glucuronoxylan, leaving the glucomannan in the residue (182). Glucomannan in the residue is then removed with aqueous sodium hydroxide containing borate (187) to obtain a glucomannan complex that is purified by fractional precipitation with barium hydroxide.

A glucomannan isolated from wood of white willow (*Salix alba*) consists of 47 β-D-glucosyl and β-D-mannosyl units in the ratio of 1:1.6 (188). The polysaccharide is slightly branched with an average of 2.5 branches per molecule. Another glucomannan from white willow is composed of a linear chain of some 20 β-D-glucosyl and β-D-mannosyl units in the ratio of 1:1.4 (189). Thus, the glucomannans of white willow are heterogeneous with respect to composition, size, and shape.

3. Minor Constituents

An α-glucan is isolated from white willow (*Salix alba*) (189). Partial hydrolysis of the polysaccharide gives D-glucose, maltose, and isomaltose. Hydrolysis of the methylated polysaccharide yields 2,3,4,6-tetra-O-methyl-D-glucose (10.9%), 2,3,6-tri-O-methyl-D-glucose (77.0%), and 2,3-di-O-methyl-D-glucose (12.1%). The average length of the polysaccharide chains is 9.1 D-glucopyranosyl units and the average length between branches is 5.1 units. Arabinogalactans are present in some hardwood species (177,190), and one has been isolated from the wood of sugar maple (*Acer saccharum*) by water extraction.

C. Hemicelluloses of Reaction Wood

When a tree is subjected to continuous physical stress, such as gravitational force on its branches or deviation from a vertical position, it tries to effect a return to its normal growth pattern by forming a specialized wood, called reaction wood. In

→4)-β-D-Glcp-(1→4)-β-D-Manp-(1→4)-β-D-Glcp-(1→4)-β-D-Manp-(1→4)-β-D-Manp-(1→

Figure 7 Representative structure of glucomannan

softwoods, reaction wood is formed on the lower side of a leaning stem or branch and is termed compression wood. In hardwoods, such abnormal wood is usually located on the upper side and is called tension wood. Compression wood differs from normal wood in having an increase in lignin and D-galactosyl units of some 9% and 7.8%, respectively, and a 10% lower cellulose content (191).

A galactan isolated from the compression wood of Norwegian spruce (*Picea abies*) (192) has a linear structure composed predominantly of (1→4)-linked β-D-galactopyranosyl units with a small proportion of (1→6) linkages. The compression wood of tamarack (*Larix laricina*) contains an acidic galactan, which accounts for 90% of the galactan present. The polysaccharide has a main chain of 200–300 (1→4) β-D-galactopyranosyl units, with an average of one (1→6)-linked D-galacturonic acid residue attached to every 20 D-galactosyl units (193). An acidic arabinogalactan in tamarack compression wood has a similar structure to the arabinogalactan in normal wood of the same species, but some of the D-arabinosyl side chains are thought to be considerably longer than two units (194).

Laricinan from the compression wood of tamarack (*Larix laricina*) is an acidic glucan of some 200 β-D-glucopyranosyl units, 6% of which are (1→4)-linked and the remainder (1→3)-linked. The average molecule contains eight branch points with D-glucuronic acid and to a lesser extent D-galacturonic acid residues attached to the glucan (195). An arabino-4-O-methyl-D-glucuronoxylan from the compression wood of tamarack is found to be similar to the xylan present in normal larch wood, with the exception that it contains only half as many L-arabinosyl side chains (196).

Tension wood differs from normal wood in its lower content of lignin and hemicellulose and higher content of cellulose (197). From the tension wood of European beech (*Fagus silvatica*), a galactan was isolated in which both (1→4)- and (1→6)-linked galactopyranosyl units are present (198).

V. LINKAGES BETWEEN HEMICELLULOSE AND LIGNIN

The difficulty in completely isolating hemicelluloses from cellulose with alkaline or acidic solutions without causing serious degradation of the cellulose and hemicelluloses led some investigators to speculate that chemical linkages exist between cellulose and hemicelluloses (199). However, it is a general belief that hemicelluloses do not bind chemically to cellulose, but are closely associated with cellulose by physically intermixing and hydrogen bondings. On the other hand, there is strong evidence that hemicelluloses are linked covalently to lignin.

A lignin-carbohydrate complex was first proposed in 1957 by Bjorkman (200,201), who isolated the complex from spruce wood. Ever since, the existence of covalent bonds between lignin and carbohydrates has been a debated issue (11). Lignin-carbohydrate complexes have been isolated from many species using a number of solvents (201–203). Hot water, dimethylformamide, methyl sulphoxide, and a dilute alkaline solution are commonly used.

A lignin-carbohydrate complex prepared from spruce wood according to Bjorkman's method (200,201) can be separated into two fractions by electrophoresis (204). One fraction consists of carbohydrates and the other contains equal amounts of carbohydrates and lignin. The lignin segments and carbohydrate moiety in the purified lignin-carbohydrate complex cannot be physically separated without causing degradation of the complex. The fact that the separation of lignin from the carbohydrate cannot be achieved by gel filtration, electrophoresis, ultracentrifugation, and hydrophobic-interaction chromatography strongly indicates the existence of chemical bonds between lignin and carbohydrates (205).

The lignin-xylan complex isolated by Bolker and Wang (206) is not separable by electrophoresis. When the lignin-xylan complex is oxidized by periodate and hydrolyzed, the hydrolyzate contains D-xylose, suggesting that some of the xylopyranosyl units bore protective groups at C-2 or C-3.

A lignin-carbohydrate complex isolated from *Pinus densiflora* forms a micell in aqueous solution that can incorporate hydrophobic substances. This indicates that hydrophobic lignin fragments co-exist with a hydrophilic polysaccharide moiety within a molecule and provides evidence for chemical bonds between the lignin and hemicelluloses (207).

A water-soluble lignin-carbohydrate complex can be isolated from a dilute alkaline extraction of *Eucalyptus obliqua* (203,208,209). The carbohydrate component of the complex consists of units of xylose, arabinose, galactose, and an unidentified compound that is an alkaline degradation product of glucose. A NMR study indicates that the carbohydrate component is attached to the lignin propane side chain possible at the α-position to the guaiacyl or syringyl unit. The polysaccharides attached to the lignin are possibly xylan and arabinan types and they are highly branched.

Borohydride reduction in situ, prior to the alkaline extraction of the xylan from aspen (*Populus tremuloids*), yields a reduced 4-O-methyl-D-glucurono-D-xylan. Treatment of the xylan with xylanase provides two similar tetrasaccharides: O-(4-O-methyl-D-glucopyranosyl)-(1→2)-D-xylotriose and O-(4-O-methyl -D-glucopyranosyluronic acid)-(1→2)-D-xylotriose. This suggests that some of the uronic acid residues may be linked to lignin through an ester bond. According to the ratio of reduced to unreduced uronic acid residues (4:5) and taking into account the competition between borohydride reduction of the esters and their alkaline saponification, we see that a high proportion of the 4-O-methyl-D-glucuronic residues of the xylan could occur in esterified form and be important in the bonding to lignin (210).

The combined actions of cellulase and hemicellulase on a lignin-hemicellulose complex from Norwegian spruce (*Picea abies*) yielded a lignin preparation containing about 20% of carbohydrate (211). Degradation of the preparation indicates that lignin is linked to an L-arabinosyl side chain in a xylan and to a D-galactosyl side chain in a galactoglucomannan. Similar work with a purified lignin-hemicellulose fraction from Pinus densiflora shows that lignin is linked glycosidically to D-galactosyl, D-xylosyl, and L-arabinosyl units in the hemicellulose, with two sugar residues linking a single lignin fragment (212).

Although the existence of covalent bonds between lignin and carbohydrates is quite certain, the nature of the linkage is still inconclusive. Four different kinds of bonding have been suggested: namely, ester (202,210), benzyl ether (213,214), glycosidic (215,216), and an acetal or hemiacetal (217). There is not sufficient evidence to fully confirm one or the other of these. It is likely that more than one type of bonding exists in the lignin-carbohydrate complex.

Besides covalent bonds between hemicelluloses and lignin, there are hydrogen bonds formed between hydroxyl and carboxyl groups of hemicelluloses and alcoholic and phenolic hydroxyls, carbonyl, and the etheric oxygen of lignin (218).

VI. APPLICATIONS OF HEMICELLULOSES

Since hemicelluloses constitute up to 20–30% of wood, it is logical to expect their emergence in commercial applications. When isolated, they have many potentially useful properties. These properties include their use as food additives, thickeners, emulsifiers, gelling agents, adhesives, and adsorbants. Some hemicelluloses have even shown value as antitumor agents. Such applications would not only involve the polymeric character of hemicellulose, but also would be based on products from chemical, biochemical, or thermal modifications of hemicelluloses. As new methods provide large quantities of low-cost hemicellulose, it is reasonable to expect that these polysaccharides will find broad industrial applications that will be based also on their abundance and essentially low price.

The presence of hemicelluloses in wheat flours improves the water binding capacity (219), mixing quality (220), and loaf volume (221). It reduces the energy requirement for dough mixing (222) and for the incorporation of added protein (220,223). Arabinogalactan may be used under the Federal Food, Drug and Cosmetic law as an emulsifier, stabilizer, binder, or bodying agent in any of a number of products such as essential oils, flavor bases, and pudding mixes (224). The addition of arabinogalactan may improve the retention of an unstable or volatile flavoring agent (225), and the polymer can be used as a bulking agent in an artificial sweetener preparation (226) to provide an acceptable "mouth feel."

A marketable, low-molecular-weight hemicellulose extract is a byproduct in the production of hardboard and insulating board made by the steam explosion of wood chips. It is the concentrated soluble product obtained from the treatment of wood chips at elevated temperatures and pressures without the use of acids, alkalis, or salts. The product was marketed in the form of feed molasses containing 65% solids and 55% digestible carbohydrates. 4-O-Methyl-D-glucuronoxylan from beech (*Fabus clenata*) suppresses the intestinal absorption of cholesterol and bile acid. Supplementation with 5% xylan to a hypercholesterolemic diet for two weeks causes 28 and 37% depression of the respective plasma and liver cholesterol levels of rats (227). The growth rate of Sarcoma-180 subcutaneously implanted in mice is significantly inhibited by 4-O-methyl-D-glucuronoxylan at a dose of 200 mg/kg (228). Glucomannan isolated from holocellulose or pulp shows antitumor action against Sarcoma 180 in mice (229).

Arabinogalactan is useful as a tablet binder and an emulsifier for pharmaceutical oil-in-water or water-in-oil emulsions (230). Arabinogalactan has found some use in the mining industry for the processing of iron (231) and copper ores. It may be used to replace gum arabic. Arabinogalactan has been obtained by hot-water extraction of Western larch chips and was marketed for a time under the trade name Stractan. It was used mainly as an additive to printing inks (232) and was also found in foods, in certain pharmaceuticals, and in mining industries.

Arabinoxylans have possible use in food, cosmetic, and pharmaceutical industries as emulsifiers, thickeners, bodying agents, or stabilizers. For example, a corn oil-in-water emulsion containing approximately 5% arabinoxylan from corn hull shows good stability. Other potential industrial applications are printing inks, paper coatings, textile sizes, and coating agents. Galactoglucomannans are not used industrially, but the addition of galactoglucomannan to a sizing composition is claimed to improve the sizing of paper (233).

Derivatives of xylans such as carboxymethylxylan have been prepared. They have properties similar to carboxymethylcellulose and have potential for use in detergent building (234). These derivatives also may prove useful in the dispersal of inorganic or organic pigments (235), paper coating (236), development of flocculents and adhesive (237), and preparation of cellophanelike film (238).

It was recognized as early as 1927 that the presence of hemicelluloses in pulp not only increases pulp yield, but also has a positive effect on paper strength (239). Hemicelluloses have been shown to contribute to tensile and bursting strength and to the folding endurance of pulp sheet (240–242). The presence of hemicelluloses in the pulp reduces the time and power required to soften and fibrillate fibers during beating. This beneficial effect can be best illustrated in the production of grease-proof paper, for which beating must be carried to an extreme. The kind and amount of hemicelluloses present in pulp are of primary importance to pulp formation and properties. Mannan has a greater effect on pulp yield than xylan (243). Although a minimum amount of hemicellulose must be present in the pulp for strength development, pulp containing too much hemicellulose hydrates too rapidly and loses freeness before adequate strength is developed. This produces a paper that is translucent, rattly, and of low strength (244). Hemicelluloses are undesirable components for pulps that are to be used as a dissolving pulp. In papermaking, hemicelluloses are partly responsible for a loss in brightness after bleaching and on storage or aging.

The bioconversion of hemicelluloses to useful chemicals often requires prior hydrolysis of the polysaccharides to their sugar constituents. Hemicelluloses are more readily hydrolyzed by acid than is cellulose, and consequently, they are more easily converted to simple sugars. Xylans that are more abundant in hardwood species yield mainly D-xylose, whereas the glucomannans found in large quantities in softwood species yield sizable amounts of D-mannose. Chemicals derived from hemicelluloses have been reviewed by Thompson (245).

D-Mannose and other hexoses can be combined with D-glucose for fermentation to ethanol. D-Xylose has been used at a rate of about 400 tons per year for the production of isomerase used in the production of high D-fructose syrup from D-glucose (246,247). Xylose and other pentoses can be converted to furfural. Furfural is produced commercially by the acid treatment of xylan in corn cobs and sugar cane bagasse. Alternatively D-xylose can be reduced to xylitol. Xylitol has attracted attention as a noncariogenic sweetener and has been tested in a variety of food products (248). Hardwoods such as sweetgum or northern birch and the existing hemicellulose extracts produced by forest industries could be a potential source for the production of xylitol. Liquid fuel, such as 2,3-butanediol and butanol, can be produced by a combined enzymic hydrolysis and fermentation process (249). Hemicelluloses can be converted by microorganisms to various products, such as methane, organic acids, sugar alcohols, solvents, animal feed, and ethanol (250).

REFERENCES

1.   J. Thomsen, *J. Prakt. Chem.*, *19*:146 (1879).
2.   H. J. Wheeler and B. Tollens, *Ber.*, *22*:1046 (1889).
3.   E. Schulze, *Ber.*, *24*:2277 (1891).
4.   T. E. Timell, *Wood Sci. Technol.*, *1*:45 (1967).
5.   R. L. Whistler and R. H. Shah, Symposium on Cellulose, Paper Textile Division, ACS, Washington, D.C., 1977 Vol. 341.
6.   G. A. Twole and R. L. Whistler, in Phytochemistry: *The Process and Products of Photosynthesis* (L. P. Miller, ed.), Van Nostrand Reinhold, New York, 1973, p. 198.
7.   G. O. Aspinall, in *Biogenesis of Plant Cell Wall Polysaccharides* (F. Loewus, ed.), Academic Press, New York, 1973, p. 95.
8.   R. L. Whistler and E. L. Richards, in *The Carbohydrates: Chemistry and Biochemistry*, 2nd ed., (W. Pigman and D. Horton, eds.), Vol. 2A, Academic Press, New York, 1970, p. 447.
9.   T. E. Timell, *Adv. Carbohydr. Chem.*, *20*:409 (1965).
10.   T. E. Timell, *Adv. Carbohydr. Chem.*, *19*:247 (1964).
11.   C. Schuerch, in *The Chemistry of Wood* (B. L. Browing, ed.), Wiley-Interscience, New York, 1963, p. 191.
12.   G. O. Aspinall, *Adv. Carbohydr. Chem.*, *14*:429 (1959).
13.   R. L. Whistler and C. L. Smart, *Polysaccharide Chemistry*, Academic Press, New York, 1953.
14.   R. L. Whistler, *Adv. Carbohydr. Chem.*, *5*:269 (1950).
15.   L. E. Wise and E. C. Jahn, *Wood Chemistry,* 2nd ed., Reinhold, New York, 1952.
16.   L. E. Wise, M. Murphy, and A. A. D'Addieco, *Paper Trade J.*, *122*:35 (1964)

17. T. E. Timell and E. C. Jahn, *Svensk Papperstidn.*, *54*:831 (1951).
18. A. Jeanes and H. S. Isbell, *J. Res. Nat. Bur. Stand.*, *A27*:125 (1941).
19. E. S. Becker, J. K. Hamilton, and W. E. Lucke, *TAPPI*, *48*:60 (1965).
20. D. Pal and O. Samuelson, *Svensk Paperstidn.*, *79*:311 (1976).
21. T. E. Timell, *TAPPI*, *44*:88 (1961).
22. G. G. S. Dutton, B. I. Joseleau, and P. E. Reid, *TAPPI*, *56*:168 (1973).
23. D. W. Clayton, *Svensk Papperstidn.*, *66*:115 (1963).
24. R. L. Whistler and J. N. BeMiller, *Advan. Carbohydr. Chem.*, *13*:289 (1958).
25. K. H. Johansson and O. Samuelson, *Carbohydr. Res.*, *54*:295 (1977).
26. G. O. Aspinall, C. T. Greenwood, and R. L. Sturgeon, *J. Chem. Soc.*, :3667 (1961).
27. J. L. Minor and N. Sanyer, *TAPPI*, *57*:109 (1974).
28. G. Katz, *TAPPI*, *48*:34 (1965).
29. T. Matsuo and T. Mizuno, *Agr. Bio. Chem.*, *38*:465 (1974).
30. H. O. Bouveng, P. G. Garegg, and B. Lindberg, *Chem. and Ind.*, *52*:1727 (1958).
31. T. E. Timell, *Svensk Papperstidn.*, *63*:472 (1960).
32. M. H. O'Dwyer, *Biochem. J.*, *20*:656 (1926).
33 . J. D. Blake, P. T. Murphy, and G. N. Richards, *Carbohydr. Res.*, *16*:49 (1971).
34. R. L. Whistler and G. E. Lauterbach, Arch. B*iochem Biophys.*, *77*:62 (1958).
35. R. E. Gramera and R. L. Whistler, *Arch. Biochem. Biophys.*, *101*:75 (1963).
36. R. L. Whistler and J. L. Sannella, *Methods Carbohydr. Chem.*, *5*:34 (1965).
37. J. D. Blake and G. N. Richards, *Carbohydr. Res.*, *17*:253 (1971).
38. B. D. E. Gaillard and N. S. Thompson, *Carbohydr. Res.*, *18*:137 (1971).
39. B. D. E. Gaillard, N. S. Thompson, and A. J. Morak, *Carbohydr. Res.*, *11*:509 (1969).
40. B. D. E . Gaillard and R. W. Bailey, *Nature*, *212*:202 (1966).
41. B. D. E. Gaillard, *Nature*, *191*:1295 (1961).
42. S. K. Chanda, E. L. Hirst, J. K. N. Jones, and E. G. V. Percival, *J. Chem. Soc.* :1289 (1950).
43. A. J. Erskine and J. K. N. Jones, *Can. J. Chem.*, *34*:821 (1956).
44. H. Meier, *Methods Carbohydr. Chem.*, *5*:45 (1965).
45. J. E. Scott, *Chem. Ind. (London)*, :168 (1955).
46. J. E. Scott, *Methods Biochem. Anal.*, *8*:145 (1960).
47. H. O. Bouveng and B. Lindberg, *Acta Chem. Scand.*, *12*:1973 (1958).
48. E. L. Hirst, D. A. Rees, and N. G. Richardson, *Biochem. J.*, *95*:453 (1965).
49. S. C. Churms, *Adv. Carbohydr. Chem. Biochem.*, *25*:13 (1970).
50. K. Kringstad and O. Ellefsen, *Papier*, *18*:583 (1964).
51. S. C. Churms and A. M. Stephen, *Carbohydr. Res.*, *21*:91 (1972).
52. A. Heyraud and M. Rinaudo, *J. Chromatogr.*, *166*:149 (1978).
53. D. Noel, T. Hanai, and M. D'Amboise, *J. Liquid Chromatogr.*, *2*:1325 (1979).
54. G. A. Adams, *Can. J. Chem.*, *38*:280 (1960).
55. D. H. Northcote, *Biochem. J.*, *58*:353 (1954).
56. R. L. Whistler and C. S. Campbell, *Methods Carbohydr. Chem.*, *5*:202 (1965).
57. R. L. Whistler and J. W. Marx, *Methods Carbohydr. Chem.*, *5*:56 (1965).
58. K. -S. Jiang and T. E. Timell, *Cell. Chem. Technol.*, *6*:493 (1972).
59. P. Albersheim, D. J. Nevins, P. D. English, and A. Karr, *Carbohydr. Res.*, *5*:340 (1967).
60. E. Fengel, M. Przyklenk, and G. Wegener, *Papier*, *30*:240 (1976).
61. R. L. Taylor and H. E. Conrad, *Biochem.*, *11*:1383 (1972).

62. C. C. Chen and G. D. McGinnis, *Carbohydr. Res.*, *90*:127 (1981).
63. R. M. Thompson, *J. Chromatogr.*, *166*:201 (1978).
64. S. Honda, Y. Matsuda, M. Takahashi, K. Kakehi, and S. Ganno, *Anal. Chem.*, *52*:1079 (1980).
65. W. A. König, I. Benecke, and H. Bretting, *Angew. Chem.*, *93*:688 (1981).
66. W. A. König, I. Benecke, and S. Sievers, *J. Chromatogr.*, *217*:71 (1981).
67. H. Schweer, *J. Chromatogr.*, *243*:149 (1982).
68. K. Leontein, B. Lindberg, and L. Lönngren, *Carbohydr. Res.*, *62*:359 (1978).
69. G. J. Gerwig, J. P. Kamerling, and J. F. Vliegenthart, *Carbohydr. Res.*, *62*:349 (1978).
70. G. O. Aspinall, *Int. Rev. Sci.*, *7*:201 (1976).
71. B. Lindberg and J. Lönngren, *Methods Enzymol.*, *50*:3 (1978).
72. P. J. Garegg and B. Lindberg, *Acta Chem. Scand.*, *14*:871 (1960).
73. H. A. Hampton, W. N. Haworth, and E. L. Hirst, *J. Chem. Soc.*, :1739 (1929).
74. T. Purdie and J. C. Irvine, *J. Chem. Soc.*, *83*:1021 (1903).
75. T. Purdie and J. C. Irvine, *J. Chem. Soc.*, *85*:1049 (1904).
76. V. D. Harwood, *Can. J. Chem.*, *29*:974 (1951).
77. R. Kuhn, H. Trischmann, and I. Löw, *Angew. Chem.*, *67*:32 (1955).
78. E. L. Falconer and G. A. Adams, *Can. J. Chem.*, *34*:338 (1956).
79. S. Hakomori, *J. Biochem.* (Tokyo), *55*:205 (1964).
80. E. J. Corey and M. Chaykovsky, *J. Amer. Chem. Soc.*, *84*:866 (1962).
81. D. R. Phillips and B. Fraser, *Carbohydr. Res.*, *90*:149 (1981).
82. T. Narui, K. Takahashi, M. Kobayashi, and S. Shibata, *Carbohydr. Res.*, *103*:293 (1982).
83. A. H. deBelder and B. Norrman, *Carbohydr. Res.*, *8*:1 (1968).
84. C. G. Hellerqvist, B. Lindberg, S. Svensson, T. Holme, and A. A. Lindberg, *Carbohydr. Res.*, *8*:43 (1968).
85. R. Oshima, A. Yoshikawa, and J. Kumanotani, *J. Chromatogr.*, *213*:142 (1981).
86. H. Björndal, C. G. Hellerqvist, B. Lindberg, and S. Svensson, *Angew. Chem.*, Int. Ed. Engl., *9*:610 (1970).
87. G. O. Aspinall and K. -G. Rosell, *Carbohydr. Res.*, *57*:C23 (1977).
88. R. L. Whistler and C. C. Tu, *J. Am. Chem. Soc.*, *75*:3609 (1952).
89. R. L. Whistler and L. Hough, *J. Am. Chem. Soc.*, *75*:4918 (1953).
90. C. L. Mehltretter, *Adv. Carbohydr. Chem.*, *8*:231 (1953).
91. G. O. Aspinall and A. Nicolson, *J. Chem. Soc.*, :2503 (1960).
92. G. O. Aspinall and I. M. Cairncross, *J. Chem. Soc.*, :3996 (1960).
93. B. Lindberg, J. Lönngren and D. A. Powell, *Carbohydr. Res.*, *58*:177 (1977).
94. P. A. J. Gorin and J. F. T. Spencer, *Carbohydr. Res.*, *13*:339 (1970).
95. I. J. Goldstein and W. J. Whelan, *J. Chem. Soc.*, :170 (1962).
96. Y. C. Lee and C. E. Ballow, *Biochem.*, *4*:257 (1965).
97. R. F. H. Dikker and G. N. Richards, *Adv. Carbohydr. Chem.*, *32*:277 (1976).
98. O. Perila and C. T. Bishop, *Can. J. Chem.*, *39*:815 (1961).
99. T. E. Timell, *Can. J. Chem.*, *40*:22 (1962).
100. J. Comtat and J. -P. Joseleau, *Carbohydr. Res.*, *95*:101 (1981).
101. J. M. Bobbitt, *Adv. Carbohydr. Chem.*, *11*:1 (1956).
102. T. Painter and B. Larsen, *Acta Chem. Scand.*, *24*:813 (1970).
103. G. D. Greville and D. H. Northcote, *J. Chem. Soc.*, :1945 (1952).

104. M. L. Wolfrom, A. Thompson, A. N. O'Neill, and T. T. Galkowaski, *J. Amer. Chem. Soc.*, *74*:1062 (1952).
105. C. T. Bishop, *Methods Carbohydr. Chem.*, *6*:350 (1972).
106. I. J. Goldstein, G. W. Hay, B. A. Lewis, and F. Smith, *Methods Carbohydr. Chem.*, *5*:361 (1965).
107. T. Painter and B. Larsen, *Acta Chem. Scand.*, *27*:1957 (1973).
108. B. Lindberg, J. Lönngren, and J. L. Thompson, *Carbohydr. Res.*, *28*:351 (1973).
109. B. Lindberg, J. Lönngren, and S. Svensson, *Adv. Carbohydr. Chem. Biochem.*, *31*:185 (1975).
110. S. J. Angyal and K. James, *Aust. J. Chem.*, *23*:1209 (1970).
111. T. Fujiwara and A. Arai, *Carbohydr. Res.*, *104*:325 (1982).
112. J. Hoffman, B. Lindberg, and S. Svensson, *Acta Chem. Scand.*, *26*:661 (1972).
113. R. K. Basak, P. K. Mandal, and A. K. Mukherjee, *Carbohydr. Res.*, *104*:309 (1982).
114. T. Usui, T. Toriyama, and T. Mizuno, *Agric. Biol. Chem.*, *43*:603 (1979).
115. G. G. S. Dutton and N. A. Funnell, *Can. J. Chem.*, *51*:3190 (1973).
116. A. S. Perlin and G. K. Hamer, *ACS Symp. Ser.*, *103*:123 (1979).
117. F. R. Seymour, *ACS Symp. Ser.*, *103*:27 (1979).
118. S. M. Bociek, M. J. Izzard, A. Morrison, and D. Welti, *Carbohydr. Res.*, *93*:279 (1981).
119. L. D. Hall and M. Yalpani, *Carbohydr. Res.*, *91*:C-1 (1981).
120. T. J. Painter, J. J. Gonzalez, and P. C. Hemmer, *Carbohydr. Res.*, *69*:217 (1979).
121. R. Albersheim, D. J. Nevins, P. D. English, and A. Karr, *Carbohydr Res.*, *5*:340 (1967).
122. A. L. Usov, K. S. Adamyants, S. V. Yarotsky, A. A. Anoshina, and N. K. Kochetkou, *Carbohydr. Res.*, *26*:282 (1973).
123. P. Kovac and J. Hirsch, *Carbohydr. Res.*, *85*:177 (1980).
124. C. C. Sweeley, R. Bentley, M. Makita, and W. W. Wells, *J. Amer. Chem Soc.*, *85*:2497 (1963).
125. F. R. Seymour, E. C. M. Chen, and S. H. Bishop, *Carbohydr. Res.*, *73*:19 (1979).
126. F. R. Seymour, S. L. Unruh, and D. A. Nehlich, *Carbohydr. Res.*, *191*:175 (1989).
127. J. H. Sloneker, *Methods Carbohydr. Chem.*, *6*:20 (1972).
128. C. Green, V. M. Doctor, G. Holzer, and J. Orò, *J. Chromatogr.*, *207*:268 (1981).
129. J. Klok, E. H. Nieberg-van Velzon, J. W deLeeuw, and P. A. Shenck, *J. Chromatogr.*, *207*:273 (1981).
130. R. Oshima, A. Yoshikawa, and J. Kumanotani, *J. Chromatogr.*, *213*:42 (1981).
131. V. Kovàcik, V. Mihàlov, and P. Kovàc, *Carbohydr. Res.*, *88*:189 (1981).
132. N. Shibuya, *J. Chromatogr.*, *208*:96 (1981).
133. H. A. Swenson, A. J. Morak, and S. Kurath, *J. Polym. Sci.*, *51*:231 (1961).
134. J. L. Minor, *J. Chromatogr.*, *2*:309 (1979).
135. H. Yamaguchi, S. Inamura, and K. Makino, *J. Biochem.* (Tokyo), *79*:299 (1976).
136. H. Yamaguchi and K. Makino, *J. Biochem.* (Tokyo), *81*:563 (1977).
137. H. Yamaguchi and K. Okamoto, *J. Biochem.* (Tokyo), *82*:511 (1977).
138. M. Tanaka, *Carbohydr. Res.*, *88*:1 (1981).
139. T. E. Timell, *TAPPI, 44*:88 (1961).
140. G. G. S. Dutton, B. I. Joseleau, and P. E. Reid, *TAPPI, 56*:168 (1973).
141. J. K. Hamilton, H. W. Kircher, and N. S. Thompson, *J. Amer. Chem. Soc.*, *78*:2508 (1956).
142. N. S. Thompson, H. H. Heller, J. D. Hankey, and O. Smith, *Pulp Pap Mag. Can.*, *67*

(12):T541 (1966).

143.   J. K. Hamilton and H. W. Kircher, *J. Amer. Chem. Soc.*, *80*:4703 (1958).

144.   E. C. A. Schwarz and T. E. Timell, *Can. J. Chem.*, *41*:1381 (1963).

145.   G. O. Aspinall, R. Begbie, and T. E. McKay, *J. Chem. Soc.*, :214 (1962).

146.   B. Lindberg, K. -G. Rosell, and S. Svensson, *Svensk Papperstidn.*, *76*:383 (1973).

147.   Y. L. Fu and T. E. Timell, *Cell. Chem. Technol.*, *6*:517 (1972).

148.   L. Kenne, K. -G. Rosell, and S. Svensson, *Carbohydr. Res.*, *44*:69 (1975).

149.   V. D. Harwood, Svensk Papperstidn., *76*:377 (1973).

150.   G. G. S. Dutton and R. H. Walker, *Cell. Chem. Technol.*, *6*:295 (1972).

151.   S. Johnson and S. Samuelson, *Anal. Chem. Acta*, *36*:1 (1966).

152.   A. Roudier and L. Eberhard, *TAPPI*, *38*:156A (1955).

153.   M. Zinbo and T. E. Timell, Svensk Papperstidn., *70:597* (1967).

154.   G. N. Richards and R. L. Whistler, *Carbohydr. Res.*, *31*:47 (1973).

155.   K. Shimizu and O. Samuelson, *Svensk Papperstidn.*, *76*:156 (1973).

156.   K. Shimizu, M. Hashi, and K. Sakurai, *Carbohydr. Res.*, *62*:117 (1978).

157.   J. Comtat and F. Barnoud, *Compt. Rend.*, *277*:61 (1973).

158.   G. G. S. Dutton and J. -P. Joseleau, *Cell. Chem. Technol.*, *11*:313 (1977).

159.   K. -G. Rosell and S. Svensson, *Carbohydr. Res.*, *42*:297 (1975).

160.   M. F. Adams and B. V. Ettling, in I*ndustrial Gums* (R. L. Whistler ed.), Academic Press, New York, 1973, p. 415.

161.   H. O. Bouveng and B. Lindberg, *Acta Chem. Scand.*, *12*:1977 (1958).

162.   B. V. Ettling and M. F. Adams, *TAPPI*, *51*:116 (1968).

163.   H. A. Swenson, H. M. Kaustinen, J. J. Bachhuber, and J. A. Carlson, *Macromolecules*, *2*:142 (1969).

164.   G. Lystad-Borgin, *J. Amer. Chem. Soc.*, *71*:2247 (1949).

165.   S. C. Churms, E. H. Merrifield, and A. M. Stephen, *Carbohydr. Res.*, *64*:C-1 (1978).

166.   O. Goldschmid and H. L. Hergert, *TAPPI*, *44*:858 (1961).

167.   N. S. Thompson and O. A. Kaustinen, *TAPPI*, *49*:83 (1966).

168.   R. A. Laidlaw and G. A. Smith, *Chem. Ind.* (London): 462 (1962).

169.   Y. L. Fu and T. E. Timell, *Cell. Chem. Technol.*, *6*:513 (1972).

170.   A. J. Roudier and L. Eberhard, *Bull. Soc. Chim.* (France), :460 (1965).

171.   J. K. N. Jones, C. B. Purves, and T. E. Timell, *Can. J. Chem.*, *39*:1059 (1961).

172.   B. Lindberg, K. -G. Rosell, and S. Svensson, *Svensk Papperstidn.*, *76*:30 (1973).

173.   H. O. Bouveng, P. J. Garegg, and B. Londberg, *Acta Chem. Scand.*, *14*:742 (1960).

174.   H. O. Bouveng, *Acta Chem. Scand.*, *15*:87 (1961).

175.   H. O. Bouveng, *Acta Chem. Scand.*, *15*:96 (1961).

176.   K. Shimizu and O. Samuelson, *Svensk Papperstidn.*, *76*:150 (1973).

177.   M. H. Johansson and O. Samuelson, *Wood Sci. Technol.*, *11*:251 (1977).

178.   R. Aurell, N. Hartler, and G. Persson, A*cta Chem. Scand.*, *17*:545 (1963).

179.   N. Hartler and I. -L. Svensson, *Ind. Eng. Chem.*, *4*:80 (1965).

180.   J. Havlicek and O. Samuelson, *Carbohydr. Res.*, *22*:307 (1972).

181.   M. Zinbo and T. E. Timell, *Svensk Papperstidn.*, *68*:647 (1965).

182.   T. E. Timell, *Svensk Papperstidn.*, *63*:472 (1960).

183.   J. E. Milks and C. B. Purves, *J. Amer. Chem. Soc.*, *78*:3738 (1955).

184.   T. Koshijima, T. E. Timell, and M. Zinbo, *J. Polym. Sci.*, Part C, *11*:265 (1965).

185.   M. J. Song and T. E. Timell, *Cell. Chem. Technol.*, *6*:67 (1972).

186. S. Karàcsonyi, M. Kubackovà, and J. Hrivnak, *Coll. Czech. Chem. Commun.*, *32*:3597 (1967).
187. J. K. N. Jones, L. E. Wise and J. Jappe, *TAPPI*, *39*:139 (1956).
188. S. Karàcsonyi, *Coll. Czech. Chem. Commun.*, *34*:3944 (1969).
189. S. Karàcsonyi, *Coll. Czech. Chem. Commun.*, *40*:1240 (1975).
190. G. A. Adams, *Svensk Papperstidn.*, *67:82* (1964).
191. W. A. Côtè, Jr., B. W. Simson, and T. E. Timell, *Svensk Papperstidn.*, *69*:553 (1966).
192. H. O. Bouveng and H. Meier, *Acta Chem. Scand.*, *13*:1884 (1959).
193. K. S. Jiang and T. E. Timell, *Svensk Papperstidn.*, *75*:592 (1972).
194. Y. Fu and T. E. Timell, *Svensk Papperstidn.*, *75*:680 (1972).
195. G. C. Hoffmann and T. E. Timell, *Svensk Papperstidn.*, *75:135* (1972).
196. G. C. Hoffmann and T. E. Timell, *Svensk Papperstidn.*, *75*:241 (1972).
197. W. Klauditz and I. Stolley, *Holzforschung*, *9*:5 (1955).
198. H. Meier, *Acta Chem. Scand.*, *16*:2275 (1962).
199. T. M. Singh, G. M. Mathur, and S. R. D. Guha, *Indian Pulp Pap.*, *6*:10 (1976).
200. A. Bjorkman, *Svensk Papperstidn.*, *59*:477 (1956).
201. A. Bjorkman, *Svensk Papperstidn.*, *60*:243 (1957).
202. H. H. Brownell, *TAPPI*, *48*:513 (1965).
203. J. W. T. Merewether and A. M. Samsuzzaman, *Holzforschung*, *26*:11 (1972).
204. B. O. Lindgren, *Acta Chem. Scand.*, *12*:447 (1958).
205. J. Azuma, N. Takahashi, and T. Koshijima, *Carbohydr. Res.*, *93*:91 (1981).
206. H. I. Bolker and P. Y. Wang, *TAPPI*, *52*:920 (1969).
207. F. Yaku, S. Tsuji, and T. Koshijima, *Holzforschung*, *33*:54 (1979).
208. J. W. T. Merewether, L. A. M. Samsuzzaman, and I. C. Calder, *Holzforschung*, *26*:180 (1972).
209. J. W. T. Merewether, L. A. M. Samsuzzaman, and R. G. Cooke, *Holzforschung*, *26*:193 (1972).
210. J. Comtat, J. P. Joseleau, C. Bosso, and F. Barnoud, *Carbohydr. Res.*, *38*:217 (1974).
211. O. Eriksson and B. O. Lindgren, *Svensk Papperstidn.*, *80*:59 (1977).
212. T. Koshijima, F. Yaku, and B. Tanake, in *Proceedings of the 8th Cellulose Conference, Applied Polymer Symposia* (T. E. Timell, ed.), 1976, Vol. 28, p. 1025.
213. K. Freudenberg, *Sci.* *148*:595 (1965).
214. B. Kosikovà, D. Joniak, and L. Kosàkovà, *Holzforschung*, *33*:11 (1979).
215. A. Hayashi, J. *Agric. Chem. Soc.* (Japan), *35*:83 (1961).
216. F. Yaku, Y. Yamada, and T. Koshijima, *Holzforschung*, *30:1*48 (1976).
217. H. I. Bolker, *Nature*, *197*:489 (1963).
218. M. Remko, *Zeitschrift fur Physikalische Chemie Neue Folge*, *126*:s-195 (1981).
219. C. G. Schwalbe, *Papier-Fabr.*, *25*:481 (1927).
220. H. L. Spiegelberg, *TAPPI*, *49*:388 (1966).
221. M. N. Fineman, *TAPPI*, *35*:320 (1952).
222. J. W. Swanson, *TAPPI*, *33*:451 (1951).
223. F. W. Klingstedt, *Svensk Papperstidn.*, *40*:412 (1937).
224. G. Jayme, *Papier-Fabr.*, *40*:137 (1942).
225. M. Glicksman and R. E. Schachat, U. S. Patent 3,264,114, 1976.
226. G. L. Stanko, U. S. Patent 3,294,544, 1966.
227. M. Hashi and T. Takeshita, *Agric. Biol. Chem.*, *39*:579 (1975).

228. M. Hashi and T. Takeshita, *Agric. Biol. Chem.*, *43*:951 (1979).
229. T. Hashimoto, *Japan. Kokai*, *75*:100 (1975).
230. L. C. Bratt, *Pulp Pap.* :102 (1979).
231. S. I. Gorlovskii, *Obogashch. Rud.*, *6*:18 (1961).
232. M. R. Nazareth, C. E. Kennedy, and V. N. Bhatia, *J. Pharm. Sci.*, *50*:560 (1961).
233. A. A. Sedov and S. A. Puzyrev, in *Lesnaya Promyshlennost* (E. Ya. Pechko, ed.) Moscow 1973, p. 116.
234. J. Schmorak and G. A. Adams, *TAPPI*, *40*:378 (1957).
235. O.Naoterv, T. Takashi, and K. Tsutomu, *Ger. Offen.*, *2*:523, 161 (1975).
236. I. Kusakabe, T. Yasui, and T. Kobayashi, *Hakko Kogaku Zasshi*, *53*:135 (1975).
237. E. Pulkkinen, M. Reintjes, and L. D. Storr, U. S. Patent 3,833,527, 1974.
238. J. A. Church, *J. Polym. Sci.*, Part A-1, *5*:3183 (1967).
239. B. L. D'appolonia and L. A. MacArthur, *Amer. Assoc. Cereal Chem.*, *52*:230 (1975).
240. S. L. Jelaca and I. Hlynka, *Amer. Assoc. Cereal Chem.*, *49*:489 (1972).
241. B. L. D'appolonia, K. A. Gilles, and D. G. Medcalf, *Amer. Assoc. Cereal Chem.*, *47*:194 (1970).
242. S. L. Jelaca and I. Hlynka, *Amer. Assoc. Cereal Chem.*, *48*:211 (1971).
243. B. L. D'appolonia and K. A. Gilles, *Amer. Assoc. Cereal Chem.*, *48*:427 (1971).
244. *Federal Register*, *30*:2430 (1965).
245. N. S. Thompson, in *Organic Chemicals from Biomass* (I. S. Goldstein, ed.), CRC Press, Inc., Boca Raton, Fla., 1981.
246. J. W. Dunning and E. C. Lathrop, U. S. Patent 2,450,586, 1948.
247. Y. Kamiyama, Y. Hirabayashi, Y. Sakai, and T. Kobayashi, *Hakko Kogaku Zasshi*, *52*:669 (1974).
248. J. R. Russo, *Food Eng.*, *48*:77 (1976).
249. E. K. C. Yu, L. Deschatelets, and J. N. Saddler, *Proceedings 5th Canadian Bioenergy R&D Seminar* (S. Hasnain, ed.), Elsevier Applied Sciences, London, 1984, p. 267.
250. C. -S. Gong, L. F. Chen, M. C. Flickinger, and G. T. Tso, *Adv. Biochem. Eng.*, *20*:93 (1981).

8

Extraneous Materials from Wood

Eugene Zavarin and Laurence Cool

University of California, Richmond, California

I. INTRODUCTION

Any comprehensive treatment of wood extractives of necessity involves an extraordinarily large number of species and an even greater number of compounds. Since such a treatment would have been impossible within the scope of this review, drastic cuts had to be made. These cuts involved reducing the number of wood species treated to include mainly those of at least some commercial importance and reducing the number of compounds to include either the most common ones or those best exemplifying the extractives from a certain source. The treatment of bark extractives was omitted for the same reason. Readers interested in a more detailed treatment are referred to the comprehensive compilations (1–12). As is well known, many differences in the properties of various woods are determined by the composition of their extractives. In addition, some extractives represent important industrial products. Unfortunately, the topic of the practical importance of extractives had to also be omitted, as this subject would itself require a full chapter.

In covering the structures of extractives, the liberal use of biosynthetic relationships was made. In the authors' opinion, any understanding of the structural relationships of natural products is impossible without at least a rudimentary understanding of their biosynthesis. All examples of the occurrence of individual structures were checked as far as possible to make sure the respective compounds were isolated from wood and not from some other plant organ; this point is unfortunately neglected by many authors. If mention is made of compounds occurring elsewhere, this is indicated in the text. It cannot be emphasized strongly enough that the compositional differences of extractives isolated from various organs can be as high as those between different species or even between higher taxonomic units.

Formulas were drawn using WIMP software (Aldrich Chemical Co., Inc.), with an IBM AT computer and IBM 7371 plotter.

II. DEFINITION AND CLASSIFICATION

A. Definition

Under extraneous or associated materials of wood one generally understands the organic compounds that do not form a part of the lignocellulosic cell wall but occur in cell lumina, cell wall interstices, or in intercellular cavities such as resin canals of wood. For the most part, these materials can be separated from the structural cellulose, hemicelluloses, and lignin by extraction with neutral organic solvents and water, and they are therefore called extractives. Strictly speaking, however, the two concepts are not synonymous. It has been shown that under some conditions lignin can be partly dissolved with polar organic solvents; on the other hand, in woods high in phenolic materials some of the extraneous materials can be so highly polymerized that they can be neither extracted with neutral organic solvents nor with water (not even with dilute alkalies). Such extraneous materials remain in the wood and are co-determined with lignin through Klason lignin analysis. Their presence can be noticed by a lower-than-usual methoxyl content of Klason lignin or by a lower yield of vanillin in the alkaline nitrobenzene oxidation of lignin, as these extraneous materials generally contain less methoxyl than actual lignin (Fig. 1). In most cases, the distinction between extractives and extraneous materials is academic, however, and the two concepts can be equated.

B. Classification

Extractives include a very large number of radically different chemical substances. This diversity places the extractives apart from the much more uniform cellulose, hemicelluloses, and lignin and necessitates treating them in groups and subgroups. These groups can be defined on the basis of various criteria, resulting in several types of classification.

Chemical classification is based on the overall similarity of the chemical structures of the extractive components. This classification is most generally used, is simple, and leads to such well-known classes of compounds as terpenoids (including subclasses such as monoterpenoids and diterpenoids), flavonoids, stilbenes, and lignans. Classification in terms of functional groups only (alcohols, carboxylic acids, hydrocarbons) does not lead very far and is useful only if one is interested in differences in chemical reactivity.

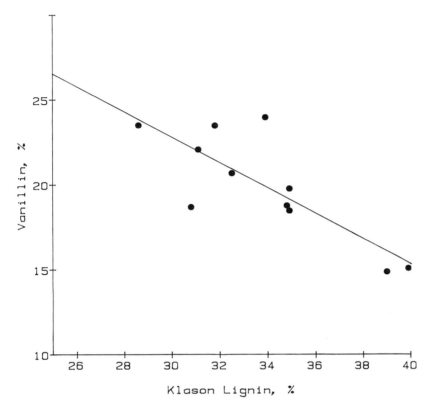

Figure 1 Relationship between percentage of Klason lignin content of various softwoods and percentage of vanillin obtained in nitrobenzene oxidation. Correlation coefficient, R=0.802, significance, <0.005. (From Ref. 312.)

Biogenetic classification is based on the biogenetic relationships of the extractive components (13). It produces groupings similar to those of chemical classification, but has the advantage of giving the classification a natural basis and is not plagued as much by exceptions. Driven to the extreme, however, it becomes too complicated and unwieldy.

Plant physiological classification subdivides the extractives into a group associated with the presence of extracellular receptacles for extractives, such as resin canals of Pinaceae (canal extractives), a group associated with heartwood formation (heartwood extractives), and a group comprising the extractives used by the plant as food (reserve materials).

Finally, the empirical classification groups the extractives in accordance with their historical role in the utilization of wood products; such groups include turpentine, tall oil, rosin, fatty materials, wax, tannin, carbohydrates, etc.

The general classification scheme adopted as a basis for discussion in this chapter is given in Fig. 2.

The individual compounds are named either systematically, using rules such as those of the International Union for Pure and Applied Chemistry, or empirically, commonly deriving the names of compounds from the botanical names of their respective plant sources (e.g. α-pinene from *Pinus* or abietic acid from *Abies*). The first approach often becomes unwieldy, however, as the names occasionally become several lines long. The second approach leads to shorter names, but greatly increases their number. Occasionally, compromise solutions are attempted, most of them only partly successful. It is probably still best to rely wherever possible on language-independent structural formulae, although they are unfortunately unpronounceable.

III. FORMATION OF EXTRACTIVES

Under formation of extractives is understood the mechanisms accounting for the accumulation of extractives in specific tissues of wood, i.e., the heartwood, the extracellular receptacles (resin canals of conifers), and, in the case of reserve

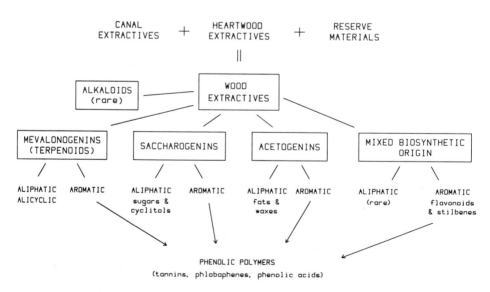

Figure 2 Classification of wood extractives as used in this chapter.

materials, in the living portion of the wood (sapwood). The mechanisms dealing with the formation of individual molecular species (molecular biosynthesis) will be covered in discussions of the individual extractive constituents.

A. Heartwood Extractives

The mechanism of the formation of heartwood extractives has been the subject of numerous investigations that led to a better understanding but by no means a complete solution of the problem (14–17). It has been demonstrated that an intensified and altered metabolic activity at the sapwood-heartwood boundary precedes the transformation of sapwood into heartwood. The enzymes implicated include peroxidase, phenol oxidases, and enzymes concerned with carbohydrate hydrolysis and metabolism. The involvement of ethylene (15), triggering a variety of transformations, has been also suggested. This increased metabolism is regarded as being responsible for the formation of heartwood extractives in their final or near-final structural form and their deposition in cell lumina and cell wall interstices. While some additional transformations (condensations, oxidations) of extractives continue to take place even after their deposition, they are generally appreciably less drastic and, since heartwood is dead tissue, must take place either by abiotic pathways (under the influence of natural acidity or oxygen) or by the involvement of extraneous organisms (fungi, insects, or bacteria).

One of the effects of the increased metabolism at the heartwood-sapwood boundary is the polymerization of lignans, the ligninlike monomeric extractives, to a material chemically difficult to distinguish from the protolignin but extractable with polar organic solvents (18) (Brauns' native lignin), as well as the formation of insoluble phenolic materials. The latter are co-determined with true lignin in the Klason lignin determination and are responsible for the higher Klason lignin content of heartwood, as well for the lower methoxyl content of the Klason lignin. The term "secondary lignification," in analogy to the primary lignification at the cambium, has been proposed for these processes (17).

Since the amount of extractives present in heartwood often greatly exceeds the amount of the reserve substances in the sapwood (starch, oligosaccharides, lipids), a question immediately arises as to the origin of the material responsible for the formation of heartwood extractives. That the reserve substances, accumulated in the parenchyma of sapwood, are partly responsible for supplying the material necessary for the formation of heartwood extractives is indicated by the disappearance of starch (19) and a marked reduction in the amount of oligomeric carbohydrates (20) and lipids (21) during the sapwood-to-heartwood transformation. Additional material is probably supplied by a centripetal material movement from the cambium inward. Whether this material is being translocated as typical primary metabolites (oligosaccharides, pyruvic acid, acetate), as secondary metabolites structurally similar to the final products, or in the form of some intermediate structures is not clear. In view of the multitude of chemical structures and biosynthetic pathways involved in heartwood formation, as well as the botanical diversity of heartwood-forming plants, it seems likely that we are dealing with a variety of transport mechanisms and that the question of heartwood formation will be ultimately answered by separately treating a multitude of individual cases.

B. Canal Extractives

Practically all plants secrete various organic materials by means of special secretory cells. In some cases, the secreted substances are stored within the cells, in others they are secreted outside. Secretory cells are often organized into a tissue that commonly assumes a characteristic structure (gland), with a central cavity where the secretion is stored. The extractives stored in special cavities in wood can be designated as secretory extractives or canal extractives if the cavities are in the form of canals. The cavities can form by dissolution of the cell walls (lysigenous cavities, such as in *Citrus*) or by pushing apart the cell walls and forming a cavity in the intercellular space (schizogenous cavities, such as resin canals in *Pinus*).

Of high practical importance is the secretion of resin by the resin canals of the wood of species of Pinaceae. In wood of *Pinus* the resin canals extend vertically as well as radially and are interconnected in the radial planes of the tree stem, with little or no tangential connections. The resin canals are surrounded by a layer of epithelial cells where the resin is synthesized (22,23). Besides *Pinus* resin canals are also found in the wood of *Pseudotsuga, Picea, Larix*, and *Keteleeria*; in the genera *Abies, Tsuga, Cedrus*, and *Pseudolarix*, resin canals are lacking (1). Resin canals can also form as a response to wood injury (traumatic resin canals) in practically all species of Pinaceae (1,24,25). Wood of the species of Cupressaceae and Taxodiaceae lacks resin canals, but they or similar secretory structures are often found in other organs of conifers such as bark, leaves, fruit, and seeds and occur occasionally in various hardwoods (4).

The resin canals of wood of *Pinus* are filled with resin composed mainly of terpenoid substances with some fatty materials. If the tree is wounded, resin exudes through the created openings, being forced out by the pressure produced by the epithelial cells that expand by absorbing water. The oleoresin exudation pressure (OEP) can reach rather high values such as 10 MPa or more and depends on many factors, including the season, time of day, and health of the tree (12,24,26,27). This resin flow includes the resin contained initially in the resin canals, as well as resin formed in the interim by the epithelial cells in response to the wound. Not much direct information is available on the composition of the resin contained in the resin canals prior to wounding, and it is generally assumed that it is the same as that produced in response to the wound. The work of Back (28,29) suggests, however, that native resin might contain an appreciably higher proportion of fatty materials.

After some time, the resin flow stops, partly by resin solidification and partly by enlargement of the epithelial cells, but resumes if the resin canals are reopened (24–26). The time of resin flow can be appreciably extended by the application of stimulants, such as sulfuric acid. This forms the basis of modern naval stores technology, which yields such products as gum rosin and gum turpentine and the various chemicals derived from them (12).

In the course of wood analysis, one meets the canal extractives either as an exudate or an extract obtained by extraction of wood using less polar organic solvents. In the latter case, the canal extractives are removed together with the fatty reserve materials contained in the ray parenchyma, as well as with some less polar heartwood extractives if heartwood is included in the material being extracted. As a result, the composition of the extract differs appreciably from the material obtained as an exudate. The difference manifests itself mainly in the increased amount of nonpolar reserve materials (fatty acids and esters) in the extract and is the reason why commercial "wood rosin," obtained by the extraction of old pine stumps, and tall oil, obtained as a byproduct of sulfate pulping, include increased amounts of fatty materials.

Not much information is available on the influence of the sapwood-to-heartwood transformation on the composition of the canal extractives. Hemingway and Hillis reported a large increase in the resin content of *Pinus radiata* wood going from the outer sapwood to the inner heartwood (21), while the composition of the resin acid fraction of the extract remained practically unchanged. It is likely that this increase is connected with the increasing number of the resin canals toward the pith (23,30). At the same time, the amount of fatty materials (reserve substances) strongly decreased in the same direction, most likely due to metabolism; the composition of the fatty materials also differed between sapwood and heartwood.

Work on the site of the biosynthesis of the canal resin is being currently conducted in several laboratories. The results demonstrate that resin acids, sesquiterpenoids, and monoterpenoids are synthesized at different sites. A remarkable achievement in this area was the histological separation of the elements responsible for the production of the monoterpenoids and sesquiterpenoids and the demonstration of the biosynthesis of these terpenoids in vitro (31,32) by Bernard-Dagan et al. using needles of *Pinus pinaster*.

C. Reserve Materials

Some materials obtained by extraction owe their presence to participation in the living processes taking place in the sapwood. These materials include substances such as fatty materials, starch and oligosaccharides, proteins, and various biological intermediates. The fatty materials are present mainly in ray parenchyma. As mentioned, their amount strongly decreases during heartwood formation (21), as well as during resin production (33). In the same way, oligosaccharides appear in appreciably higher amounts in the living portions of a tree and both decrease in amount and change composition as the tissues die (20). Starch is present exclusively in sapwood and living bark and is absent in heartwood and the outer, dead bark; its amount also decreases during resin production (34).

IV. CHEMICAL STRUCTURE AND OCCURRENCE OF EXTRACTIVE COMPONENTS

The discussion in this section of the chemical structures of the various compounds occurring in wood extracts follows the scheme outlined in Fig. 2. According to this scheme, the extractive components are separated into mevalonogenins or terpenoids formed through the mevalonic acid intermediate; acetogenins forming by linear polymerization of acetate units; saccharogenins including carbohydrates, cyclitols, and their derivatives as well as aromatics forming from carbohydrate-type precursors*; compounds of multiple biosynthetic origin; and alkaloids. Each class is then subdivided further according to biosynthetic or structural similarity. In each case, characteristic examples are given for the occurrence of the most important compounds or classes of compounds.

A. Mevalonogenins or Terpenoids

Terpenoids can be chemically defined as the group of compounds whose carbon skeleta can be dissected into "isoprene units" (carbon skeleton of isoprene or 2-methyl-1,3–butadiene) linked head to tail (Fig. 3). Some authors use the name terpenes for this class of compounds. In organic chemistry, however, the suffix "ene" denotes an unsaturated hydrocarbon, so the name terpene should be reserved for unsaturated terpene hydrocarbons. It should be stressed that the terpenoid group includes a number of compounds that do not strictly adhere to the above definition, but are classified as terpenoids on the basis of their predominant structural characteristics and biosynthetic considerations. These exceptions include compounds such as abietic acid, triterpenoids, tetraterpenoids, and particularly steroids; the latter bear a very close structural similarity to tetracyclic triterpenoids, differing from them by the absence of several methyl groups.

The terpenoids are further subdivided according to the number of isoprene units forming their carbon skeleton or by the number of biosynthetic C_5 intermediates making up the terpenoid skeleton into hemiterpenoids (one isoprene unit),

---

*Including sugars, cyclitols, and shikimic-acid-derived materials into one category might appear a bit forced. For one, sugars represent the materials leading not only to shikimic acid but also to acetogenins and mevalonogenins. Still, the biosynthetic link between sugars and the shikimic-acid-derived materials appears to be appreciably stronger. An alternative classification including two separate categories—saccharogenins sensu stricto (sugars and closely related aliphatics) and shikimigenins—represents another possibility.

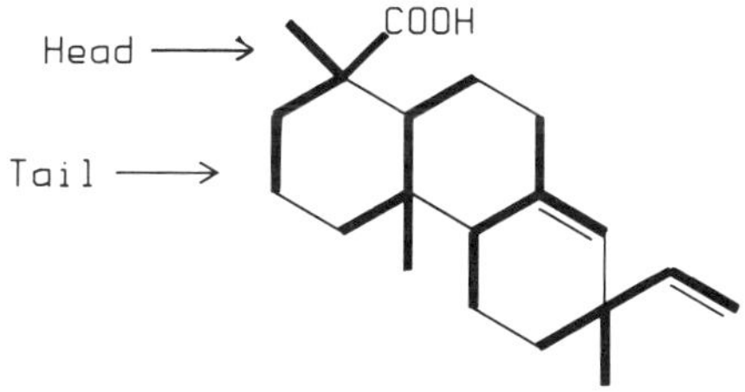

Figure 3   Dissection of the carbon skeleton of pimaric acid (a diterpenoid) into isoprene units linked head to tail.

monoterpenoids (two isoprene units), sesquiterpenoids (three isoprene units), diterpenoids (four isoprene units), sesterterpenoids (five isoprene units), triterpenoids (six isoprene units), tetraterpenoids (eight isoprene units), and polyterpenoids with more than eight isoprene units (Fig. 4). Terpenoids with seven or nine isoprene units are rare (components of acyclic terpenoid alcohols, "prenols").

Of the above classes, the monoterpenoids and diterpenoids are of major importance in wood chemistry. Sesquiterpenoids and triterpenoids are of secondary importance, hemiterpenoids and tetraterpenoids (carotenoids) are of marginal importance, and sesterterpenoids are of no importance. Although terpenoid polymers such as rubber are of great industrial importance (exudates of *Hevea* species such as *Hevea braziliensis*), the discussion of this subject belongs in treatises on organic polymers and will be omitted from this chapter.

Biosynthesis of terpenoids begins with two molecules of acetyl coenzyme A that react to give acetoacetyl coenzyme A. The latter can react with a third molecule of acetyl coenzyme A at C_3 to produce mevalonic acid (MVA). Several additional transformations of MVA lead to isopropenyl pyrophosphate (IPP) and dimethylallyl pyrophosphate (DMAPP), the two immediate precursors of terpenoids. Since the carbon skeleta of IPP and DMAPP are identical with that of isoprene, this provides the biological basis for the isoprene rule. The oligomerization of IPP and DMAPP leads to the open-chain terpenoid pyrophosphates, differing in the number of C_5 isoprene units, each type representing a parental intermediate for a particular terpenoid class. The terpenoid final products occurring as secondary metabolites in wood are ultimately formed from the parental intermediates by a variety of biological transformations, analogous to in vitro carbonium ion transformations, oxidations, and dehydrations. The biosynthesis of terpenoids is sketched in Figs. 5 and 6 and provides a natural basis for the chemical definition and classification of terpenoids. The biosynthesis of terpenoids has been reviewed many times (35–37).

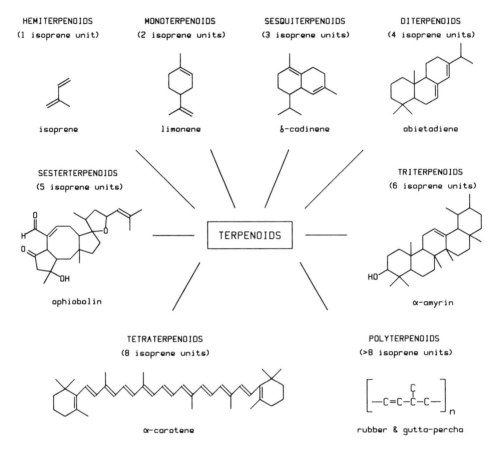

Figure 4   General classification of terpenoids (mevalonogenins).

1. Monoterpenoids

Monoterpenoids can be defined chemically as compounds constructed from two isoprene units attached head to tail, or biochemically as compounds formed from two IPP/DMAPP units. Depending on the number of rings within a structure, the monoterpenoids are separated into acyclic, monocyclic, bicyclic, and the rare tricyclic monoterpenoids. The bicyclic monoterpenoids are further subdivided into carane, pinane, thujane, and camphane subgroups on the basis of their carbon skeleta (Fig. 7). Monoterpenoids can be aliphatic, alicyclic, or aromatic hydrocarbons or can include various oxygen functional groups in their structure. In some

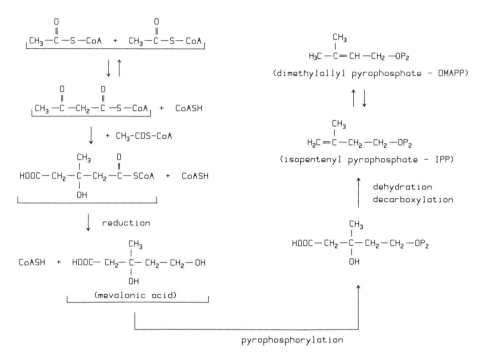

Figure 5 Biosynthesis of dimethylallyl (DMAPP) and isopentenyl pyrophosphate (IPP) from acetyl coenzyme A.

cases, subsequent rearrangement and degradation reactions can lead to compounds with different carbon skeleta and/or compounds containing less than 10 carbon atoms, both resulting in structures not fully obeying the isoprene rule.

Geranyl (GPP) and neryl pyrophosphate (NPP) (and according to some, also linalyl pyrophosphate-LPP) are the immediate biosynthetic precursors of monoterpenoids. The difference between GPP and NPP resides in its stereochemistry at the C_2 double bond. It has been assumed that only NPP is capable of forming cyclic compounds due to the *cis* arrangement of the substituents at the C_2 double bond. This assumption has been challenged recently (11). Be that as it may, the structures of a large portion of natural monoterpenoids can be explained by assuming the formation of the nonclassical C_1 carbonium ion from GPP, NPP, or LPP by hydrolysis of the pyrophosphate group. This carbonium ion can generate other carbonium ions by intramolecular double-bond attack, hydride shifts, or attack of the saturated carbon atoms α to double bonds. Each carbonium ion can, in turn, stabilize itself by splitting an α proton under formation of a double bond, or by the addition of water or acetate under formation of an alcohol or ester. The

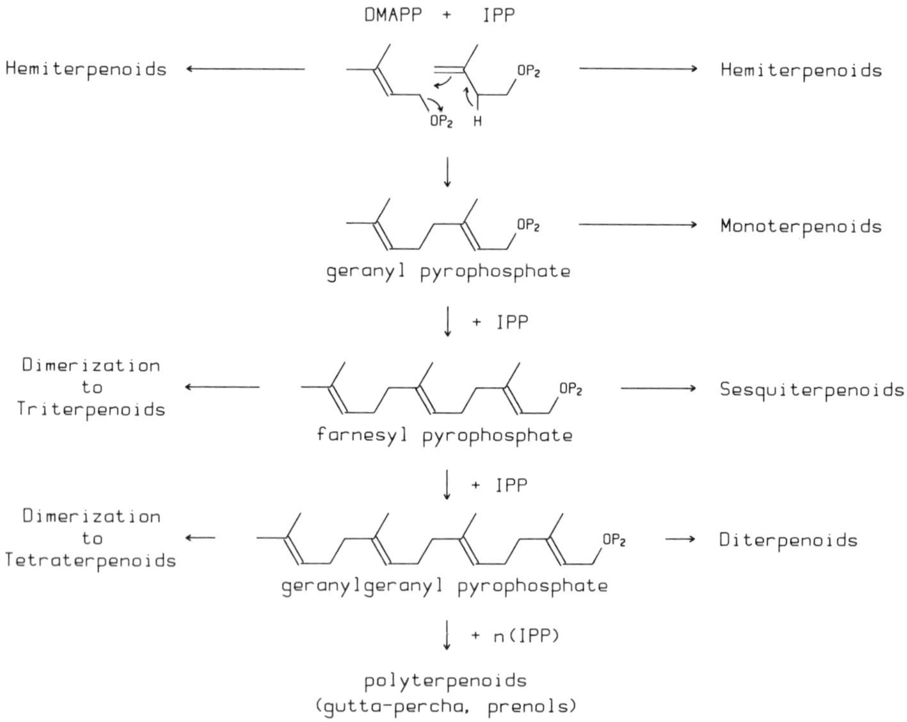

Figure 6   Polymerization of DMAPP and IPP to the immediate precursors of the various classes of terpenoids.

"stabilized" compounds can further change by oxidation, reduction, or aromatization to other monoterpenoid structures. An important step represents the cyclization under formation of the monocyclic 1-p-menthene-8-carbonium ion; the latter can exist in two enantiomeric forms and through additional transformation can give rise to the D and L series of monoterpenoids. The above biosynthetic transformations are sketched in Fig. 8 and can be used for classifying monoterpenoids according to biosynthetic affinity. This classification is particularly useful in biological problems involving monoterpenoids.

Monoterpenoid hydrocarbons represent one of the most important constituents of the canal extractives and exudates of the wood of conifers, where they are found together with diterpenoid resin acids and some fatty acids and their glycerides, as well as smaller quantities of related aldehydes, alcohols, and hydrocarbons. Aliphatic and alicyclic monoterpenoid hydrocarbons comprise the bulk of the canal monoterpenoids. (-)α-Pinene or (+)α-pinene and (-)β-pinene are the main

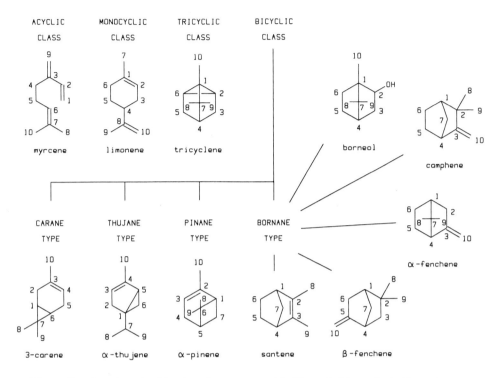

Figure 7   Chemical classification of monoterpenoids. Here, the biosynthetically equivalent carbon atoms often carry different numbers, e.g., C-3 in myrcene , C-1 in limonene, C-4 in thujene, C-2 in a-pinene. The bornane group includes compounds of somewhat different carbon skeleta. (From Ref. 313.)

monoterpenoids. (+)3-Carene, (-)limonene, and (-)β-phellandrene are occasionally found in large amounts. (-)Sabinene as well as the biosynthetically related terpinolene, thujene, α- and γ-terpinene are rarely found as major constituents. Myrcene is generally found as a trace to secondary constituent. Fenchenes, santene, α-phellandrene, and *cis*-and *trans*-ocimenes are only trace constituents (47). Camphene is ubiquitous in small amounts in *Pinus* wood, but has been occasionally found as one of the main constituents of the cortical canal resin of *Abies* (310,311). Aromatic *p*-cymene is relatively ubiquitous in the canal resin; it is present generally in small amounts, although it is one of the main constituents of the monoterpenoids obtained by extraction of the wood of *Pinus discolor* (38) (Figs. 7 and 8). Oxygenated monoterpenoids are found in resin canals in small amounts only; their structures derive from those of monoterpenoid hydrocarbons or their immediate precursors by the introduction of hydroxy, acetoxy, or carbonyl groups without

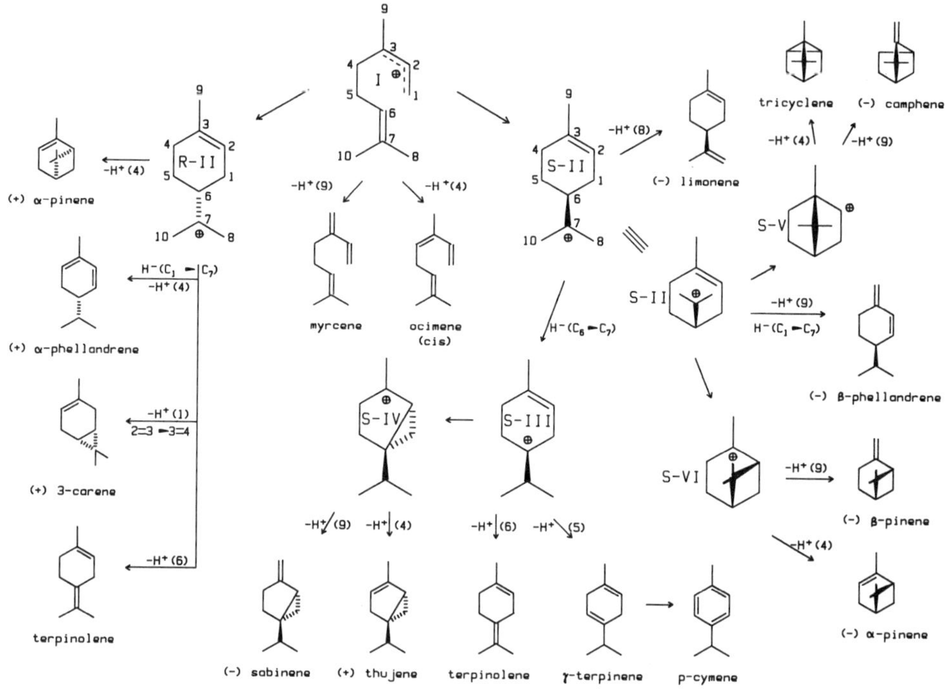

Figure 8 Formation of monoterpenoid hydrocarbons from geranyl/neryl carbonium ion. For clarity's sake, we allowed the biosynthetically equivalent carbon atoms to retain their numeration throughout the various transformations.

major skeletal degradations.

Apart from their occurrence in the resin canals, monoterpenoids as constituents of the heartwood extractives of the Pinaceae are apparently rare, as only a few have been reported such as andirolactone (Fig. 11), an irregular monoterpenoid from the wood of *Cedrus libani* that carries an "extra" lactonized carboxyl (39). At the same time, the Cupressaceae species, which lack resin canals, tend to produce larger amounts of monoterpenoids as heartwood deposits. Such monoterpenoids generally include relatively small amounts of hydrocarbons (α-pinene, limonene, camphene, α- and γ-terpinene, terpinolene) and are mainly composed of aromatized and oxygenated materials, occasionally with less than 10 carbon atoms (40), including monoterpenoid phenols such as chamenol, *p*-cresol, carvacrol, *p*-methoxycarvacrol, *p*-methoxythymol, thymoquinone, and *o*-methylcarvacrol; and monoterpenoid carboxylic acids such as thujic acid, shonanic acid, citronellic acid, dehydrogeranic acid, chamic acid, chaminic acid, and derivatives of *p*-isopropylbenzoic acid (Fig.

9). The aromatic tropolones, with a seven-member ring, are also commonly present and include α-, β-, and γ-thujaplicins, thujaplicinol and dolabrinol isomers and their methylation products, dimeric utahine and the related but nontropolonic nezukone (Fig. 10). Phenolic monoterpenoids can undergo the oxidative coupling reaction and appear as dimers such as 6-*p*-methoxythymoxy-*p*-methoxythymol (libocedrol), 6-*p*-methoxycarvacroxy-*p*-methoxythymol (heyderiol), trimers (3-libocedroxythymoquinone), or phlobaphenic polymers (41) (Fig. 9). Occasionally, oxygenated monoterpenoid alcohols, esters, ethers, and ketones are found (cineole, bornyl acetate, camphor, α-fenchyl alcohol, borneol, α-terpineol, citronellol, citronellyl acetate) (Fig. 11). Monoterpenoids are usually absent from the wood of Taxodiaceae with the exception of the *Cunninghamia* species where α- and β-pinene, camphene, β-phellandrene, limonene, sabinene, 1,8–cineole, α-terpineol, and borneol were found (42).

The monoterpenes are rare in the wood of other conifers and the more common hardwood species. It has been reported that the wood of *Phebalium nudum* (hardwood) produces the monoterpenoid citral (=mixture of neral and geranial) and that the wood of *Santalum album* produces a small amount of santene, a C9 monoterpene (43) (Fig. 7).

2. Sesquiterpenoids

Sesquiterpenoids are chemically defined as compounds constructed of three isoprene units linked head to tail. Biosynthetically, the immediate precursor of

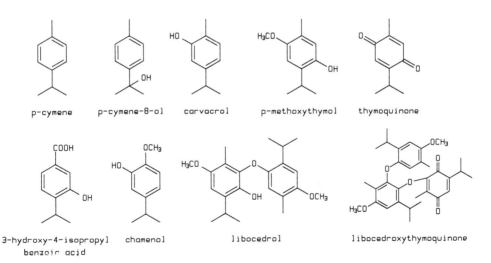

Figure 9 Aromatic, benzenoid monoterpenoids.

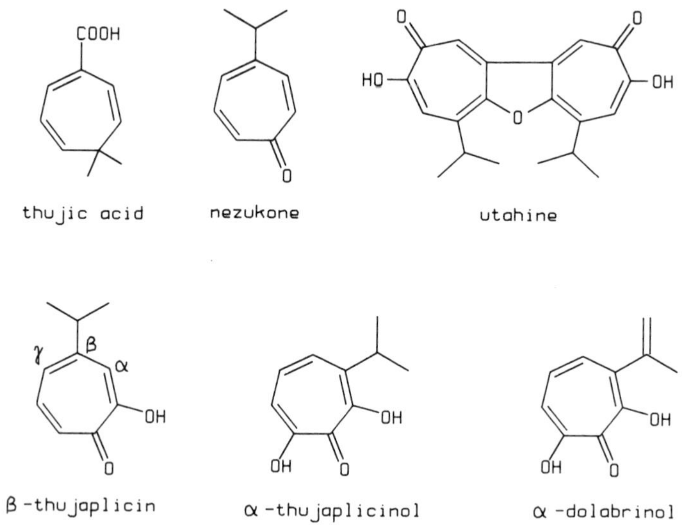

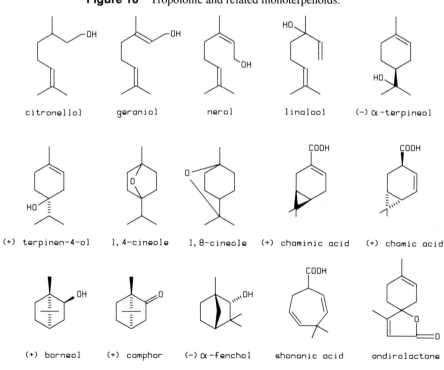

thujic acid   nezukone   utahine

β-thujaplicin   α-thujaplicinol   α-dolabrinol

Figure 10   Tropolonic and related monoterpenoids.

citronellol   geraniol   nerol   linalool   (−) α-terpineol

(+) terpinen-4-ol   1,4-cineole   1,8-cineole   (+) chaminic acid   (+) chamic acid

(+) borneol   (+) camphor   (−) α-fenchol   shonanic acid   andirolactone

Figure 11   Oxygenated aliphatic and alicyclic monoterpenoids

sesquiterpenoids is farnesyl pyrophosphate that forms from one DMAPP and two IPP units. Farnesyl pyrophosphate can exist in *trans-trans* and *trans-cis* isomeric forms corresponding to geranyl and neryl pyrophosphates of monoterpenoids. Both forms can cyclize, however (Fig. 12); and with the longer chain and presence of three instead of two double bonds, a greater diversity of structures results. The sesquiterpenoids are classified chemically according to the basic structures of their carbon skeleta and biosynthetically according to the manner of their formation. A complete classification of sesquiterpenoids would transcend the limits of the present review, however, and the reader is referred elsewhere (35–37,44–46).

In wood species, the sesquiterpenoids are found as components of canal resin as well as of the heartwood deposits of wood. In the gum oleoresins of *Pinus* (resin canal exudates from wood), the sesquiterpenoids commonly represent a minor portion of the volatile materials (gum turpentines). Here as in the case of monoterpenoids, the hydrocarbons predominate. In the case of hard pines (subgenus *Pinus*), the percent of sesquiterpenoids in turpentines is generally small, below 10%, whereas in soft pines (subgenus *Strobus*), the percentage is higher, occasionally about 20%. In *Pinus albicaulis* and *P. parviflora* gum turpentines, the sesquiterpenoids occur in quantities as high as 58 and 48%, respectively (47). The relative amount of sesquiterpenoids appears to be appreciably higher in turpentines obtained by the solvent extraction of wood, although their nature remains roughly the same. Thus, turpentine obtained by the extraction of the wood of *P. edulis* includes sesquiterpenoids and monoterpenoids in a 1.6:1.0 ratio, whereas gum turpentines from the same species exhibit a 0.06:1 ratio (48). Some important sesquiterpenoids of the canal extractives of *Pinus* are shown in Fig. 13.

The wood of genera *Abies*, *Tsuga*, *Pseudolarix*, and *Cedrus* of Pinaceae is devoid of resin canals, and in these cases, the sesquiterpenoids obtained by wood extraction should come from the heartwood deposits. Heartwood of the *Cedrus* species con-

trans-cis-farnesyl pyrophosphate
(8-dimethylallylneryl pyrophosphate)

trans-trans-farnesyl pyrophosphate
(8-dimethylallylgeranyl pyrophosphate)

Figure 12 Farnesyl pyrophosphate precursors of sesquiterpenoids.

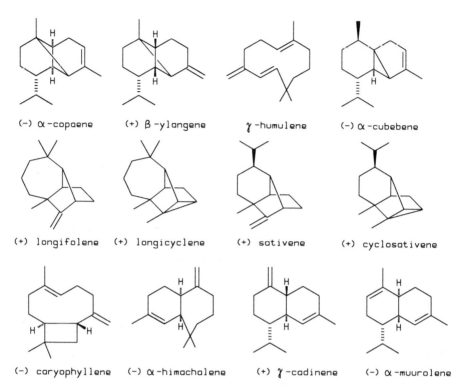

Figure 13 Important sesquiterpenoid hydrocarbons of the resin canals of *Pinus*.

tains relatively large amounts of volatile extractives (2.5–5.0%), composed mainly of sesquiterpenoids including atlantones, himachalol, allohimachalol, (+)-longiborneol, deodardione, deodarone, limonene carboxylic acid, todomatsuic acid derivatives (Δ-10-dehydrotodomatsuic acid, Δ-7-dehydrotodomatsuic acid, and 7-hydroxytodomatsuic acid), and himachalene hydrocarbons (49,50) (Figs. 13–15). Recently, the oxygenated himachalene derivatives, α-and β-torosol, *trans*-atlantone-6-ol, and several atlantone-derived carboxylic acids were isolated from the wood of *Cedrus libani* (39,51). The wood of *Abies* is appreciably poorer in volatile oils; it contains juvabione (todomatsuic acid methyl ester) and related materials (todomatsuic acid, dehydrojuvabione, and two stereoisomers of tetrahydrotodomatsuic acid), i.e., compounds similar to those from Cedrus (52). Little information is available on the sesquiterpenoids from *Tsuga* and *Pseudolarix*.

The wood of the other genera of Pinaceae includes resin canal systems that, however, are less developed than in *Pinus*. The sesquiterpenoids of the respective gum oleoresins, i.e., of the canal resins, do not basically differ from those of *Pinus*

(46,53–55). The sesquiterpenoids obtained by the solvent extraction of wood should, however, derive from both canal resins and heartwood deposits. This is possibly the reason that the wood extract of *Pseudotsuga menziesii* contains several juvabionelike materials that are characteristic of the heartwood extratives of the resin canal free wood of *Abies* and *Cedrus* (56).

Wood of several genera of Taxodiaceae produces essential oils containing sesquiterpenoids *(Metasequoia, Cryptomeria, Cunninghamia, Sciadopytis,* and *Taiwania);* identified were eudesmol, cedrenes, cedrol, cadinenes, cadinols, muurolols, β-humulene, caryophyllene, and calamenene (42,57,58) (Figs. 13,14). "Cadinene" was reported in the wood of *Dacrydium franklinii* (59). Oxygenated sesquiterpenoids dendrolasin and (+)nuciferol and their derivatives have been identified in the wood of *Torreya nucifera* (Taxaceae) (60) (Fig. 16).

The heartwood of the Cupressaceae species (resin canals absent) follows the trend mentioned under monoterpenoids and includes aromatic and oxygenated sesquiterpenoids in preference to hydrocarbons. Thus, the sesquiterpenoid tropolone

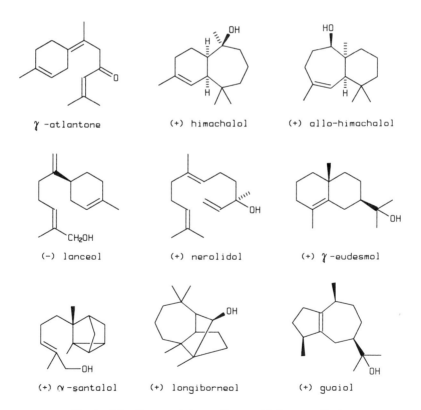

γ-atlantone   (+) himachalol   (+) allo-himachalol

(−) lanceol   (+) nerolidol   (+) γ-eudesmol

(+) α-santalol   (+) longiborneol   (+) guaiol

Figure 14 Oxygenated sesquiterpenoids from conifers.

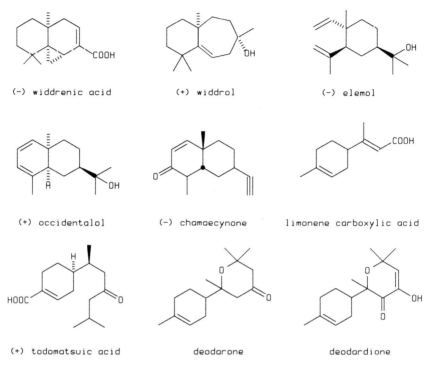

Figure 15   Oxygenated sesquiterpenoids from conifers.

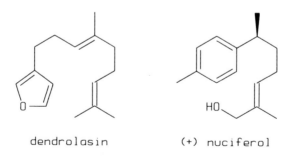

Figure 16   Sesquiterpenoids from the wood of *Torreya*.

nootkatin and related compounds are common; additional constituents include
aromatic cuparene and cuparenic acid and oxygenated sesquiterpenoids cedrol,

guaiol, widdrenic (=hinokiic) acid, various cadinols, eudesmol, widdrol, occidentalol, and others, including the interesting acetylenic ketone, chamaecynone (Figs. 15–17). The most common sesquiterpenoid hydrocarbons include cedrene and its isomers, thujopsene, nootkatene, and α-alaskene (1,39) (Fig. 18). The sesquiterpenoid hydrocarbons δ-cadinene and *trans*-calamenene and oxygenated sesquiterpenoids including 15-copaenol, cubenol, *epi*-cubenol, torreyol (= pilgerol, albicaulol, or δ-cadinol), caryophyllene-4,5-epoxide, and humulene-1,2-epoxide were recently isolated from the heartwood of *Pilgerodendron univerum* (68). The wood of *Ginkgo biloba* produces about 5% of essential oil containing sesquiterpenoid elemol, γ-eudesmol, and atlantone derivatives (61) (Figs. 14 and 15).

Sesquiterpenoids are also rare, albeit a bit more common than monoterpenoids, in the hardwoods. They represent the main constituents of the essential oil from sandalwood (Santalaceae and Myoporaceae) and include the open chain nerolidol and farnesol and cyclic lanceol and α- and β-santalols (62,63) (Fig. 14). Oxygenated, aromatic sesquiterpenoids related to cadalene and calamenene have been found in the heartwood of the *Ulmus* species (64–66) (Fig. 17). A variety of sesquiterpenoids has been found in the wood of *Copaifera* and related species (family Leguminosae) (67), including α-bergamotene, β-bisabolene, cadinenes, muurolenes, copaenes, cubebenes, elemenes, calamenene, caryophyllene, and selinenes (Figs. 13 and 18).

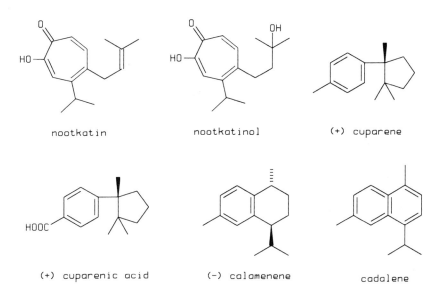

nootkatin   nootkatinol   (+) cuparene

(+) cuparenic acid   (–) calamenene   cadalene

Figure 17 Aromatic sesquiterpenoids characteristic of Cupressaceae and *Ulmus*

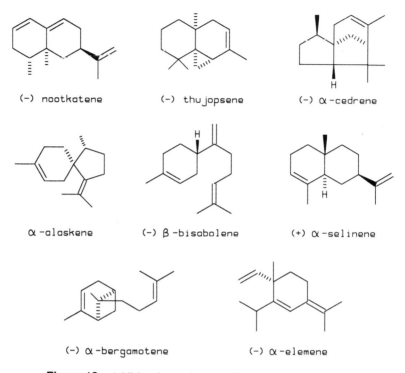

(−) nootkatene   (−) thujopsene   (−) α-cedrene

α-alaskene   (−) β-bisabolene   (+) α-selinene

(−) α-bergamotene   (−) α-elemene

Figure 18   Additional sesquiterpenoid hydrocarbon structures.

3.   Diterpenoids

Diterpenoids can be defined chemically as compounds whose carbon skeleta are composed of four isoprene units attached head to tail and are classified as acyclic, mono-, bi-, tri-, and tetracyclic, depending on the number of medium size-rings in a structure, and as macrocyclic in the case of large rings (10–15 carbons). Biosynthesis of the diterpenoids begins with geranylgeranyl pyrophosphate. In the case of the bicylic labdane and tricyclic pimarane and abietane diterpenoids, which are the most important in wood chemistry, the biosynthesis starts with the electrophilic attack on the terminal double bond of geranylgeranyl pyrophosphate, followed by a concerted closure of two rings and formation of a cationic pyrophosphate. A variety of reactions without additional cyclizations leads to the labdane-type diterpenoids. Hydrolysis of the pyrophosphate group of the labdane pyrophosphate leads to the carbonium ion that forms the tricyclic ion by ring closure between carbons 13 and 17 and finally stabilizes to one of the pimarane structures. The abietane-type diterpenoids form from pimarane structures by electrophilic attack on the 13,14 double bond, followed by methyl migration from C_{13} to C_{14} and final stabilization of the resulting carbonium ion (35,36) (Fig. 19).

Tricyclic diterpenoids, particularly the diterpenoid acids ("resin acids"), are the main components of the canal extractives of the wood of conifers, especially of *Pinus* , amounting to roughly 70% of the total in the latter. The abietane structures predominate; pimarane structures are also present in sizable quantities (Fig. 20). The bicyclic labdane structures (Fig. 21) appear only occasionally in amounts above 10% of the total diterpenoids and include (+)elliotinic acid from *P. elliottii* and (+)lambertianic acid from *P. lambertiana*. In addition to the monocarboxylic resin acids, the corresponding aldehydes, alcohols, and hydrocarbons are also commonly found in smaller quantities (69,70). Macrocyclic diterpenoids, such as cembrene and cembrol from the canal resin of *P. albicaulis* and *P. sibirica*, are occasionally found in the form of hydrocarbons and alcohols in relatively small amounts (47). Tetracyclic diterpenoids such as (+)phyllocladene and related compounds from certain *Picea* species are generally rare (71) (Fig. 21).

Diterpenoids also occur as components of the heartwood extractives of conifers. As in the case of mono- and sesquiterpenoids, one often encounters here diterpenoids that have been modified by aromatization, hydroxylation, oxidation, rearrangement,

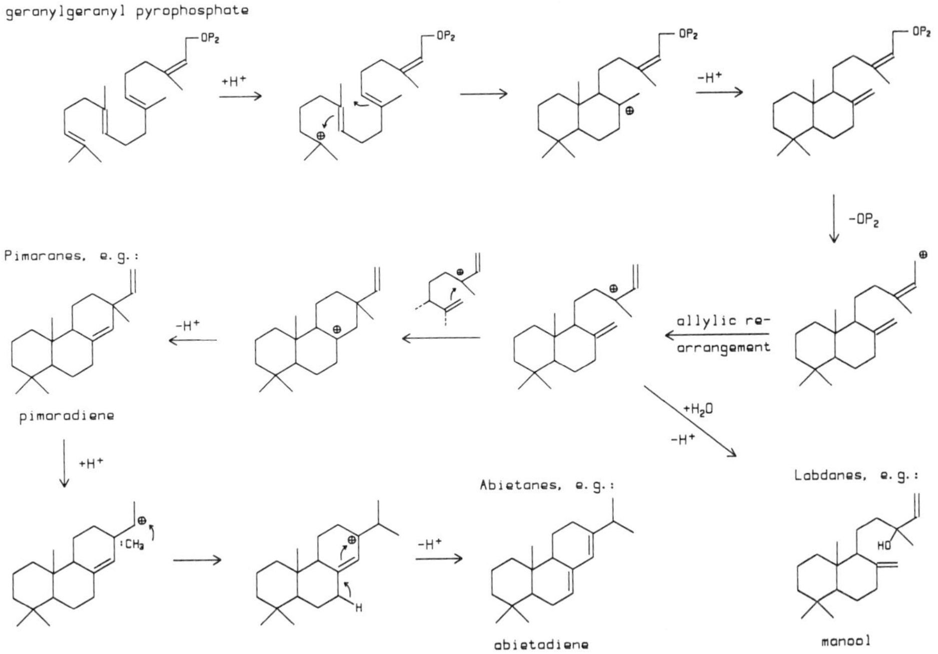

Figure 19 Biosynthesis of diterpenoids from geranylgeranyl pyrophosphate.

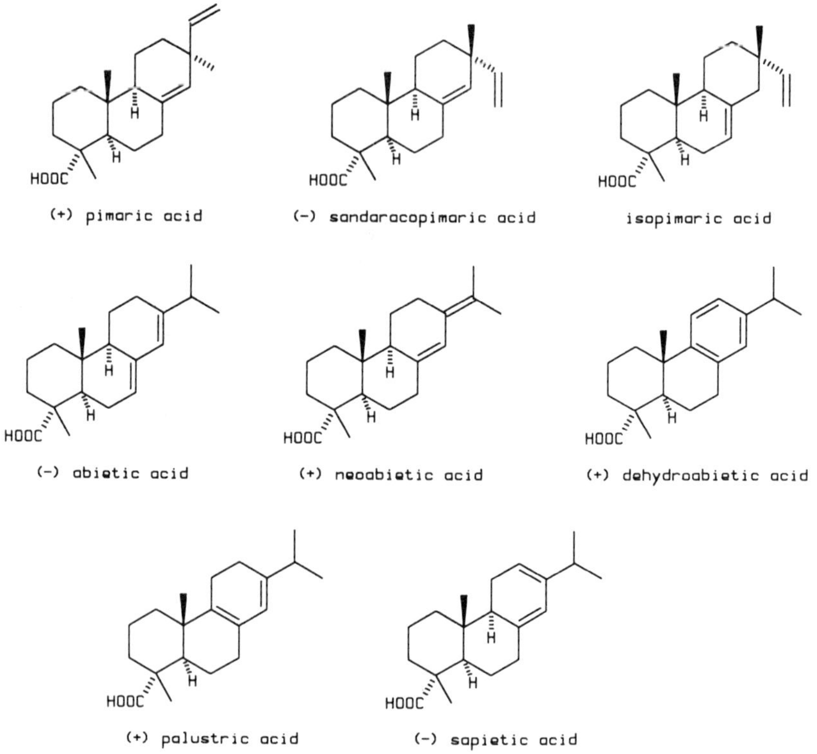

Figure 20 Most important diterpenoid acids (resin acids) from the resin canals of Pinaceae. Sapietic acid is more commonly known as levopimaric acid. The former name has an advantage, however, as it avoids confusion with pimarane structures. (From Ref. 314)

and loss of carbon atoms. The compounds include a great variety of di-, tri-, and tetracyclic diterpenoids comprising alcohols, aldehydes, ketones, carboxylic acids, and particularly phenolics. In Cupressaceae, labdanic structures are quite common, although other types of diterpenoids also occur. Thus, the labdanic alcohols manool, torulosol, and agathadiol; the aldehydes torulosal and agatholal; and *cis*-communic acid were isolated from the heartwood of several Cupressaceae species. These were accompanied by sandaracopimaric acid, by the abietanic phenols sugiol, ferruginol, hinokiol, hinokione and quasi-abietanic (isopropyl group at C-14, instead of at C-13) phenols sempervirol, totarol, totarolone, and totarolenone (72–76). In the

cembrene   cembrol   (+) manool

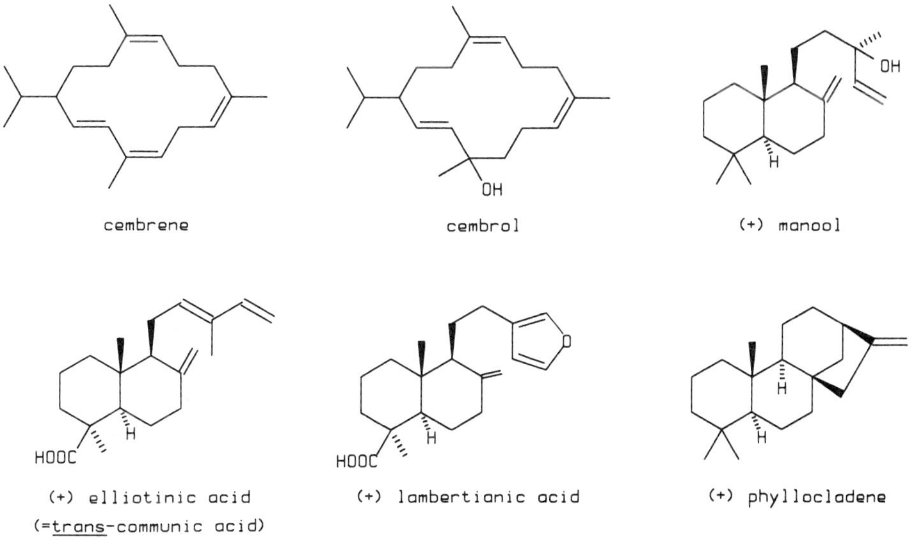

(+) elliotinic acid   (+) lambertianic acid   (+) phyllocladene
(=trans-communic acid)

Figure 21   Some of the more important diterpenoid cembrane, phyllocladane, and labdane structures.

heartwood of Taxodiaceae was found isopimaric acid, together with sugiol and ferruginol as well as the tetracyclic alcohol phyllocladanol (77) (Fig. 22). Podocarpaceae heartwood extractives include particularly interesting diterpenoids. Besides isopimaric acid and the labdanic alcohol manool and its derivatives, the wood also includes a variety of phenolic diterpenoids such as totarol, sugiol, and ferruginol as well as their dimers formed by oxidative coupling [e.g., (+)podototarin]. In addition, a series of much degraded diterpenoid lactones (podocarpuslactones) including compounds carrying the unusual sulfoxide and sulfone groups has been identified (78–82) (Fig. 23). A number of very interesting, strongly oxygenated taxan-type diterpenoids, including structures containing eight-member rings, occur in the wood of the *Taxus* species (83,84). Some of these structures include substituted amino or amide groups (taxan alkaloids) (Fig. 24).

Diterpenoids are apparently absent or rare in the wood of common hardwoods. The heartwood of the African sandalwood, *Spirostachys africana*, contains 12–14% of resin, composed mainly of tetracyclic diterpenoids, ketonic stachenone and its derivatives (85). Wood of certain species of the subfamilies Caesalpinoideae, Mimosoideae, and Lotoideae of the family Leguminosae produces diterpenoids. Over 50 different structures have been identified so far with labdane structures predominating (67) (Fig. 25).

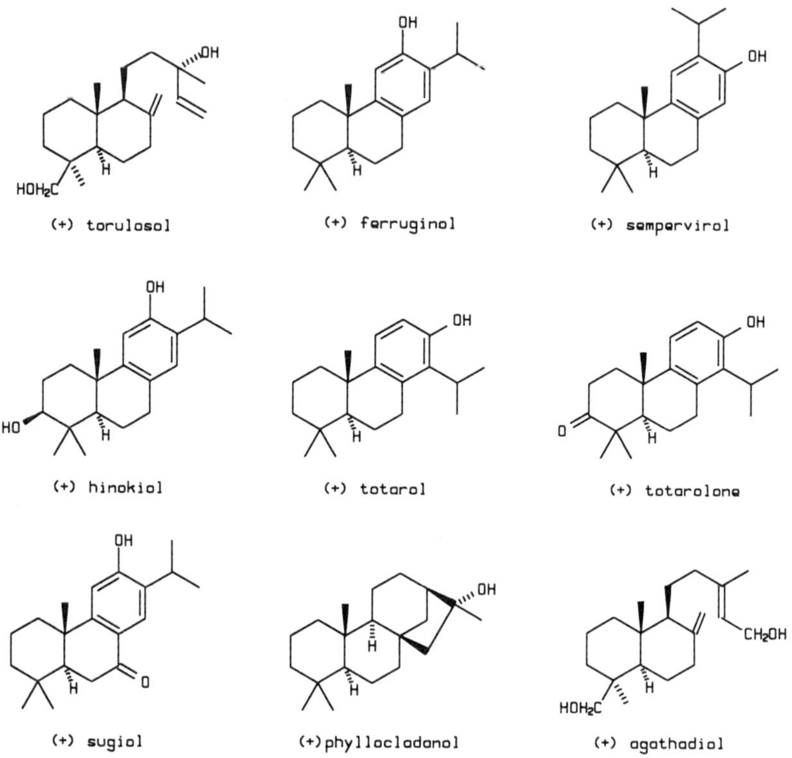

Figure 22   Diterpenoids from Cupressaceae and tetracyclic diterpenoids.

4.   Triterpenoids

Carbon skeleta of triterpenoids are composed of six isoprene units linked head to tail, with the exception of a central tail-to-tail linkage. The biosynthesis of triterpenoids begins with the joining of two units of farnesyl pyrophosphate with loss of the pyrophosphate groups; the biochemistry of this step is not completely clear. The resulting hydrocarbon, squalene (Fig. 26), cyclizes in a concerted fashion, leading to several carbon structures, of which those including four and five rings are most common. Postcyclization modifications lead to the multitude of known structures. In earlier times, and even occasionally now, the triterpenoids were treated as two classes of compounds: triterpenoids *sensu stricto* and steroids.

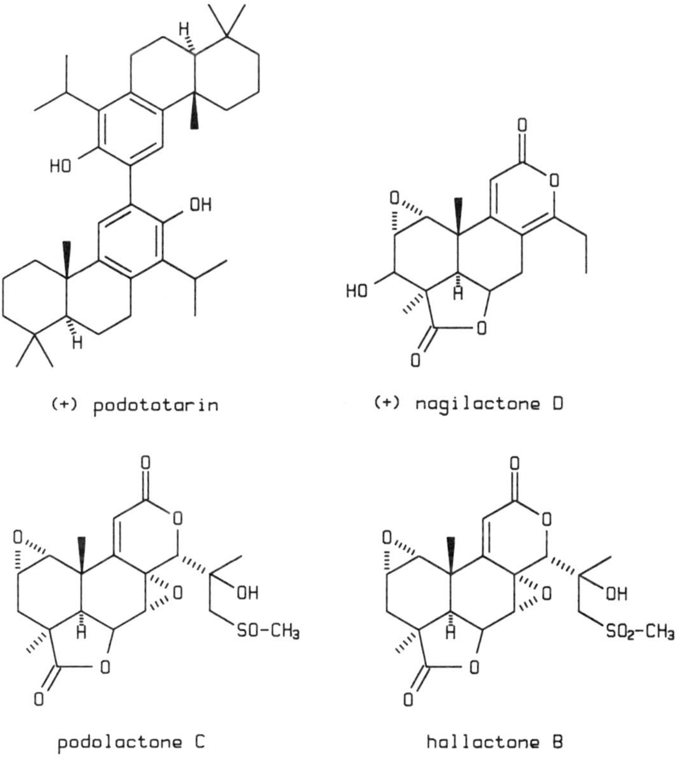

Figure 23   Diterpenoids of Podocarpaceae.

From a biosynthetic point of view, there is little reason for this distinction, as the steroids differ from some of the tetracyclic terpenoids only by the postcyclization loss of several methyl groups (35,36).

Triterpenoids are common in the wood of softwoods, although they are rarely found in large amounts. The most common compound is β-sitosterol (69); many other structures have been also identified, generally in smaller amounts, occasionally esterified with fatty or other acids. Thus, the sapwood of *Pseudotsuga menziesii* contains the steroids β-sitosterol, stigmastanol, campesterol, cycloartenol,

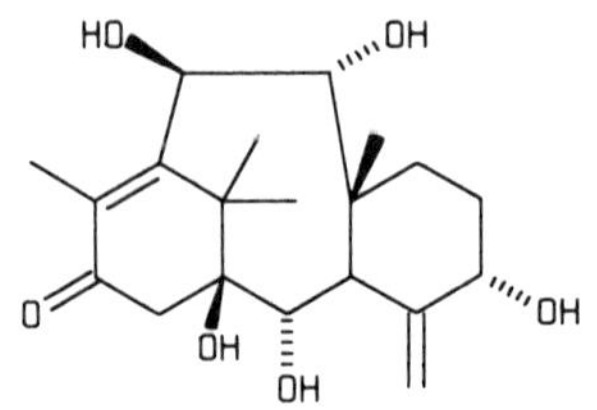

taxicin I

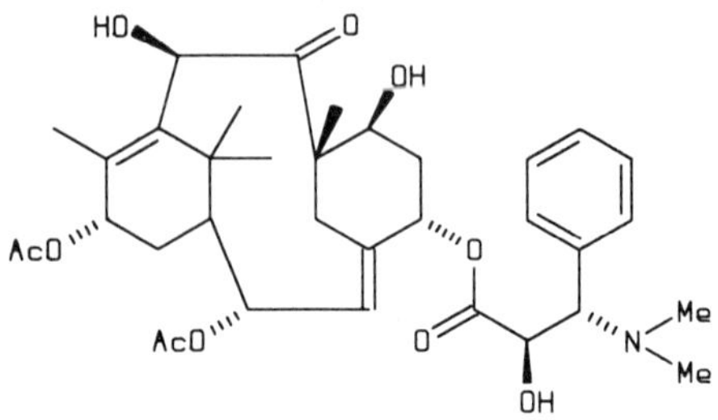

taxin A

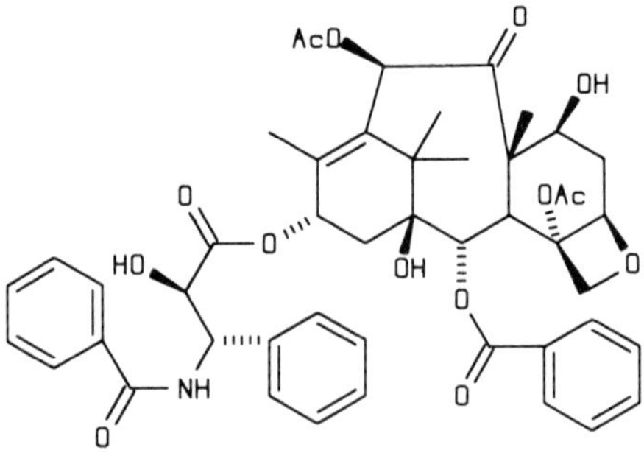

taxol A

Figure 24   Diterpenoids from Taxaceae.

cycloeucalenol, 24-*R*-cycloeucalenol, and 24-*R*-methyllophenol (86). β-Sitosterol, 24-methylenecycloartanol, and 24-methylenecycloartanone were found in the wood of *Picea obovata* (87). Seventeen steroids, with β-sitosterol as the major constituent, some stigmastanol and the rest in traces, and five trace triterpenoids *sensu stricto* (squalene and the rest of the serratane type) were identified in the neutral fraction of the tall oil from the wood of "southern pines" (pines from the southeastern United States) (70). The wood of *Picea sitchensis* was found to contain β-sitosterol and its derivative, 2-octyl-β-sitosteryl phthalate (88). β-Sitosterol was identified in the wood of several Podocarpaceae species. Steroid phytoecdysones appear to be also common in wood of this family (89). β-Sitosterol and other triterpenoids are also common in the wood of Taxaceae (1,90).

Triterpenoids are also ubiquitous in the wood of hardwoods, but in small amounts, β-sitosterol being again the most common. *Liquidambar styracyflua* wood contains β-sitosterol and stigmastanol (91). The pentacyclic triterpenoid saponins, acerotin, and related compounds, together with the ubiquitous β-sitosterol, are present in the wood of *Acer negundo* (92,93). The wood of the *Betula* species contains β-

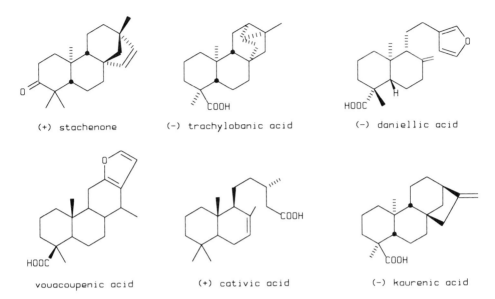

(+) stachenone   (-) trachylobanic acid   (-) daniellic acid

vouacoupenic acid   (+) cativic acid   (-) kaurenic acid

Figure 25   Some hardwood diterpenoids.

sitosterol and the lupane triterpenoids, with the latter also responsible for the white color of the birch bark. Triterpenoids of the same type have also been identified in the wood of the *Platanus* species (8). The wood of the *Salix* and *Populus* species contains triterpenoids (α- and β-amyrin, lupeol, and others) and steroids (mainly, β-sitosterol) (8). The steroids β-sitosterol and stigmasterol were identified in the wood of *Robinia pseudoacacia,* and β-sitosterol in the wood of the *Acacia modesta* and *Dalbergia* species. The pentacyclic triterpenoid oleanolic acid and its derivatives have been found in the wood of some species of the same family (Leguminosae) (67). Triterpenoid structures are summarized in Figs. 26 and 27.

5.   Tetra and Polyterpenoids

Information on tetraterpenoid carotenoids (eight isoprene units) in wood is practically nonexistent. Polymeric terpenoids with many isoprene units also appear to be rare, and when they occur, they generally do so in small amounts. Alcoholic polymeric terpenoids ("prenols") composed of 6–13 isoprene units and carrying *cis* and *trans* double bonds (betulaprenols) have been isolated from the wood of *Betula verrucosa* in 0.0275% yield. The wood of Pi*cea, Pinus, Populus,* or *Aesculus hippocastanum* yielded no prenols (29).

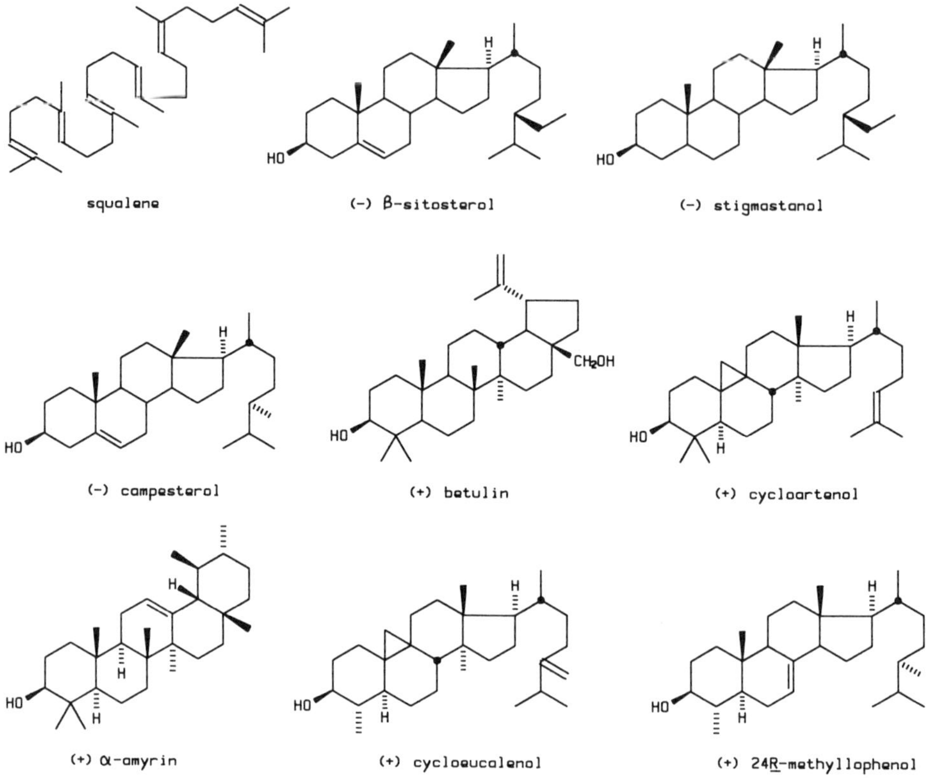

Figure 26   Some of the more important triterpenoids.

Exudates of some Euphorbiaceae species, such as *Hevea brasiliensis*, are industrially very important since they yield polyisoprenoid natural rubber and gutta-percha.

B. Acetogenins

Acetogenins comprise aliphatic and aromatic compounds. Aliphatic acetogenins include a variety of long straight-chain compounds generally designated as "fatty," as they constitute the main components of various oils and fats. Among these compounds, the acidic materials, fatty acids, are most important. The aromatic materials comprise certain phenolic compounds characteristically tending to carry oxygen functions on alternate carbons. In most common woods, the aromatic compounds derived entirely by the acetogenic pathway are relatively rare. Acetogenic aromatic structures are very important, however, as components of the molecular

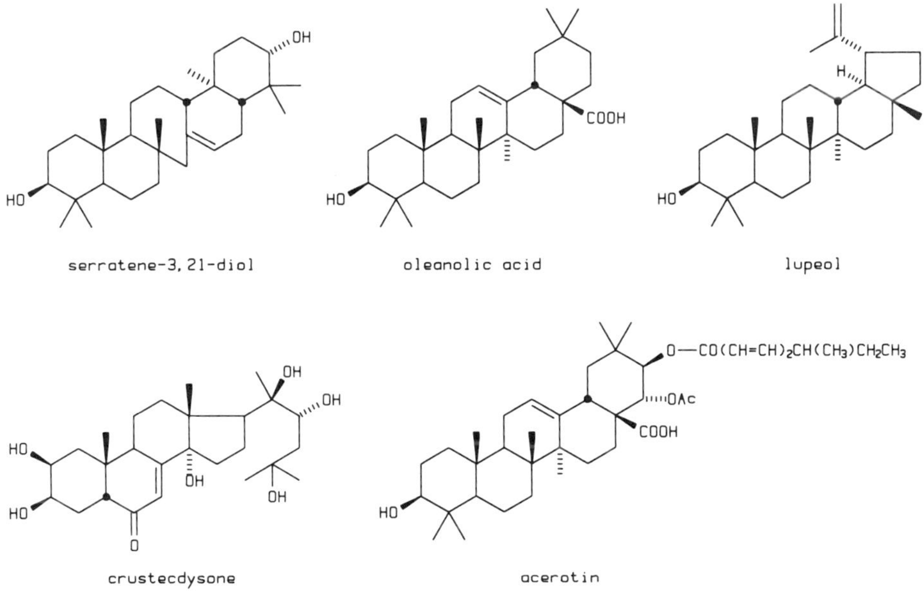
HO
OH
serratene-3, 21-diol
HO
COOH
oleanolic acid
HO
H
lupeol
HO
HO
OH
OH
OH
OH
O
crustecdysone
HO
O—CO(CH=CH)2CH(CH3)CH2CH3
OAc
COOH
acerotin

Figure 27 Some of the more important triterpenoids.

structures of flavonoids, stilbenes, and related polymers, such as tannins and phlobaphenes.

Biosynthetically, the formation of acetogenins involves the repeated Claisen-like condensation of acetylcoenzyme A units to long chains. In actuality, the acetyl coenzyme A unit being added is transformed by a reaction with carbon dioxide into a malonyl coenzyme A unit that then loses carbon dioxide during the condensation process. Thus, carbon dioxide acts as a kind of condensation catalyst.

In the case of aromatic acetogenins, the process results in the formation of long polyketide chains that cyclize intramolecularly by aldol or Claisen-like mechanisms forming aromatic rings after appropriate enolizations and dehydrations. In the case of aliphatic acetogenins, a reaction sequence involving the reduction of the β-keto group to methylene occurs after each addition of an acetyl unit. This results in the formation of aliphatic saturated long-chain acyl coenzyme A compounds and leads to saturated fatty acids after coenzyme A hydrolysis (35). The formation of unsaturated fatty acids generally follows the same sequence, although the mechanism of the introduction of double bonds into the chains is not yet well understood. The biosynthesis of acetogenins is summarized in Fig. 28.

1. Aliphatic Acetogenins

Aliphatic acetogenins are by far the most important components of waxes, as well as of fats and oils.

Waxes commonly occur in plants on the leaf surface, which reduces evaporation of water. They are also found in some barks, but are rare in wood. One of the difficulties in wax chemistry stems from the definition of a wax; the term tends to be applied to anything looking waxy, and since almost every relatively nonpolar aliphatic compound of sufficiently high molecular weight looks "waxy," wax can refer to many chemically unrelated things. Thus, mixtures of triterpenoids and steroids, occasionally esterified with fatty or other acids, and mixtures of saturated fatty acids and fatty alcohols, as well as long-chain hydrocarbons, have been referred to as waxes. The most common constituents of waxes, however, are long-chain (up to about C-50) saturated acids and alcohols, often esterified to each other.

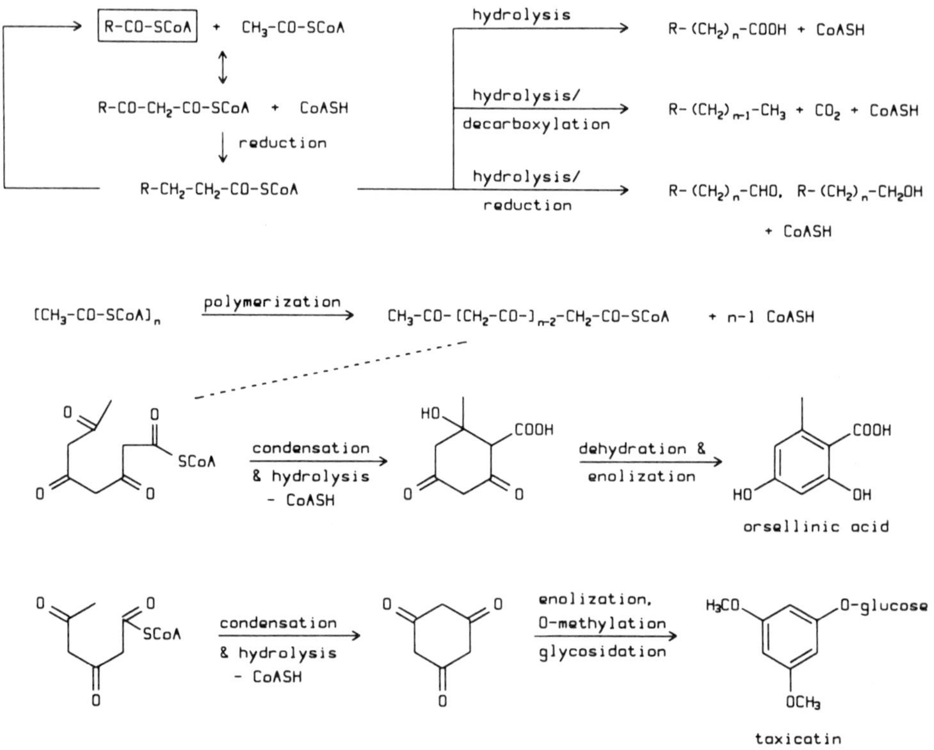

Figure 28 Biosynthesis of aliphatic and aromatic acetogenins. Chain lengthening of R—CO—SCoA (R=CH$_3$(CH$_2$CH$_2$)n for the nth reaction cycle).

Not uncommon are long-chain compounds carrying a carboxyl and hydroxyl at the ends, found in the form of polymeric esters (estolid waxes). Long-chain hydrocarbons are also common. Reports of the occurrence of waxes in wood, however, are very rare and deal with relatively small amounts of material.

Fats and oils, on the other hand, are very common in the wood of both hard- and softwoods. Commonly, the term "fat" is applied to materials that are solid or near-solid at ambient temperatures, whereas the corresponding liquid materials are designated as oils. Long-chain saturated and unsaturated monocarboxylic acids ("fatty acids") are the most important constituents of fats and oils. These acids are commonly esterfied with glycerol and thus are often found in forms of glycerides. Corresponding fatty alcohols, aldehydes, and hydrocarbons are occasionally present, but usually in much smaller amounts.

Fatty acids are predominantly compounds with even-numbered carbon atom chains, which is an obvious consequence of their biosynthesis. Fatty alcohols and aldehydes generally follow the same rule, being formed by the reduction of the corresponding fatty acids. Hydrocarbons, on the other hand, include predominantly odd-numbered compounds, since they are formed by the decarboxylation of fatty acids. The most common fatty acids include chains of 16 and 18, or less often 20 or 22, carbon atoms. Shorter and longer chains, chains with an odd number of carbons, or compounds with branched chains are also common, but usually only in small amounts.

Fatty acids and related compounds can be saturated or unsaturated. The common unsaturated acids carry one, two, or three double bonds, predominantly in *cis* configuration. These double bonds are generally in the 9; 9,12; or 9,12,15 positions in respect to the carboxyl, i.e., in that they are unconjugated, being separated by methylene groups (Fig. 29).

Whereas fatty materials are present in rather small percentages in pine oleoresins (a few percent at most and usually less) (95), they are major constituents occasionally representing half or more of the wood extracts of *Pinus* and Pinaceae and can amount to a few percent of the dry wood weight (96–105). Although many reports are available on the fatty acid fractions of wood, unfortunately they are often incomplete and occasionally useless, as it is often unclear whether only free acids or also the glycerides were analyzed and whether sapwood, heartwood, phloem, or cortex was investigated. These omissions make it unclear whether any differences encountered are genetic or physiological.

Unsaturated fatty acids generally predominate (roughly 80% of the total) and commonly consist of oleic and linoleic acids as the main constituents, with palmitoleic and linolenic acids in secondary quantities. The saturated fatty acids comprise mainly palmitic and stearic acids, and occasionally also arachidic, behenic, and lignoceric acids. In some instances, other unsaturated acids, acids with odd-numbered carbon chains, and branched saturated acids with methyl side groups near the end of the carbon chain were reported (95,105).

$$CH_3-(CH_2)_n-COOH$$

n=14	palmitic acid
n=16	stearic acid
n=18	arachidic acid
n=20	behenic acid
n=22	lignoceric acid

$$CH_3-(CH_2)_m-(CH_2-CH=CH)_n-(CH_2)_7-COOH$$

m=4, n=1	palmitoleic acid
m=6, n=1	oleic acid
m=3, n=2	linoleic acid
m=0, n=3	linolenic acid

Figure 29   Structures of the most important fatty acids; all double bonds *cis*.

Fatty acids of the softwood families other than Pinaceae are poorly known. A few papers on the fatty acids and their glycerides from the wood of hardwood species indicate that the composition of fatty acids and their content in wood are not much different from what is known for softwoods (106,107). The structures of the most common fatty acids are summarized in Fig. 29.

Another source of aliphatic acetogenins are volatile components of the gum and wood oleoresins of some *Pinus* species where acetogenic odd-numbered hydrocarbons, especially *n*-heptane and/or *n*-undecane, replace monoterpenes, sometimes almost completely (*Pinus jeffreyi, P. sabiniana*). Other hydrocarbons such as *n*-pentane and *n*-nonane also often occur in small quantities. These hydrocarbons are commonly accompanied by small amounts of normal aldehydes, such as C_8, C_9, C_{10}, C_{12}, C_{14} and possibly higher ones (47).

2.   Aromatic Acetogenins

Pure aromatic acetogenins are encountered rather rarely as components of the extraneous materials from wood. The wood from the tropical *Diospyros* species has been shown to contain naphthoquinones and closely related aromatic compounds such as isodiospyrin, methyljuglone, and shinanolone (Fig. 30), with structures suggestive of their formation from acetate units and polyketide chains (108–111). Other references to such compounds can be found in (8). The wood of *Grevillea robusta* contains compounds that include a long paraffinic chain, to which are attached units of resorcinol or hydroxybenzoquinone (112–114); a similar compound, betulachrysoquinone, is occasionally found (as a hemiketal) in the wood of the *Betula* species where it is produced by a fungus (115) (Fig. 30). These compounds are structurally related to the active toxic principles of poison-oak and poison-ivy and likewise cause dermatitis. They are biosynthesized from polyacetate chains partly in polyketide and partly in paraffin form (35).

C. Saccharogenins

The saccharogenins, like the mevalonogenins and acetogenins, comprise aliphatic and aromatic compounds (see the footnote on p. 328). Aliphatic compounds include mono-, oligo-, and polysaccharides, as well as the closely related cyclitols (cyclohexane polyols). The aromatic materials comprise various phenolic compounds that characteristically carry a three-carbon chain on the aromatic ring and one, two, or three oxygen functions in the 4; 3,4; or 3,4,5 ring positions relative to the chain. Some of the aliphatic saccharogenins represent reserve materials and are found mainly in sapwood (mono- and oligosaccharides, starch), whereas others are typical heartwood materials (cyclitols and arabinogalactans). Aromatic saccharogenins are important constituents of heartwoods. They are also occasionally found in the resin of softwood resin canals, but only in small amounts. They represent, however, the major constituents of certain exudates (balsams, gums) produced by the wood of some hardwood species. Parts of the carbon skeleta of flavonoids and stilbenes (and derived tannins and phlobaphenes) are saccharogenic in nature, with the latter materials being ubiquitous heartwood extractives. Low-molecular-weight saccharogenic compounds are occasionally volatile and are responsible for the

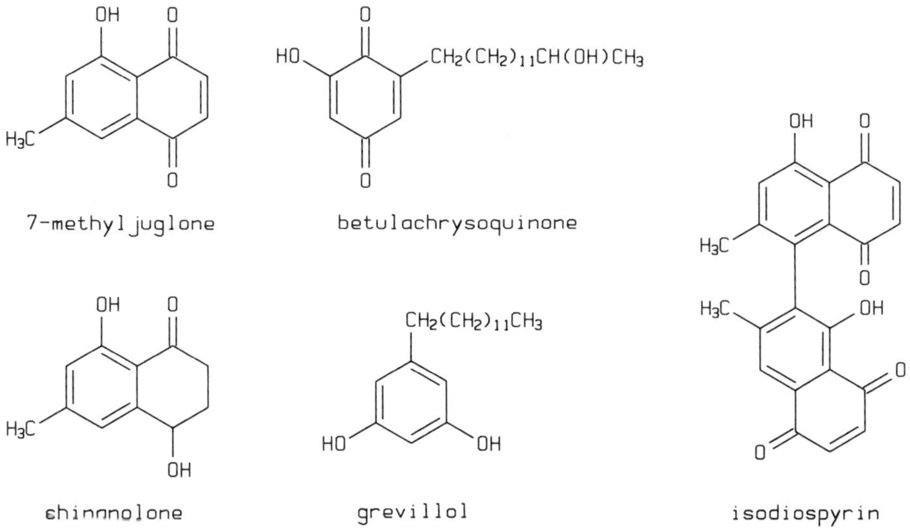

Figure 30   Some aromatic acetogenins.

balsamic odor of some woods.

Biosynthetically, polysaccharides form by the polymerization of monosaccharide units. Cyclitols also derive from monosaccharides (116,117). It seems that *myo*-inositol forms first by oxidation of *D*-glucose-6-phosphate to 5-keto-*D*-glucose-6-phosphate, aldolisation to 1-phospho-*myo*-inosose-2, and reduction-hydrolysis to *myo*-inositol-1-phosphate. Isomerization by oxidation–reduction of the appropriate hydroxyl leads to *D*-inositol, whereas methylation of *myo*- and (+)inositols gives sequoyitol and pinitol, respectively (Fig. 31).

Aromatic saccharogenins form from monosaccharides by a series of transformations involving 5-dehydroquinic acid, 5-dehydroshikimic acid, shikimic acid, prephenic acid, phenylpyruvic acid, phenylalanine, and cinnamic acid. Starting with cinnamic acid, a large variety of natural products is formed by (1) reduction of the carboxylic group of the side chain to the aldehyde or alcohol group or transformation of the side chain into *n*-propenyl group, (2) degradation of the three-carbon side chain to two- and one-carbon chains, (3) hydroxylation of the aromatic nucleus in the 4; 3,4; or 3,4,5 positions and occasionally in still others, (4) methylation of the hydroxyls, preferentially in the 3 or 3,5 positions, (5) formation of methylenedioxy rings between neighboring hydroxyls, or (6) oligo- and polymerization of the initially formed monomers.

In addition to their formation from cinnamic acid, some aromatic saccharogenins are also derived by the direct aromatization of the 5-dehydroshikimic acid intermediate (35). The biosynthetic sequence leading to cinnamic acid is depicted in Fig. 32.

1. Aliphatic Saccharogenins

Aliphatic saccharogenins include mono- and oligosaccarides, and cylitols as well as polymeric starch, arabinogalactans, and pectins. While all of these materials are rather ubiquitous, they generally occur in small amounts.

Mono- and oligosaccharides are found in the heartwood and sapwood of all wood species. They occur in higher amounts in the living sapwood (0.1-1.0% vs. traces to 0.3% in heartwood), which is understandable in view of their metabolic role The composition of the heartwood and sapwood sugars is also different. Sapwood contains monohexoses (glucose and fructose), and oligosaccharides (sucrose, raffinose, and stachyose), the latter in traces only. Heartwood, on the other hand, contains monohexoses (glucose and galactose), pentoses (arabinose and xylose), and methylpentoses (rhamnose and fucose), with arabinose always predominating (118) (Fig. 33).

The ubiquitous cyclitols, although generally minor components, do become quite abundant in a few species. Thus, in *Pinus lambertiana* (sugar pine), their amount in heartwood fluctuates between 1.5 and 9.5% and between 2.4 and 26.7% in old stumpwood (119, 120). In the heartwood of *Sequoia sempervirens* and *Sequoiadendron giganteum*, cyclitols were found in 1.0 and 0.58% amounts,

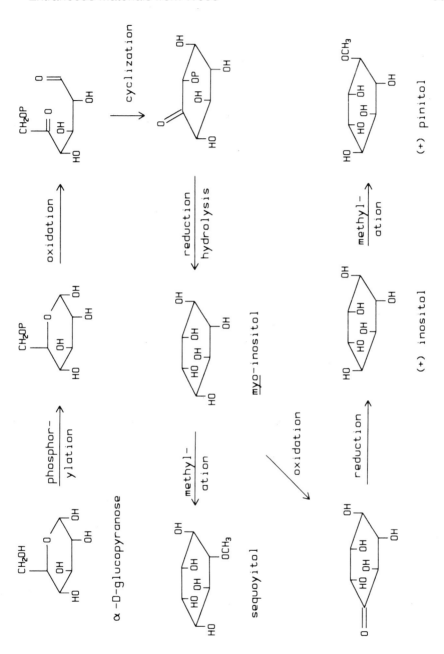

Figure 31 Biosynthesis and structures of cyclitols occurring in wood.

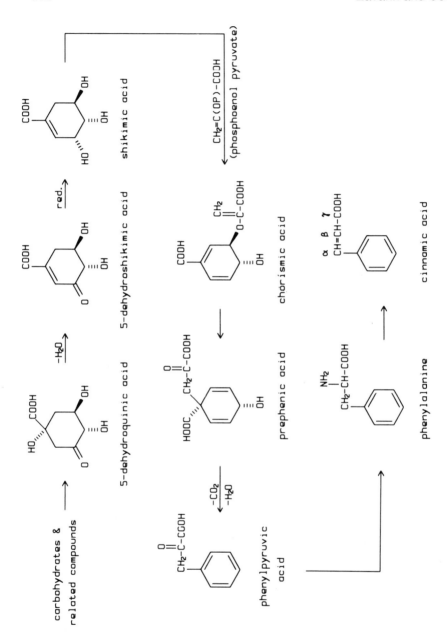

Figure 32 Biosynthesis of 5-dehydroshikimic and cinnamic acids, the precursors of aromatic saccharogenins.

respectively(121,122). *Pinus ayacahuite* heartwood has also been reported to produce cyclitols in sizable amounts (123). The most common cyclitols isolated were
pinitol, sequoyitol, *myo*-inositol and (+) inositol (Fig. 31), with pinitol dominating
in *P. lambertiana* and *S. sempervirens* and sequoyitol in *S. giganteum*.

Most of the polymeric carbohydrates of wood (cellulose, hemicelluloses) are cell
wall components and are insoluble in water or organic solvents. Exceptions are the
starches, arabinogalactans, and pectic substances. Starches are α-D-(1-4) glucans,
composed of the straight-chain amylose of 200-350 DP and of the branched, high-
molecular-weight amylopectin; they are ubiquitous in relatively small amounts in
the sapwood of soft- and hardwoods as metabolic reserve materials. Pectic substances are α-D-(1-4) straight-chain polymers of galacturonic acid, of a widely

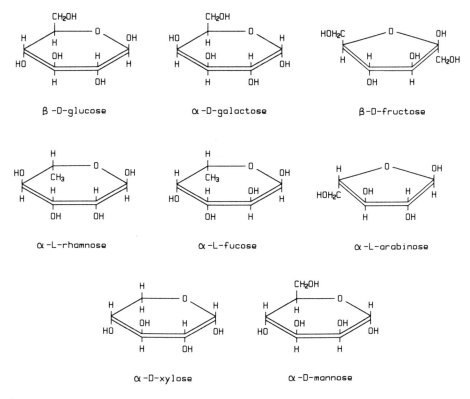

Figure 33 Mono- and oligosaccharides from wood (Sucrose=α-D-glycosyl-β-D-fruc-
tose. Raffinose=α-D-galactosyl-α-D-glucosyl-β-D-fructose. Stachyose=α-D-galactosyl-α-
D-galactosyl-α-D-glucosyl-β-D-fructose.)

varying DP, with carboxylic groups present in part as methyl esters. They are also ubiquitous minor extractives. Arabinogalactans occur predominantly in heartwood and thus represent genuine heartwood extractives. Generally, their amount is not very high, varying in the softwoods around 1%. *Larix* heartwood is an exception; they have been reported in amounts between 5 and 25%, or even 40% (124). Arabinogalactans are highly branched polysaccharides composed of two polymeric components with a DP of about 60 and 400, respectively, and are constituted of a (1-3)-β-D-galactopyranose backbone to which are attached (1-6)-β-D-galactose and α-L-arabinofuranose units, as well as a few β-D-glucuronic acid units. To some α-L-arabinofuranose units are additionally attached (3-1)- β-L-arabinopyranose units. The ratio between the arabinose and galactose units is about 1:6.

2. Aromatic Saccharogenins

Aromatic saccharogenins include a great variety of predominantly phenolic compounds and with a few exceptions represent typical heartwood extractives. Chemically, they can be divided into monomeric saccharogenins, oligomeric saccharogenins (lignans, sequirins), and polymeric saccharogenins; the last category will be discussed in the section on polymeric phenols.

a. Monomeric Aromatic Saccharogenins. Monomeric aromatic saccharogenins comprise a wide variety of materials from nonpolar aromatic ethers to highly polar polyhydroxy aromatic acids.

The balsamic odor of many essential oils and resins is due to the presence of saccharogenins such as eugenol, *O*-methyl eugenol, safrole, myristicin, and elemicin. Such oils are derived in the majority of cases from the foliage, fruits, or bark of various plants, and the fragrant aromatic saccharogenins are rather rare in wood. One of these compounds, methyl chavicol, does occur in the gum turpentine of many *Pinus* species, but in rather small amounts, usually 1–2%; the highest amount (5%) was reported for the Mexican *P. patula* turpentine (47). The distinctive balsamic odor of *P. lambertiana* heartwood is derived from cinnamic acid present at a maximum of 0.8% of dry wood weight (125). Methyl eugenol, elemicin, *O*-methyl coniferyl alcohol, and methyl coniferylaldehyde were identified in the heartwood of *Dacrydium franklinii* (126). Vanillin and veratraldehyde were isolated from the wood of *Abies sibirica* in very small amounts (127). For structures see Fig. 34.

The heartwood of the species of *Salix* and *Populus* contains a wide variety of monomeric saccharogenins attached as esters or glycosides to a sugar molecule. The aglycones and aromatic acids include salicylic acid, benzoic acid, vanillic acid, syringic acid, *p*-hydroxybenzoic acid, *p*-hydroxycinnamic acid, caffeic acid, ferulic acid, syringaldehyde, vanillin, acetosyringone, acetovanillone, and salicyl alcohol; the sugar moities include glucose, galactose, mannose, and xylose (128). Some of these aromatics apparently also exist as free compounds and include *p*-hydroxybenzoic acid, benzyl alcohol, *p*-ethylphenol, phenol, and β-phenyl ethanol (129–131) (Fig. 35). Similar compounds have been isolated recently from the

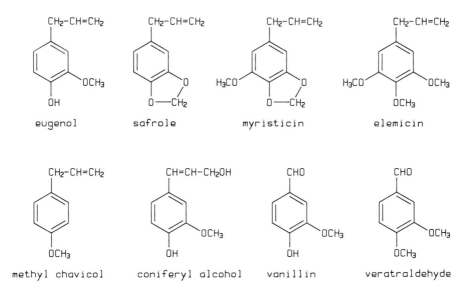

Figure 34   Nonpolar, monomeric, aromatic saccharogenins.

heartwood of *Prunus* (132).

The heartwoods of Fagaceae (*Fagus* and *Quercus*) contain a large variety of simple, free, polar saccharogenic phenolics including gallic acid, sinapaldehyde, coniferaldehyde, syringaldehyde, propiovanillone, vanillin, *p*-hydroxybenzaldehyde, resorcinol, ferulic acid, sinapic acid, salicylic acid, gentisic acid, pyrogallol, and eugenol (Figs. 34 and 35). The presence of glucosides has been also indicated (133,134). The above compounds have also been identified in the heartwood of many other tannin-rich hardwoods. Gallic acid and its dimer, ellagic acid, are important components of the hydrolyzable tannins present in these and other species and will be discussed later in more detail.

Another class of saccharogenins commonly occurring in wood and including a single aromatic nucleus are the coumarins, the lactones of 2-hydroxy-*cis*-cinnamic acids. Coumarins, such as esculetin, scopoletin, and fraxetin, are present in the wood of Hippocastanaceae (135), *Fraxinus* (136) and Fagaceae (137,138) (Fig. 36).

b. Oligomeric Aromatic Saccharogenins.   The great majority of oligomeric aromatic saccharogenins form by oxidative coupling of the respective monomers. Most of the monomeric aromatic saccharogenins are phenolic in nature, with the hydroxy group attached in the *para* position of the aliphatic side chain. Enzymatic oxidation of the phenolic hydroxyl gives rise to free radicals that can exist in several mesomeric forms, depending on the structure of the compound oxidized. Dimerization of the

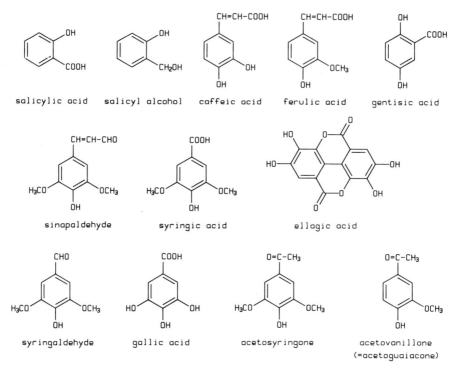

Figure 35 Polar, monomeric, aromatic saccharogenins.

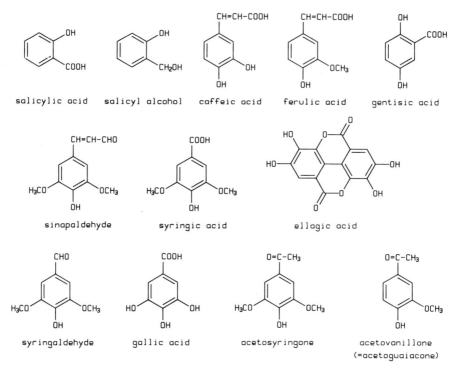

Figure 36 Coumarins from wood.

free radicals under formation of a single bond can give rise to a variety of products, the structures of which depend on the location of the unpaired electron in the respective mesomers. Such initially formed products are generally rather unstable and react further either by prototropy (cyclohexadienones) or the addition of

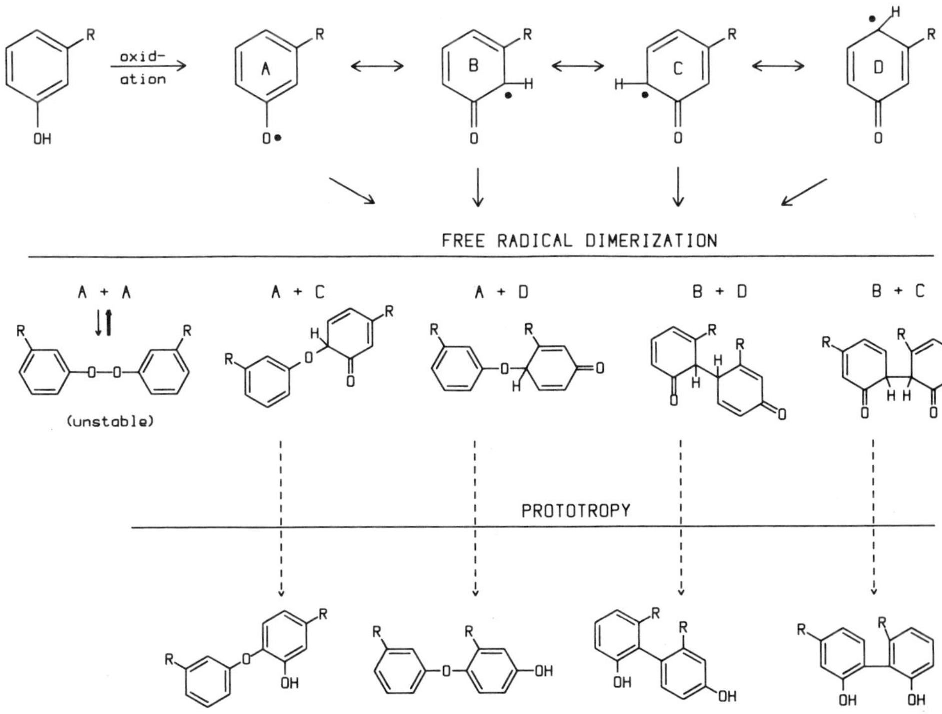

Figure 37 Oxidative coupling of phenolic materials.

elements of water or of organic hydroxyls (quinonemethides) to give the final compounds (35) (Fig. 37).

As mentioned, the structure of the final materials depends on the monomers involved and the position of the single electron during dimerization. One of the common dimerization patterns is the oxidative dimerization of the *p-n*-propenylphenol structures, common to compounds such as *p*-coumaryl alcohol, coniferyl alcohol, and sinapyl alcohol. Dimerization of the respective free radicals via the mesomeric form, with the lone electron located at the β position of the side chain, results in a diquinonemethide, which is further stabilized by a variety of pathways, all involving the addition of a hydroxyl-carrying entity, either the extramolecular addition of water or intramolecular addition of a hydroxyl group (Fig. 38). The compounds so produced, which include a carbon skeleton of two phenylpropane units linked by a bond between β-carbons, are called lignans. Under some conditions, certain lignans can rearrange by the substitution of an α-β-linkage for the β-β-linkage and the loss of one carbon. Such compounds are here denoted

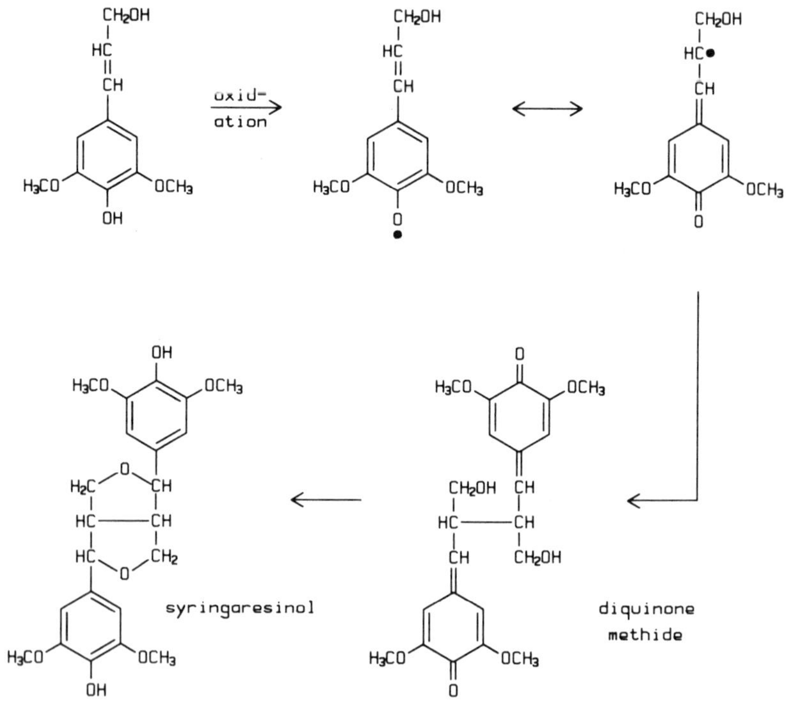

Figure 38   Biosynthesis of lignans.

sequirins. The dimerization of other types of saccharogenic monomers or alternative dimerizations of the above monomers are also known.

1. Lignans.   Lignans include a large number of compounds with the common feature of two 1-phenyl-*n*-propane units linked β-β, but differing from each other in substitution at the 3,4, and 5 positions of the benzenoid ring, in the nature of the linked side chains, and in the nature of the additional bonds linking the two units. Some authors have recently extended the definition of lignans to include any oligomeric products formed by the coupling of *n*-propylbenzene units, but in this review we will adhere to the earlier definition. The additional bonds between the two *n*-propylbenzene units can include linkages between side chains, lactone linkages between side chains, bonds between the α-position of the side chains and position 2 of the benzene ring, etc. As the result of alicyclic or heterocyclic ring closures, lignans can exist in several stereomeric forms. The classification and chemistry of lignans were discussed by Hathway (137). Structures of some lignans are depicted in Figs. 39a and 39b.

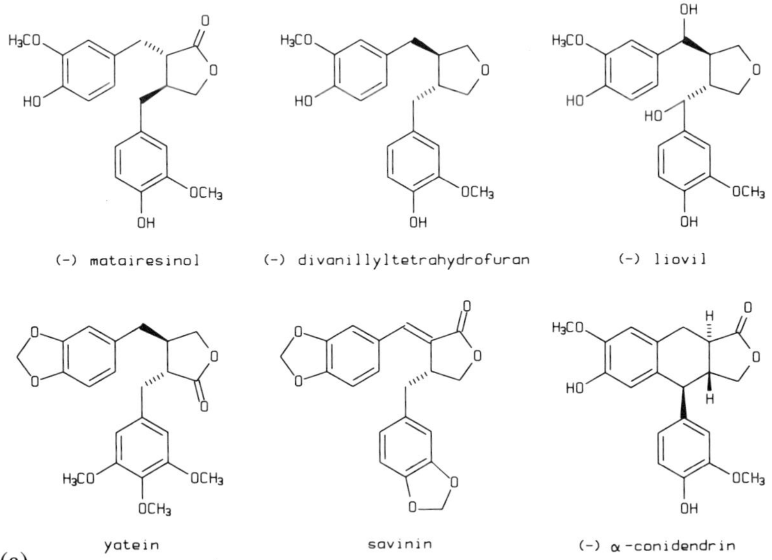

(a)

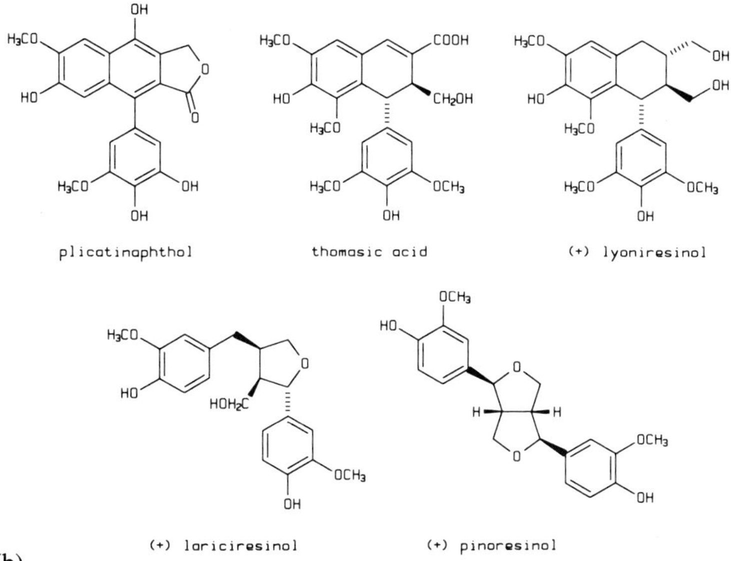

(-) matairesinol

(-) divanillyltetrahydrofuran

(-) liovil

yatein

savinin

(-) α-conidendrin

(b)

plicatinaphthol

thomasic acid

(+) lyoniresinol

(+) lariciresinol

(+) pinoresinol

Figure 39 Structures of some important lignans.

Lignans have been isolated from the heartwood of practically all genera of conifers, usually in amounts below 1%. Thus the heartwood of *Picea abies* (Pinaceae) yielded 0.5% of lignans, including (-)matairesinol, (-)hydroxymataircsinol, (+)oxomatairesinol, (-)allohydroxymatairesinol, (-)α-conidendrin, pinoresinol, (+)lariciresinol, (-)liovil, and 3,4 vanillyltetrahydrofuran (138). The wood of *Abies amabilis* contains matairesinol (139). *Larix decidua* wood produces liovil and secoisolariciresinol (140). *Tsuga mertensiana* and *T. heterophylla* wood also contain matairesinol (139). The wood of *Cedrus deodara* contains lariciresinol, isolariciresinol, and mesosecoisolariciresinol (141). Species of *Pinus* produce pinoresinol and lariciresinol (142). *Thuja plicata* (Cupressaceae) heartwood contains several lignans, thujaplicatin and its methyl ether, plicatin, plicatic acid, plicatinaphtol, and plicatinaphthalin (143–145); *Libocedrus yateensis* produces yatein; *Chamaecyparis pisifera* produces savinin (154,159). *Fitzroya cupressoides* wood produces isotaxiresinol, its 6-methylether, isolariciresinol, and secoisolariciresinol (147). The wood of some Taxodiaceae (*Taiwania cryptomerioides*), Podocarpaceae, and Taxaceae species has been reported to contain lignans (148).

Lignans are also common in the wood of hardwoods. The wood of *Quercus rubra* (northern red oak) was found to contain syringaresinol, (5-5'-dimethoxypinoresinol) and lyoniresinol and its glycoside (134). *Liriodendron tulipifera* (yellow poplar) heartwood produces (+)syringaresinol and its dimethyl ether (149). The heartwood of the *Ulmus* species (elms) contains the lignans thomasic acid, thomasidioic acid, lyoniresinol, and its glycoside in various amounts (150–152). Although (-)eudesmin (*o*-dimethyl pinoresinol) was isolated from the kino (exudate) of *Eucalyptus hemiphloia*, the lignans seem to be rare in the otherwise phenol rich *Eucalyptus* heartwoods (153).

2. Sequirins.   In relatively recent years, another class of dimeric saccharogenins has been discovered. The characteristic feature of these compounds is a phenylpropane unit linked by its α-carbon on the side chain to the β-carbon of a phenylethane unit. The substitution of the benzene nuclei of these compounds corresponds, however, to that of lignans. Besides the α-β-linkage, the side chains can additionally join to form heterocyclic rings. This class of compounds has been named conioids by Erdtman, but no explanation for the choice of the name could be found. More recently, the same compounds have been named norlignans. We prefer to give this class of compounds the independent trivial name *sequirins* (from *Sequoia sempervirens* and the names of the most important compounds), mainly because the linkage between the two saccharogenic monomers is α-β, rather than β-β as in true lignans. The biosynthetic sequence leading to the sequirins has been postulated to involve side chain rearrangement of the appropriate lignan precursors (Fig. 40) (154).

Sequirins are typical of the heartwood of Taxodiaceae. Thus, *Sequoia sempervirens* heartwood contains sequirins A (=sugiresinol), B, and C and agatharesinol.

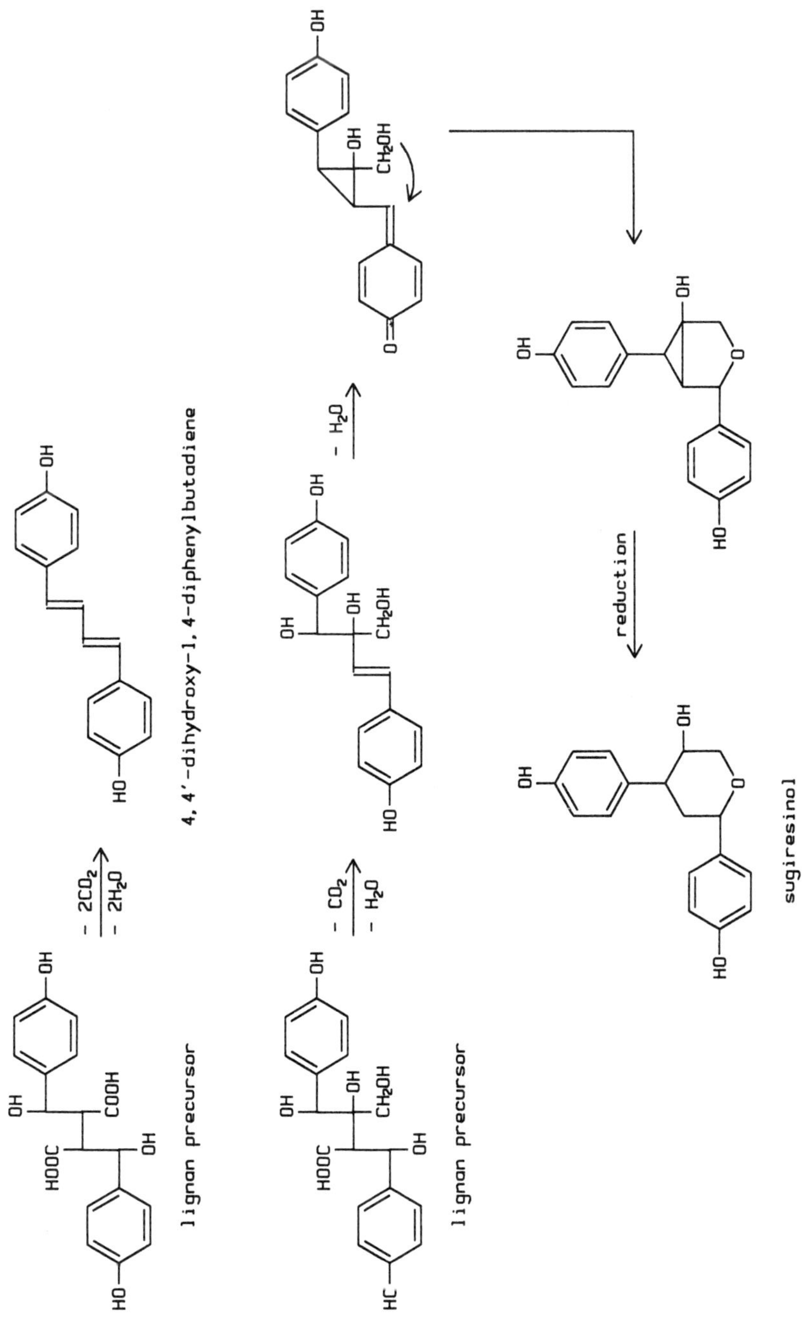

Figure 40 Biosynthesis of 4,4'-dihydroxy-1,4-diphenylbutadiene and of sequirins.

Sequoidendron giganteum heartwood produces agatharesinol and sequirins E, F, and G, (155). *Athrotaxis selaginoides* wood contains hinokiresinol, agatharesinol, and sugiresinol, (156). A sequirin, cryptoresinol, has been isolated from the heartwood of *Cryptomeria japonica* (157). *Metasequoia glyptostroboides* heartwood contains six sequirins: metasequirins A and B, hydroxymetasequirin A, athrotoxin, agatharesinol, and hydroxyagatharesinol (158). Sequirins have also been identified in some species of Cupressaceae heartwood; thus, hinokiresinol and dehydroagatharesinol have been isolated from *Libocedrus yateensis* (154). Agatharesinol has also been isolated from the wood of *Agathis australis* (160), and hinokiresinol from *Chamaecyparis obtusa* (146) (Fig. 41).

   3. Oligomeric Benzoic Acid Derivatives.   This class of compounds derives mainly from gallic acid. Gallic acid forms predominantly by the aromatization of the 5-dehydroshikimic acid intermediate of the cinnamic acid biogenetic pathway (35). It is found free in nature and as a monomeric or oligomeric component of the so-called hydrolyzable tannins, attached to a monosaccharidic moiety by an ester linkage. The oligomerization of gallic acid takes place essentially in two ways: by the formation of an ester linkage (depsidic units) or by oxidative coupling occasionally followed by additional transformations. Oxidative coupling leads to a sizable number of compounds and apparently takes place during the attachment of the gallic acid to the monosaccharide moieties. As a result, the esters of acids such as hexahydroxydiphenic acid, dehydrodigallic acid, valoneaic acid, flavogallonic acid, dehydrohexahydroxydiphenic acid, and the peculiar isohexahydroxydiphenic acid are formed. Upon acid or base hydrolysis, these materials split and are generally found in altered form, such as lactones (ellagic acid) or rearrangement products

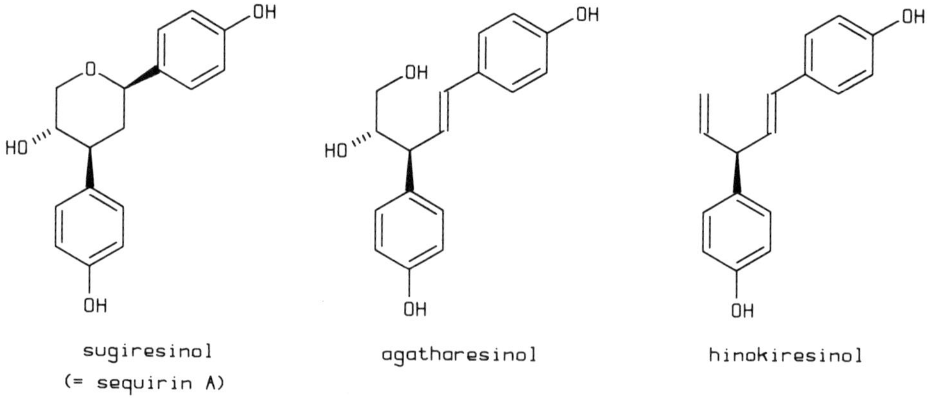

Figure 41   Represenatative sequirins.

(brevifolin carboxylic acid, chebulic acid) (161–164). These materials will be further discussed under "Tannins." For formulas, see Fig. 42a and 42b.

4. Other Saccharogenic Oligomers. The oxidative coupling of the phenylpropane monomers based on other free radical mesomers, coupling of the other type of saccharogenic monomers, or degradation of lignans and sequirins can lead to additional types of extractive structures, although such compounds have been more often reported for bark and needles than for wood. Of the few available example, one can mention three compounds derived by oxidative coupling, followed by quinonemethide stabilization, of the coumarin fraxetin with either coniferyl, 5-

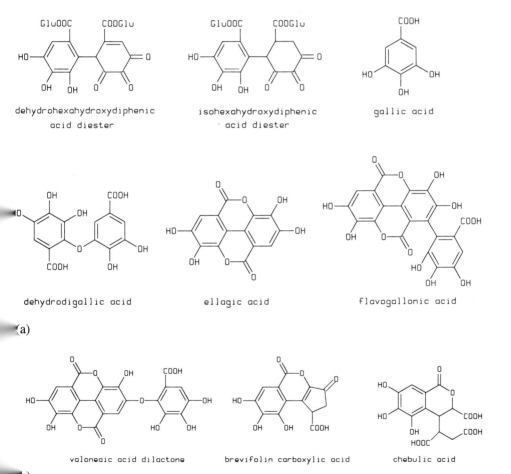

dehydrohexahydroxydiphenic
acid diester

isohexahydroxydiphenic
acid diester

gallic acid

dehydrodigallic acid

ellagic acid

flavogallonic acid

(a)

valoneaic acid dilactone

brevifolin carboxylic acid

chebulic acid

(b)

Figure 42 Phenolic components of hydrolyzable tannins and products of their hydrolysis (a and b). Glu=glucose (ester).

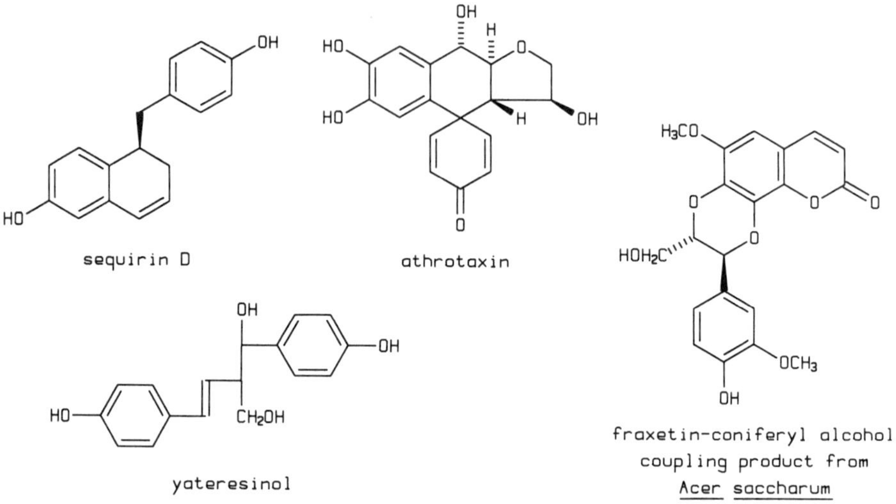

Figure 43 Lignan- and sequirin-related saccharogenins and a phenolic coupling product from *Acer saccharum* wood.

hydroxyconiferyl, or sinapyl alcohol. These compounds have been identified in the stressed wood of *Acer saccharum* (165) and exhibit antifungal properties (Fig. 43).

Compounds biosynthetically related to lignans and sequirins are occasionally found in the heartwood of the Cupressaceae and Taxodiaceae species. Thus, 4,4'-dihydroxy-1,4-diphenylbutadiene and yateresinol have been isolated from the wood of *Libocedrus yateensis* and the former compound also from *Taxodium distichum* (Figs. 40 and 43). Most likely, 4,4'-dihydroxy-1,4-diphenylbutadiene forms by the decarboxylation of both γ-carbons of the respective lignan precursor (154) (Fig. 40). Sequirin D has been found in the wood of *Sequoia sempervirens* and athrotaxin has been isolated from wood of *Athrotaxis selaginoides* (156) (Fig. 43).

Recently, 42 phenylpropane dimers, trimers, and tetramers were isolated from the wood and bark of *Larix leptolepis* and *Picea jezoensis*. These were partly true lignans like lariciresinol and pinoresinol, with β-β linkages between *n*-propyl side chains, and partly compounds based on other modes of coupling. Examples are given in Fig. 44. It seems likely that the isolated structures are quite ubiquitous in conifers and broad-leaved species (18).

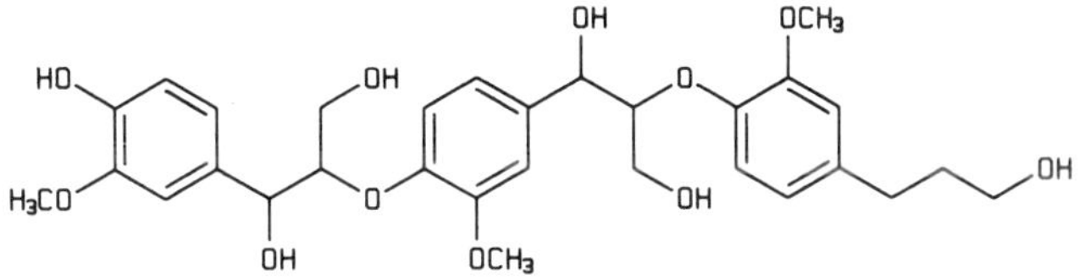

leptolepisol B

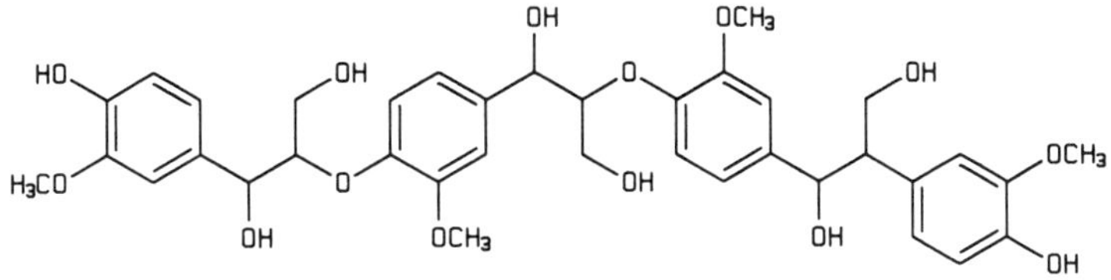

leptolepisol E

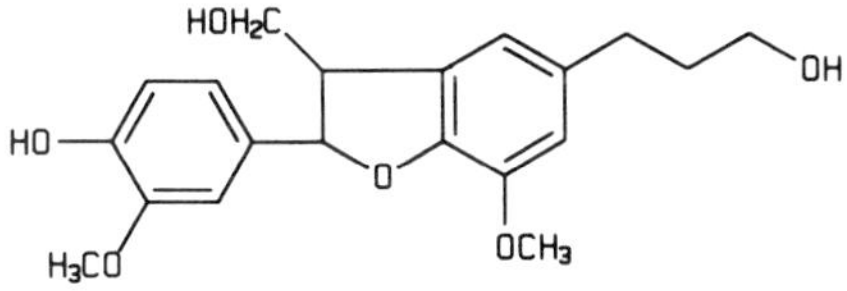

compound X

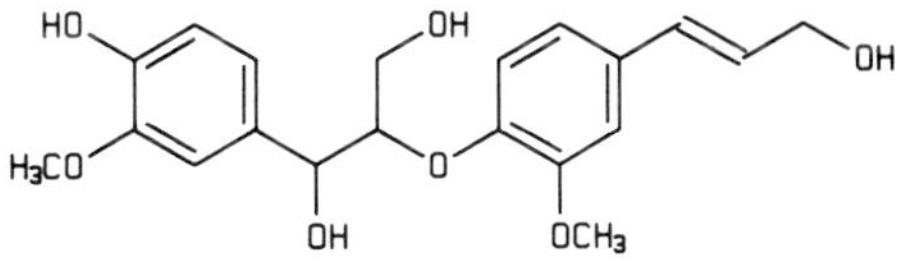

compound XVIII

Figure 44 Lignan-related dimers, trimers, and tetramers from the wood of *Larix leptolepis* and *Picea jezoensis*.

D. Compounds Formed by Several Biosynthetic Pathways

In addition to the compounds formed by a single biosynthetic route, many types of wood extractives form by a combination of several of the aforementioned biosynthetic sequences. By far the most important of these are the flavonoids and stilbenes, both of which form by a combination of acetogenic and saccharogenic pathways. Compounds formed by mevalonogenic/acetogenic, mevalonogenic/saccharogenic, or all three routes are also known, but are less common and will be discussed only briefly.

The biosynthesis of both stilbenes and flavonoids is based on saccharogenic cinnamic acid-type "starter" units that are acetogenically extended by three $CH_2C{=}O$ units into a 15-carbon polyketomethylene precursor. This precursor can undergo two types of cyclizations, leading to either a stilbene or chalcone, and the latter can cyclize to give a flavonone, from which various types of flavonoids can form by additional structural changes (35,166) (Fig. 45).

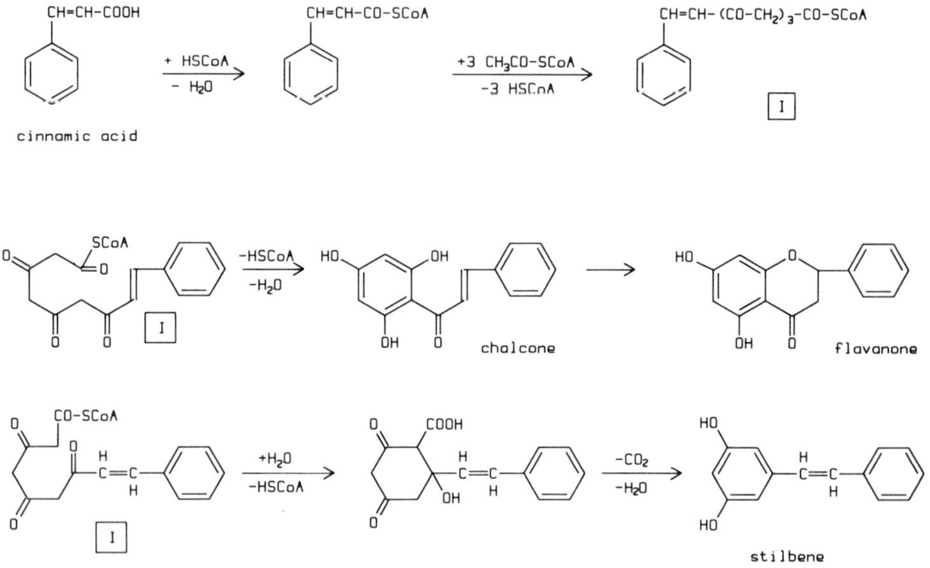

Figure 45 Biosynthesis of flavonoids and stilbenes.

1. Monomeric Flavonoids

Flavonoids are compounds based on the fundamental 2-phenylchroman skeleton and include an aromatic ring A to which is fused an oxygen heterocyclic ring C, substituted in the 2 position by phenyl (ring B). In accordance with its acetogenic derivation, ring A commonly carries hydroxy groups in meta positions to the heterocyclic oxygen, i.e., on carbons 5 and 7. In accordance with its saccharogenic derivation, ring B tends to carry the oxygen functions on carbon 4', carbons 3' and 4', or carbons 3',4', and 5'; if methylated, the hydroxyls at carbons 3' and 5' are methylated preferentially in the majority of cases. Ring C exists in several modifications and can carry hydroxyls, carbonyls, and double bonds. The structure of ring C is the basis for the empirical classification and nomenclature of the flavonoids. Some classes of flavonoids derive their names from the root "flavan" and include flavan-3-ols (catechins), flavan-3,4-diols (leucoanthocyanidins or leucoanthocyanins if glycosides), flavanones, and flavanonols. Other flavonoids derive their name from the root "flavon" (flavan-2-ene-4-one) and include flavones and flavonols. To these classes should be added anthocyanidins (anthocyanins if glycosides), those peculiar oxonium compounds with an aromatic oxygen heterocycle.

Anthocyanidins, although important constituents of flower and leaf pigments, have apparently never been found in wood and will not be discussed in this review.

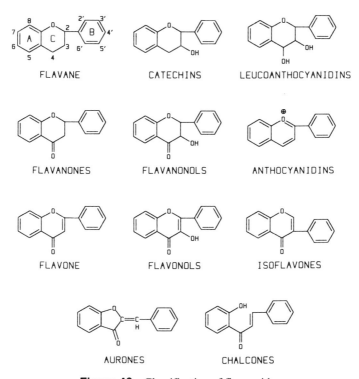

Figure 46   Classification of flavonoids.

Closely related to the flavonoids and usually considered with them are the isoflavonoids, aurones, and chalcones, all of which are rare in wood. Classification of the flavonoids is shown in Fig. 46

Flavonoids are possibly the most ubiquitous monomeric natural products and occur in the heartwood and in lesser amounts in the sapwood of many wood species. They can be separated into the "reduced" group comprising flavan-3-ols and flavan,-3,4-diols, i.e., compounds lacking a carbonyl group and formed by the reduction of flavanonol precursors, and into unreduced flavonoids that carry a carbonyl in ring C at position 4, i.e. flavanones and compounds formed by the oxidation of ring C of flavanones.

a. Flavonoids Lacking Ring C Carbonyl (Reduced Flavanonols).   Relatively little information is available on the occurrence of the reduced flavanonols in the wood of softwoods. These are generally found in other organs or tissues, such as bark or foliage. Hergert mentions (without reference) the chromatographic identification of the flavan-3-ols catechin and gallocatechin and four flavan-3,4-diols, two isomeric leucocyanidins and two isomeric leucodelphinidins, in the wood of 12 *Pinus* species (167). The occurrence of (+) and (-) catechin, (+) and (-) afzelechin, and (+) and (-) gallocatechin has been reported for the wood of *Juniperus communis* (168,169). In *Abies concolor*, the total amount of catechins has been determined to be 0.12% in wood vs. 3.8% in bark (170). The sapwood of *Pseudotsuga menziesii*, *Picea sitchensis*, *Tsuga heterophylla*, and *Pinus elliottii* was shown to contain catechin, leucocyanidin, and occasionally epicatechin, whereas the heartwood only contained the corresponding polymers (171). (-)Epicatechin was found in the wood of *Metasequoia glyptostroboides* (172). Structures are given in Fig. 47.

Considerably more is known of the occurrence of flavan-3-ols and flavan-3,4-diols in the wood of hardwoods. Thus, catechin has been identified in the wood of *Quercus rubra* (134) and can generally be expected in the wood of oaks and related species. Various flavan-3-ols and flavan-3,4-diols [e.g., (+)-catechin, (-)-epicatechin, leucofisetinidin, melacacidin, mollisacacidin, leucorobinetidin, teracacidin, guibourtacacidin, leucocyanidin] have been identified in the heartwood of the *Acacia* and *Robinia* species (173–175), *Prunus* (167), *Acer* (176–179), *Anacardiaceae* (*Rhus*, *Schinopsis*, *Cotinus*, *Melanorrhoea*) (180), *Nothofagus*

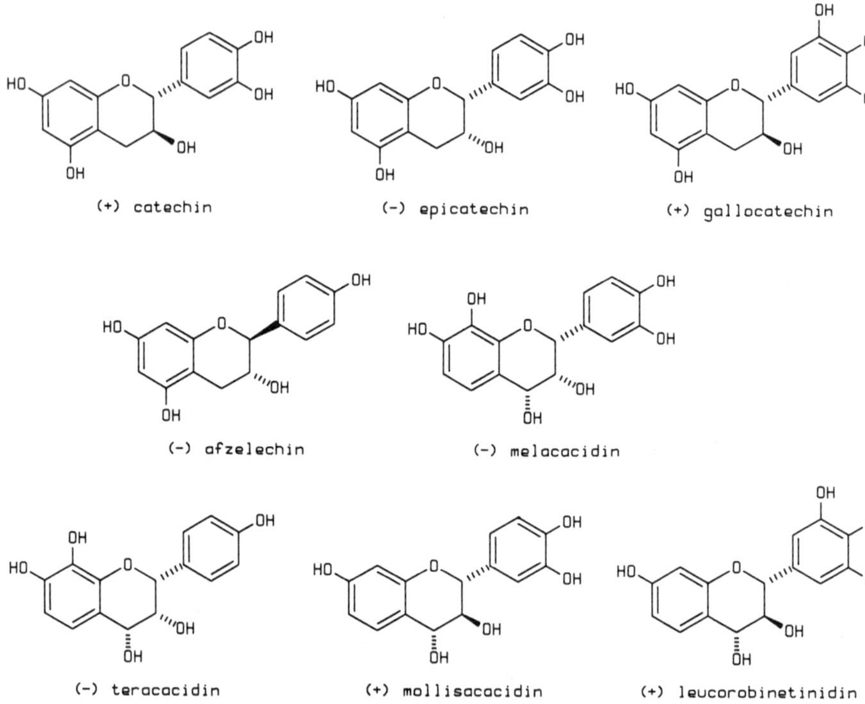

Figure 47 Flavan-3-ols and flavan-3,4-diols from wood.

(181), and others. The wood of the *Eucalyptus* species has been also shown to contain these compounds [(+)-catechin, (-)-epicatechin, leucoanthocyanidin] that appear to polymerize partly or fully during the transformation of sapwood into heartwood (182). In general, flavan-3-ols and flavan-3,4-diols as such or as corresponding glycosides are likely to be found in trees that produce dark heartwood and larger amounts of the so-called condensed tannins, with which they are biogenetically linked (see Sec. IV. E. 1(b)).

b. Flavonoids with Ring C Carbonyl (Flavanones and Oxidized Flavanones). Flavanones and their oxidation products (flavanonols, flavones, flavonols) appear to be more widespread and are especially common in the heartwood of Pinaceae. Flavonoids and stilbenes from the heartwood of about 80 pine species have been investigated by Erdtman and his co-workers. The flavanone (-) pinocembrin, flavanonol (+) pinobanksin, and stilbene pinosylvin and its monomethylether have been identified in the wood extracts of all species of both subgenera *Strobus* (soft pines) and *Pinus* (hard pines) investigated (183,184). The heartwood of soft pines also contained four additional flavanones: one flavononol, three flavones, and one flavonol; stilbenes included dihydropinosylvin and its monomethyl ether. All the above materials had a common trait of a B ring lacking hydroxyl substituents. Later, however, compounds were discovered in *Pinus* wood that carried hydroxyls in ring B, such as the flavanonols aromadendrin and dihydroquercetin (taxifolin) (185), the flavone apigenin (186), and the flavonols galangin and izalpinin (187) as well as 2,4-dimethoxy-6-hydroxychalcone.

The flavanonol dihydroquercetin has also been found in the heartwood of the other genera of Pinaceae: *Pseudotsuga* (1.0–2.2%) (179), *Larix*, *Picea* (188), and *Cedrus* (189,190). The flavonols quercetin, aromadendrin, pinobanksin, and pinocembrin were also reported in *Pseudotsuga* (188); aromadendrin, quercetin, and dihydrokaempferol in *Larix* (below 1%) (188,191,192); and aromadendrin, pinobanksin, dihydroquercetin, dihydromyricetin, and their 6-methyl adducts and glucosides in *Cedrus* (189,190). Tiwari and Minocha reported the occurrence of several chalcones and a flavanone, mostly as glucosides, in the "stem" of *Abies pindrow*; it is unclear whether these compounds came from wood, bark, or cortical tissue (193).

Carbonyl-carrying flavonoids, just like their reduced counterparts, also occur in the wood of other families of conifers. Aromadendrin and dihydroquercetin have been reported for *Biota orientalis* (194), dihydroquercetin for *Austrocedrus chilensis* (195), 4,4'-dihydroxychalcone for *Chamaecyparis obtusa* heartwood (196), and a flavanone for *Pilgerodendron uviferum* (Cupressaceae) (196); the wood of *Podocarpus spicatus* has been shown to produce aromadendrin, dihydroquercetin, kaempferol, quercetin, and the isoflavones genistein and podospicatin (Podocarpaceae) (192,198,200). Isoflavonoid-derived pterocarpan flavonoids such as (6aR,11aR)-3,8-dihydroxy-g-methoxypterocarpan (Fig. 48b) have been identified in the heartwood of several *Podocarpus* species (199). Isoflavones are apparently characteristic of Podocarpaceae wood (201). For structures, see Figs. 48a and 48b.

Ketonic flavonoids are also not uncommon in the wood of hardwoods, although most of the attention has been given to foliage and bark, rather than wood. They seem to be rather common in Leguminosae heartwood. Thus, Tindale and Roux (173) reported the occurrence of chalcones (butin, butein), flavanonols [(+) fustin], and flavonols (fisetin) in the heartwood of the Australian *Acacia* species; the compounds had the same substitution pattern as the previously mentioned flavan-3-ols and flavan-3,4-diols isolated from the same source. The heartwood of *Robinia* also appears to be a good source of ketonic flavonoids: the chalcones robtein, butein, and liquiritigenin; the flavanonols dihydrorobinetin and fustin; and the flavonols (4–5%) robinetin and fisetin have been identified in the heartwood of *Robinia pseudoacacia* (174,175,202). The ketonic flavonoids naringenin, aromadendrin, kaempferol, and 7-methylaromadendrin have been reported in *Eucalyptus* (*E. calophylla*, *E. corymbosa*, and *E. maculata*) (203–206), although reduced flavonoids seem to be more common in the heartwood of this genus. A few occurrences of ketonic flavonoids in Salicaceae have been reported: naringenin in the wood of *Populus tremuloides* (207) and 3-methyldihydroquercetin (or its stereoisomer) in the heartwood of *Salix hultenii* (208,209). The relatively large body of information on flavonoids of *Prunus* heartwood (24 species) has been reviewed by Hergert (167); over a dozen ketonic flavonoids and their glucosides have been identified (quercetin, dihydroquercetin, aromadendrin, naringenin, pinocembrin, chrysin, tectocrysin, dihydrotectochrysin, their glucosides, as well as other flavonoids). Flavonoids have been reported in the heartwood of some Fagaceae; thus, quercetin has been found in the wood of *Castanea sativa* (210), and quercetin, dihydroquercetin, naringenin, aromadendrin, and kaempferol in the wood of the *Nothofagus* species (181). On the other hand, we could find no reports of ketonic flavonoids in the wood of the *Quercus, Fagus, Lithocarpus,* or *Castanopsis* species. In Juglandaceae, 5-methylquercetin and 3,5 dimethylquercetin have been reported for *Carya pecan* heartwood (211). Many tropical trees also produce ketonic flavonoids in their heartwood, e.g., the *Melanorrhea* species (Malaya) and *Ocotea usumbarensis* contain dihydroquercetin (212).

Aniba rosaedora of Guaiana contains pinocembrin (213). *Zelkova serrata* produces glycoflavonoids (flavonoids attached to a sugar molecule by a carbon-to-carbon bond), in particular a large amount of keyakinin and keyakinol -(3,5,4'-trihydroxy-7-methoxyflavonol and its dihydro derivative) attached to glucose in the 6 position (214). Orobol-7-methyl ether (5,7,3',4'-isoflavone=orobol) has been found in *Pterocarpus santalinus* wood (215). Ketonic flavonoids have been identified in the wood of other *Pterocarpus* species and other genera of Leguminosae (216). Besides various common flavonoids, *Caesalpinia sappan* wood contains an interesting group of homoisoflavonoids such as 8-methoxybonducellin (Fig. 48b) (217).

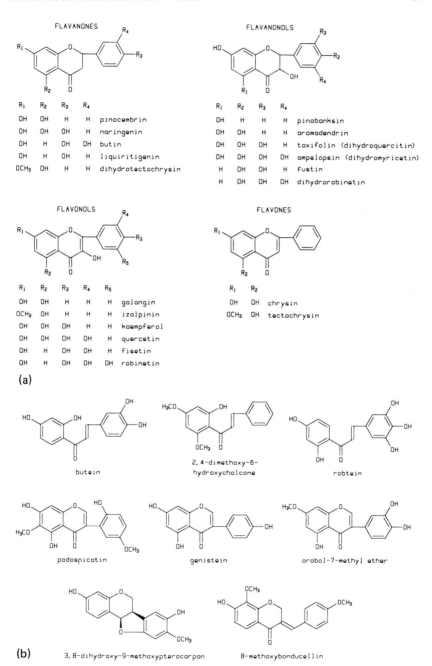

Figure 48 Flavanones, flavanonols, flavones, flavonols, isoflavones, and chalcones from wood (a and b).

2. Oligomeric Flavonoids

Oligomeric flavonoids are rather widespread in nature and commonly co-occur with the monomeric flavonoids. Several types of oligomeric flavonoids can be distinguished on the basis of their biosynthesis and structure.

Oligomeric *Proanthocyanidins** represent oligomeric flavonoids based on the polymerization of flavan-3,4-diols and flavan-3-ols. The polymerization takes place at a slightly acidic pH, most likely with the participation of enzymes, but it is nonoxidative in nature. The basic mechanism, depicted in Fig. 49, involves the activation of the 4 positions of flavan-3,4-diols under the formation of quinonemethides, 4-carbonium ions, or 3-ene-3-ol structures, which condense with ring A of another flavan-3,4-diol or flavan-3-ol unit at positions 6 or 8. As a result, a straight or branched chain of flavan-3-ol units is generated. Since flavan-3-ols are incapable of forming quinone methides by this mechanism, they act only as chain-terminating units. The main linkage between units has been found to be 4-8 or occasionally 4-6, although other bonds, including ether linkages, have also been discovered (218–224). These oligomers will be further discussed under tannins.

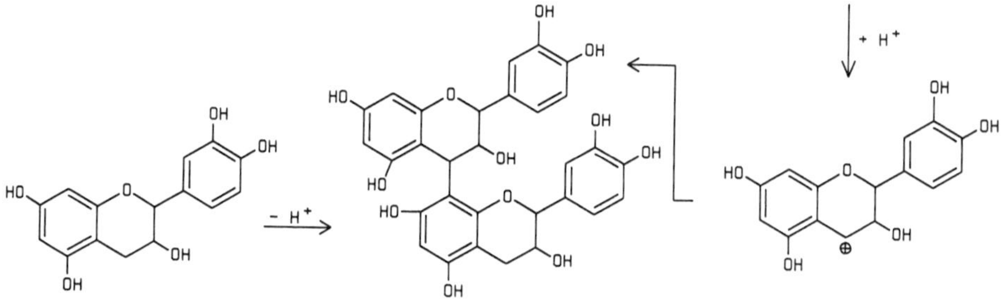

Figure 49   Nonoxidative, acidic polymerization of flavan-3-ols and flavan-3,4-diols.

---

*Proanthocyanidins are defined as colorless plant materials forming anthocyanidins upon heating with acid. They include flavan–3,4–diols and their polymers.

Although flavan-3-ols are incapable of nonoxidatively forming quinone methides (or 4-carbonium ions) and polymerizing in the absence of flavan-3,4-diols, they can polymerize by oxidative coupling to produce oligomers linked in a different way. Following oxidative attack on the hydroxy groups of rings A and B of the flavan-3-ols, free radicals or orthoquinones can form, which couple to produce oligo- and polymeric materials. This reaction has been demonstrated in vitro using enzymatic oxidation, and several dimeric products linked 8-6' have been isolated. It is suspected that this or a similar reaction is responsible for the loss of brightness of *Tsuga heterophylla* wood pulp (225–229). Recently, dehydrodicatechin A (Fig. 50), apparently formed by oxidative coupling, has been isolated from *Prunus grayana* heartwood (132).

Oligoflavonoids (biflavonoids) denotes the materials, mainly of a dimeric nature, that form by the oxidative coupling of ketonic flavonoids. Their structures involve carbon-to-carbon bonds at positions 3,6,8, or 3' or ether linkages involving

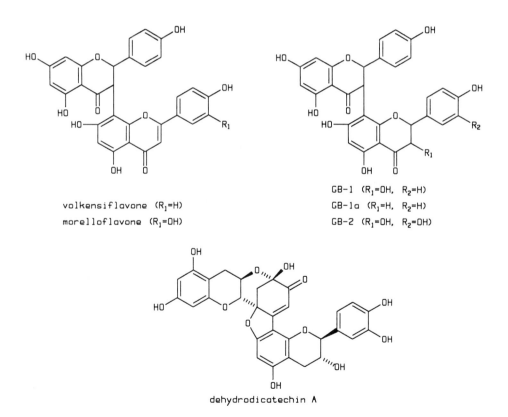

volkensiflavone (R₁=H)
morelloflavone (R₁=OH)

GB-1 (R₁=OH, R₂=H)
GB-1a (R₁=H, R₂=H)
GB-2 (R₁=OH, R₂=OH)

dehydrodicatechin A

Figure 50 Biflavonoids from the wood of *Garcinia* and *Prunus grayana*.

the 4' hydroxyl of one flavonoid molecule and carbons at 6,8, or 3' of the other, and they are easy to understand as coupling products of the respective free radicals, with ultimate stabilization of the resultant dimer (228,229). A relatively large number of such dimeric oligoflavonoids (biflavonoids) has been isolated in recent years, predominantly from the leaves of Gymnospermae (Ginkgoaceae, Taxaceae, Araucariaceae, Cephalotaxaceae, Cupressaceae, Podocarpaceae, Taxodiaceae, but none in Pinaceae), or from the leaves of certain hardwood families. Oligoflavonoids are apparently rare in heartwood, having been identified only in a few tropical species such as *Garcinia* (Guttiferae) where several biflavonoids were found (230–235). These represented flavone, flavanone, and flavanonol units linked 3-8 by carbon-to-carbon bonds and were found both as aglycones and glycosides (Fig. 50).

3. Stilbenoids

Stilbenoids include various derivatives of *trans*-stilbene (although *cis*-stilbenes are also known) and its hydrogenation product bibenzyl or dihydrostilbene. Stilbenes are biosynthetically close to flavonoids and tend to occur with them. They are commonly hydroxyl-substituted, with the substitution pattern conforming to their biosynthetic derivation: Ring A is commonly substituted in positions 3 and 5 and ring B in positions 4'; 3',4'; or 3',4',5'. The hydroxyls can be methylated or glycosidated. Exceptions to the above pattern have occasionally been found, either with hydroxyls at the "wrong" locations or arising due to various transformations of the initially formed compounds (236). Stilbenoids are much less diversified than flavonoids and include less than 100 structures; a few important ones are given in Fig. 51.

In conifers, stilbenoids are found in the foliage of *Abies* and *Picea*, in the bark of *Picea* and *Pinus*, and in the heartwood of *Pinus*, particularly in subgenus *Strobus* (soft pines). In *Pinus* heartwood, stilbenoids include 3,5-dihydroxystilbene (pinosylvin), its mono- and dimethyl ethers (183,184), and 4-hydroxystilbene and its methyl ether (237), as well as dihydropinosylvin (3,5-dihydroxybibenzyl) and its mono- and dimethyl ethers, the latter three generally in smaller amounts than their double-bond counterparts (183,184,238). Recently, seven *cis*- and *trans*-pinosylvin dimethyl ether derivatives were isolated from the heartwood of *Pinus armandi*, including two epoxides (186). The taxonomic value of stilbenoids in *Pinus* was mentioned earlier.

Among the hardwoods, stilbenes are common constituents of the heartwood of Moraceae. Thus, resveratrol, 2'-hydroxyresveratrol, piceatannol, and the bibenzyls 3,4'-dihydroxydihydrostilbene, 3,2',4'-trihydroxydihydrostilbene, and dihydroresveratrol together with 6,3',5'-trihydroxy-2-phenylbenzofuran were identified in the heartwood of several *Morus* species (239); hydrogenated stilbenes were generally found in smaller quantities. Resveratrol was the main constituent, occurring in amounts of from 0.6 to as high as 7.3%. Six *Artocarpus* species (the same family) produced resveratrol and oxyresveratrol in the heartwood; that of

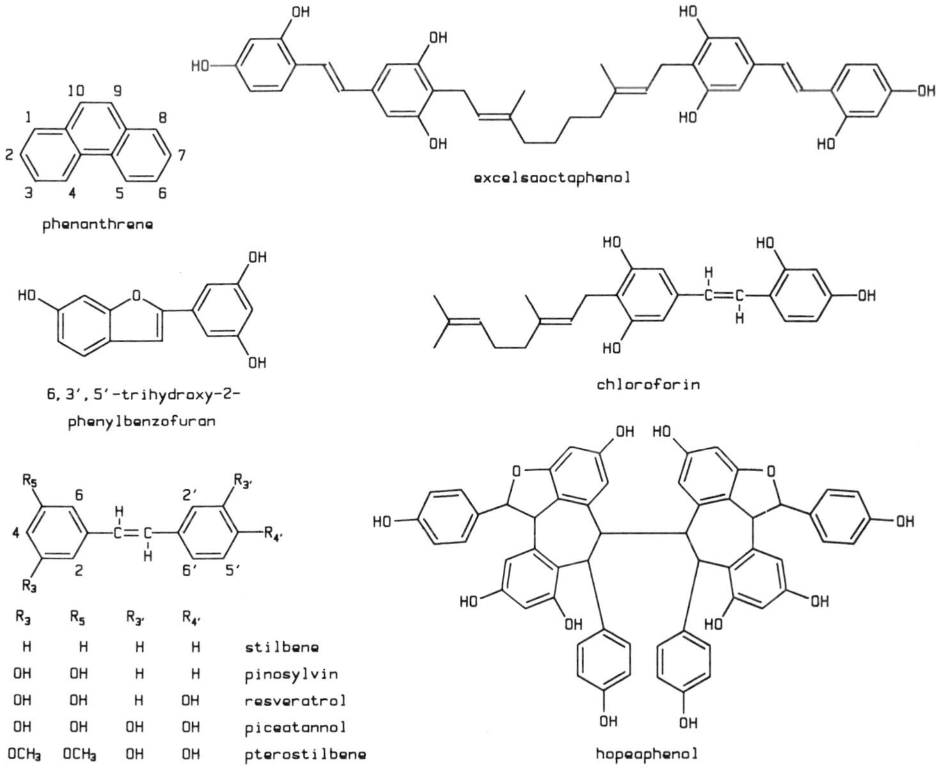

Figure 51 Stilbenoids and related compounds from wood.

Maclura pomifera contains 1% of 2' hydroxyresveratrol (240).

Resveratrol, free and as its glucoside, was found in the heartwood of many *Eucaluptus* species (241,242). Among Fagaceae, resveratrol was identified in *Nothofagus fusca* (243). Heartwoods of the tropical Leguminosae commonly contain stilbenes. Thus, piceatannol and 3,4,3',4',5'-pentahydroxystilbene were isolated from *Vouacapoua macropetala* and *V. americana* (244); the *Intsia* species contain resveratrol and 3,5,3'4'-tetrahydroxystilbene (245); and *Centrolobium robustum* contains piceatannol (246). Several *Pterocarpus* species contain pterostilbene in their heartwood (247,248). *Schotia brachypetala* heartwood produces 3,5,3',4',5'-pentahydroxystilbene in an amount of 7.2% (249). *Chlorofora regia* and *C. excelsa* (Moraceae) contain 2,4,3',5'-tetrahydroxystilbene and chloroforin (oxyresveratrol with a chain of two isoprene units at 4'). The heartwood of *Haplormosia monophylla* contains resveratrol and the *Afrormosia* species contain 3,4,3',5'-tetrahydroxystilbene (250,251) (Fig. 51).

The heartwood of the tropical genus *Combretum* (Combretaceae) produces bibenzyls (4'-hydroxy-3,4,5-trimethoxybibenzyl, 4,4'dihydroxy-3,5-dimethoxybibenzyl, and 3,4'dihydroxy-4,5-dimethoxybibenzyl) and various hydroxy- and methoxyphenanthrenes and their 9,10-dihydroderivatives oxygenated in 2,3,4,6, and 7 (rarely also in the 5) positions (252). Phenanthrenes and their dihydro derivatives most likely form by an internal oxidative coupling reaction, from corresponding *cis*-stilbenes or dihydrostilbenes, followed by electron redistribution, prototropy, and in some cases methylation.

Oligomers of stilbenes (Fig. 46) are relatively rare. Hopeaphenol (Fig. 51), a tetramer of resveratrol, has been found in the heartwood of the Dipterocarpaceae species (*Hopea odorata*, *Balanocarpus heimii*, *Shorea robusta*, and *S. talura*) (253,254).

4. Other Compounds of Mixed Biosynthetic or Unclear Origin

Compounds resulting from other combinations of biosynthetic pathways are relatively rare and are found mainly in the wood of exotic species. *Chlorofora excelsa* produces geranylated stilbenes such as chloroforin and 4-geranyl resveratrol, as well as the more complicated excelsaoctaphenol, a compound composed of two stilbene units and a geranyl chain extended by two carbons (Fig. 51) (255–259). Prenylated flavonoids have been identified in the *Artocarpus* species (239). Various benzoquinones, synthesized in part by mevalonic acid pathways, have been identified in west African and South American *Cordia* species (cordiachromes) (Fig. 52a) (260,261). Naphthoquinones and napthalenes carrying terpenoid units have been isolated from the South American *Tabebuia avellanedea* and *Paratecoma peroba*, as well as from East Asian *Tectona gradis* (lapachole derivatives) (262,263), whose heartwood also contains derivatives of anthraquinone, such as 9,10-dimethoxy-2-methylanthra-1,4-quinone (Fig. 52a) (264). Prenylated quinones related to lapachol and their dimers, such as tecomaquinone I, were identified in the heartwood of *Tababuia pentaphylla* (Fig. 52b) (265). The African, South Asian, and South American *Dalbergia* species and South American *Machaerium* species contain compounds in which a benzoquinone unit or related phenolic unit is attached to the side chain of a saccharogenic *n*-propylphenol unit (mucroquinone, dalbergione derivatives). Several xanthones (e.g., euxanthone) formed by a combination of acetogenic and saccharogenic biosynthetic pathways (35) have been identified in the heartwood of the *Garcinia* species. Mangostin, a diprenylated xanthone formed by a combination of all three biosynthetic pathways, was identified in the same source (232). Some relatively simple 1,4-benzoquinone derivatives have been identified in the *Dalbergia* (2,5-dimethoxy-1-4-benzoquinone) and *Acacia* species (2,6-dimethoxy-1,4-benzoquinone and acamelin) (9) (Fig. 52a). The heartwood of *Chamaecyparis pisifera* produces the "ascorbigen" sawaranin (Fig. 52b) (266).

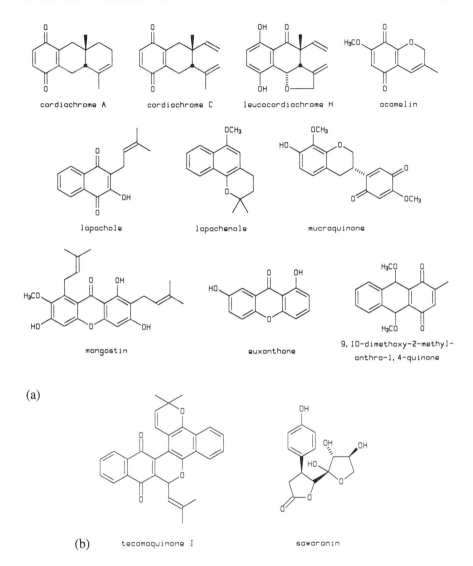

cordiachrome A   cordiachrome C   leucocordiachrome H   acamelin

lapachole   lapachenole   mucroquinone

mangostin   euxanthone   9, 10-dimethoxy-2-methyl-
                                                anthra-1, 4-quinone

(a)

(b)   tecomaquinone I   sawaranin

Figure 52 Benzoquinones, naphthoquinones, xanthones, and ascorbigens from the wood of some hardwood species (a and b).

E. Polymeric Phenols

Polymeric phenolic compounds are the most widespread and abundant heartwood extractives. The dark color of heartwood depends primarily (but not entirely) on the presence of these polymers. The polymeric phenols can be divided into several groups on the basis of their solubility.

1.   *Tannins* are the designated polymeric phenolic materials that are soluble in water as well as in polar organic solvents. Of all phenolic polymers, they have the lowest DP (molecular weights between 500 and 3000) and are predominantly oligomeric rather than truly polymeric. Occasionally, the name "tannins" is used in a narrower sense to designate polyphenolic materials that can combine with proteins, a reaction forming the basis of the tanning of animal hides. We prefer the broader definition, however, since tannins are also used for other purposes, such as control of the viscosity of oil-drilling muds, preparation of adhesives, etc., in which the precipitation of proteins plays no role.

2.   Natural *phlabaphenes* are polymeric phenolic materials that are soluble in polar organic solvents, but insoluble in water.

3.   *Phenolic acids* are phenolic polymers that are insoluble in either water or polar organic solvents, but are dissolved by dilute alkalies. This name is rather unfortunate since it leads to confusion with unrelated compounds such as gallic acid or salicylic acid that bear a carboxylic and phenolic group.

4.   Even after extracting some heartwoods with all the above solvents, some extraneous phenolic polymers remain undissolved. These are usually co-determined with Klason lignin and increase the Klason lignin content of certain woods. One can generally recognize their presence by the abnormally low methoxyl content of the Klason lignin or by abnormally low vanillin yields in the alkaline nitrobenzene oxidation of lignin (Fig. 1).

1. Tannins

There is a very large body of information, albeit of varying quality, on the distribution, structure, and chemical reactions of plant tannins. Most of the information, however, relates to bark and foliage rather than wood, an imbalance that is no doubt a reflection of the relative commercial importance of these tannin sources. Still, it is probably safe to say that tannins are present in most heartwoods, particularly in those having a darker color. Heartwood tannin content has been shown to vary between rather wide limits: It can be as high as 25% (the *Schinopsis* species) (164), but is usually much less, often below 1%. Most of the detailed chemical information available on wood tannins is from studies of those species that are commercially important sources of these materials, so our knowledge of the chemistry and structure of wood tannins is somewhat incomplete.

Traditionally, tannins are divided into (1) hydrolyzable tannins and (2) condensed tannins.

a. Hydrolyzable Tannins.   In hydrolyzable tannins, the individual units of the tannin oligomers are attached to each other by ester bonds, which can be broken by acid or alkaline hydrolysis. The resulting monomers are of two types: carbohydrate monomers, predominantly D-glucose and aromatic compounds, predominantly gallic acid and the product of its oxidative coupling, 2,2'-dicarboxy-4,5,6,4',5',6'-hexahydroxybiphenyl, which lactonizes during isolation to its dilactone, ellagic acid. The hydrolyzable tannins are subdivided into gallotannins and ellagitannins, depending on whether hydrolysis produces gallic acid alone, or also yields ellagic acid and related materials.

The structure of most gallotannins includes a unit of D-glucose (or less commonly other aliphatic or alicyclic materials like quinic acid) to which are attached by ester linkages the gallyl units. In some cases, additional gallyl units are attached to the hydroxy groups of the primary gallyl units (depside linkages), forming oligomeric gallyl chains.

Ellagitannins can be considered products of the enzymatic oxidation of gallotannins, resulting in the oxidative coupling of the gallyl units attached to D-glucose. As a result, carbon-carbon or carbon-oxygen-carbon bonds are formed. In some cases, the coupled products are transformed to produce orthoquinoid or 1,2,3-triketocyclohexane units and further modified products. Upon hydrolysis, the freed oligomers often undergo additional changes that further contribute to the difficulties of structure determination. Typical tannin structures are given in Fig. 42. For review, see (162–164).

Hydrolyzable tannins are rare or nonexistent in conifer wood, but are more common in hardwoods, although even here, most information does not distinguish between hydrolyzable and condensed tannins. In some cases, the occurrence of gallic and/or ellagic acid has been reported and the presence of hydrolyzable tannins has been assumed on that basis. The botanical genera that produce substantial amounts of hydrolyzable tannins in wood include *Quercus*, *Castanea*, *Eucalyptus*, and related genera. Thus, the wood of *Castanea crenata* contains 8.6% of hydrolyzable tannins (267). *C. sativa* wood produces hydrolyzable tannins from which four different compounds were isolated and the structure determined (268). The wood of the *Nothofagus* species such as *N. procera* contains hydrolyzable tannins (269), as does the heartwood and sapwood of *Quercus rubra* (134), the heartwood of *Q. alba* (270), and the wood of almost all *Eucalyptus* species (1) e.g. *E. astringens*, *E. gigantea*, *E. regnans*, and *E. sieberiana* (271–273). The presence of hydrolyzable tannins or their building blocks (e.g., gallic and ellagic acids) has also been reported for the wood of *Juglans sieboldiana* and *Platycarya strobilacea* (274–275). Hydrolyzable tannins apparently occur in the heartwood of some *Celtis* species, as gallic and 3,3'-di-*O*-methylellagic acid have been found in *Celtis australis* (276). The wood of some species of Tiliaceae, such as *Elaeocarpus dentatus*, *E. hookerianus*, and *Aristotelia serrata*, contain hydrolyzable tannins (277, 278).

b. Condensed Tannins.   Condensed tannins are those that upon treatment with hot mineral acids condense to water-insoluble, brown materials termed synthetic phlobaphenes. Condensed tannins are generally more widespread in wood than are hydrolyzable tannins. The commercially most important condensed tannins obtained from wood are those extracted from the *Schinopsis* species of South America and the *Acacia* species of South Africa. These condensed tannins and a few others have been the subject of numerous investigations and their chemical constitution is thus relatively well known. The precursors of these tannins are flavan-3-ols (catechins) and flavan-3,4-diols (leucoanthocyanidins) that, as mentioned earlier, condense between carbons 4 and 8 (less frequently between carbons 4 and 6) to form di-, tri-, and higher oligomers with flavan-3-ols as the repeat unit (Fig. 49). These oligomers are designated as proanthocyanidins. It would be a mistake, however, to equate condensed tannins with proanthocyanidins. Convincing indirect evidence, although not yet supported by the structural determination of the pertinent oligomers, points to the existence of other types of condensed tannins, based on different units. Thus, Grassmann and Endres proposed that the condensed tannins from spruce bark derive from the polymerization of stilbene monomers (279), which are rather ubiquitous in the various plant organs. Other examples will be dealt with later.

Condensed tannins are widespread in conifers, not only in bark, where they are plentiful, but also in wood, where they are usually found only in small amounts. These tannins are apparently not only of a proanthocyanidin nature. Unfortunately, the published information on the occurrence and nature of conifer wood tannins is restricted to a few papers, so the above generalizations must be considered tentative.

Juniperus communis, J. nana, and *J. phoenicea* wood has been found to contain a proanthocyanidin tannin (168,169,280–282). It has been demonstrated that the heartwood of *J. occidentalis* contains 1% of a phlobaphene (283). The heartwood of *Libocedrus decurrens* contains 0.15% of a tannin (purified), which on pyrolysis gives carvacrol and *p*-methoxythymol, as well as the more common products of phenolic pyrolysis (phenol, catechol, pyrogallol, and their methyl adducts). This tannin thus appears to be structurally different from proanthocyanidins (284). *Sequoia sempervirens* heartwood contains appreciable amounts of tannin, generally about 7%, that has been the subject of several investigations. It is generally classified as a condensed tannin, and a polycatechin structure was proposed for it (286). The evidence for this structure was weak, however, and since neither catechins nor leucoanthocyanidins were isolated from redwood, it seems more reasonable to assume that the tannin has a polysequirin structure, since sequirins are also present in sizable amounts and are known to polymerize under acidic or oxidative conditions (285,287,288).

More information on condensed tannins is available for the hardwoods. Hillis lists the heartwood of the South American *Schinopsis balansae* and *S. lorentzii,* the South African *Acacia catechu,* and the Australian *Eucalyptus wandoo* as major sources of commercial condensed tannins, with tannin contents of 20–25%, 16–17%, 15%, and 13–15%, respectively. All these tannins appear to be entirely or predominantly of the proanthocyanidin type (164, p. 325). The heartwood of other species of *Acacia,* such as *A. mearnsii* (289–291), also represents a rich source of tannins. As for the other *Eucalyptus* species, direct evidence from the isolation of individual tannin components as well as indirect evidence from the isolation of the co-occurring monomers point to the presence of both hydrolyzable and condensed tannins, in varying proportions and amounts, possibly including tannins based on stilbene units (292–294). The heartwood of *Quercus* contains condensed tannins, accompanied by hydrolyzable tannins, with the latter generally predominating. The heartwood of *Prunus* contains sizable amounts of tannin, as does *Maclura pomifera* (9.4%), and *Robinia pseudoacacia* (1.5%) (1, Vol. IV, p. 144; 295–298). Proanthocyanidins have also been identified in the heartwood of *Acer rubrum* (14,299), the species of Tiliaceae, Myrtaceae, and *Nothofagus* (301), and in small amounts some species of the Lauraceae family (300).

2. Phlobaphenes, Phenolic Acids, Insoluble Polymeric Phenols

These three classes of materials could be called the pariahs among natural products in that they are little studied despite their widespread occurrence. For example, although phlobaphenes can occur in heartwood in sizable amounts, only their percentage is occasionally reported, while in most cases they are simply ignored. The same is true for the phenolic acids and insoluble polymeric phenolics that occur in barks in large quantities, although their amount in heartwood is usually small. As a result, little is known about these classes of compounds.

All three classes are regarded as closely related to the co-occurring condensed tannins. Their insolubility in water is generally explained by higher molecular weight or a higher degree of condensation involving the loss of hydrophilic hydroxyl groups, but this is most likely an oversimplification. Thus, Swan, working with a phlobaphene from *Thuja plicata* bark, identified fatty acids in its alkaline hydrolyzate and concluded that they must have been originally bound by ester linkages (302). The work of Foo and Karchesy on the bark phlobaphene of *Pseudotsuga menziesii* indicated that this component was a complex mixture derived from condensed tannins, carbohydrates, and dihydroquercetin and contained a sizable number of methoxyls (306). Thus, it seems likely that in many cases not only differences in molecular weight and degree of condensation but also inclusion of some hydrophobic units in the structure are responsible for the insolubility of phlobaphenes.

Phlobaphene from the sapwood of *Sequoia sempervirens* was obtained in a 0.7% yield, from the heartwood in a 6.2% yield, and from the roots in a 22.1% yield. The methoxyl content of the heartwood phlobaphene was 6.9% vs. 2.8% for the co-occurring tannin (291). *Calocedrus decurrens* heartwood contains 11% of a phlobaphene and 3% of a tannin (crude); both materials are present in sapwood in much smaller quantities. This phlobaphene is composed of a polar and a nonpolar fraction with a methoxyl content of 6.6 and 13.3%, respectively; the corresponding tannin has a methoxyl content of 12.2% (303). Later Zavarin and Snajberk investigated the phlobaphenes from the heartwood of *Calocedrus decurrens*, *Cupressus pygmaea*, *Abies magnifica*, *Pinus albicaulis*, and *Sequoia sempervirens* (yields of 6.2, 1.2, 0.25, 0.2, and 5.5%, respectively). The methoxyl content of the phlobaphenes varied between 6 and 14%. On pyrolysis, 27 low-boiling phenols were obtained, most of them methyl derivatives of phenol, pyrocatechol, and pyrogallol and their 4-ethyl and 4-*n*-propyl homologues. In addition, carvacrol and *p*-methoxythymol (Fig. 9) were obtained from the phlobaphenes from *Calocedrus decurrens* and *Cupressus pygmaea* (304). Much of these data are inconsistent with the picture of phlobaphenes as compounds derived by the polymerization and condensation of proanthocyanidins.

A few words are necessary here on the relationship of the phlobaphenes to so-called Braun's native lignin (BNL). This material was first isolated from *Picea mariana* by alcohol extraction and was thought to represent the soluble portion of the true lignin (305). Later, this assumption was repeatedly challenged for a variety of reasons. The recent careful studies of Hergert and of Sakakibara et al., which included the distribution of BNL in wood, isolation of intermediates, and comparison with milled wood lignin using NMR spectroscopy, IR spectroscopy, gel filtration, chemical degradation, and elemental and functional group analysis, left little doubt that BNL is a polymer of lignans and related phenylpropane coupling products (17,18) and is therefore just a special form of phlobaphene.

F. Alkaloids

With the exception of *Taxus,* in which, as mentioned earlier, a number of diterpenoids (taxan alkaloids) with amine or amide groups have been identified (83) (Fig. 24), alkaloids are absent in the wood of softwoods.

Some hardwoods, however, particularly those from the tropics, contain alkaloids in appreciable amounts. Thus, alkaloids were identified in the wood of *Pericopsis elata* from western Africa, *Dicorynia quaianesis* from South America, *Gonioma kamassi* from South Africa, *Ocotea rodiaei* from South America, *Prosopis juliflora* from Central America, *Aspidosperma peroba* and *A. vargasii* from India and Sri Lanka, and *Cryptocarya pleurosperma* from Australia (9). Of North American woods containing alkaloids, mention should be made of *Liriodendron tulipifera* that contains aporphine alkaloids of which about 30 have been identified (307–309) (Fig. 53).

Additional references to the occurrence of alkaloids in wood can be found in (8,9,11).

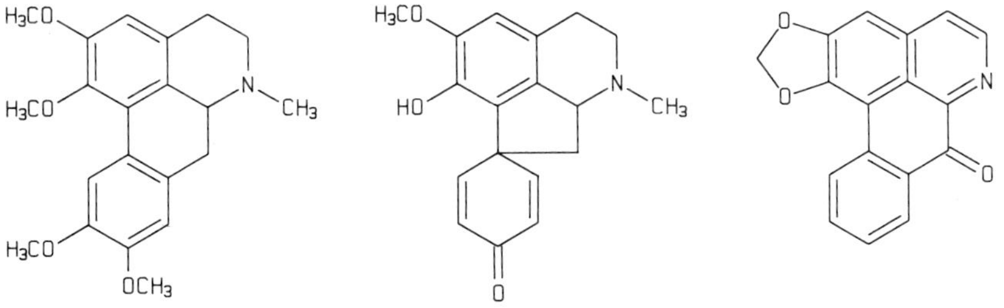

(+) glaucine   N-methylcrotsparine   liriodenine

Figure 53   Alkaloids from *Liriodendron tulipifera*.

APPENDIX

Latin and Common Names for Botanical Taxa[a]

Families

Latin Name	Common name
Araucariacea	Araucaria family
Cephalotaxaceae	Plum yew family
Combretaceae	Combretum family
Cupressacae	Cypress family
Dipterocarpaceae	[b]
Euphorbiaceae	Spurge family
Fagaceae	Beech family
Ginkgoaceae	Ginkgo family
Guttiferae	Mangosteen family
Hippocastanaceae	Horsechestnut family
Juglandaceae	Walnut family
Lauraceae	Laurel family
Leguminosae	Legume family
Moraceae	Mulberry family
Myoporaceae	[b]
Myrtaceae	Myrtle family
Pinaceae	Pine family
Podocarpaceae	Fern pine or yew pine family
Salicaceae	Willow family
Santalaceae	Sandaltree family
subfamily Mimosoideae	[b]
subfamily Caesalpinoideae	[b]
subfamily Lotoideae	[b]
Taxaceae	Yew family
Taxodiaceae	Redwood family
Tiliaceae	Linden family

Genera

Latin name	Common name	Family
Abies Mill.	Firs	Pinaceae
Acacia Mill.	Acacias	Leguminosae
Acer L.	Maples	Acaraceae
Afrormosia Harms[b]	Kokrodua, assamela	Leguminosae
Artocarpus L.	Keledang	Moraceae
Betula L.	Birches	Betulaceae
Carya Nutt.	Hickories	Juglandaceae
Castanea Mill.	Chestnuts	Fagaceae
Castanopsis (D. Don) Spach	Chinkapins	Fagaceae
Cedrus Link	True cedars	Pinaceae
Celtis L.	Hackberries	Ulmaceae
Chamaecyparis Spach	White cedars	Cupressaceae
Citrus L.	Citruses	Rutaceae
Combretum L.	[b]	Combretaceae
Copaifera L.	Copaiba	Leguminosae
Cordia L.	Cordias: bohari, thanat, anonang	Boraginaceae
Cotinus Mill.	Smoke trees	Anacardiaceae
Cryptomeria D. Don	Japanese cedar, sugi	Taxodiaceae
Cunninghamia (Richard) R.Brown	China firs	Taxodiaceae
Dacrydium Solander	[b]	Podocarpaceae
Dalbergia L.	Indian rosewood	Leguminosae
Diospyros L.	Persimmons, ebonies	Ebenaceae
Eucalyptus L´. Hèr.	Eucalypts	Myrtaceae
Fagus L.	Beeches	Fagaceae
Fraxinus L.	Ashes	Oleaceae
Garcinia L.	Kandis	Guttiferae
Intsia Thou.	Ipil, merbau	Leguminosae
Juglans L.	Walnuts	Juglandaceae
Keteleeria Carr.	David fir	Pinaceae
Larix Mill.	Larches	Pinaceae
Lithocarpus Blume	Tanoaks	Fagaceae
Machaerium Pers	Caviuna, pau ferro	Leguminosae
Melanorrhoea Wall.	Roble	Anacardiaceae
Metasequoia S. Miki	Dawn redwood	Taxodiaceae
Nothofagus L.	Antarctic beeches	Fagaceae

Latin name	Common name	Family
Picea A. Dietr.	Spruces	Pinaceae
Pinus L.	Pines	Pinaceae
subgenus *Strobus* Lemm.	Soft or white pines	Pinaceae
subgenus *Pinus*	Hard or yellow pines	Pinaceae
Platanus L.	Sycamores	Platanaceae
Podocarpus (Pers.) L'Herit.	[b]	Podocarpaceae
Populus L.	Cottonwoods, poplars	Salicaceae
Prunus L.	Plums, cherries, peaches	Rosaceae
Pseudolarix Gord.	Golden larch	Pinaceae
Pseudotsuga Carr.	Douglas firs	Pinaceae
Pterocarpus L.	Padauk, narra	Leguminosae
Quercus L.	Oaks	Fagaceae
Rhus L.	Sumacs	Anacardiaceae
Robinia L.	Locusts	Leguminosae
Salix L.	Willows	Salicaceae
Schinopsis Endl.	Quebracho	Anacardiaceae
Sciadopytis Sieb. & Zucc.	Japanese umbrella pine, koya-maki	Taxodiaceae
Taiwania Hayata	[b]	Taxodiaceae
Taxus L.	Yews	Taxaceae
Tsuga (Endl.) Carr.	Hemlocks	Pinaceae
Ulmus L.	Elms	Ulmaceae

Species

Latin name	Common name	Family
Abies amabilis (Dougl.) Forbes	Pacific silver fir	Pinaceae
Abies concolor (Gord. & Glend.) Lindl.	White fir	Pinaceae
Abies magnifica A. Murr.	California red fir	Pinaceae
Abies pindrow (Royle) Spach	West Himalayan fir	Pinaceae
Abies sibirica Ledeb.	Siberian fir	Pinaceae
Acacia catechu Willd.	African catechu, cutch	Leguminosae
Acacia mearnsii De Wild.	Black wattle	Leguminosae
Acacia modesta Wall.	Palosa	Leguminosae

Latin name	Common name	Family
Acer negundo L.	Boxelder	Aceraceae
Acer rubrum L.	Red maple	Aceraceae
Acer saccharum Marsh.	Sugar maple	Aceraceae
Aesculus hippocastanum L.	Horsechestnut	Hippocastana-ceae
Agathis australis (D. Don) Salisb.	Kauri pine	Araucariaceae
Aniba rosaeodora Ducke	Brazilian rosewood, louro	Lauraceae
Aristotelia serrata W.R.B. Oliver	Makomako, wine berry	Tiliaceae
Aspidosperma peroba Fr. Allem.	Red peroba, peroba amarello	Apocynaceae
Aspidosperma vargasii A. DC.	Satinwood, amarillo	Apocynaceae
Athrotaxis selaginoides Don	King William pine	Taxodiaceae
Austrocedrus chilensis (D. Don) Florin & Boulelje	Chilean cedar	Cupressaceae
Balanocarpus heimii King	Takien-chan, chengal	Dipterocarpa-ceae
Betula verrucosa Ehrh.	Silver birch	Betulaceae
Biota orientalis Endl.	Chinese arbor-vitae	Cupressaceae
Caesalpinia sappan L.	Sappan	Leguminosae
Calocedrus decurrens (Torr.) Florin[d]	Incense-cedar	Cupressaceae
Carya pecan (Marsh.) Engl. & Graebn.	Pecan	Juglandaceae
Castanea crenata Sieb. & Zucc.	Japanese chestnut, Kuri	Fagaceae
Castanea sativa Mill.	Sweet chestnut	Fagaceae
Cedrus deodara (Roxb.) Loudon	Deodar cedar	Pinaceae
Cedrus libani Loudon	Lebanon cedar	Pinaceae
Celtis australis L.	Celtis	Ulmaceae
Centrolobium robustum Mart.	Porcupine wood, canary wood	Leguminosae
Chamaecyparis obtusa (Sieb. & Zucc.) Endl.	Hinoki, Hinoki cypress	Cupressaceae
Chamaecyparis pisifera Endl.	Sawara	Cupressaceae
Chlorophora excelsa (Welw.) Benth. & Hook.	Iroko, African teak	Moraceae
Chlorophora regia A. Cheval.	Odoum	Moraceae
Cryptomeria japonica D. Don	Sugi	Taxodiaceae

Latin name	Common name	Family
Cryptocarya pleurosperma White & Francis	Poison walnut	Lauraceae
Cupressus pygmaea (Lemm.) Sarg.	Pygmy cypress	Cupressaceae
Dacrydium franklinii Hook.	Huon pine	Podocarpaceae
Dicorynia guaianesis Amshoff	Basralocus	Leguminosae
Elaeocarpus dentatus Vahl	Hinau	Tiliaceae
Elaeocarpus hookerianus Raoul	Pokaka	Tiliaceae
Eucalyptus astringens Maiden.	Brown mallet	Myrtaceae
Eucalyptus calophylla (Lindl.) R. Br.	Red gum	Myrtaceae
Eucalyptus corymbosa Cav.	Bloodwood	Myrtaceae
Eucalyptus gigantea Hook.	Gum-top stringybark	Myrtaceae
Eucalyptus hemiphloia (Benth.) F. Muell.	Bark	Myrtaceae
Eucalyptus maculata Hook.	Spotted gum	Myrtaceae
Eucalyptus regnans F. Muell.	Giant gum	Myrtaceae
Eucalyptus sieberiana F. Muell.	Mountain ash	Myrtaceae
Eucalyptus wandoo Blakely	Wandoo	Myrtaceae
Fitzroya cupressoides (Mol.) Johnst.	Alerce	Cupressaceae
Ginkgo biloba L.	Maidenhair tree	Ginkgoaceae
Gonioma kamassi E. Mey	African boxwood	Apocynaceae
Grevillea robusta Cunn.	Southern silky oak	Proteaceae
Haplormosia monophylla Harms.	Liberian black gum, idewa	Leguminosae
Hevea brasiliensis *M*uell.	Seringa, Para rubbertree	Euphorbiaceae
Hopea odorata Roxb.	Thingan, merawan	Dipterocarpaceae
Juglans sieboldiana Maxim.	Japanese walnut, Oni-gurumi	Juglandaceae
Juniperus communis L.	Common juniper	Cupressaceae
Juniperus occidentalis Hook.	Western juniper	Cupressaceae
Juniperus phoenicea L.	Phoenician juniper	Cupressaceae
Larix decidua Mill.	Common larch	Pinaceae
Larix leptolepis Sieb. & Zucc.	Japanese larch, Karamatsu	Pinaceae
Libocedrus decurrens Torr.[d]	Incense cedar	Cupressaceae
Libocedrus yateensis Guillaumin	[b]	Cupressaceae
Liriodendron tulipifera L.	Yellow poplar	Magnoliaceae

Latin name	Common name	Family
Maclura pomifera (Raf.) Schneid.	Osage-orange	Moraceae
Metasequoia glyptostroboides Hu & Cheng	Dawn redwood, shui-hsa	Taxodiaceae
Nothofagus fusca Oerst.	Red beech	Fagaceae
Nothofagus procera Oerst.	Rauli	Fagaceae
Ocotea rodiaei (Schomb.) Mez	Greenheart	Lauraceae
Ocotea usumbarensis Engl.	East African camphor wood	Lauraceae
Paratecoma peroba Kuhlm.	White peroba	Bignoniaceae
Pericopsis elata (Harms) Van Meuven[c]	Kokrodua, assamela	Leguminosae
Phebalium nudum Hook.	Tasmanian lancewood, satinbox	Rutaceae
Picea abies (L.) Karsten	Common spruce	Pinaceae
Picea jezoensis (Sieb. & Zucc.) Carr.	Yezo spruce, Hondo spruce	Pinaceae
Picea mariana (Mill.) B.S.P.	Black spruce	Pinaceae
Picea sitchensis (Bong.) Carr.	Sitka spruce	Pinaceae
Pilgerodendron uviferum (D. Don) Florin	White alerce	Cupressaceae
Pinus albicaulis Engelm.	White bark pine	Pinaceae
Pinus armandi Franchet	Armand's pine	Pinaceae
Pinus ayacahuite Ehrenb.	Mexican white pine	Pinaceae
Pinus discolor Bailey & Hawksworth	Border pinyon	Pinaceae
Pinus edulis Engelm.	Pinyon	Pinaceae
Pinus ellottii Engelm.	Slash pine	Pinaceae
Pinus radiata D. Don	Monterey pine	Pinaceae
Pinus jeffreyi Grev. & Balf.	Jeffrey pine	Pinaceae
Pinus lambertiana Dougl.	Sugar pine	Pinaceae
Pinus parviflora Sieb. & Zucc.	Japanese white pine	Pinaceae
Pinus patula Schl. & Cham.	Jelicote pine	Pinaceae
Pinus pinaster Ait.	Cluster pine, pin maritime	Pinaceae
Pinus radiata D. Don	Monterey pine	Pinaceae
Pinus sabiniana Dougl.	Digger pine	Pinaceae
Pinus sibirica Mayr.	Siberian white pine, Siberian cedar	Pinaceae
Platycarya strobilacea Sieb. & Zucc.	Nobunoki	Juglandaceae

Latin name	Common name	Family
Podocarpus spicatus R. Brown	Black pine, red pine	Podocarpaccae
Populus tremuloides Michx.	Quaking aspen	Salicaceae
Prosopis juliflora (Sw.) DC.	Mesquite	Leguminosae
Prunus grayana Maxim.	[b]	Rosaceae
Pseudotsuga menziesii (Mirb.) Franco	Douglas fir	Pinaceae
Pterocarpus marsupium Roxb.	Bija	Leguminosae
Pterocarpus santalinus L.	Sandalwood	Leguminosae
Quercus alba L.	White oak	Fagaceae
Quercus rubra L.	Northern red oak	Fagaceae
Robinia pseudoacacia L.	Black locust	Leguminosae
Salix hultenii Floderus	Hulten willow	Salicaceae
Santalum album L.	Genuine sandalwood	Santalaceae
Schinopsis balansae Engl.	Quebracho	Anacardiaceae
Schinopsis lorentzii Engl.	Quebracho	Anacardiaceae
Schotia brachypetala Sond.	Boerboen	Leguminosae
Sequoia sempervirens (D. Don) Endl.	Sequoia, California redwood	Taxodiaceae
Sequoiadendron giganteum (Lindl.) Buchh.	Big tree, giant sequoia	Taxodiaceae
Shorea robusta Gaertn.	Philippine mahagony	Dipterocarpaceae
Shorea talura Roxb.	Philippine mahagony	Dipterocarpaceae
Spirostachys africana Sond.	African sandalwood	Euphorbiaceae
Tabebuia avellanedea Lorentz	White cedar	Bignoniaceae
Tabebuia pentophylla (L.) Hemse	Roble	Bignoniaceae
Taiwania cryptomerioides Hayata	Taiwania	Taxodiaceae
Taxodium distichum (L.) Richards	Deciduous cypress	Taxodiaceae
Tectona grandis (L.)	Teak	Verbenaceae
Thuja plicata Don	Westem red cedar	Cupressaceae
Torreya nucifera Sieb. & Zucc.	Kaya	Taxaceae
Thuja heterophylla (Raf.) Sarg.	Western hemlock	Pinaceae
Tsuga heterophlla (Bong.) Carr.	Mountain hemlock	Pinaceae
Vouacapoua americana Aubl.	Acapu, wacapou, brownheart	Leguminosae
Vouacapoua macropetala Sandwith	Sarabebeballi	Leguminosae
Zelkova serrata Mak.	Sawleaf zelkova	Ulmaceae

aList of families and genera includes only the names mentioned as such. Common names preferentially included the botanical common names for the species; if such could not be found, the name or names for the corresponding wood were used. The authors are deeply grateful to Ms. N. Rem for checking this list.
bNo common name could be found.
cPericopsis elata (Harms) Van Meuven = Afrormosia elata Harms.
dCalocedrus decurrens (Torr.) Florin = Libocedrus decurrens Torr.

REFERENCES

1. R. Hegnauer, *Chemotaxonomie der Pflanzen*, Birkhäuser Verlag, Basel, 1962–1989, 8 vols.
2. J. B. Harborne, D. Boulter, and B. L. Turner, eds., *Chemotaxonomie of the Leguminosae*, Academic Press, London, 1971.
3. C. Wehmer, *Die Pflanzenstoffe*, Verlag von Gustav Fischer, Jena, 1929–1935, Germany, 3 vols.
4. A. Tschirch and E. Stock, *Die Harze*, Verlag von Gebrüder Borntraeger, Berlin, 1935, 2 vols. (photo reproduction 1943 by Edwards Brothers, Ann Arbor, Mich.).
5. E. Guenther, *The Essential Oils*, van Nostrand, New York, 1948–1952, 6 vols.
6. W. Karrer, *Konstitution und Vorkommen der Organischen Pflanzenstoffe*, Birkhäuser Verlag, Basel, 1958.
7. W. Karrer, E. Cherbuliez, and C. H. Eugster, *Konstitution und Vorkommen der Organischen Pflanzenstoffe—Ergänzungsband 1*, Birkhäuser Verlag, Basel, 1977.
8. J. W. Rowe and A. H. Conner, "Extractives in Eastern Hardwoods—A Review," Forest Products Lab., U.S. Dep. Agr., Madison Wis., 1979.
9. B. M. Hausen, *Woods Injurious to Human Health*, W. de-Gruyer, Berlin, 1981.
10. R. W. Hemingway and J. J. Karchesy, eds., *Chemistry and Significance of Condensed Tannins*, Plenum Press, New York, 1989.
11. J. W. Rowe, ed., *Natural Products of Woody Plants*, Vols. I and II, Springer Verlag, New York, 1989.
12. D. F. Zinkel and J. Russel, eds., *Naval Stores—Production—Chemistry—Utilization*, Pulp Chemicals Assoc., New York, 1989.
13. Ch. Mentzer and O. Fatianoff, *Actualities de Phytochimie Fondamentale*, Masson et Cie, Paris, 1964.
14. W. E. Hillis, *Wood Sci. Technol.*, 5:272 (1971).
15. W. E. Hillis, "Occurrence of Extractives in Wood Tissue," in *Biosynthesis and Biodegradation of Wood Components* (T. Higuchi, ed.), Academic Press, New York, 1985, p. 209.
16. W. E. Hillis, *Phytochem.*, 11:1207 (1972).
17. H. L. Hergert, "Secondary Lignification in Conifer Trees," in *Cellulose Chemistry and Technology* (J. C. Arthur, Jr., ed.), ACS Symposium Series 48, Washington, D.C., 1977, p. 227.
18. A. Sakakibara, T. Sasaya, K. Miki, and H. Takahashi, *Holzforschung*, 41:1 (1987).
19. A. Frey-Wyssling and H. H. Bosshard, *Holzforschung*, 13:129 (1959).
20. L. V. Smith and E. Zavarin, *TAPPI*, 43:218 (1960).
21. R. W. Hemingway and W. E. Hillis, *APPITA*, 24:439 (1971).

22.   A. Fahn, "Morphological and Anatomical Changes Related to Resin Stimulation," Final Rep., Dep. Botany, Hebrew Univ. of Jerusalem, Israel, 1970, Project A10-FS 15.

23.   P. Koch, "Utilization of the Southern Pines," *Agricultural Handbook*, Vol. 1, No. 420, U.S. Dept. Agr. Forest Service, Washington, D.C., 1972, p. 128.

24.   A. Tschirch and E. Stock, *Die Harze,* Verlag von Grüder Borntraeger, Berlin, 1935 Vol. 1, p. 20.

25.   W. Sandermann, *Naturharze, Terpentinöl, Tallöl*, Springer-Verlag, Belin, 1960, p. 70.

26.   E. Münch, "Naturwissenschaftliche Grundlagen der Kiefernharznutzung," in *Arb. Biol. Reichsant. Land-und Forstwirtsch.*, Springer-Verlag, Berlin, 1921, Vol. 10.

27.   F. W. Cobb, Jr., D. L. Wood, R. W. Stark, and P. R. Miller, *Hilgardia*, *39*:127 (1968).

28.   E. Back, *Svensk Papperstidn.*, *63*:647 (1960).

29.   E. Back, *Svensk Papperstidn.*, *63*:793 (1960).

30.   F. Mergen and R. M. Echols, *Sci.*, *121*:306 (1955).

31.   C. Bernard-Dagan, G. Pauly, A. Marpeau, M. Gleizes, J.-P. Carde, and Ph. Baradat, *Physiol. Veg.*, *20*:775 (1982).

32.   C. Bernard-Dagan, "Biosyntheses des Terpenes du Pin Maritime," in *Recueil des Conferences*, La Journee sur les Produits Resineux, Utilisations et Perspectives d'Application, Universite de Bordeaux, Institut du Pin, Talence, 1982.

33.   A. Ya. Kalninsh and E. E. Lazda, *Izv. Akad. Nauk Latv. SSR,(9)*:80(1974).

34.   E. E. Lazda, *Khim. Drev.* (1):108(1977).

35.   T. A. Geissman and D. H. G. Crout, *Organic Chemistry of Secondary Plant Metabolism,* Freeman, Cooper & Co., San Francisco, Calif., 1969.

36.   J. W. Porter and S. L. Spurgeon, eds., *Biosynthesis of Isoprenoid Compounds*, Wiley, New York, 1981.

37.   D. V. Banthorpe and B. V. Charlwood, "Biogenesis of Terpenes," in *Chemistry of Terpenes and Terpenoids*, (A. A. Newman, ed.), Academic Press, London, 1972, p. 337.

38.   E. Zavarin and K. Snajberk, *Biochem. Syst. Ecol.*, *14*:1 (1986).

39.   H. Avcibasi, H. Anil, and M. Toprak, *Phytochem.*, *26*:2852 (1987).

40.   H. Erdtman, 4th International Congress of Biochemistry, *Biochemistry of Wood* (K. Kratzl and G. Billek, eds.), Pergamon Press, London, 1959, Vol. II, p. 1.

41.   A. B. Anderson and E. Zavarin, *J. Inst. Wood Sci.*, *15*:3 (1965).

42.   K. Kafuku and R. Kato, *Bull. Chem. Soc. Japan*, *6*:65 (1931).

43.   F. Mueller, *Arch. Pharm.*, *238*:366 (1900).

44.   W. Parker and J. S. Roberts, "Sesquiterpene Biogenesis," *Chem. Soc.* (London), *Quar. Rev.*, *21*(1):331 (1967).

45.   N. H. Andersen, Y. Ohata, and D. D. Syrdal, "Studies in Sesquiterpene Biogenesis: Implications of Absolute Configuration, New Structural Types, and Efficient Chemical Simulation of Pathways," in *Bioorganic Chemistry* (E. E. van Tamelen, ed.), *Academic Press*, New York, 1978, Vol. II, Chap. 1, p. 1.

46.   Y. Hirose, Paper 28, Div. of Agr. and Food Chem., *158th Nat. Amer. Chem. Soc. Meeting*, Symp. Chemistry of Essential Oils, New York, 1969.

47.   N. T. Mirov, "Composition of Gum Turpentines of Pines," *Tech. Bull. 1239*, U.S. Dept. Agr., Washington, D.C., 1961.

48.   K. Snajberk and E. Zavarin, personal observations (1985).
49.   S. Krishnappa and S. Dev, *Tetrahedron*, *34*:599 (1978).
50.   P. Bahn, B. S. Pande, R. Soman, N. P. Damodaran, and S. Dev, *Tetrahedron*, *40*:2961 (1984).
51.   H. Avcibasi, H. Anil, and M. Toprak, *Phytochem.*, *27*:3967 (1988).
52.   J. F. Manville, *Can. J. Chem.*, *53*:1579 (1975).
53.   Zh. V. Dubovenko, M. A. Chirkova, N. K. Kashtanova, E. N. Shmidt, V. A. Babkin, and V. A. Pentegova, "Seskviterpenoidy Zhivits Khvoinykh Sibiri," in *Sinteticheskie Produkty iz Kanifoli i Skipidara* (I. I. Bardyshev, ed.), Volgo-Viatskoe Kn. Izd., Gorkii, 1970, p. 45.
54.   E. N. Shmidt, V. A. Khan, Z. A. Isaeva, T. D. Drebushchak, Zh. V. Dubovenko, E. P. Kemertelidze, and V. A. Pentegova, *Khim. Prir. Soed.* (2):189 (1982).
55.   V. I. Bolshakova, L. I. Demenkova, V. A. Khan, Zh. V. Dubovenko, E. N. Shmidt, and V. A. Pentegova, *Khim. Prir. Soed* (6):790 (1985).
56.   I. H. Rogers and J. F. Manville, *Can. J. Chem.*, *50*:2380 (1972).
57.   Y. Fujita, Sh. Fujita, Y. Iwamura, and Sh. Nishida, *J. Pharm. Soc. Japan* (Yakugaku Zasshi), *95*:349(1975); *CA*, *83*:151483 (1975).
58.   T. Norin, S. Sundin, B. Karlsson, P. Kierkegaard, A.-M. Pilotti, and A.-Ch. Wiehager, *Tetr. Lett.*, (1):17 (1973).
59.   H. G. Smith, *J. Soc. Chem. Ind.*, *30*:1353 (1911).
60.   T. Sakai, K. Nishimura, andf Y. Hirose, *Bull. Chem. Soc. Japan*, *38*:381 (1965).
61.   H. Irie, K. Ohno, Y. Ito, and Sh. Uyeo, *Chem. Pharm. Bull.*, *23*:1892 (1975).
62.   A. J. Birch, K. M. C. Mostyn, and A. R. Penfold, *Austr. J. Chem.*, *6*:391 (1953).
63.   G. Brieger, *Tetr. Lett.* (30):2123 (1963).
64.   M. Fracheboud, J. W. Rowe, R. W. Scott, S. M. Fanega, A. J. Buhl, and J. K. Toda, *For. Prod. J.*, *18*(2):37 (1968).
65.   B. O. Lindgren and C. M. Svahn, *Phytochem.*, *7*:1407 (1968).
66.   J. W. Rowe and J. K. Toda, *Chem. and Ind.*, (27):992 (1969).
67.   J. H. Langenheim, "Terpenoids in the Leguminosae," in *Advances in Legume Systematics*, (R. M. Polhill and P. H. Raven, eds.), Royal Botanical Garden, Kew, Richmond, England, 1981, Vol. 2, p. 627.
68.   M. L. Oyarzun and J. A. Garbarino, *Phytochem.*, *27*:1121 (1988).
69.   D. F. Zinkel, "Turpentine, Rosin and Fatty Acids form Conifers," in *Organic Chemicals from Biomass* (I. S. Goldstein, ed), CRC Press, Boca Raton, Fla., 1981, Chap. 9, p. 163.
70.   A. H. Conner and J. W. Rowe, *J. Amer. Oil Chem. Soc.*, *52*:334 (1975).
71.   E. N. Shmidt, N. K. Kashtanova, M. A. Chirkova, A. I. Lisina, and V. A. Pentegova, "Diterpenovye Soedineniya Zhivits Khvoinykh Sibiri," in *Sinteticheskie Produkty iz Kanifoli i Skipidara* (I. I. Bardyshev, ed.), Volgo-Viatskoe Kn. Izd., Gorki, 1970, p. 55.
72.   Y.-L. Chow and H. Erdtman, *Acta Chem. Scand.*, *14*:1852 (1960).
73.   J. B. -son Bredenberg and J. Gripenberg, *Acta Chem. Scand.*, *8*:1728 (1954).
74.   L. Mangoni and M. Belardini, *Gazz. Chim. Ital.*, *96*:206 (1966).
75.   L. Mangoni and R. Caputo, *Gazz. Chim. Ital.*, *97*:908 (1967).
76.   L. J. Gough and J. S. Mills, *Phytochem.*, *9*:1093 (1970).

77. T. Kondo, H. Imamura, and M. Suda, *Bull. Agr. Chem. Soc. Japan*, *23*:233(1959); *CA*, *54*:887 (1960).
78. C. W. Brandt and B. R. Thomas, *Nature*, *170*:1018 (1952).
79. L. H. Briggs, *Tetrahedron*, 7:270 (1959).
80. R. C. Cambie, W. R. J. Simpson, and L. D. Colebrook, *Tetrahedron*, *19*:209 (1963).
81. S. M. Bocks, R. C. Cambie, and T. Takahashi, *Tetrahedron*, *19*:1109 (1963).
82. S. Ito and M. Kodama, *Heterocycles*, *4*:595 (1976).
83. R. W. Miller, *J. Natural Prod.* (Lloydia), *43*:425 (1980).
84. M.-K. Yeh, J.-Sh. Wang, L.-P. Lin, and F.-Ch. Chen, *Phytochem.*, *27*:1534 (1988).
85. W. H. Baarshers, D. H. S. Horn, and L. R. F. Johnson, *J. Chem. Soc.*, (London), :4046 (1962).
86. A. H. Conner and D. O. Foster, *Phytochem.*, *20*:2543 (1981).
87. A. I. Lisina, L. N. Vol'skii, G. A. Mamontova, and V. A. Pentegova, *Izv. Sib. Otd. Akad. Nauk SSSR, Ser. Khim. Nauk*:, (6):98(1969); *CA*, *72*:68403 (1970); see also A. I. Lisina, L. N. Vol'skii, V. G. Leont'eva, and V. A. Pentegova, *Izv. Sib. Otd. Akad. Nauk SSSR, Ser. Khim. Nauk* (6):102(1969); *CA*, *72*:68404 (1970).
88. J. Kohlbrenner and C. Schuerch, *J. Org. Chem.*, *24*:166 (1959).
89. G. B. Russel and P. G. Fenemore, *New. Zeal. J. Sci.*, *13*:61 (1970).
90. H. Erdtman and K. Tsuno, *Phytochem.*, *8*:931 (1969).
91. D. O. Foster and D. F. Zinkel, *Wood Fiber Sci.*, *16*:298 (1984).
92. T. Kondo, H. Ito, and M. Suda, *J. Jap. Wood Res. Soc.*, *3*:151 (1957).
93. S. M. Kupchan, M. Takasugi, R. M. Smith, and P. S. Steyn, *J. Org. Chem.*, *36*:1972 (1971).
94. F. W. Hemming, "Polyisoprenoid Alcohols (Prenols)," in *Terpenoids in Plants* (J. B. Pridham, ed.), Academic Press, London, 1967, Chap. 12, p. 223.
95. I. I. Bardyshev and A. L. Pertsovskiy, "Izuchenie Sostava Smesei Smolianykh Kislot Metodom Gazo-Zhidkosnoi Khromatografii," in *Sinteticheskie Produkty iz Kanifoli i Skipidara* (I. I. Bardyshev, ed.), Volgo- Viatskoe Kn. Izd., Gorki, 1970, p. 103.
96. Yu. G. Sannikov and V. A. Samoilov, *Gidrol. i Lesokhim. Prom.* (7):15 (1976).
97. A. I. Lisina, V. K. Finogenova, L. N. Vol'skii, and V. A. Pentegova, *Izv. Sib. Otd. Akad. Nauk SSSR, Ser. Khim. Nauk* (2):122 (1967).
98. H. G. Daessler, *Holz als Roh- und Werkstoff*, *18*:162 (1960).
99. G. V. Nair and E. von Rudloff, *Can. J. Chem.*, *38*:177 (1960).
100. R. W. Hemingway, R. W. Hillis, and L. S. Lau, *Svensk Papperstidn.*, *76*:371 (1973).
101. E. P. Swan, Inform. Rept. VP-X-115, *Can. Dept. Forestry*, 1973.
102. M. A. Buchanan, R. V. Sinneff, and J. A. Jappe, *TAPPI*, *42*:578 (1959).
103. I. I. Bardyshev, S. I. Kriuk, B. G. Udarov, and N. G. Yaremchenko, *Khim. Prir. Soed.*, (5):650 (1974).
104. J. A. Lloyd, *Phytochem.*, *14*:483 (1975).
105. H. Hafizoglu, *Holzforschung*, *37*:321 (1983).
106. L. P. Clermont, *Pulp Paper Mag. Can.*, *62*:(12):T-511 (1961).
107. D. O. Foster and D. F. Zinkel, *Wood Fiber Sci.*, *16*:298 (1984).
108. T. J. Lillie and O. C. Musgrave, *J. Chem. Soc.*, *Perkin Trans. I*:355 (1977).
109. J. Tannock, *Phytochem.*, *12*:2066 (1973).
110. M. Tezuka, C. Takahashi, M. Kuroyanagi, M. Satake, K. Yoshihira, and S. Natori, *Phytochem.*, *12*:175 (1973).

111. L. M. van der Vijver and K. W. Gerritsma, *Phytochem.*, *13*:2322 (1974).

112. D. D. Ridley, E. Ritchie, and W. C. Taylor, *Austr. J. Chem.*, *21*:2979 (1968).

113. E. Ritchie, W. C. Taylor, and S. T. K. Vautin, *Austr. J. Chem.*, *18*:2015 (1965).

114. J. L. Occolowitz and A. C. Wright, *Austr. J. Chem.*, *15*:858 (1962).

115. C.-L. Chen, H.-M. Chang, and T. K. Kirk, *Phytochem.*, *16*:1983 (1977).

116. H. Kindl, "Biosynthesis of Meso-inositol in Microorganisms and Higher Plants," in *Proceedings of 2nd Meeting of the Federation of European Biochemistry Society*, (H. Kindl, ed.), Vienna, April 21–24, 1965, Pergamon Press, Oxford, 1966, Vol. 2, p. 15.

117. O. Hoffman-Ostenhof, "The Biosynthesis of the Cyclitols Other than Mesoinositol," in *Proceedings of 2nd Meeting of the Federation of European Biochemistry Society*, (H. Kindl, ed.), Vienna, April 21–24, 1965, Pergamon Press, Oxford, 1966, Vol. 2, p. 23.

118. L. V. Smith and E. Zavarin, *TAPPI*, *43*:218 (1960).

119. A. B. Anderson, *TAPPI*, *35*:198 (1952).

120. A. B. Anderson, *Ind. Engin. Chem.*, *45*:593 (1953).

121. A. B. Anderson, R. Riffer, and A. Wong, *Phytochem.*, *7*:1867 (1968).

122. A. B. Anderson, R. Riffer, and A. Wong, *Phytochem.*, *7*:1367 (1968).

123. H. Erdtman, 4th International Congress of Biochemistry, *Biochemistry of Wood* (K. Kratzl and G. Billek, eds.), Pergamon Press, London, 1959, Vol. II, p. 468.

124. T. E. Timell, *Compression of Wood in Gymnosperms*, Springer-Verlag, Berlin, 1986, Vol. 1, p. 353.

125. A. B. Anderson, *J. Amer. Chem. Soc.*, *74*:6099 (1952).

126. K. H. Baggaley, H. Erdtman, N. Y. McLean, T. Norin, and G. Eriksson, *Acta Chem. Scand.*, *21*:2247 (1967).

127. N. A. Tyukavkina, S. A. Medvedeva, and L. N. Ermolaeva, *Khim. Prir. Soed.*, (1):131 (1970).

128. J. W. Rowe and A. H. Conner, "Extractives in Eastern Hardwoods—A Review," Tech. Rept. FPL 18, Forest Products Lab., U.S. Dept. Agri., Madison, Wis., 1979, p. 39.

129. R. A. Abramovitch and R. G. Micetich, *Can. J. Chem.*, *44*:2913 (1963).

130. I. A. Pearl, D. L. Beyer, and D. Whitney, *TAPPI*, *44*:656 (1961).

131. I. A. Pearl, D. L. Beyer, S. S. Lee, and D. Laskowski, *TAPPI*, *42*:61 (1959).

132. H. Shimomura, Y. Sashida, and K. Yoshinari, *Phytochem.*, *28*:1499 (1989).

133. J. W. Rowe and A. H. Conner, "Extractives in Eastern Hardwoods—A Review," Tech. Rept. FPL 18, Forest Products Lab., U.S. Dept. Agri., Madison, Wis., 1979, p. 15.

134. M. K. Seikel, F. D. Hostettler, and G. J. Niemann, *Phytochem.*, *10*:2249 (1971).

135. L. Reppel, *Planta Medica*, *4*:199 (1956).

136. R. Hegnauer, *Chemotaxonomie der Pflanzen*, Birkhaüser Verlag, Basel, 1962–1989, Vol. V, p. 238.

137. D. E. Hathway, "The Lignans," in *Wood Extractives* (W. E. Hillis, ed.), Academic Press, New York, 1962, p. 159.

138. K. Freudenberg and L. Knof, *Chem. Ber.*, *90*:2857 (1957).

139. G. M. Barton and J. A. F. Gardner, *J. Org. Chem.*, *27*:322 (1962).

140. K. Freudenberg and K. Weinges, *Tetr. Lett.*, (17):19 (1959).

141. P. K. Agrawal, S. K. Agarwal, and R. P. Rastogi, *Phytochem.*, *19*:893,1260 (1980).

142. R. Hegnauer, *Chemotaxonomie der Pflanzen*, Birkhaüser Verlag, Basel, 1962–1989, Vol. I, p. 335.

143. J. A. F. Gardner, E. P. Swan, S. A. Sutherland, and H. MacLean, *Can. J. Chem.*, *44*:52 (1966); see also, H. MacLean and K. Murakami, *Can. J. Chem.*, *44*:1541,1827 (1966); see also, H. MacLean and K. Murakami, *Can. J. Chem. 45*:305, (1967).

144. H. MacLean and B. F. MacDonald, *Can. J. Chem.*, *45*:305, 739 (1967).

145. H. MacLean and B. F. MacDonald, *Can. J. Chem.*, *47*:457, 4495 (1969).

146. Y. Hirose, N. Oishi, H. Nagaki, and T. Nakatsuka, *Tetr. Lett.* (41):3665 (1965).

147. H. Erdtman and K. Tsuno, *Acta Chem. Scand.*, *23*:2021 (1969).

148. R. Hegnauer, *Chemotaxonomie der Pflanzen*, Birkhaüser Verlag, Basel, 1962–1989, Vol. 7, pp. 531, 537, 544.

149. C. L. Chen, H. M. Chang, and E. B. Cowling, *Phytochem.*, *15*:547 (1976).

150. F. D. Hostettler and M. K. Seikel, *Tetrahedron, 25*:2325 (1969).

151. M. K. Seikel, F. D. Hostettler and D. B. Johnson, *Tetrahedron, 24*:1475 (1968).

152. A. F. A. Wallis, *Tetr. Lett.*, (51):5287 (1968).

153. W. E. Hillis and A. Carle, *Austr. J. Chem.*, *16*:147 (1963).

154. H. Erdtman and J. Harmatha, *Phytochem.*, *18*:1495 (1979).

155. P. Henley-Smith and D. A. Whiting, *Phytochem.*, *15*:1285 (1976).

156. P. Daniels, H. Erdtman, K. Nishimura, T. Norin, P. Kierkegaard, and M. Pilotti, *J. Chem. Soc.* (London), *Chem Commun.*: 246 (1972).

157. K. Takahashi, M. Yasue, and K. Ogiyama, *Phytochem.*, *27*:1550 (1988).

158. A. Enoki, S. Takahama, and K. Kitao, *Phytochem.*, *23*:579,587 (1977).

159. T. Takashima, M. Ogushi, Y. Kiuchi, and F. Akabane, *Mokuzai Gakkaishi*, *14*:391 (1968).

160. C. R. Enzell and B. R. Thomas, *Tetr. Lett.*, (22):2395 (1966).

161. L. Jurd, "The Hydrolyzable Tannins," in *Wood Extractives* (W. E. Hillis, ed.), Academic Press, New York, 1962, p. 229.

162. E. Haslam, *Chemistry of Vegetable Tannins*, Academic Press, London, 1966.

163. E. Haslam, "Vegetable Tannins," in *Recent Advances in Phytochemistry, Biochemistry of Plant Phenolics,* Vol. 12, (T. Swain, J. B. Harborne, and Ch. F. Van Sumere, eds.), Plenum Press, New York, 1977, p. 475.

164. W. E. Hillis, "Biosynthesis of Tannins," in *Biosynthesis and Biodegradation of Wood Components* (T. Higuchi, ed.), Academic Press, Orlando, Fla., 1985, p. 325.

165. J. F. Manville and N. Levitin, *Bimon. Res. Notes*, Environ. Can. For. Serv., *30* (1):3 (1974).

166. H. Grisebach, "Biosynthesis of flavonoids," in *Biosynthesis and Biodegradation of Wood Components*, Academic Press, Orlando, Fla., 1985, p. 291.

167. H. L. Hergert, "Economic Importance of Flavonoid Compounds: Wood and Bark," in *Chemistry of Flavonoid Compounds* (T. A. Geissman, ed.), MacMillan, New York, 1962, p. 566.

168. H. Friedrich and R. Engelshowe, *Planta Medica*, *33*:251 (1978).

169. R. Engelshowe and H. Friedrich, *Planta Medica*, *49*:170 (1983).

170. H. L. Hergert and E. F. Kurth, *J. Org. Chem.*, *18*:521 (1953).

171. H. L. Hergert, *Forest Prod. J.*, *10*:610 (1960).

172. A. Sato, M. Senda, T. Kakutani, Y. Watanabe, and K. Kitao, *Mokuzai Kenkyo*, *39*(11):13 (1966); *CA*, *68*:41228 (1968).

173. M. D. Tindale and D. G. Roux, *Phytochem.*, *8*:1713 (1969).
174. T. Kubota and T. Hase, *Nippon Kagaku Zasshi*, *87*(11):1201 (1966).
175. D. G. Roux and E. Paulus, *Biochem. J.*, *82*:324 (1962).
176. T. A. Tatta and A. E. Rich, *Phytopathol.*, *63:*167 (1973).
177. G. H. N. Towers and R. D. Gibbs, *Nature*, *172*:25 (1953).
178. T. R. Seshadri, 1973 as quoted in J. W. Rowe and A. H. Conner, "Extractives in Eastern Hardwoods—A Review," Tech. Rept. FPL 18, Forest Products Lab., U.S. Dept. Agri., Madison, Wis., 1979, p. 10.
179. J. C. Pew, *J. Amer. Chem. Soc.*, *70*:3031 (1948).
180. R. Hegnauer, *Chemotaxonomie der Pflanzen,* Birkhäuser Verlag, Basel, 1962–1989, Vol. 3, pp. 100–110.
181. W. E. Hillis and H. R. Orman, *J. Linn. Soc.* (London), *58*:175 (1962).
182. R. Hegnauer, *Chemotaxonomie der Pflanzen,* Birkhäuser Verlag, Basel, 1962–1989, Vol. 5, pp. 181–182.
183. T. Norin, *Phytochem.*, *11*:1231 (1972).
184. H. Erdtman, "Conifer Chemistry and Taxonomy of Conifers," in *Biochemistry of Wood* (K. Kratzl and G. Billek, eds.), Pergamon Press, London, 1959.
185. J. Chopin and G. Grenier, *Chemie et Industrie* (Paris), *79*:605 (1958).
186. J.-M. Fang, W.-Ch. Su, and Y.-Sh. Cheng, *Phytochem.*, *27*:1395 (1988).
187. J. V. B. Mahesh and T. R. Seshadri, *J. Sci. Ind. Research* (India), *13B*:835 (1954).
188. H. L. Hergert, "Economic Importance of Flavanoid Compounds: Wood and Bark", in *The Chemistry of Flavanoid Compounds*, (T. A. Geissman, ed.), MacMillan, New York, 1962, p. 558.
189. P. K. Agrawal, J. K. Agrawal, and R. P. Rastogi, *Phytochem.*, *19*:893 (1980).
190. P. K. Agrawal and R. P. Rastogi, *Biochem. Syst. Ecol.*, *12*:133 (1984).
191. N. A. Tyukavkina, K. I. Lapteva, and V. A. Pentegova, *Khim. Prir. Soed.*, (3):278 (1967).
192. M. F. Shostakovskii, N. A. Tyukavkina, N. G. Devyatko, and K. I. Lapteva, *Izv. Sib. Otd. Akad. Nauk SSSR, Ser. Biol. Nauk*, (3):77 (1969).
193. K. P. Tiwari and P. K. Minocha, *Phytochem.*, *19*:2501 (1980).
194. H. Erdtman and Z. Pelchowicz, *Chem. Ber.*, *89*:341 (1956).
195. H. Erdtman and Z. Pelchowicz, *Acta Chem. Scand.*, *9*:1728 (1955).
196. H. Ohashi, Y. Ido, T. Imai, K. Yoshida, and M. Yasue, *Phytochem.*, *27*:3993 (1988).
197. H. Erdtman, Z. Pelchowicz, and J. Topliss, *Acta Chem. Scand.*, *10*:1563 (1956).
198. L. H. Briggs, R. C. Cambie, and J. L. Hoare, *Tetrahedron*, *7*:262 (1959).
199. B. C. B. Bezuidenhoudt, E. V. Brandt, and D. Ferreira, *Phytochem.*, *26*:531 (1987).
200. H. W. Brewerton, *New Zeal. J. Sci.*, *1*:220 (1958).
201. L. H. Briggs, and B. F. Cain, *Tetrahedron*, *6*:143,145 (1959).
202. D. G. Roux and E. Paulus, *Biochem. J.*, *84*:416 (1962).
203. A. K. Ganguly and T. R. Seshadri, *J. Sci. Ind. Research* (India), *17B*:168 (1958).
204. A. K. Ganguly and T. R. Seshadri, *Tetrahedron*, *6*:21 (1959).
205. W. E. Hillis, *Aust. J. Sci. Res.*, *A5*:379 (1952).
206. G. J. Gell, J. T. Pinkley, and E. Richie, *Austr. J. Chem.*, *11*:372 (1958).
207. I. A. Pearl and D. L. Beyer, *TAPPI*, *46*:502 (1963).
208. M. Hanzawa and T. Sasaya, *Mok. Gakkaishi*, *4·*125 (1958).

209.   T. Sasaya, *Hokkaido Daigaku Nogakubu Enshurin Kenkyu Hokoku*, *24*(1):177 (1965); *CA*, *65*:5435e (1966).

210.   C. Mentzer, *Bull. Assoc. Franc. Chimistes Inds. Cuir et Dec. Sci. et Tech. Inds. Cuir*, *22*:180; *CA*, *55*:5943 (1961).

211.   T. Sasaki and M. Mikami, *Yakugaku Zasshi*, *83*:879 (1963).

212.   F. E. King, T. J. King, and D. W. Rustidge, *J. Chem. Soc.* (London):1192 (1962).

213.   R. Hegnauer, *Chemotaxonomie der Pflanzen*, Birkhäuser Verlag, Basel, 1962–1989, Vol. 4, p. 369.

214.   W. E. Hillis and D. H. S. Horn, *Austr. J. Chem.*, *19*:705 (1966).

215.   A. Robertson, C. W. Suckling, and W. B. Whalley, *J. Chem. Soc.* (London):1571 (1949).

216.   J. B. Harborne, "Distribution of Flavonoids in the Leguminosae," in *Chemotaxonomy of the Leguminosae* (J. B. Harborne, D. Boulter, and B. L. Turner, eds.), Academic Press, London, 1971, p. 31.

217.   M. Namikoshi, H. Nakata, and T. Saitoh, *Phytochem.*, *26*:1831 (1987).

218.   E. Haslam, "Proanthocyanidins," in *The Flavonoids: Advances in Research* (J. B. Harborne and T. J. Mabry, eds.), Chapman and Hall, London, 1982, p. 417.

219.   E. Haslam, "Natural Proanthocyanidins," in *The Flavonoids, Vol. 1* (J. B. Harborne, T. J. Mabry, and H. Mabry, eds.), Academic Press, New York, 1975, p. 505.

220.   Z. Czochanska, L. Y. Foo, R. H. Newman, and L. J. Porter, *J. Chem. Soc.* (London), *Perkin Trans. I*:2278 (1980).

221.   R. Hemingway, L. Y. Fee, and L. J. Porter, *J. Chem. Soc.* (London), *Perkin Trans. I*:1209 (1982).

222.   R. Hemingway and P. E. Laks, *J. Chem. Soc.* (London): Chem. Commun., 746 (1985).

223.   R. W. Hemingway, G. W. McGraw, J. J. Karchesy, L. Y. Foo, and L. J. Porter, *J. Appl. Polym. Sci.*, *37*:967 (1983).

224.   R. W. Hemingway and G. W. McGraw, *J. Wood Chem. Technol.*, *3*:421 (1983).

225.   K. Weinges and W. Ebert, *Phytochem.*, *7*:153 (1968).

226.   K. Weinges and D. Huthwelker, *Liebigs Ann. Chem.*, *731*:161 (1970).

227.   K. Weinges, H. Mattauch, C. Wilkins, and D. Frost, *Liebigs Ann. Chem.*, *754*:124 (1971).

228.   K. Weinges, E. Ebert, D. Huthwelker, H. Mattauch, and J. Perner, *Liebigs Ann. Chem.*, *726*:114 (1969).

229.   B. F. Hrutfiord, R. Luthi, and K. F. Hanover, *Wood Chem. Tech.*, *5*:451 (1985).

230.   H. Geiger and C. Quinn, "Biflavonoids," in *The Flavonoids: Advances in Research* (J. B. Harborne and T. J. Mabry, eds.), Chapman and Hall, London, 1982, p. 505.

231.   H. Geiger and C. Quinn, in *The Flavonoids* (J. B. Harborne, T. J. Mabry, and H. Mabry, eds.), Chapman and Hall, London, 1975.

232.   M. W Bandaranayake, S. S. Selliah, M. U. S. Sultanbawa, and D. Ollis, *Phytochem.*, *14*:1878 (1975).

233.   F.-C. Chen, Y.-M. Lin, and J.-C. Huang, *Phytochem.*, *14*:300 (1975).

234.   F.-C. Chen, Y.-M. Lin, and J.-C. Huang, *Phytochem.*, *14*:818 (1975).

235.   P. J. Cotterill, F. Scheinmann, and G. S. Puranik, *Phytochem.*, *16*:148 (1977).

236.   J. Gorham, "The Stilbenoids," in *Progress in Phytochemistry*, (J. Reinhold, J. B. Harborne, and T. Swain, eds.), Pergamon Press, Oxford, 1980, Vol. 6, p. 203.

237.   V. B. Mahesh and T. R. Seshadri, *J. Sci. Ind. Research*, *13B*:835 (1954).

238. A. I. Lisina, N. K. Kashtanova, A. K. Dzizenko, and V. A. Pentegova, *Izv. Sib. Otd. Akad. Nauk SSSR, Ser. Khim. Nauk.*, (1):165 (1967).

239. K. Venkataraman, *Phytochem.*, *11*:1571 (1972).

240. S. Mongolsuk, A. Robertson, and R. Towers, *J. Chem. Soc.* (London):2231 (1957).

241. W. E. Hillis, J. H. Hart, and Y. Yazaki, *Phytochem.*, *13*:1591 (1974).

242. W. E. Hillis and K. Isoi, *Phytochem.*, *4*:541 (1965).

243. W. E. Hillis and T. Inoue, *Phytochem.*, *6*:59 (1967).

244. F. E. King T. J. King, D. H. Godson, and L. C. Manning, *J. Chem. Soc.* (London):4477 (1956).

245. W. E. Hillis and Y. Yazaki, *Phytochem.*, *12*:2491 (1973).

246. A. A. Craverio, A. da Costa Prado, O. R. Gottlieb, and P. C. W. de Albuquerque, *Phytochem.*, *9*:1869 (1970).

247. F. E. King, C. B. Cotterill, D. H. Godson, L. Jurd, and T. J. King, *J. Chem. Soc.* (London):3693 (1953).

248. E. Spaeth and T. Schlaeger, *Chem. Ber.*, *73*:881 (1940).

249. S. E. Drewes, *Phytochem.*, *10*:2837 (1971).

250. Y. Hayashi, K. Sakurai, T. Takahashi, and K. Kitao, *Mokusai Gakkaishi*, *20*:595 (1974).

251. J. W. W. Morgan and R. J. Orsler, *Holzforschung*, *22*:11 (1968).

252. J. Gorham, "The Stilbenoids", in *Progress in Phytochemistry*, (J. Reinhold, J. B. Harborne, and T. Swain, eds.), Pergamon Press, Oxford, 1980, Vol. 6, pp. 220 and 229.

253. P. Coggon, N. F. Janes, F. E. King, T. J. King, R. J. Molyneux, J. W. W. Morgan, and K. Sellars, *J. Chem. Soc.* (London):406 (1965).

254. R. Madhav, T. R. Seshadri, and G. B. V. Subramanian, *Phytochem.*, *6*:1155 (1967).

255. M. F. Grundon and F. E. King, *Nature*, *163*:564 (1949).

256. F. E. King and M. F. Grundon, *J. Chem. Soc.* (London):3348 (1949).

257. F. E. King and M. F. Grundon, *J. Chem. Soc.* (London):3547 (1950).

258. L. P. Christensen, J. Lam, and T. Sigsgaard, *Phytochem.*, *27*:3014 (1988).

259. L. P. Christensen and J. Lam, *Phytochem.*, *28*:917 (1989).

260. M. A. Moir and R. H. Thomson, *J. Chem. Soc., Perkin Trans. I*:1352,1556 (1973).

261. G. D. Manners and L. Jurd, *J. Chem. Soc., Perkin Trans. I*:405 (1977).

262. A. R. Burnett and R. H. Thomson, *J. Chem. Soc.* (London):2100 (1967).

263. W. Sandermann, M. H. Simatupang, and W. Wendeborn, *Naturwiss.*, *55*:38 (1968).

264. P. Singh, S. Jain, and S. Bhargava, *Phytochem.*, *28*:1258 (1989).

265. P. K. Sharma, R. N. Khanna, B. K. Rohatgi, and R. H. Thomson, *Phytochem.*, *27*:632 (1988).

266. Sh. Hasegawa, Y. Hirose, and H. Erdtman, *Phytochem.*, *27*:2703 (1988).

267. K. Sakata, J. Kishimoto, H. Sakai, and N. Kosaka, *Tottori Nagakkaiho*, *15*:52 (1963); *CA*, *60*:3192 (1964).

268. W. Mayer, K. Lauer, W. Gabler, U. Panther, and A. Riester, *Naturwiss.*, *46*:669 (1959).

269. N. F. Janes and J. W. W. Morgan, *J. Chem. Soc.* (London):2560 (1960).

270. C.-L. Chen, *Phytochem.*, *9*:1149 (1970).

271. W. E. Hillis and A. Carle, *Holzforschung*, *12*:136 (1958).

272. W. E. Hillis and A. Carle, *Biochem. J.*, *74*:607 (1960).

273. W. E. Hillis and A. Carle, *Biochem. J.*, *82*:435 (1962).

274. T. Kondo, H. Ito, and M. Suda, *Mokusai Gakkaishi*, *2*:221 (1956); *CA 51*:9152 (1957).

275. T. Kondo, H. Ito, and M. Suda, *Nippon Nogei-Kagaku Kaishi*, *30*:281 (1956); *CA*, *52*:12395 (1958).

276. V. M. Chari, S. Neelakantan, and T. R. Seshadri, *Indian J. Chem.*, *6*:231 (1968).

277. R. C. Cambie and J. C. Parnell, *New Zeal. J. Sci.*, *12*:457 (1969).

278. R. C. Cambie, *New Zeal. J. Sci.*, *2*:257 (1959).

279. W. A. Grassmann and H. Endres, *Leder*, *10*:237 (1959).

280. Ph. Lebreton, *Candollea*, *34*:211 (1979).

281. Ph. Lebreton, *Naturalia Monospeliensia*, *Ser. Bot.*, *47*:1 (1981).

282. Ph. Lebreton, *Agronomia Lusitana*, *42*:55 (1983).

283. E. F. Kurth and H. B. Lackey, *J. Amer. Chem. Soc.*, *70*:220 (1948).

284. E. Zavarin and K. Snajberk, *TAPPI*, *48*:612 (1965).

285. A. B. Anderson, *J. Inst. Wood Sci.*, (8):14 (1961).

286. C. C. Scalione and D. R Merrill, *J. Ind. Eng. Chem.*, *11*:643 (1919).

287. M. A. Buchanan, H. F. Lewis, and E. F. Kurth, *Ind. Eng. Chem.*, *36*:97 (1944).

288. Anonymous, "Redwood", *Res. Bull. Inst. Paper Chem.*, Appleton, Wis., 1945.

289. D. G. Roux and D. Ferreira, *Prog. Chem. Org. Nat. Prod.*, *41*:47 (1982).

290. L. Y. Foo, *Phytochem.*, *23*:2915 (1984).

291. D. Ferreira, I. C. duPreeez, J. C. Wijnmaalen, and D. G. Roux, *Phytochem.*, *24*:2415 (1985).

292. W. E. Hillis and A. Carle, *Holzforschung*, *12*:136 (1958).

293. W. E. Hillis and A. Carle, *Biochem. J.*, *74*:607 (1960).

294. W. E. Hillis and A. Carle, *Biochem. J.*, *82*:435 (1962).

295. A. Russel, C. R. Vanneman, and W. E. Waddey, *J. Amer. Leather Chem. Assoc.*, *40*:422 (1945).

296. A. Russel, E. A. Kaczka, W. G. Tebbens, C., R. Vanneman, and S. Cody, *J. Amer. Leather Chem. Assoc.*, *39*:173 (1944).

297. A. Russel, E. A. Kaczka, W. G. Tebbens, C. R. Vanneman, and S. Cody, *J. Amer. Leather Chem. Assoc.*, *38*:235 (1943).

298. G. Jayme and H. Semmler, *Das Papier*, *11*:396 (1957).

299. V. Narayanan and T. R. Seshadri, *Indian J. Chem.*, *7*:213 (1969).

300. L. L. Prado and E. Ricci, "Systematic Study of the Tannin Content of Various Indigenous Species", Anales Admin. Nac. Bosques, Rep. Arg., No. 7, 1956; *CA*, *51*:9803 (1957).

301. R. C. Cambie, B. F. Cain, and S. LaRoche, "A New Zealand Phytochemical Survey, II Dicotyledons," *New Zeal. J. Sci.*, *4*:604 (1961).

302. E. P. Swan, *Forest Prod. J.*, *13*:195 (1963).

303. A. B. Anderson and E. Zavarin, *J. Inst. Wood Sci.*, *15*:3 (1965).

304. E. Zavarin and K. Snajberk, *TAPPI*, *48*:574 (1965).

305. F. E. Brauns, *J. Amer. Chem. Soc.*, *61*:2120 (1939).

306. L. Y. Foo and J. J. Karchesy, "Chemical Nature of Phlobaphanes," in *Chemistry and Significance of Condensed Tannins* (R. W. Hemingway and J. J. Karchesy, eds.), Plenum Press, New York, 1989, p. 109.

307. C.-L. Chen, H.-M. Chang, and E. B. Cowling, *Phytochem.*, *15*:547 (1976).

308. C.-L. Chen, H.-M. Chang, E. B. Cowling, C.-Y. Huang Hsu, and R. P. Gates, *Phytochem.*, *15*:1161 (1976).

309. P. D. Senter and C.-L. Chen, *Phytochem.*, *16*:2015 (1977).
310. E. Zavarin and K. Snajberk, *Phytochem.*, *4*:141 (1965).
311. L. A. Smedman, K. Snajberk, E. Zavarin, and T. R. Mon, *Phytochem.*, *8*:1471 (1969).
312. R. H. J. Creighton, R. D. Gibbs, and H. Hibbert, *J. Amer. Chem. Soc.*, *66*:32 (1944).
313. Anonymous, "System of Nomenclature for Terpene Hydrocarbons," Advances in Chemistry Series 14, ACS, Washington, D.C., 1955, pp. 1–98.
314. T. K. Devon and A. I. Scott, *Handbook of Naturally Occurring Compounds, Vol. II. Terpenes*, Academic Press, New York, 1972, p. 220.

9

Bark

Murray L. Laver

Oregon State University, Corvallis, Oregon

I. INTRODUCTION

Bark is that material formed by a tree to the outside of the vascular cambium. The vascular cambium forms wood (xylem) to the inside and inner bark (phloem) to the outside. It is a remarkable feat that this essentially monomolecular layer of living tissue should form two materials that are so different, one on each side. The physiology and chemical mechanisms of how this is done are certainly not completely understood.

Bark has several functions for the living tree. It protects the tree from mechanical injury, from attack by insects, from disease, and from chemicals in the atmosphere, to name only a few. The inner bark, or phloem, also performs the important physiological function of transporting the products of photosynthesis synthesized in the foliage to other parts of the tree.

The chemical composition of bark is complex. There are a great number of different types of chemicals in bark. Bark would thus appear to be a renewable source for chemicals. Indeed it is, but the problems of isolating, separating, and purifying the chemical entities have, in general, proven prohibitive from an economic standpoint. However, there have been some materials used commercially such as polyflavonoid polymers from Western hemlock [*Tsuga heterophylla* (Raf.) Sarg], tannins from the bark of the black wattle tree (*Acacia mearnsii* De Wild), and wax from the bark of Douglas fir [*Pseudotsuga menziesii* (Mirb.) Franco].

The processing of bark for chemicals involves the handling of large volumes of solid matter. Although it is harvested and transported to centralized locations to obtain the wood to which it is attached, the methods often result in the embedding of dirt and stones that makes the chemical processing of bark extra difficult. The recovery of selected chemicals from bark invariably leaves large residues of processed bark that have to be disposed. Although burning the residues is usually feasible, additional handling is involved.

Bark is also a variable material. It varies with the age of the tree, the height of the tree, soil conditions, moisture conditions, and other factors. The yields of a particular chemical are thus unpredictable. All of these reasons contribute to the difficulties of obtaining chemicals from bark, especially commodity chemicals.

Bark today is primarily used as a fuel by direct burning in boilers. It is still a byproduct of wood products' manufacture. There are vast tonnages of bark available for fuel. Corder (1) estimated that in 1972 some 46,000,000 m^3 (or about 15,000,000 metric tons, dry weight basis) of bark was delivered to mills in the United States as part of sawlogs, pulpwood bolts, and wood chips. Worldwide in 1972, the volume of bark arriving at mill sites was 319,000,000 m^3. It has been calculated that this amount of bark would fill a train of rail cars 70,000 km long, which would extend almost twice around the world at the equator. This certainly represents a lot of chemicals. Although these data were collected in 1972, the figures are still relevant today.

II. BARK ANATOMY

This chapter is concerned primarily with the chemistry of tree barks, but a brief description of bark anatomy is presented to define and clarify the specific materials that produce and contain many of the chemicals described. Three descriptions of bark anatomy have been recently published (2–4) that along with the references contained there, provide access to original and detailed information. Bark is said to consist of inner bark and outer bark. There are thus essentially five major classifications of bark that should be considered when the chemistry of bark is discussed: (1) hardwood inner bark; (2) hardwood outer bark; (3) softwood inner bark; (4) softwood outer bark;(5) cork. The inner bark (phloem) is the portion from the vascular cambium to the innermost periderm, or innermost cork layer. The outer bark (rhytidome) is everything to the outside of the innermost cork layer.

Hardwood inner bark contains longitudinal and ray parenchyma, phloem fibers, and sieve cells. Companion cells are always paired with sieve cells and regulate the functions of the sieve cells. Parenchyma and sieve cells remain alive in the living tree as long as they are components of the inner bark.

Hardwood outer bark is composed of old periderms and crushed phloem tissue. With the exception of the cork that is produced by the cork cambium, all cells in the outer bark were once cells in the inner bark.

Softwood inner bark contains longitudinal and ray parenchyma cells, longitudinal and ray epithelial cells, and albuminous cells. There are also sieve cells and, in most species, phloem fibers and stone cells (sclereids or brachysclereids). Phloem fibers are long, slender, thick-walled, and often heavily lignified cells. These cells serve as structural elements. The sclereids or brachysclereids have thick and heavily lignified cell walls. They are usually found in clusters. In the early literature of *Pseudotsuga menziesii* bark, these sclereids were termed bast fibers.

A single cork layer is called a periderm. In hardwood and softwood barks, the periderms are comprised of three morphological tissues, the phellem (outer bark side), the phellogen (cork cambium), and the phelloderm (inner bark side). The phellogen of the innermost cork layer is meristematic tissue that divides to form phellem to the outside and phelloderm to the inside. New cork phellogen cells form in the inner bark and cut away part of the inner bark, which now becomes part of the outer bark. All cells outside the innermost phellogen layer (cork cambium) are dead because no food supply can pass through this layer of cork cells. This then results in an outer bark composed of cork cells and dead phloem cells, which were once inner bark. Litvay (5) and Litvay and Krahmer (6,7) describe these cells in detail for *Pseudotsuga menziesii* bark.

These cork cells are very important to the chemistry of bark because they contain different chemicals in different amounts than the cells of the inner and outer barks. For example, in *Pseudotsuga menziesii*, the cork is said to be rich in wax and suberin. Some trees, such as *Pseudotsuga menziesii*, and the *Quercus*, *Abies*, and *Betula* species contain considerable amounts of cork, whereas others do not. It is important when investigating the chemistry of bark to know which portion of the bark is being studied, and if whole bark is being studied, to know the amounts of inner bark, cork, and outer bark.

III. BARK CARBOHYDRATES

The carbohydrates of bark have proven difficult to isolate. Thus, their structures have not been as thoroughly elucidated as those from most of the woods. The difficulty is that bark contains many noncarbohydrate constituents such as lignins, suberins, tannins, and numerous other compounds in lesser quantities. These materials make the isolation of carbohydrates, especially polysaccharides in unaltered form, difficult. Segal and Purves (8) discussed these difficulties in a review of the early literature on the chemistry of bark. The so-called "extractives" were known to interfere unless they were removed by exhaustive, successive extractions with ethanol, water, or other neutral, chemically inert liquids, such as diethyl ether or petroleum ether. After all of the solvent-soluble materials had been removed by exhaustive extraction, some 70–90% remained undissolved residue. Segal and Purves (8) reported that this residue material still contained considerable noncarbohydrate material, including suberins, a name derived from cork oak (*Quercus suber L.*). The residue also contained large amounts of complex, insoluble bodies of low or zero methoxyl content that appeared to be closely allied chemically to the soluble phlobaphenes and tannins. These bodies grossly increased the apparent Klason lignin content of barks. However, it was found that they could be extracted with dilute alkali, and thus an improved lignin value could be obtained on the remaining residue.

After removing all of the above-mentioned material, it was possible to acid-hydrolyze the remaining residue and investigate some of the monosaccharides released. It appeared (8) that the polysaccharides in tree barks were based predominantly on glucose, and to a lesser extent on galactose, mannose, xylose, arabinose, and rhamnose.

The problems of isolating pure, unaltered carbohydrate fractions from bark are still with us. Various researchers have developed procedures for isolating and purifying specific fractions, particularly specific hemicelluloses, depending on the nature of the starting material and the accompanying impurities. Experience has led, however, to a more or less general procedure for the systematic isolation of bark carbohydrates. The procedure involves sequential extraction with selected solvents and the preparation of a holocellulose material.

The initial step in the isolation is essential. When collecting live biological material, the species, age of the tree, its location, and where the sample is taken from in the tree are important. It is best to obtain the bark sample as soon after the tree is cut as possible. Enzyme action can alter the carbohydrates considerably, and if the bark sample is old, there may be little relation between the carbohydrates in the sample and what was in the living tree. When collecting live biological material, it is wise to immerse the material immediately in 95% ethanol to inactivate the enzymes (9).

After ethanol extraction, the bark can be extracted in sequence with benzene-ethanol 2:1, hot water, and 0.5% aqueous ammonium oxalate. The residue from these extractions can then be delignified. The ethanol-water extraction not only denatures the enzymes, but also solubilizes simple sugars. The benzene-ethanol extraction removes waxes and lipids and opens the tissues to penetration by hydrophilic solvents (8,10). After removal of the waxes and lipids, it is possible to remove the hot-water-soluble material. Water extraction generally removes simple sugars, water-soluble polysaccharides, and varying amounts of noncarbohydrates. The ammonium oxalate extraction removes pectin and pectic substances prior to delignification (9,11). Each of the extracts should be investigated for carbohydrate materials in any systematic study of bark carbohydrates.

The presence of lignin in plant tissues, in general, presents an obstacle to the removal and purification of polysaccharides. There is a special problem with tree barks because of the added presence of additional polyphenolics such as the condensed tannins. It is preferable to remove all of these phenolic polymers prior to polysaccharide extraction. In fact, Timell (12) showed that it was essential to completely delignify the bark of gymnosperms if a successful fractionation of the polysaccharides was to be achieved.

Numerous methods have been described for the delignification of plant materials with the aim of minimum alteration of the carbohydrates present (9,11). However, the method now in general use is the acidified sodium chlorite method. It involves the generation of chlorine dioxide in situ through a reaction of dilute acetic acid with sodium chlorite. An exact procedure is described by Green (13) and by Whistler and BeMiller (14) in *Methods in Carbohydrate Chemistry* and the references contained there. The resulting residue has generally been known over the years as "holocellulose" after that proposed by Ritter and Kurth (15).

The holocellulose preparation, however, does not cleanly delignify the bark material and yield a quantitative recovery of polysaccharides. According to Timell (16), holocellulose isolation by any means can result in losses of polysaccharides, as well as alterations in their structures. The solubilization of bark carbohydrates during acidified sodium chlorite delignification was demonstrated by Lai (17) and Laver, et al. (18) in their investigations of the inner bark of *Pseudotsuga menziesii*. They isolated the solids dissolved by the acidified sodium chlorite treatment and after acid hydrolysis determined the ratio of monosaccharide residues by the gas-liquid chromatography of their alditol acetates as glucose, 59.1; arabinose, 11.9; galactose, 3.9; mannose, 3.7; xylose, 1.0; rhamnose, 1.0. Rhamnose had not been previously reported in *Pseudotsuga menziesii* bark. The results of this work emphasize that care must be exercised in isolating the holocellulose material to keep carbohydrate solubilization to a minimum.

After the isolation of the holocellulose, or delignified material, it is necessary to separate and purify the various polysaccharides that comprise the holocellulose because studies of the structures require homogeneous polymers. Alkaline extraction has been the principal method employed for the removal of the hemicellulose group. Relatively dilute alkalies suffice to dissolve xylans and galactoglucomannans, but higher concentrations are required for the extraction of glucomannans (9).

A great improvement in the isolation of homogeneous polysaccharides resulted from the work of Beélik et al. (19). These workers developed a procedure based on the principle of selective extraction for the isolation of the three main hemicelluloses from softwood holocelluloses. Selectivity was assured by impregnating the holocellulose with a 1–2% barium hydroxide solution in the first of three extraction steps. The barium hydroxide formed complexes with polysaccharides containing mannose units and rendered them insoluble in aqueous alkali. Extraction of the impregnated medium with 10% potassium hydroxide dissolved the xylans. After removal of the barium hydroxide, the easily soluble galactoglucomannan was extracted from the holocellulose residue in the second step of the sequence with 1% sodium hydroxide, and the glucomannan in the last step with 15% sodium hydroxide. All three hemicelluloses obtained in this fashion were quite homogeneous, and thus tedious secondary purifications by precipitations were kept to a minimum. The residue from these extractions was predominantly a glucan, usually with cellulose character.

Impure fractions have usually been purified by precipitation, using acids, organic liquids, or complexing reagents such as Fehling's solution. None of these agents is specific for a given hemicellulose; separation depends on differing rates of precipitation, is usually lengthy and laborious, and is rarely complete (9). Drying of the polysaccharide fractions is best accomplished by lyophilization (9).

The early work by Segal and Purves (8) assisted considerably in isolating the carbohydrates from tree barks, but very little structural elucidation was presented. Timell and co-workers defined the structures of several polypaccharides from tree barks, and although fine structural details may exist in polysaccharides from other barks, Timell and his co-workers described the basic framework of bark polysaccharide structures. Timell (20) isolated a xylan from the bark of amabilis fir [*Abies amabilis* (Dougl.) Forbes] that consisted of a framework of at least 124 (1→4)-linked β-D-xylopyranose units. To these were directly attached single, terminal (1→2)-linked side chains of 4-O-methyl-α-D-glucuronic acid, one per 6 xylose units, and (1→3)-linked L-arabinofuranose residues, one per 10 xylose units. Ramilingam and Timell (21) isolated a similar xylan from the bark of Engelmann spruce (*Picea enqelmanni* Parry). Timell (22) also isolated a galactoglucomannan from the bark of *Abies amabilis*. The polymer contained a slightly branched framework of at least 80 β-(1→4)-linked D-mannopyranose and D-glucopyranose units, every tenth of which, on the average, carried a (1→6)-linked D-galactopyranose unit probably attached directly by an α-glycosidic bond. A very similar galactoglucomannan was isolated by Ramalingam and Timell (21) from the bark of *Picea enqelmanni*. Timell (23) also isolated a glucomannan from the bark of *Abies amabilis*. The polysaccharide consisted of a minimum number of 70 (1→4)-β-linked D-glucopyranose and D-mannopyranose units, constituting a linear framework to which a few single side chains of galactose units were probably attached. The difference between the naming of these polysaccharides, whether glucomannan or galactoglucomannan, depends on the amount of galactose obtained on hydrolysis. The name galactoglucomannan usually indicates that there is considerably more galactose in the polysaccharide than there is in one named glucomannan.

Ramalingam and Timell (21) isolated a polysaccharide from the bark of *Picea engelmanni* that they referred to as a "glucan." The glucan contained small amounts of galactose and xylose in addition to glucose residues. The monosaccharide ratio was not altered by numerous efforts at purification. They proposed that the structure consisted, in part, of a sequence of β-(1→4)-linked glucopyranose units frequently branched through C-6.

Timell (24) isolated cellulose from extractive-free barks of lodgepole pine (*Pinus contorta* Dougl. ex Loud.) and *Ginkgo biloba* L. in yields of 38.1, 30.9, 30.4, and 37.6%, respectively. The number-average degrees of polymerization varied between 216 and 702. The weight-average degrees of polymerization ranged from 7100–8800. These celluloses were similar to the cellulose (25) from the inner bark of white birch (*Betula papvrifera* Marsh.). These above-mentioned polysaccharides from bark are remarkably similar to the polysaccharides isolated from woods, except perhaps in molecular weight. The polysaccharides in woods have been reviewed by Timell (16,26).

There are some gums and mucilages in tree barks that possess very complex structures. Slippery elm (*Ulmus fulva* Michx.) mucilage is a polysaccharide that has been extensively investigated. Beveridge et al. (27) concluded that the polysaccharides contained chains of 3-O-methyl-D-galactose residues attached to the C-4 positions of certain L-rhamnose residues and that 3-O-methyl-D-galactose residues occur in some cases as nonreducing end groups. D-galactose is attached as single residues or as 4-O-substitute residues to the C-3 positions of some L-rhamose units. The structure was thus highly branched.

The barks of hardwoods have been less thoroughly investigated than the barks of the conifers. Jabbar et al.(28) isolated a xylan from the inner bark of *Betula papyrifera* that contained a linear framework of a minimum number of 230 (1→4)-linked β-D-xylopyranose residues with, on the average, every tenth residue possessing a single, terminal side chain of a (1→2)-linked 4-O-methyl-α-D-glucuronic acid residue. Jiang and Timell (29,30) also isolated a xylan and a galactoglucomannan from the inner bark of quaking aspen (*Populus tremuloides* Michx.) that showed similar structural characteristics to previously described xylans and galactoglucomannans. These workers also isolated a unique arabinan from the *Populus tremuloides* bark by direct extraction with 70% methanol (31). The arabinan was highly branched with an α-(1→5)-linked arabinofuranose framework with branch points at C-2, C-3, and sometimes both C-2 and C-3. Toman and co-workers (32–34) isolated a xylan, galactan, and arabinan from the bark of white willow (*Salix alba* L.). The above descriptions definitely show that many different types of polysaccharides are present in tree barks.

Fu et al. (35) reported the isolation and characterization of callose from the bark of Scots pine (*Pinus svlvestris* L.).Callose is a β-D-(1→3)-glucan, a rather unique polysaccharide. It is widely distributed in the plant kingdom and occurs especially in the sieve elements of tree barks, where it fills the sieve pores of nonfunctioning or dead sieve cells (gymnosperms) and sieve tubes (angiosperms). It had previously been shown to be a β-D-(1→3)-glucan by Aspinall and Kessler (36). Litvay (5) and Litvay and Krahmer (6,7) reported that a substance that stained and reacted like callose was present as a plugging material in pit-like areas in the walls of mature cork (phellem) cells of *Pseudotsuga menziesii* bark.

The carbohydrates of the inner bark of Douglas-fir have been systematically investigated by Laver and co-workers (17,18,37,38,39,40). The inner bark contained 44.3% (on a dry-weight basis) of a yellowish holocellulose material that, on further treatment, resulted in a white holocellulose in 30.6% yield. These same researchers isolated and characterized a xylan, galactoglucomannan, glucomannan, and a glucan-rich fraction. These polysaccharides were similar in structures and physical characteristics to corresponding polysaccharides isolated from other barks and wood.

There has been a continuing interest in carbohydrates from tree barks (41), but the polysaccharides isolated from different tree barks have proven to be quite similar to those reported earlier. Dietrichs et al. (42) reported some galactoglucomannans and xylans that contained some acetyl groups. The polysaccharides that were characterized earlier possibly contained acetyl groups also, but they were hydrolyzed away during isolation.

Yields of holocellulose from tree barks vary considerably, depending on the conditions of isolation. However, the holocellulose content represents about all of the insoluble carbohydrates that can be realized by a chemical treatment. Hemingway (2) shows tables that give the holocellulose contents of several tree barks. The quantities for conifer barks range from 65.0% of the inner bark of *Pinus sylvestris* to 37.4% of the bark of *Pinus ponderosa* Dougl. ex Laws. on an extractive-free basis. Hardwood barks (2) range from 75.0% for *Betula papyrifera* to 41.5% for *Betula platyphylla* Suk. on an extractive-free basis. Clearly, researchers wanting to know the holocellulose content of tree barks should determine the data themselves.

Although tree barks contain considerable quantities of carbohydrate materials, bark alone has not been pulped commercially as a source of carbohydrates. Cram et al. (43) in an early study of the pulping of Western red cedar (*Thuja plicata* Donn) bark essentially demonstrated that special techniques would have to be developed for bark. These techniques have not yet been developed. The pulping of bark requires considerably more chemical usage than does the pulping of wood because of the increased content of polyphenolics and other chemical components that have to be solubilized. The pulp from bark is also of inferior quality to wood pulp for present commercial products. Embedded dirt and stones that are often present in bark after logging operations also create problems for the processing of bark. Increasing amounts of bark are accepted at pulp mills with wood chips, but it will be some time before bark supplies any significant amount of commercial carbohydrates.

IV. BARK LIGNINS

The lignins of bark have not been thoroughly characterized. There are a myriad of publications about the lignins of woods, but not nearly so many about bark lignins. Wood lignins, of course, are of the utmost importance to the recovery of pulp carbohydrates, but since there is no comparable commercialization of bark, the interest has not been so great.

One of the reasons bark lignins have not been completely investigated in their own right is because of the difficulty of isolating material pure enough, or homogeneous enough, for structural elucidations. Bark lignins and bark tannins both possess phenolic groups and the similarity of these functional groups has made it difficult to separate them. Cram et al. (43) in their early efforts to pulp *Thuja plicata* bark outlined difficulties with the so-called lignin materials. Segall and Purves (8) reviewed the early literature on bark lignins. They reported lignin values ranging from 9.4–53.7%, depending on the fraction of bark analyzed, method of analysis, and species. Thus, there is little to be gained in trying to evaluate the lignin content in any particular bark species from the early literature. The early reports also presented essentially no structural characteristics.

The advent of new experimental techniques and new instrumentation allowed researchers to obtain insights into bark lignins, but still with limited success as the following descriptions illustrate. Hergert and Kurth (44) and Kiefer and Kurth (45) obtained lignin from the bark of *Pseudotsuga menziesii*. They solubilized a fraction with 1% sodium hydroxide that represented 49% of what appeared to be Klason lignin based on the extractive-free bark. This easily removed lignin material was found by infrared spectroscopy and chemical analysis to resemble a high-molecular-weight phenolic acid rather than material that was generally designated as lignin. In an article that demonstrated the uncertainties of bark lignin research at the time, Kurth and Smith (46) showed that bark lignins of *Pseudotsuga menziesii* were composed of at least two fractions. One contained carboxylic groups and was low in methoxyl content. Kurth and Smith (46) termed this material bark phenolic acid. The material was readily soluble in dilute alkalies, insoluble in mineral acids and diethyl ether, but when freshly prepared was somewhat soluble in acetone, ethanol, dioxane, and water. The second lignin product was extracted with a dioxane-hydrochloric acid reagent and possessed a methoxyl content comparable to lignins isolated from conifer woods. However, the bark lignin products did not give the typical wood lignin color reaction with a phloroglucinol-hydrochloric acid reagent.

Swan (47) isolated a dioxane acidolysis lignin from both the inner and outer bark fractions of *Thuja plicata*. He concluded that the inner bark lignin was similar to softwood lignins, but that the outer bark lignin was different. The outer bark lignin contained fewer methoxyl groups, fewer aromatic protons, more catechol groups, and more aliphatic protons per C_9 unit as measured by proton nuclear magnetic resonance.

Sogo and Hata (48) in one of their numerous publications on bark commented on the isolation of lignin fractions from both barks of gymnosperm and angiosperm trees. They demonstrated the presence of condensed-type guaiacyl nuclei and smaller amounts of open-type guaiacyl and syringyl nuclei. They also obtained a dioxane lignin that was similar in analyses and characteristics to wood lignins.

Sarkanen and Hergert (49) published a comprehensive review of the lignins in tree barks. They commented that the most convenient method for isolating lignins from bark was by extraction with dioxane-hydrochloric acid. Lignin may also be isolated by Björkman's method, although yields are lower. Bark lignins were said to have a 1–2% lower methoxyl content than wood lignins. However, the comparison of ultraviolet and infrared spectra of bark lignins isolated by dioxane-hydrochloric acid and by the Björkman technique with spectra of corresponding wood lignins confirmed the essential identity of these materials. Sarkanen and Hergert (49) used the amount of the 72% acid-insoluble residue after alkaline extraction as representing the true lignin content of barks. They recalculated the data of Chang and Mitchell (50) on these bases and showed lignin contents ranging from 9.9% for *Pinus contorta* pine bark to 31.7% for slash pine (*Pinus elliottii* Engelman.) bark on an extractive-free basis. This represents a wide range of values indeed.

Hergert (51) reported that no Braun's native lignin was detected in the living inner bark of longleaf pine (*Pinus palustris* Mill.), but it was obtained in 1.1% yield from outer bark. The reason given was that on the death of the parenchymatic cells (outer bark formation), the substantial deposition of lignin occurred. Preliminary chemical analyses of the pine bark lignin fractions suggested a somewhat higher parahydroxyphenyl to guaiacyl ratio than in wood.

Hemingway (2) compiled a thorough review of the lignins of bark in 1981. Since that time, little or no work has appeared, at least none that made major advances in the understanding of bark lignins. Lignins from bark probably will not become commercially important in the near future. Bark lignins would have to be isolated for their own value and, in this respect, differ from the lignin preparations that are so readily available as byproducts of wood pulping.

V. BARK WAXES AND SUBERINS

Wax materials have been solubilized from a number of tree barks, but possibly the wax of *Pseudotsuga menziesii* has been the most thoroughly investigated. The wax has been of commercial interest for some 40 years (52). The presence of wax in *Pseudotsuga menziesii* bark was recognized as early as 1923 (53), but attracted little attention until the late 1940s when Clark et al. (54,55) extracted the bark with benzene and reported wax-like brown and black substances.

Kurth and co-workers (44,56–58) published pioneering studies on *Pseudotsuga menziesii* bark wax. Extracting bark from 80- to 95-year-old trees with n-hexane gave a light-colored "hexane wax" (5.47% yield), and extracting the residue with benzene gave a light brown "benzene wax" (an additional 2.52% yield). Patents were granted in the early 1950s for the extraction and refining of the wax (59–61), and, more recently, for extraction and utilization with new solvent systems and processes (62,63). The wax has been refined to the point at which it is a high-quality vegetable wax that can be used in formulating polishing waxes and manufacturing carbon paper (64,65). A commercial plant (Bohemia Inc., Eugene, Oregon) for extracting *Pseudotsuga menziesii* bark wax was founded in 1975 and operated for 6–7 years.

Considerable research has been done on the chemical composition of the wax from *Pseudotsuga menziesii* bark. Fang (66) and Laver et al. (67) reported the presence of uncombined sitosterol and campesterol in the n-hexane-soluble fraction, as well as some unidentified terpenes and "steroid-like" compounds. Loveland and Laver (68,69) showed that on chemical fragmentation by saponification, the n-hexane-insoluble but benzene soluble (benzene wax) yielded n-fatty acids, ω-dicarboxylic acids, ω-hydroxy fatty acids, and some fatty alcohols. However, these studies involved the identification of components that existed in the free state in the original wax and of components that resulted from degradation when the wax was saponified. Fang and Laver and Fang (70,71) isolated sterol and wax esters as they existed in their combined forms in the n-hexane wax. The sterol esters were composed of sitosterol and campesterol esterified to fatty acids, ranging in chain length from 13 carbons to 24 carbons. The wax esters were composed of 1-docosanol and 1-tetracosanol, also esterified to fatty acids, ranging in chain length from 13 carbons to 24 carbons. This work established that most of the wax from *Pseudotsuga menziesii* as extracted with n-hexane was a true wax in the chemical sense i.e., a material comprised mostly of esters of long-chain fatty acids and long-chain fatty alcohols.

Fang and Laver and Fang (70,72) also isolated chemically intact ferulic acid esters from n-hexane wax. Kurth (56) first reported the presence of ferulic acid in the n-hexane wax of *Pseudotsuga menziesii* bark and commented that the presence of ferulic acid did not appear to have been established previously in a natural wax. Kurth (56) speculated that the n-hexane wax contained ferulic acid esters, but he did not isolate any in their chemically intact forms.

Brooker (73) actually reported the first isolation of wax alcohol ferulates from a tree bark (*Phyllocladus glaucus* Carr.). Since then, numerous reports indicating the presence of ferulic acid esters in tree barks have appeared. For example, Rowe et al. (74) isolated ferulic acid esters from the benzene extract of jack pine (*Pinus banksiano* Lamb.) bark. Saponification of the esters yielded 1-docosanol and 1-tetracosanol and trace amounts of 1-octadecanol, 1-eicosanol, 1-tricosanol, and 1-hexacosanol. Adamovics et al. (75) using low-pressure liquid chromatography also isolated a mixture of ferulic acid esters from a chloroform extract of a cork fraction from the bark of *Pseudotsuga menziesii*. Saponification of the esters yielded 1-docosanol and 1-tetracosanol and ferulic acid

Adamovics et al. (75) were prompted to study the phenolic compounds of *Pseudotsuga menziesii* bark by the work of Mizicko et al. (76) who had demonstrated that cork-rich fractions of bark had enhanced the natural healing process of suberization and wound periderm initiation in cut seed potatoes. Bohemia Inc., Eugene, Oregon, sells a fine grind of *Pseudotsuga menziesii* bark, marketed as "Douglas fir bark 100," to potato growers for the specific purpose of treating freshly cut seed potatoes to enhance the wound healing of the potatoes.

The wax of *Pseudotsuga menziesii* bark exists primarily in the cork cells. Litvay (5) and Litvay and Krahmer (6,7) investigated the anatomical and chemical characteristics of the cork layers of *Pseudotsuga menziesii* bark. Their results showed that the cell wall structure was comprised of four zones or layers as follows: (1) a primary wall with randomly oriented microfibrils, (2) a secondary suberin layer composed of alternating lamella of phenolics and waxes that have the wax molecules oriented perpendicular to the cell surface, (3) a wax extractives layer, and (4) a layer of cellular debris such as cytoplasm and old membranes lining the lumen. A middle lamella composed of phenolic and pectic substances holds the cells together. These anatomical studies demonstrated that in order to obtain a decent yield of wax, bark from older trees that had formed outer bark and hence cork cells had to be used, and that the bark had to be quite finely ground to crush the cell walls and expose the wax layers to the extracting solvent.

Tree barks other than *Pseudotsuga menziesii* have been investigated for their wax content. Early work by Zellner, reviewed by Segall and Purves (8), reported results on European hardwood barks. Jensen et al. (77) and later Hemingway (2) have reviewed research reported on many bark waxes. Although the yields from each species might be somewhat different, the findings on *Pseudotsuga menziesii* bark wax are quite typical.

Suberin seems not to be clearly defined. In fact, the structure and composition of suberin is still incomplete (78). Suberin was first described by Priestly (79) as an aggregate of various modified forms of certain organic acids that can be in combination with glycerine as true fats. The properties of suberin are described by Priestly (79) as (1) insoluble and impermeable to water, (2) considerably insoluble in organic solvents, (3) greatly resistant to concentrated sulfuric acid, (4) readily oxidized by nitric acid or chromic acid, and (5) readily soluble in warm alkali. Hergert and Kurth (44) commented that in *Pseudotsuga menziesii* cork, the hydroxy acids were not only esterified to the phenolic acids but also to each other in an etholide-type structure. Jensen et al. (77) considered suberin as being in the cork cell walls in the form of a polyestolid of hydroxy acids and that it could be dissolved by saponification. The acids isolated from the saponification solution contained two or more functional groups that made it possible for the acids to form a network bound by ether or ester bonds. The exact chemical structure is thus ill-defined.

Litway (5) and Litvay and Krahmer (6,7) investigated suberin in the cork of *Pseudotsuga menziesii*. They stripped away the suberin layer by a saponification reaction. Paper chromatography indicated that the saponified material was phenolic and that "ferulic-acid-like" material was present. They commented that the phenolics in the suberin layer may be composed of cinnamic acid derivatives.

Suberin has been studied only sparingly. The early work consisted of Zetzsche's (80) studies on suberin from *Quercus suber* and Jensen's (81,82) reports on *Betula verrucosa* Enrh. The chemistry of the suberins of the *Betula* and *Quercus* species has been reviewed by Jensen et al. (77). Hemingway (2) has reviewed more recent work on suberin and lists the following fatty acids that have been identified on the saponification of suberin-containing material: ω-hydroxybehenic acid (83); eicosanedicarboxylic acid (77); 9,10,18-trihydroxystearic acid (77);18-hydroxy-9-acetadeceonic acid (77); *cis*-9,10-epoxy-18-hydroxystearic acid (83); 9-hydroxy-

1,18-octadecanedioic acid (84); *cis*-9,10-epoxyoctadecanedioic acid (83); 8,9-dihydroxy-1,16-hexadecadioic acid (77) and 8-hexadecene-1,16-dicarboxylic acid (77). These acids compare very well with the type of acids that Kolattukudy (78) later described as being part of the suberin polymer.

New and modern instruments are being utilized in attempts to better define the structure of suberin. Zimmermann et al. (85) used proton and carbon-13 nuclear magnetic resonance spectroscopy to study extracts from the cork of *Quercus suber*. They attempted to elucidate the involvement of phenolic compounds similar to lignin that Kolattukudy (78,86,87) had reviewed as being part of the suberin polymer. Zimmermann et al. (85) milled preextracted cork from *Quercus suber* in a vibrational ball mill. The material released was extracted with aqueous dioxane and N,N-dimethylformamide. Gel permeation chromatography of the extracts resulted in high- and low-molecular-weight fractions. This is a method widely used for the isolation of lignin from wood samples. The isolated fractions were characterized by proton and carbon-13 nuclear magnetic resonance spectroscopy to obtain information on the presence of lignin-like components and linkages in suberin. The nuclear magnetic resonance spectroscopic examination revealed that these fractions consisted mainly of saturated and unsaturated aliphatic compounds, alcohols, acids, and esters. Signals corresponding to guaiacyl, syringyl, and dilignol units were not found, indicating that lignin was not present in the examined cork extracts. Thus, it would appear that the suberin polymer contains little or no phenolic-like material, but is mostly aliphatic in nature, consisting of long-chain saturated and unsaturated aliphatic components, fatty alcohols, and acids. The increased use of instrumentation on undegraded samples of the suberin polymer will surely elucidate important components of the structure.

The commercial utilization of wax and suberin materials from tree barks does not appear promising at this time. Bohemia Inc., Eugene, Oregon, no longer extracts wax from *Pseudotsuga menziesii* bark, although the factory still makes bark products for potato growers. The waxes from tree barks have lower melting points and are softer than the waxes usually marketed in commerce. The bark waxes are also dark in color and efforts to bleach them have not been very successful. The dark colors are the result of polyphenolic contaminants and the removal of the contaminants is expensive and difficult.

VI. BARK VOLATILES

Tree barks naturally emit volatile materials, but steam distillation may be used to better remove the readily volatile components. The major volatile components recovered by the steam distillation of whole *Pseudotsuga menziesii* bark were furfural, terpinene-4-ol, α-terpineol, guaiacol, 2,5-dimethyl-3-acetylfuran, and β-cyclocitral (70,88). The presence of α-pinene, β-pinene, limonene, 1,5-p-menthadien-7-ol, β-citronellol, and geraniol were also tentatively established. Steam distillation may have altered certain compounds in the bark, and furfural and 2,5-dimethyl-3-acetylfuran may have been the result. The work on *Pseudotsuga menziesii* bark demonstrates the great number of compounds that can be volatilized from tree barks by simple steam distillation.

The majority of readily volatile materials from tree barks are terpenes and terpenoids. Terpenes and terpenoids have always been of interest to researchers studying the chemistry of bark, perhaps because there are so many different types in barks. Hemingway (2) has reviewed the work of Zavarin and his co-workers who investigated terpenes and terpenoids from many, many tree barks, based on their importance as taxonomic indicators, particularly with regard to their genetic and geographical variation. Zavarin and Snajberk (89) continue work on these important taxonomic questions, although much of the research involves terpenes and terpenoids in wood, as well as in bark.

Tree barks have long been a source of new and unusual compounds. Rowe and co-workers, in particular, have isolated and elucidated the structures of many new terpenes and terpenoids (2,90–92). Several of these compounds are not too volatile and were isolated by extraction with selected solvents, but they are included here because of their structural similarities. Several resin acids were first isolated from tree barks and their structures shown. Communic acid was first isolated from the bark of *Juniperous communis* L. by Arya et al. (93). Zinkel and Spaldina (94) found the new resin acid, strobic acid, in the cortical oleoresin of *Pinus strobus* L. These and other unique terpenes and terpenoids from both conifers and hardwoods have been reviewed by Hemingway (2).

Zinkel (95) has reviewed the chemistry and use of terpenes and terpenoids. Some of the uses include synthetic flavors and fragrances, raw materials for the synthesis of fine chemicals, and polyterpene resins. The use of bark as a meaningful supply of terpenes and terpenoids does not appear imminent.

VII. BARK PHENOLIC COMPOUNDS

Bark phenolic compounds are those possessing free phenolic functional groups in their structure. Although bark lignins possess free phenolic functional groups, they have been given separate treatment in Sec. IV of this chapter. Hemingway (2,96) has compiled two reviews on natural phenolic materials. The first review in 1981 (2) was restricted to tree barks. The second review (96) focuses on biflavonoids and proanthocyanidins in woody plants. These two reviews and references cited there are recommended reading for those interested in natural phenolic materials, as is a recent book of contributed chapters edited by Hemingway and Karchesy (97) on the chemistry and significance of condensed tannins.

Tree barks contain many unusual and unique phenolic compounds. One of the more novel compounds was isolated from the bark of red alder (*Alnus rubra* Bong.) by Karchesy (98) and Karchesy et al. (99). The compound was a novel diarylheptanoid xyloside that the researchers named Oregonin because red alder had previously been named *Alnus oregona* Nutt. Oregonin was assigned the structure 1,7-bis-(3,4-dihydroxyphenyl)heptan-3-one-5-xylopyranoside on the basis of the nuclear magnetic resonance, infrared, ultraviolet, and mass spectrometric data of its derivatives and on partial synthesis. The aglycone of Oregonin, 1,7-bis-(3,4-dihydroxyphenyl)heptan-3-one-5-ol, has also been found in the bark of *Alnus hirsuta* Turez. (100). Terazawa et al. (100) refer to the aglycone as hirsutanonol after the name of the tree in which it was found. Terazawa et al. (101) later isolated the xyloside of hirsutanonol and named it hirsutoside. Oregonin from the bark of *Alnus rubra* and hirsutoside from the bark of *Alnus hirsuta* are one and the same compound. Perhaps, these compounds are widespread in the *Alnus* species.

Oregonin is the precursor to the red color that develops in *Alnus rubra* bark and wood when they are cut and exposed to the light and atmosphere. The first reference to the fact that a xyloside might be involved was by Kurth and Becker (102) who reported pioneering work on the extractives of *Alnus rubra*. This work on the *Alnus* barks demonstrates that chemical studies on barks often are informative for investigations of commercial woods. The studies on Oregonin have been most helpful to researchers trying to understand and prevent the formation of color in the *Alnus* species (103–105).

A common class of phenolic compounds in tree barks are the flavonoids and their glycosides. Quercetin (3,5,7,3',4'-pentahydroxyflavone) and dihydroquercetin (taxifolin) (3,5,7,3',4'-pentahydroxyflavanone) are the most common flavonoids in conifer barks. Dihydroquercetin in *Pseudotsuga menziesii* bark received a great deal of attention in the 1950s and 1960s as reviewed by Hall (52) and by Hemingway (2). Patents were granted for the commercial extraction and utilization of dihydroquercetin from *Pseudotsuga menziesii* bark (2,52). However, markets never really developed for the compound and very little commercial work has been done on it since about 1970. However, interest still continues in the chemistry of flavonoids and their glycosides. Tsukamato et al. (106) have isolated two new lignan glucosides from the bark of *Olea europaea* L., a bark long used for its medicinal properties in southern Africa. Examples of novel biphenyl-linked derivatives of dihydroquercetin have been recently reported by Kolodziej (107) and by Foo and Karchesy (108). It is expected that chemical interest in these varied materials will continue.

Flavonoids have a long history of use as commercial dyes, food colorings, and pharmaceuticals. However, cheaper substitutes have replaced many of them and others have been banned by the U.S. Food and Drug Administration as carcinogenic or ineffective as pharmaceuticals. Hergert (109), Ryan (110), McClure (111), and Hemingway (2) have reviewed the extensive literature on bark flavonoids.

The development of red colors from colorless compounds present in plant tissues by treatment with hot mineral acid has long been known, as mentioned by Haslam (112). As a result of work over the years by many investigators, it is now accepted that two groups of substances are principally responsible for the red colors produced on mineral acid hydrolysis (113). These precursors of the red colors are flavan-3,4, -diols and flavan-3-ol dimers and higher oligomers. It was proposed (114) that flavan-3,4-diols be called leucoanthocyanidins and the flavan-3-ol products be named proanthocyanidins. Condensed tannins, which are now generally recognized as flavanoid polymers, are included in the proanthocyanidin classification (115).

The leucoanthocyanidins (flavan-3,4-diols) have received considerable interest because they have long been proposed as intermediates in the biosynthesis of the condensed tannins. Hemingway (96) and others have reviewed biosynthesis studies.

There has been considerable work done on the proanthocyanidin oligomers from plants in general in an effort to clarify their exact structures, particularly their stereochemistry (2, 96–97). Because of the chiral carbon centers at C-2, C-3, and C-4 of each monomer unit, the stereochemistry becomes quite complex. However, much of the stereochemistry has been carefully resolved, as explained by Hemingway (2,96) and others (97). The proanthocyanidin oligomers are conceptualized as small-molecular-weight oligomers that are really short-chain condensed tannins. Therefore, their exact structures are very important. Recent work in this field is exemplified by that of Foo and Karchesy (116) and Kolodziej (117).

The correlation between condensed tannins and anthocyanidin production was not obvious to early workers because most of the higher oligomers condense into "tannin reds" without the formation of anthocyanidins when treated with aqueous acid. It was only through the introduction by Pigman et al. (118) of improved reaction conditions (high dispersion in an alcoholic hydrochloric acid solution under pressure) that facilitated the generation of hydrolysis products that condensed tannins were found to yield anthocyanidins.

The primary approach to understanding the structures of condensed tannins has been to degrade them to small units and resolve the chemistry of those small units. Betts et al. (119) advanced the field greatly when they showed that tannin from common heather (*Calluna vulgaris* Salisb.) could be degraded with thioglycolic acid and that crystalline products of degradation could be isolated. The procedure called for treating the methylated tannin with thioglycolic acid and then reacting the products with diazomethane to produce the methyl ester of the thioglycolic acid derivative. The methyl esters could be crystallized and characterized.

The procedure represented a milestone in condensed tannin chemistry since for the first time not only could a position of interflavanoid linkage be established, but the stereochemistry at C-2 and C-3 of the pyran ring was also preserved in a simple derivative. Betts et al. (119) proposed that the tannin from heather was linked by ether linkages from C-4 of one unit to a phenolic oxygen at C-5 or C-7 of a second unit because thioglycolic acid was known to cleave benzylic ethers (120). Sears and Casebier (121), however, in the very next year showed that thioglycolic acid could also cleave carbon-carbon benzylic bonds in a model procyanidin that they had synthesized.

Sears and Casebier (122) applied the thioglycolysis reaction to methylated polyflavanoid fractions from the bark of *Tsuga heterophylla*. They obtained approximately equal amounts of the methyl 2,3-*cis* and 2,3-*trans*-(3-hydroxy-3',4',5,7-tetramethoxyflavan-4-ylthio)acetates after esterification. These results indicated that the tannin was comprised of approximately equal amounts of catechin and epicatechin flavan-3-ol units, linked together by carbon bonds via C-4 of the pyran ring of a flavan unit to C-6 or C-8 of the phloroglucinol ring of the adjacent unit. Since the original work with thioglycolic acid, additional sulfur nucleophiles including toluene-α-thiol and benzenethiol (123) have been used to cleave condensed tannins, as has phloroglucinol (124,125).

Karchesy (98) and Karchesy et al. (126) investigated the condensed tannins from the barks of *Alnus rubra* and *Pseudotsuga menziesii*. Thioglycolysis and subsequent permethylation of the methylated condensed tannin from *Alnus rubra* gave only the epicatechin thioglycolate derivative, methyl 2,3-*cis*-3,4-*trans*-(3-hydroxy-5,7,3',4'-tetramethoxyflavan-4-ylthio)acetate. However, the methylated condensed tannin from *Pseudotsuga menziesii* yielded methyl 2,3-*cis*-3,4-*trans*-(3-hydroxy-5,7,-3',-4'-tetramethoxyflavan-4-ylthio)acetate and methyl 2,3-*trans*-(3-hydroxy-5,7,3',4'-tetramethoxyflavan-4-ylthio)acetate in approximately a 3:1 ratio. The authors concluded that the condensed tannin from *Alnus rubra* was based on epicatechin units, whereas the condensed tannin from *Pseudotsuga menziesii* consisted of both catechin and epicatechin units in an undetermined sequence. Foo and Karchesy (125,127) have continued the work on the polyphenolic materials in *Pseudotsuga menziesii* bark.

The condensed tannins of many barks have been investigated as reviewed by Hemingway (2). Hemingway and McGraw (128) investigated the condensed tannins and extractive-free barks of *Pinus taeda* L. and *Pinus echinata* Mill. The authors showed that the condensed tannins and phenolic acid fractions were linked C-4 to C-8 or C-4 to C-6. Karchesy and Hemingway (129) showed that the inner bark of *Pinus taeda* contained (+)-catechin, a C-4 to C-8 linked (-)-epicatechin to (+)-catechin dimer, and three polymeric procyanidins that had different solubility and chromatographic properties. Despite differences in their physical properties, thiolysis with benzenethiol and carbon-13 nuclear magnetic resonance spectra indicated that all three polymeric procyanidins were composed of C-4 to C-8 or C-4 to C-6 linked (-)-epicatechin upper units (chain extenders) and that the lower unit (chain initiator) was (+)-catechin.

Markham and Porter (130) investigated the tannins of *Pinus radiata* D. Don bark and found that the hydroxylation pattern was more complex than in other pine barks (2). The yield of thiolysis products from all of these investigations was very low, less than 15%. Thus, the complete picture of these extracts that are referred to as condensed tannins is far from complete. The only complete parts are the thiolysis products that are identified, but there are considerable solids, over 85% that are unaccounted for.

The elucidation of the precise structures of the condensed tannins continues to be of interest (97,131). Hemingway et al. (132) demonstrated the heterogeneity of the interfavanoid linkages in trimeric procyanidins by the isolation of an epicatechin-4,8-epicatechin-4,8-catechin trimer, epicatechin-4,8-epicatechin-4,6-catechin trimer, and epicatechin-4,6-epicatechin-4-8-catechin trimer from the bark of *Pinus taeda*. These configurational isomers demonstrated the possible linkages that had been previously proposed (2) for condensed tannins. Hemingway et al.(133) further demonstrated the heterogeneity of interflavanoid bond location by additional studies of the procyanidins from the bark of *Pinus taeda*. Hemingway et al. (134) showed that configurational isomers in the polymeric procyanidins existed in plants, in addition to *Pinus taeda*, by isolating isomers from the bark of *Pinus palustris* and the leaves of *Photinia glabrescens*. These researchers also proposed a new system of nomenclature for proanthocyanidins analogous to the system used for oligo- and polysaccharides.

Hemingway and McGraw (135) investigated the kinetics of the acid-catalyzed cleavage of procyanidins. Comparisons of the rates of cleavage of isomeric procyanidin dimers in the presence of excess phenylmethane thiol and acetic acid showed that C-4 to C-8 interflavanoid linkages were cleaved more rapidly than C-4 to C-6 linkages, that 2,3-*cis* isomers with an axial flavan substituent were cleaved more rapidly than a 2,3-*trans* isomer with an equatorial substituent, and that the cleavage rate was independent of the stereochemistry in the terminal unit. These kinetic data are important in the isolation of specific types of dimers and trimers from the degradation of polymeric condensed tannins and hence provides considerable insight into chemical structures. Of additional assistance to better understanding the structures of the condensed tannins has been the work of Mattice and Porter (136) who have demonstrated that molecular-weight averages and carbon-13 nuclear magnetic resonance intensities provide evidence for branching in proanthocyanidin polymers.

The condensed tannins are polymers and so the molecular sizes and molecular weights of the polymers have been of interest. The phenolic proanthocyanidin polymers are too polar to be separated on gel permeation columns and therefore must be derivatized. Two studies of bark tannins used methyl ethers for gel permeation studies of pine bark tannins (129,138). Williams et al. (139) reported the molecular-weight profiles of condensed tannin polymers from a wide range of plant tissues of many different species. The molecular-weight profiles were obtained by gel permeation chromatography of the peracetate derivatives. The tannins varied widely in molecular weights, with number-average molecular-weight values for the peracetates in the range 1600–5500. This type of work and the more recent work of Mattice and Porter (136) aids considerably in understanding the overall properties of condensed tannin polymers. Research on the molecular sizes and weights of condensed tannins continues to be reported (97,137).

The synthesis of polymers such as the proanthocyanidins is difficult. However, work is progressing as evidenced by some recent reports. Hemingway and Foo (140) showed that condensations between epicatechin-(4β)-phenyl sulfide and leucocyanidin proceeded more rapidly at alkaline than at acidic pH. The authors suggest that the biosynthesis of condensed tannins may occur through a quinone methide rather than a carbocation intermediate. Foo and Hemingway (141) later reported the synthesis of the first "branched" procyanidin trimer, epicatechin-(4β→8)-catechin-(6→4β)-epicatechin. The synthesis was accomplished in higher yield than the linear analogues, suggesting that procyanidin polymers may be highly branched.

Hemingway and Laks (142) have recently proposed a biogenetic route to 2R,3R-(2,3-cis)-proanthocyanidins. Their route comes from 2R,3R-(2,3-trans)-dihydroflavonols and can be accounted for by tautomerism between quinone methide and flav-3-en-3-ol intermediates. Roux and co-workers have been exceptionally active in the synthesis of various condensed tannins. In a recent publication (143), the synthesis of 2,3-trans-3,4-cis-procyanidins by the condensation of (+)- leucocyanidin with (-)-epicatechin is reported. The condensation initiated a succession of substitutions, leading mainly to the introduction of (4,8)-2,3-trans-3,4-trans-procyanidin units, but also to the incorporation of "terminal" moieties that possess a unique 3,4-cis-procyanidin configuration. Now that various experimental techniques have been developed for the condensation of products leading to condensed tannins, it is expected that many more syntheses will be reported. Such synthesis work is important because the structures of the synthetic compounds are unambiguous in most cases, and they can be compared to the natural products to better define the structures of the materials from natural sources.

Condensed tannins represent only a part of the tannin picture. Generally, natural tannins are structurally divided into two groups: the above-mentioned condensed tannins and a second group referred to as hydrolyzable tannins (12). Historically, the main distinction was made on the basis of the two groups' actions toward hydrolysis. As discussed previously, the condensed tannins do not readily break down under mild conditions of hydrolysis. However, the hydrolyzable tannins, in which ester and depside links predominate, are readily hydrolyzed into a sugar or related polyhydric alcohol and a phenol carboxylic acid. Depending on the nature of the latter, a further subdivision is made in the literature. Gallotannins are those that yield gallic acid, and ellagitannins are those that yield ellagic acid.

The hydrolyzable tannins of tree barks have not received a great deal of attention. Hemingway (2) has ably reviewed work on hydrolyzable tannins from tree barks. Hydrolyzable tannins have been chemically investigated from the barks of some of the following species (2): *Acer*, *Quercus*, *Hamamelis*, *Castanea*, *Eucalyptus*, and

Schinopsis. Roux et al. (144) discussed the hydrolyzable tannins in regard to structural considerations in predicting the utilization of tannins, but in general, the hydrolyzable tannins from tree barks have not been chemically studied in great detail.

The barks of trees contain a variety of phenolic materials in what seem to be an endless array of compound classes. It is not possible to present a thorough review of all of these compounds here. Some of these compounds include the phlobaphenes (soluble in alcohol-insoluble in water) and the phenolic acids (soluble in alkali-insoluble in neutral solvents) (2) that are somehow related in structure to the condensed tannins. Other materials are stilbenes, salicins, lignans, coumarins, alkaloids, and naphthoquinones to name only a few. Hemingway (2) has surveyed the literature on much of the work on these classes.

The most recent efforts to utilize chemicals from tree barks has emphasized the phenolic compounds, particularly condensed tannins. Hemingway (2) and Porter and Hemingway (145) have reported on the utilization prospects for these materials, as have other authors (97). Some of the uses covered include reaction with formaldehyde to make adhesives, as additives for drilling muds, as grouting agents, as complexes for metal ions, and as polyols for reaction with diisocyanates to make adhesives and urethane foams. A recent report (146) describes the use of condensed tannins as a substitute for resorcinol in bonding polyester and nylon cord to rubber. This is by no means a complete list, and for further information, those interested should consult Hemingway's review (2), the book edited by Hemingway and Karchesy and a historical perspective by Hergert (147).

In recent work Foo et al. (148) reacted bark tannins from *Pinus taeda* with sodium hydrogen sulfite to yield sulfonated products. Condensed tannins used in leather tanning, mud additives for oil well drilling, and wood adhesives are often sulfonated to increase their solubility and reduce their viscosity in water. The authors (148) carefully investigated the sulfonation reactions and obtained a high yield of sulfonated products. Foo and Hemingway (149) later investigated the reaction of phloroglucinol and catechin with furfuryl alcohol and furfuraldehyde in an effort to better understand the reactions of phenolic materials with aldehydes other than formaldehyde. A considerable amount of the phenolic materials remained unreacted under the conditions used. Kreibich and Hemingway (150) prepared a condensed tannin-resorcinol adduct by coreaction of an extract from *Pinus taeda* bark with resorcinol at a 2:1 weight ratio. A laminating resin was prepared in which the entire amount of resorcinol normally used was replaced by the above adduct. The conclusions recorded indicate that over 60% of the resorcinol requirement in a room-temperature cure wood laminating adhesive can be replaced by extracts of *Pinus taeda* bark through the use of a condensed tannin-resorcinol adduct. These latest experiments and applications demonstrate that phenolic materials from tree barks can be used in commercial products.

VIII. CONCLUSIONS

Tree barks are composed of an awesome array of chemicals. Chemists from the earliest times have employed bark for a myriad of uses and investigated the chemicals involved for their own interests. Tree barks will continue to be explored.

Research into the chemical utilization of tree barks has unfortunately slowed. During the 1950s, 60s, and early 70s, there was almost feverish activity in bark utilization. This was, in part, due to the solid-waste disposal problems associated with wood products' manufacturing. The petroleum shortages in the 1970s spurred the engineering of boiler systems to utilize wood and bark wastes for energy. Although problems still remain, that engineering was quite successful, and at present most waste bark is burned for energy. This may represent a low-value end use for a great number of interesting chemicals, but it is a use that most manufacturing firms are comfortable with, and they remain unwilling to invest the research money necessary to separate commercially valuable chemicals from bark. However, some elegant and very definitive work is presently being reported on bark chemicals, most notably in the area of phenolics. These works will surely result in some valuable products. There also exists the potential utilization of bark for additional uses, as shown by recent work that demonstrated the use of bark as a raw material for charcoal and carbon pellets (151,152).

ACKNOWLEDGMENTS

The author would like to acknowledge Dr. Robert L. Krahmer, Department of Forest Products, Oregon State University, Corvallis, Oregon, for his consultation on bark anatomy. Acknowledgment is also made to Dr. Joseph J. Karchesy, Department of Forest Products, Oregon State University, Corvallis, Oregon, and to Dr. Richard W. Hemingway, Southern Forest Experiment Station, U.S. Department of Agriculture, Pineville, Louisiana, for consultation with them during the preparation of the manuscript.

REFERENCES

1. S. E. Corder, For. Res. Lab., Oregon State Univ., Corvallis, Ore., paper 31, 1972.
2. R. W. Hemingway, in *Organic Chemicals From Biomass* (I. S. Goldstein, ed.), CRC Press, Inc., Boca Raton, Fla., 1981, Chap. 10.
3. J. G. Haygreen and J. L. Bowyer, *Forest Products and Wood Science: An Introduction,* The Iowa State University Press, Ames, Iowa, 1982.
4. D. Fengel and G. Wegener, *Wood, Chemistry, Ultrastructure, Reactions*, Walter de Gruyther, New York, 1984.
5. J. D. Litvay, "Anatomical and Chemical Characteristics of the Douglas-fir [*Pseudotsuga menziesii* (Mirb.),Franco] Phellem Cell," Ph.D. Thesis, Oregon State Univ., Corvallis, Ore., 1977.
6. J. D. Litvay and R. L. Krahmer, *Wood Fiber, 8*:146 (1976).
7. J. D. Litvay and R. L. Krahmer, *Wood Sci.*, *9*:167 (1977).
8. G. H. Segall and C. B. Purves, *Pulp Pap. Mag. Can.*, 47(3):149 (1946).
9. R. L. Whistler and C. L. Smart, *Polysaccharide Chemistry*, Academic Press, New York, 1953.
10. E. F. Kurth, *Ind. Enq. Chem., Anal. Ed., 11*:203 (1939) .
11. R. L. Whistler and E. L. Richards, in *The Carbohydrates Chemistry and Biochemistry* (W. Pigman and D. Horton, eds.), Academic Press, New York, 1970, Vol. II A.
12. E. Timell, *Svensk Papperstidn., 64*:651 (1961).
13. J. W. Green, in *Methods in Carbohydrate Chemistry* (R. L. Whistler, ed.), Academic Press, New York, 1963, Vol. III p. 9.
14. R. L. Whistler and J. N. BeMiller, in *Methods in Carbohydrate Chemistry* (R. L. Whistler, ed.), Academic Press, New York, 1963, Vol. III p. 21.
15. G. J. Ritter and E. F. Kurth, *Ind. Eng. Chem.*, *25*:1250 (1933).
16. T. E. Timell, in *Advances in Carbohydrate Chemistry* (M. L. Wolfrom, ed.), Academic Press, New York, 1964, Vol. 19.
17. Y. C. L. Lai, "Douglas-fir Bark; Carbohydrates Solubilized by the Acidified Sodium Chlorite Delignification Reaction," M. S. Thesis, Oregon State Univ., Corvallis, Ore., 1972.
18. M. L. Laver, C.-H. Chen. J. V. Zerrudo, and Y.-C. L. Lai, *Phytochem., 13*:1891 (1974).
19. A. Beélik, R. J. Conca, J. K. Hamilton, and E. V. Partlow, *Tappi, 50*:78 (1967).
20. T. E. Timell, *Svensk Papperstidn., 64*:748 (1961).
21. K. V. Ramilingam and T. E. Timell, *Svensk Papperstidn., 67*:512 (1961).
22. T. E. Timell, *Svensk Papperstidn., 65*:843 (1962).
23. T. E. Timell, *Svensk Papperstidn., 64*:744 (1961).
24. T. E. Timell, *Svensk Papperstidn., 64*:685 (1961).
25. A. Jabbar Mian and T. E. Timell, *Can. J. Chem.*, *38*:1191 (1960).
26. T. E. Timell, in *Advances in Carbohydrate Chemistry* (M. L. Wolfrom, ed.), Academic Press, New York, 1964, Vol. 20.
27. K. J. Beveridge, J. F. Stoddart, W. A. Szarek, and J. K. N. Jones, *Carbohydr. Res.*, *9*:429 (1969).
28. A. Jabbar Mian and T. E. Timell, *TAPPI, 43*:775 (1960).
29. K. S. Jiang and T. E. Timell, *Cell. Chem. Technol.*, *6*:493 (1970).
30. K. S. Jiang and T. E. Timell, *Cell. Chem. Technol.*, *6*:503 (1972).

31. K. S. Jiang and T. E. Timell, *Cell. Chem. Technol.*, *6*:499 (1972).

32. R. Toman, *Cell. Chem. Technol.*, *7*:351 (1973).

33. R. Toman, S. Karacsonyi, and V. Kovacik, *Carbohydr. Res.*, *25*:371 (1972).

34. S. Karacsonyi, R. Toman, F. Janecek, and M. Kubackova, *Carbohydr. Res.*, *44*:285 (1975).

35. Y.-L. Fu, P. J. Gutmann, and T. E. Timell, *Cell. Chem. Technol.*, *6*:507 (1972).

36. G. O. Aspinall and G. Kessler, *Chem. Ind.* (London):1296 (1957).

37. C.-H. Chen, "Douglas-fir Bark: Isolation and Characterization of a Holocellulose Fraction," Ph.D. Thesis, Oregon State Univ., Corvallis, Ore., 1973.

38. J. V. Zerrudo, "Douglas-fir Bark: Water-Soluble Carbohydrates and Alkaline Degradation of a Xylan," Ph.D. Thesis, Oregon State Univ., Corvallis, Ore., 1973.

39. E. C. Fernandez, "Douglas-fir Bark: Structure and Alkaline Degradation of a Glucomannan," Ph.D. Thesis, Oregon State Univ., Corvallis, Ore.,

40. E.C. Fernandez and M.L. Laver, *Wood Fiber Sci.*, *18*(3):436 (1986).

41. H. H. Dietrichs, *Holz Roh-Werkst.*, *33*:13 (1975).

42. H. H. Dietrichs, K. Garves, D. Behrensdorf, and M. Sinner, *Holzforschung*, *32*:60 (1978).

43. K. H. Cram, J. A. Eastwood, F. W. King, and H. Schwartz, *Pulp Pap. Mag. Can.*, *48*(10):85 (1947).

44. H. L. Hergert and E. F. Kurth, *TAPPI*, *35*:59(1952).

45. H. J. Kiefer and E. F. Kurth, *TAPPI*, *36*:14(1953).

46. E. F. Kurth and J. E. Smith, *Pulp Pap. Mag. Can.*, *55*(12):125 (1954).

47. E. P. Swan, *Pulp Pap. Mag. Can.*, *67*(10):T456 (1966).

48. M. Sogo and K. Hata, *J. Jpn. Wood Res. Soc.*, *14*:334 (1968).

49. K. V. Sarkanen and H. L. Hergert, in *Lignins, Occurrence, Formation, Structure and Reactions* (K. V. Sarkanen and C. H. Ludwig, eds.), Wiley-Interscience, New York, 1971, p. 81.

50. Y. Chang and R. L. Mitchell, *TAPPI, 38*:315 (1955).

51. H. L. Hergert, in *Cellulose Chemistry and Technology* (J. C. Arthur, Jr., ed.), ACS Symposium Series 48, Washington, D.C., 1977, p. 240.

52. J. A. Hall, "Utilization of Douglas-fir Bark," Tech. Rev., U.S.D.A. For. Serv., Pac. N.W. For. Range Exp. Station, Portland, Ore., 1971.

53. G. C. Howard, *West Coast Lumberman*, *44*:22 (1923).

54. I. T. Clark, J. R. Hicks, and E. E. Harris, *J. Amer. Chem. Soc.*, *69*:3142 (1947).

55. I. T. Clark, J. R. Hicks, and E. E. Harris, *J. Amer. Chem. Soc.*, 70:3729 (1948).

56. E. F. Kurth, *J. Am. Chem. Soc.*, *72*:685 (1950).

57. E. F. Kurth and J. H. Kiefer, *TAPPI, 33*:183 (1950).

58. E. F. Kurth and J. H. Kiefer, *TAPPI, 50*:253 (1967).

59. E. F. Kurth, "Waxes and Their Production From Wood Waste," U.S. Patent No. 2,526,607, 1950.

60. E. F. Kurth, "Extraction of Valuable Products From Bark," U.S. Patent No. 2,662,893, 1953.

61. E. F. Kurth, "Process of Beneficiating Douglas-fir Wax and Product Thereof," U.S. Patent No. 2,697,717, 1954.

62. F. S. Trocino, "Extender for Thermosetting Resin," U.S. Patent No. 3,616,201, 1971.

63.   F. S. Trocino, "Methods of Separating Bark Components," U.S. Patent No . 3,781,187, 1973.

64.   R. D. Good and F. S. Trocino, *Chem. Eng.*, 81 (1974).

65.   F. S. Trocino, in *Forest Products Residuals* (U.K. Lautner, ed.), AIChE Symp. Ser. 71, 1975, p. 46.

66.   H. H.-L. Fang, "Douglas-fir Bark; n-Hexane Soluble Fraction," M.S. Thesis, Oregon State Univ., Corvallis, Ore., 1971.

67.   M. L. Laver, H. H.-L. Fang, and H. Aft, *Phytochem.*, *10*:329 (1971).

68.   P. M. Loveland and M. L. Laver, *Phytochem.*, *11*:430 (1972).

69.   P. M. Loveland and M L. Laver, *Phytochem.*, *11*:3080 (1972).

70.   H. H.-L. Fang, "Douglas-fir Bark; n-Hexane-Soluble and Volatile Materials," Ph.D. Thesis, Oregon State Univ., Corvallis, Ore., 1974.

71.   M. L. Laver and H. H.-L. Fang, *Wood Fiber Sci.*, *18*(4):553 (1986).

72.   M. L. Laver and H. H.-L. Fang, *J. Agric. Food Chem.*, *27*:114 (1989).

73.   E. G. Brooker, *New Zeal. J. Sci.*, *2*:212 (1959).

74.   J. W. Rowe, C. L. Bower, and E. R. Wagner, *Phytochem.*, *8*:235 (1969).

75.   J. A. Adamovics, G. Johnson, and F. R. Stermitz, *Phytochem.*, *16*:1089 (1977).

76.   J. Mizicko, C. H. Livingston and G. Johnson, *Am. Potato J.*, *51*:216 (1974).

77.   W. Jensen, K. E. Fremer, P. Sierila, and V. Wartiovaara, in *The Chemistry of Wood* (B. L. Browning, ed.), Wiley-Interscience, New York, 1963, Chap. 12.

78.   P. E. Kolattukudy, in *Annual Review of Plant Physiology* (W. R. Briggs, P. D. Green, and R. L. Jones, eds.), Annual Reviews Inc., Palo Alto, Calif., 1981, Vol. 32, p. 539.

79.   J. H. Priestly, *New Phytol.*, *20*:17 (1921).

80.   F. Zetzschie, in *Handbuch der Pflanzenanalyse* (G. Klein, ed.), Springer Wein, 1932, Vol. 3, Part 1, p. 205.

81.   W. Jensen, *Pap. Puu*, *B32*:261 (1950).

82.   W. Jensen, *Pap. Puu*, *B32*:291 (1950).

83.   E. Seoane, M. C. Serra, and C. Aquillo, *Chem. Ind.* (London) :662 (1977).

84.   A. Gonzales, *Acta Cient. Compostelana*, *5*:57 (1968).

85.   W. Zimmermann, H. Nimz, and E. Seemuller, *Holzforschung*, *39*:45 (1985).

86.   P. E. Kolattukudy, in *Recent Advances in Phytochemistry*, (F. A. Loewus and V. C. Runeckies, eds.), Plenum Press, New York, 1977, Vol. II, p. 185.

87.   P. E. Kolattukudy, in *Biochemistry and Plants* (P. K. Stumpf, ed.), Academic Press, New York, 1980, Vol. 4, p. 571.

88.   M. L. Laver, P. M. Loveland, C.-H. Chen, H. H.-L. Fang, J. V. Zerrudo, and Y.-C. L. Liu, *Wood Sci.*, *10*:167 (1977).

89.   E. Zavarin and K. Snajberk, *Biochem. Syst. Ecol.*, *13*:89 (1985).

90.   A. H. Conner, B. A. Nagasampagi, and J. W. Rowe, *Phytochem.*, *19*:1121 (1980).

91.   J. P. Kutney, G. Eigendorf, B. R. Worth, J. W. Rowe, A. H. Conner, and B. A. Nagasampagi, *Helv. Chim. Acta*, *64*:1183 (1981).

92.   A. H. Conner, B. A. Nagasampagi, and J. W. Rowe, *Tetrahedron*, *40*:4217 (1984).

93.   V. P. Arya, L. Enzell, H. Erdtman, and T. Kuboto, *Acta Chem. Scand.*, *15*:225 (1961).

94.   D. F. Zinkel and B. P. Spalding, *Tetr. Lett.*, *27*:2459 (1971).

95.   D. F. Zinkel, in *Organic Chemicals from Biomass* (I.S. Goldstein, ed.), CRC Press, Boca Raton, Fla., 1981, Chap. 9 .

96. R. W. Hemingway, in *Natural Products Extraneous to the Liqnocellulosic Cell Wall of Woody Plants* (J.W. Rowe, ed.), Springer-Verlag, New York, 1989, p. 571.
97. *Chemistry and Significance of Condensed Tannins* (R W Hemingway and J. J. Karchesy, eds.), Plenum Press, New York, 1989.
98. J. J. Karchesy, "Polyphenols of Red Alder: Chemistry of the Staining Phenomenon," Ph.D. Thesis, Oregon State Univ., Corvallis, Ore., 1975.
99. J. J. Karchesy, M. L. Laver, D. F. Barofsky, and E. Barofsky, *J. Chem. Soc. Chem. Commun.*:649 (1974).
100. M. Terazawa, H. Okuyama, and M. Miyake, *Mokuzai Gakkaishi*, *19*:45 (1973).
101. M. Terazawa, H. Okuyama, M. Miyake, and M. Sasaki, *Mokuzai Gakkaishi*, *30*:587 (1984).
102. E. F. Kurth and E. L. Becker, *TAPPI*, *36*:461 (1953).
103. B. F. Hrutfiord and R. Luthi, *Preprints of the International Symposium on Wood and Pulping Chemistry (Ekman Days)*, Stockholm, Sweden, 1981, p. I:95.
104. M. Terazawa, M. Miyake, and H. Okuyama, *Mokuzai Gakkaishi*, *30*:601 (1984).
105. M. Terazawa and T. Kayama, *Preprints of the International Symposium on Wood and Pulpinq Chemistry*, Vancouver, B. C., Canada, 1985.
106. H. Tsukamoto, S. Hisada, S. Nishibe, and D. G. Roux, *Phytochem.*, *23*:2839 (1984).
107. H. Rolodziej, *J. Chem. Soc., Perkin Trans I*: 219 (1988).
108. L. Y. Foo and J. Karchesy, *J. Chem. Soc.: Chem. Commun.*, 217 (1989).
109. H. L. Hergert, in *The Chemistry of Flavonoid Compounds* (T. A. Geissman, ed.), Macmillan, New York, 1962, Chap. 17.
110. A. S. Ryan, in *Utilization of Douglas-fir Bark* (T. A. Hall, ed.), U.S.D.A. For. Serv., Pac. N.W. For. Range Exp. Station, Portland, Ore., 1971, p. 109.
111. J. W. McClure, in *The Flavonoids*, (J. B. Harborne, T. J. Mabry, and H. Mabry, eds.), Academic Press, New York, 1975, Chap. 18.
112. E. Haslam, *Chemistry of Vegetable Tannins*, Academic Press, New York, 1966.
113. R. S. Thompson, D. Jacques, E. Haslam, and R. J. N. Tanner, *J. Chem. Soc., Perkin I*:1387 (1972).
114. K. Weinges, W. Bahr, W. Ebert, K. Goritz, and H. Marx, *Fortschr. Chem. Org. Naturst.*, *27*:158 (1969).
115. T. A. Geissman and D. H. G. Crout, *Organic Chemistry of Secondary Plant Metabolism*, Freeman, Cooper and Co., San Francisco, Calif., 1969.
116. L. Y. Foo and J. J. Karchesy, *Phytochem.*, *28*:1743 (1989).
117. H. Kolodziej, *Phytochem.*, *29*:955 (1990).
118. W. Pigman, E. Anderson, R. Fischer, M. Buchanan, and B. L. Browning, *TAPPI*, *36*:4 (1953).
119. M. J. Betts, B. R. Brown, P. E. Brown, and W. T. Pike, *J. Chem Soc.: Chem. Commun.*, 1110 (1967).
120. B. O. Lindgren, *Acta Chem. Scand.*, *4*:1365 (1950).
121. K. D. Sears and R. L. Casebier, *J. Chem Soc.: Chem. Commun.*, *1437* (1968).
122. K. D. Sears and R. L. Casebier, *Phytochem.*, *9*:1589 (1970).
123. B. R. Brown and M. R. Shaw, *J. Chem. Soc., Perkin Trans. I*:2036 1974).
124. L. Y. Foo and L. J. Porter, *J. Chem. Soc., Perkin Trans. I*:1186 (1978).
125. L. Y. Foo and J. J. Karchesy, *Phytochem.*, *28*:3185 (1989).
126. J. J. Karchesy, P. M. Loveland, M. L. Laver, D. F. Barofsky, and E. Barofsky,

Phytochem., *15*:2009 (1976).

127. L. Y. Foo and J. J. Karchesy, *Phytochem.*, *28*:1237 (1989).
128. R. W. Hemingway and G. W. McGraw, *Appl. Polym. Symp.*, *28*:1349 (1976).
129. J. J. Karchesy and R. W. Hemingway, *J. Agri. Food Chem.*, *28*:222 (1980).
130. K. R. Markham and L. J. Porter, *New Zeal. J. Sci*, *16*:751 (1973).
131. G. Nonaka, *Pure Appl. Chem.*, *61*(3):357 (1989).
132. R. W. Hemingway, L. Y. Foo, and L. J. Porter, *J. Chem. Soc.: Chem.Commun.*, 320 (1981).
133. R. W. Hemingway, J. J. Karchesy, G. W. McGraw, and R. W. Wielesek, *Phytochem.*, *22*:275 (1983).
134. R. W. Hemingway, L. Y. Foo, and L. J. Porter, *J. Chem. Soc., Perkin Trans. I.*:1209 (1982).
135. R. W. Hemingway and G. W. McGraw, *J. Wood Chem. Technol.*, *3*:421 (1983).
136. W. L. Mattice and L. J. Porter, *Phytochem.*, *23*:1309 (1984).
137. D. Cho, W. L. Mattice, L. J. Porter, and R. W. Hemingway, *Polym.*, *30*(11):1955 (1989).
138. M. Samejima and T. Yoshimoto, *Mokuzai Gakkaishi*, *27*:491 (1981).
139. V. M. Williams, L. J. Porter, and R. W. Hemingway, *Phytochem.*, *22*:569 (1983).
140. R. W. Hemingway and L. Y. Foo, *J. Chem. Soc.: Chem. Commun.*, 1035 (1983).
141. L. Y. Foo and R. W. Hemingway, *J. Chem. Soc.: Chem. Commun.*, 85 (1984).
142. R. W. Hemingway and P. E. Laks, *J. Chem. Soc.: Chem. Commun.*, 746 (1985).
143. J. A. Delcour, E. J. Serneels, D. Ferreira, and D. G. Roux, *J. Chem. Soc. Perkin Trans. I.*: 669 (1985).
144. D. G. Roux, D. Ferreira, and J. J. Botha, *J. Agri. Food Chem.*, *28*:216 (1980).
145. L. J. Porter and R. W. Hemingway, in *Natural Products Extraneous to the Ligno-cellulosic Cell Wall of Woody Plants* (J. W. Rowe, ed.), Springer-Verlag, New York, 1989, Vol. 2, p. 988.
146. G. R. Hamed, K. H. Chung, and R. W. Hemingway, in *Adhesives from Renewable Resources* (R. W. Hemingway, A. H. Connor, and S. J. Branham, eds.), ACS Symposium Series 385, Washington, D.C., 1989, p. 242.
147. H. L. Hergert, in *Adhesives from Renewable Resources* (R. W. Hemingway, A. H. Connor, and S. J. Branham, eds.), ACS Symposium Series 385, Washington, D.C., 1989, p. 155.
148. L. Y. Foo, G. W. McGraw, and R. W. Hemingway, *J. Chem. Soc.: Chem. Commun.*, 672 (1983).
149. L. Y. Foo and R. W. Hemingway, *J. Wood Chem. Technol.*, *5*:135 (1985).
150. R. E. Kreibich and R. W. Hemingway, *For. Prod. J.*, *35*(3):23 (1985).
151. A. M. Aslam, R. D. Sproull, M. L. Laver, and C. J. Biermann, *Appl. Biochem. Biotech.*, *20/21*:135 (1989).
152. M. A. Ali, M. L. Laver, C. J. Biermann, R. L. Krahmer, and R. D. Sproull, *Appl. Biochem. Biotech.*, *24/25*: xxx (1990).

10

The Composite Nature of Wood

Arno P. Schniewind and Harald Berndt

University of California, Berkeley, California

I. INTRODUCTION

The physical, mechanical, and chemical behavior of wood cannot be fully understood without reference to its physical organization, including that of its chemical constituents. Each of the major constituent groups, the cellulose, the hemicelluloses, the lignin, and the extractives, make unique contributions to the characteristic properties and behavior of wood. How each component exerts its influence depends on each of the other components as well. It is the purpose of this chapter to examine the composite action of wood constituents and its effect on the physical and mechanical behavior of wood.

   Wood can best be understood as a fiber-reinforced composite material. Composite interactions in wood can occur on several levels of physical organization and structure. At its most elementary level, composite action takes place between oriented, fibrous, framework material and amorphous matrix material. These interactions take place within a cell wall layer. The next level is the cell wall, a layered system where individual layers differ in both composition and structure. Further interaction takes place in the double cell wall of two neighboring cells. Although their overall structure and composition may be nearly the same, there will usually be differences in the orientation of the framework of adjacent walls. Composite action also takes place between tissues or aggregates of similar cells, but within the same growth increment. Finally, composite interaction can develop between growth zones, representing different stages of growth in the life of a tree, each of which encompasses a number of annual growth increments.

Physical organization and structure will also affect the chemical behavior of wood, principally by the way it determines accessibility to reactive matter. Since wood is a porous material with a very large internal surface [about 800 cm^2 in a cube of wood 1 cm on a side with a relative density of 0.4 (1) not counting possible pore space within the cell wall], permeability to reactive fluids is determined by structural details. Similarly, encrustation by matrix material will affect the accessibility of the framework. However, discussion of chemical behavior is outside the scope of this chapter.

II. ORGANIZATION AND STRUCTURE

A mid-nineteenth century view of the cell wall had it consisting of nonswelling, crystalline particles and a swelling, gellike substance. The organization was pictured as similar to that of bricks and mortar. The crystalline particles were called micelles, derived from their resemblance to mica (2). This is not unlike today's ideas of composite materials with oriented reinforcing material embedded in an isotropic matrix.

Today's concept of the cell wall resembles that of a fiber-reinforced composite. The anisotropic cellulose microfibrils form the fiber reinforcement and the isotropic lignin the matrix. The partially oriented hemicelluloses are intimately connected to the other two components. The cellulose fiber reinforcement is arranged in a random pattern in the primary wall and in parallel alignment in the secondary wall, with angles of orientation that differ from one layer to the next (Fig. 30, Chap. 2).

A. Physical Structure of Cellulose as Framework

Cellulose forms the structural framework of plant cell walls. It is naturally crystalline, exhibits crystalline polymorphism, and possesses a highly ordered, fibrous morphology (3). It consists of anhydroglucopyranose residues, linked via β-1-4 glycosidic bonds (Chap. 4). The β-glucopyranose conformation with the lowest energy content is the chair form with equatorially arranged substituents. It is shown in Fig. 1, together with its spatial dimensions. As a consequence of the linkage via the C_1 OH-group in the β position, adjacent anhydroglucose residues are rotated around the C_1-C_4 axis by 180°. The repeating unit of cellulose is therefore cellobiose rather than glucose. The degree of polymerization (DP), however, is given as the number of anhydroglucose units (AGU). Goring and Timell (5) found that "..celluloses in ferns, gymnosperms, and angiosperms have approximately the same molecular size, amounting to a minimum weight-average degree of polymerization of 9,000 to 10,000." More recently, the DP of native spruce wood cellulose was estimated as 12,000 (6).

Studies of the kinetics of cellulose formation indicate that higher plants produce

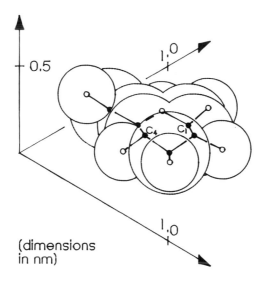

Figure 1   Spatial dimensions of the minimum-energy conformation of glucose. (From Ref. 4.)

cellulose by a matrix mechanism, similar to protein synthesis, resulting in molecules of uniform DP (7–9). The kinetics of cellulose hydrolysis suggest the presence of four weak links in each native molecule, evenly separated by approximately 3500 AGU. The probable cause is a chemical modification, an AGU being replaced by a 2-deoxyglucoside, fructoside, or xyloside (10). The DP of native wood cellulose was found to decrease with the age of the cell, from the cambium toward the pith (11). This decrease and the polydispersity of DP typically observed in technical cellulose preparations are most likely due to degradation by hydrolysis or oxidation.

Because of the β-1-4 bond and the minimum energy content of the conformation shown in Fig. 1, the cellulose molecule is elongated, with the AGU's forming a ribbonlike structure. Each AGU contains three OH groups as substituents and one oxygen atom in the pyranose ring. These functional groups are hydrophilic and allow formation of hydrogen bonds. Various schemes of intra- and intermolecular hydrogen bonding have been proposed (see Fig. 2). Cellulose coils randomly only at DP > 2000, whereas xylan chains, which are identical to cellulose except for the absence of the CH_2OH group, do so at DP > 500 (15). This behavior indicates that the C_6 hydroxyl group plays an important role in stiffening the molecule and would strengthen the argument for two intramolecular hydrogen bonds in cellulose. Raman spectral studies indicate differences between adjacent AGU's, suggesting that two alternating hydrogen-bonding patterns exist in the chain molecule (16,17) In our opinion, the currently most accepted model (Figs. 2a and 2b) is an arrangement of cellulose molecules in sheets that have hydrophilic edges and hydrophobic faces. These sheets are thought to be intermediates in microfibril formation and are bonded into larger units via London (van der Waals) forces (18).

Its conformation and configuration give the cellulose molecule a stiff, ribbonlike shape. Viscosity measurements on cellulose solutions showed that the chains have a persistence length, i.e., a length over which they will always be straight, of 7.1 nm (approximately 10–13 AGU's) (15). The persistence length needs to be considered in modeling transitions between ordered and less ordered regions. It casts doubt on the existence of truly amorphous states of native cellulose.

Present models of the crystalline unit cell of native cellulose have evolved from the structure proposed by Meyer and Misch (19). The lattice is monoclinic, i.e., it has three axes of unequal lengths and one non-90° angle, which is described by the notation C_2^2. Only the corners of the crystalline unit are occupied (P) and its translation is 1/2. i.e., the chains are screwed around the longitudinal axis by 180° (2_1), leading to the complete description of the space unit of cellulose as $C_2^2 P 2_1$. The pertinent literature has been reviewed by Marchessault and Sundararayan (20–22).

If we assume that the AGU is present in the chair conformation, only a few degrees of freedom are left for a complete description of the unit cell. These include the rotational position of the C_6 hydroxyl group and of the AGU around the glycosidic linkage. There is also the question of whether the two chains of the unit cell have the same or opposite polarity, i.e., whether they are parallel or antiparallel. Attempts have been made to find the most likely values for these free parameters by conformational and packing energy calculations (12,13). The results favor a parallel arrangement, whereas the Meyer-Misch unit cell assumed antiparallel chains. The lattice positions are nearly equal for the two models, but some effect on intermolecular bonding, by hydrogen bonds or van der Waals forces, would have to be expected.

Solid-state NMR studies have shown resonance multiplets for the C_1, C_4, and C_6 signals, indicating the presence of two distinct crystalline forms, I_α and I_β, the cellulose of higher plants consisting of 60–70% of the latter (23,24). The two forms differ mainly in their hydrogen bonding pattern, and less in molecular conformation (25).

The details of crystalline structure probably have no effect on the chemical reactivity of the crystallites (26), but one would expect some influence on such physical properties as dielectric and elastic constants, and possibly specific heat. Hydrogen bonding inside the crystallites has been shown to have a major effect on the elastic constants of cellulose (27).

Several approaches to defining the degree of lateral order are possible, all of which will furnish somewhat different results. Most methods divide cellulose into

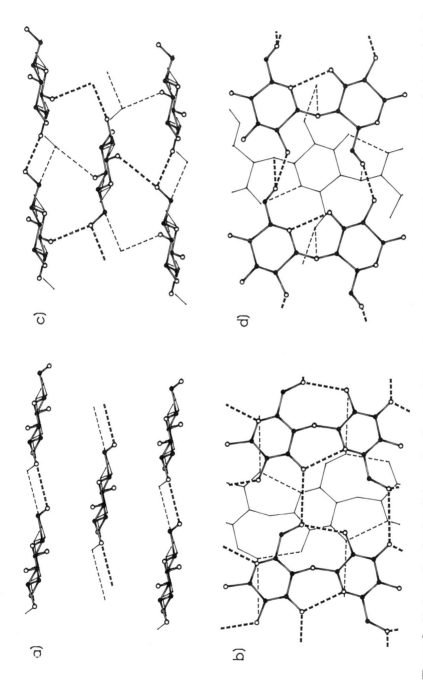

Figure 2 Hydrogen bonding in cellulose unit cells. a), c): Cross section of unit cell. b), d): Side view. [a), b): From Ref. 12 and 13; c), d): From Ref. 14.].

two phases, termed either crystalline and amorphous, or inaccessible and accessible. Data obtained for cellulose I by various methods are summarized in Table 1. The terms accessible and amorphous are not always used synonymously, as some researchers assume that an outer layer of unit cells around crystallites is ordered but accessible (30,35).

A two-phase description of molecular order in cellulose has to be considered a first approximation, since the ordered regions are not truly crystalline, the disordered regions not truly amorphous (36), and the transition between them is gradual rather than abrupt. Continuous degree-of-order spectra have been proposed and determined (37,38).

It has been shown that the theory of paracrystallinity (39) provides a valid description of cellulosic materials (40). In a paracrystalline material, lattice vectors are not equal in all unit cells, but fluctuate statistically around a mean. The magnitude of this fluctuation determines the distortion factor, another measure of molecular order. A model has been proposed to characterize the degree of order in

Table 1   Degrees of Crystallinity of Cellulose I Determined by Various Methods

Method of determination	Fraction of cellulose in ordered regions (%)		Reference
	Cotton	Wood pulp	
Ethylation	100		(28)
Hydrolysis and oxydation	95	88–92	(29)
D_2O exchange	54–82	36–64	(30)
X-ray diffraction	69	67	(31)
D_2O exchange and IR spectroscopy	64–74		(32)
Near infrared spectroscopy	62		(33)
X-ray diffraction	72		(33)
Heat of crystallization	91–95	57	(34)
X-ray diffraction	80–82	72	(34)

cellulose by parameters based on a three-phase model, i.e., degree of crystallinity, degree of paracrystallinity, degree of amorphity and distortion factor (41). It has been observed that these parameters are slightly anisotropic with respect to the crystallographic planes (42).

The differences between measuring molecular order qualitatively and quantitatively have been discussed by Ray and Bandyopadyhay (43). They observed that the perfection of crystalline regions, measured by the sharpness of the x-ray diffraction pattern, increased with increasing moisture content, while the quantitative degree of crystallinity decreased.

Sizes of ordered regions in cellulose, usually referred to as crystallites or micelles, have been determined by many researchers using various methods. Some results are summarized in Table 2. Crystallite size seems to be a function of the angle between the fiber axis and the longitudinal axis of the crystallite, which was explained as being due to the restrictions imposed by the cell wall curvature (44).

Cellulose molecules aggregate into microfibrils (sometimes called fibrils) with a width on the order of 1–2 nm in primary walls and 10–20 nm in secondary walls, and of indefinite length (45). It had been proposed that microfibrils are formed by the lateral fasciculation of so-called elementary fibrils with a uniform diameter of 3.5 nm (46), but this concept is no longer accepted (47–49). Microfibrils can be split into units of nearly molecular dimensions (50), with diameters of 1–5 nm or more (51). Valonia microfibrils have cores of single crystallites with diameters of approximately 20 nm (48,49). Microfibril cross-sections have been reported as rectangular to nearly square in poplar tension wood (52). Hexagonal cross-section geometry has been proposed to explain variations in observed width (53).

Small-angle x-ray diffraction data (Table 2) indicate an alternating pattern of

Table 2   Crystallite Sizes of Cellulose I

Material	Method	Diameter(nm)	Length(nm)	Reference
Ramie	WAXD	5.0		(40)
Ramie	SAXS	5.2, 10.6, 21.7, 40.1	48.6	(40)
Cotton	AH, EM		36–54	(35)
Cotton	DP_nSL		103	(35)
Alpine fir	SAXS		4.3–31.1	(44)
Western hemlock	SAXS		4.2–28.2	(44)
Trembling aspen	SAXS		4.2–28.2	(44)
White birch	SAXS		4.6–26.0	(44)
Bigleaf maple	SAXS		4.5–22.6	(44)

WAXD: wide-angle x-ray diffraction; SAXS: small-angle x-ray scattering; AH, EM: acid hydrolysis, electron microscopy; DP_nSL: DP_n at complete fiber strength loss.

ordered and less ordered regions along the microfibril axis. The existence of periodic, less ordered regions longitudinally provides the microfibril with the flexibility to follow the curvature of the cell wall (51,44). Several proposed models of molecular arrangement are summarized in Fig. 3. The folded-chain model can probably be rejected because it is incompatible with mechanical behavior (54) and electron microscopic observations (50,51). A cellulose unit cell with parallel chains also rules out chain folding.

B. Lignin and Hemicelluloses as Matrix

Lignin and hemicelluloses, in combination, form the matrix in which the cellulose microfibrils are embedded. As a first approximation, at least, both constituents can be considered amorphous, as is evident from the detection of distinct glass transition temperatures for each one in situ (55). This indicates the presence of two distinct phases in the cell wall, but a slight smearing of the loss tangent peaks suggests some mixing between phases. A combined treatment of these constituents is necessary and justified, however, because it is well established that they are intimately connected (56). From x-ray diffraction analyses, it appears that there are two types of amorphous material: one with random and one with preferred orientation. The preferred orientation is parallel to the cellulose microfibril directions (57).

1. Lignin

Lignin is a crosslinked, phenolic polymer built from three basic phenyl propane monomers connected via a large number of different linkages (Chap. 6). It is thought that lignin macromolecules are formed by a largely random polymerization process.

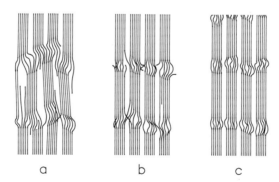

a   b   c

Figure 3   Possible arrangements of cellulose molecules in microfibrils, by increasing order from a) oriented fringed-micellar system and b) intermediate system to c) individual fibrillar units with alternating ordered and less ordered regions. (From Ref. 4.)

Protolignin, i.e., lignin in situ is considered an insoluble, essentially infinite network (58), although lignin repeat units have been proposed (59). Molecular-weight determinations are extremely difficult and their interpretation questionable, since all isolation processes significantly change its constitution. Molecular weights for various lignin preparations range from 324–150,000 (58). Isolation of lignin from eastern hemlock at 100% yield by enzymatic cellulose degradation and ball-milling resulted in fractions with a weight-average molecular weight of 77,000 (60).

Several model structures have been built based on evidence from various analytical techniques, most recently by employing computer optimization (61). Since, on the one hand, it is difficult to assemble a structural model of lignin that includes all relevant details in correct proportions, and, on the other hand, it is arduous to gather all analytically important details from a hypothesized lignin structure, a different approach to lignin description has been proposed. It uses graphs that draw attention to such analytically important details as elemental composition, functional groups, and frequency of linkage types. It leaves the sequence of the basic units unconsidered, which is said to be less important in amorphous polymers (62,63).

One main characteristic of lignins from different origins is the relative content of building blocks. According to their content of guayacyl-(G), syringyl-(S), and p-hydroxyphenyl-(H) groups, lignins from softwoods (G-lignins), hardwoods (GS-lignins), grasses (GSH-lignins), and compression wood (GH-lignins) can be distinguished (64). Any description of the lignin of a given plant must be considered as an average, since it was found that lignin characteristics are different in different morphological units of the same plant (65,66).

Conformational energy calculations on lignin showed that the most stable conformation of the β-O-4 dimer, the most common linkage in lignins, is a sandwich-type structure (67). A predominance of such an arrangement would lead to a planar organization of most benzene rings. Polarized Raman microspectroscopy has supplied evidence for a preferred orientation of the benzene rings in the plane of the cell wall surface (68).

2. Hemicelluloses

The hemicelluloses include a diverse group of cell wall carbohydrates (Chap. 7). They have low degrees of polymerization, ranging from a few dozen to a few hundred monomeric units. Most hemicelluloses are branched, or have at least some side groups. They are soluble in alkali or even in water. Their structure inside the wood cell wall is thought to be that of a water-swelling gel. Polarized infrared spectra indicate that the xylan (69,70) and glucomannan (69) components are oriented. The same results were obtained from birefringence measurements (71).

The straight-chain hemicelluloses can, especially after removal of substituents, form crystalline aggregates. Unsubstituted xylan can crystallize in various forms depending on water content, the dihydrate crystalline form showing the highest degree of order (72). Even xylan diacetate shows crystalline features in x-ray diffraction (73). Glucomannan can exhibit crystalline diffraction similar to that of a pure mannan I (74).

It has been shown that O-acetyl-4-O-methyl glucuronoxylan crystallizes inside the cell wall after mild deacetylation (75). The xylan crystallites were closely parallel to the cellulose microfibrils. Similarly, alkali-soluble beechwood xylan formed fibrous structures on precipitation, as did sprucewood xylan (76). Thin fibrils of glucomannan were found in the same manner (77). The ability of hemicelluloses to crystallize onto the surfaces of cellulose microfibrils is thought to contribute to their observed stabilization in some pulping processes (78–80).

C. Nature of Association of the Components

1. The Lignin-Carbohydrate Complex (LCC)

Isolated wood carbohydrate fractions always contain some lignin, and lignin preparations invariably include some carbohydrates. This intimate connection of the wood constituents has been known since the first attempts at wood analysis and led to the proposal of covalent bonds between the lignin and carbohydrate components. The pertinent literature has been reviewed repeatedly (81–83). Many of the observations can be explained by assuming incrustation and/or entanglement of the carbohydrates in a three-dimensional lignin network (81), also known as "snake cage" structures (58), but there seems to be good evidence for the existence of covalent bonds.

In studies modeling the formation and aging of lignin via quinone methide intermediates, it was found that these intermediates react with carbohydrates present to form ether bonds (84). A similar ether bond was formed between a monomeric lignin model compound, vanillyl alcohol, and polyols or sugars in a neutral aqueous solution (85). Viscosimetry and analysis of milled-wood lignin from spruce led to the proposal that the LCC consists of one hemicellulose molecule of DP_n 18 linked to single molecules of lignin of M_n 3500, and an estimate of one lignin-carbohydrate linkage per 44 hemicellulose sugar units (86). For milled-wood enzyme lignin, a lignin-carbohydrate bond frequency of 0.028 per phenyl propane unit has been determined experimentally (87).

Eriksson et al.(88) found evidence for linkages between xylan and lignin via ether bonds to the 2- or 3-positions of arabinose and, to a lesser extent, xylose units, and between galactoglucomannan and lignin via ether bonds to the 3-positions of galactose units. These authors also found evidence for ester bonds of lignin to 4-O-methyl glucuronic acid units and suggest that even cellulose and lignin form covalent bonds. Mild alkaline treatment cleaved 10–20% of the lignin-carbohydrate bonds in milled-wood enzyme lignin, and it was proposed that these bonds are mainly uronic acid ester linkages (87). Glycosidic linkages of the arabinofuranic type were found after in vitro synthesis of LCC with a crude enzymatic mixture from actively lignifying poplar stems (89). The authors consider the possibility that arabinose-predominated sugar residues may only be attached to lignin, without being incorporated in a polysaccharide chain.

Electron microscopic investigations showed that in fractions with high lignin content, polysaccharide fibrils were embedded in the lignin, whereas in fractions with low lignin content, the fibrils were coated with a thin layer of lignin (90,91). These studies showed that carbohydrate fibrils are coiled in the LCC, probably due to many covalent bonds between individual carbohydrate fibrils and large lignin domains. Other researchers found evidence only for very thin fibrils studded with small spherical lignin aggregates of the dimensions typical of Brauns native lignin, with single bonds between the fibrillar material and separate lignin domains (92).

2. Morphological Organization

The distribution of the three wood components in the various layers of the cell wall has been studied, e.g., by microspectrophotometry for lignin (93–95) and for carbohydrates (96), and by analysis of the fractions of developing xylem obtained by microdissection (97). Table 3 shows the average values of component distribution in the cell wall layers. Similar determinations were made by Mark (99).

Resistance to ozonization indicates an intimate association of hemicelluloses with the ordered regions, i.e., cellulose, of native plant cell walls (100). On the other hand, changes in pore size distribution during the hemicellulose removal from chlorite delignified wood demonstrate that hemicelluloses are also distributed throughout the three-dimensional lignin network (101). This strongly indicates that the hemicelluloses form the link between the cellulose framework and lignin matrix.

When the shrinking behavior of wood cells was followed during delignification (102) and hemicellulose removal (103), the results pointed toward a concentric lamellar arrangement of these components in the cell wall. A lamellar arrangement of lignin has also been observed by the staining and microdensitometry of the obtained electron micrographs (104). Such an arrangement would be in line with the layered structure of the cell wall, which can readily be explained by growth mechanisms.

Figure 4 summarizes the experimental evidence regarding the ultrastructural association of cell wall components. This is, in effect, a combination of the interrupted lamellae model of Kerr and Goring (105) with the detailed view of molecular orientation and interfaces presented by Fengel and Wegener (4). The latter already included a variation of the proposal of oriented hemicelluloses by Marchessault (70,106). An interrupted lamellae model that allows for the limited variability of sizes and orientation of the structural elements would appear to most readily accommodate all reported observations. The honeycomb model proposed by Scallan (107) is essentially similar to the interrupted lamellae model and can be considered a special case or modification thereof. The above-mentioned models, as

Table 3 Distribution of Wood Components in Softwood Tracheids[a]

Morphological unit	Fraction of total material (vol %)[b]	Cellulose (vol. %)[b]	Hemi-cellulose (vol. %)[b]	Lignin (vol. %)[b]
Compound middle lamella	12.3	12.4	25.6	62.0
S1	10.0	34.5	35.5	30.0
S2 + S3	77.7	55.7	14.3	30.0

[a]Based on data collected by Fengel (98).
[b]Assuming the following densities: cellulose = 1.55 g/cm^3 , hemicellulose = 1.50 g/cm^3, lignin = 1.30 g/cm^3.

well as the proposals by Stone and Scallan (108) and Page (71), were compared to electron microscopic observations on enzymatically and cytochemically modified wood cell walls (109). None of the models could be rejected based on the new evidence.

D. Role of the Extractives

The extractives derive their name from the fact that they can largely be removed from wood with neutral solvents. They are also referred to as the extraneous substances because they are not thought of as being an integral part of the cell wall structure. They can probably best be described as encrusting substances. Nevertheless, extractives can have significant effects on the properties and behavior of wood (110–112). Extractives encompass a very broad spectrum of chemical compounds, and their composition varies according to species as well as from sapwood to heartwood in a given stem (Chap. 8). Extractives are more abundant in the outer than the inner heartwood (113).

Perhaps the most obvious effect of extractives is that they add to the weight of wood. Many wood properties are related to wood density, but such relationships may be obscured by high extractive contents whenever the extractives do not contribute to a particular property, as for instance, pulp yield (114). In this respect, extractives can truly be considered as extraneous materials. On the other hand, extractives are responsible for the characteristic color and odor of various species, or in some woods for resistance to decay and insect attack, which makes them essential components even if they may be extraneous in a structural sense.

The effect of extractives may be purely chemical, as when they interfere with the setting of gypsum or cement in composite board manufacture (115,116), or purely

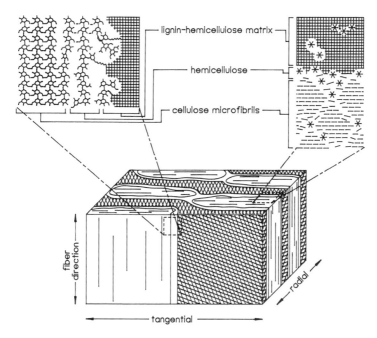

Figure 4 Ultrastructural association of cell wall components, with detailed views (tangential and cross section) of molecular association. In the magnified tangential view, covalent and intermolecular hydrogen bonds are shown. Wavy lines represent lignin-polysaccharide bonds. (From Refs. 4 and 105.)

physical, as when they occlude avenues for the passage of fluids (117,118), or physico-chemical, as in their effect on the shrinkage and hygroscopic properties of wood (119,120).

With regard to such properties as shrinkage and hygroscopicity, the effect of the extractives will depend on their specific location within the wood. Those merely deposited in the cell lumen would not be expected to affect shrinkage, e.g., whereas extractives located within the microcapillary structure of the cell wall could occupy sorption sites, making the latter inaccessible to water molecules (121). Extractives in the cell cavities may be found either lining the wall surfaces in the lumen or as incrustations in the pit membranes where they can have a major effect on permeability to fluids (118).

The relationship between extractive content and hygroscopicity has been used to determine the proportion of extractives that are contained within cell walls, as compared to cell cavities. Tarkow and Krueger (122) investigated this for redwood, assuming that the removal of extractives from within the cell wall would result in a volume change (additional shrinkage) of equal volume. They determined that 73% of the extractives were in the cell wall and 27% in the lumen of the redwood heartwood specimens investigated. A detailed study of extractive distribution in redwood and incense cedar cell cavities and cell walls was made by Kuo and Arganbright (123,124). Major percentages of extractives were found to be in cell walls of the heartwood of both species, amounting to 50.6, 77.6, and 74.1% of total extractives in the outer, intermediate, and inner heartwood of redwood, respectively, and similar results were obtained for incense cedar. In Japanese larch, the major component of the water-soluble extractives is arabinogalactan. In the heartwood, 30% of these were determined to be inside the cell wall, whereas in sapwood, the percentage rose to 60 (125).

Water-soluble extractives, not being an integral part of wood structure, are free to migrate along with any movement of liquid water within wood. On a gross scale, extractive migration to the surface of lumber during drying leads to a so-called seasoning stain in some species, such as redwood, that have deeply colored extractives(126). Kuo and Arganbright (124) found that the extractives were concentrated in the S_2 layer with the highest concentrations in its outer and inner portions, but that there was a general tendency for the concentration to decrease from the lumen toward the middle lamella. This suggests that the extractives originally diffused into the cell wall from the lumen. Hart (127), in fact, observed swelling during the early stages of drying, which he attributed to the entry of extractives from the lumen into the cell wall. Extractives must therefore be recognized as potentially mobile constituents that are subject to migration during wood processing, depending on their solubility.

III. MECHANICAL PROPERTIES

Attempts to elucidate the contribution of the constituents of wood to its mechanical properties and behavior have followed three general approaches. One is to examine a series of specimens of the same or different species that are similar in all respects, except for the characteristic under investigation, as, e.g., degree of lignification or microfibril angle. There are many difficulties associated with this method, principally since the compositional and structural characteristics of wood are usually interrelated. The second approach, rather than relying on natural variation, introduces changes by degrading or removing certain constituents preferentially. This also has limitations, except when examining the effect of extractives as they can be removed with solvents without changing the structure of what remains. The third approach involves the use of models that represent an idealized abstraction of the real structure and are amenable to mathematical analysis.

Some of the early work following these approaches has been reviewed by

Schniewind (128). The evidence then available suggested that, other structural factors being equal, a high cellulose content leads to increased tensile strength (129), whereas a high lignin content improves the compression strength parallel to grain (130). The importance of microfibril orientation, which is often taken to mean microfibril angle in the S_2 layer, was recognized, but it was also suggested that an integral microfibril angle based on the orientation in all cell wall layers may be the more important factor (131). Models were limited to representing the wood cell as a system of helical springs in analogy to the helically wound microfibrillar structure (132,133). Although the mechanics of a single helical spring are well understood, the interaction of several such springs is another matter, especially if it is taken into consideration that they are embedded in a matrix.

The modern era of modeling wood as a fiber-reinforced composite, where the cellulose microfibrils possibly in association with some of the hemicellulose represent the fiber reinforcement or framework, and the lignin and remaining hemicellulose the matrix, began with the shrinkage model of Barber and Meylan (134). This was a very simple model, which considered the cell wall a single layer of matrix reinforced with a framework of microfibrils of uniform inclination to the cell axis throughout. It was followed shortly thereafter by the model of Mark (135,99) that was far more detailed and ambitious in attempting to predict the mechanical behavior of cell walls from the basic properties of its constituents, taking into account the layered structure of the cell wall and the differing composition and microfibril orientation in each layer.

Most investigations have been limited to the mechanical properties and behavior parallel to grain, with the focus on coniferous woods. Tracheids make up as much as 95% of the volume of the latter and can be treated as thin-walled tubes as a first approximation. This allows the assumption that stresses parallel to the grain are confined to the plane of the cell wall, so that the behavior of the cell can be modeled by a small element of cell wall subjected to a state of plane stress. The cell wall element can then be modeled as a layered system, and various investigators have used schemes ranging from a single layer (134) to a continuous distribution of layers (136), but more often they have attempted to represent the recognized layers of the primary and secondary cell walls (137). Each layer, in turn, is a composite of framework and matrix, but with angles of orientation with respect to the main cell axis that vary from one layer to the next.

A. Mechanical Properties of Wood Constituents

Because of the tenacious association between wood constituents in situ, it is virtually impossible to isolate any of them without altering their characteristics somewhat. For that reason, the necessary basic property values of wood constituents are not always available as such and must be inferred or estimated.

1. Glass Transition Temperatures

The most fundamental parameter for the physical characterization of amorphous polymeric materials is the glass transition temperature T_g, which locates the temperature region where materials change from the glassy to rubbery state. The change to the rubbery state is accompanied by a significant reduction in stiffness, so that materials well below their T_g at ambient conditions will have rather different mechanical properties and behavior than materials above it.

Glass transition temperatures are of particular significance in thermomechanical pulping and have been studied extensively in that context (138–141). They are often referred to as softening temperatures, which in the case of some measuring methods may actually represent the onset of rubbery flow rather than the glass transition (142). Glass transitions exist for lignin, hemicellulose, and the noncrystalline regions of cellulose (143). Published data have been reviewed by Salmén (144) and Back and Salmén (141). Values vary widely, but in the dry state T_g of cellulose is most likely about 230°C, as compared to about 180°C for hemicellulose and 200°C for lignin.

Glass transition temperatures of polymers are significantly affected by plasticizers, of which water is particularly important for wood components. Reductions in softening temperatures of lignin and hemicellulose due to water vapor sorption were already observed by Goring (143). Salmén and Back (145) used the Kaelble equation to calculate the effect of water content on T_g of cellulose and obtained values as low as about -50°C for the completely swollen state, depending on the degree of crystallinity, which agreed with rather limited experimental data. Similar projections were made for hemicellulose (141). Lignin is less affected, reaching T_g values ranging from 50–100°C in milled-wood lignin and 60°C for lignin in situ in the water swollen state (141,142). Irvine's differential thermal analysis measurements could not be made at temperatures less than 40°C, and no values for in situ hemicellulose could be found. Irvine considered this as possible evidence for an otherwise unlikely degree of miscibility of lignin and hemicellulose. Kelley et al. (55) have been able to show by means of dynamic mechanical thermal analysis and differential scanning calorimetry measurements of whole wood that in situ lignin and hemicellulose are immiscible. They also obtained T_g as a function of moisture content for both lignin and hemicellulose. Lignin values agreed well with the data of Irvine (142) and hemicellulose values reached -20°C in the water-saturated state, in general conformance with the predictions of Back and Salmén (141). Hillis and Rozsa (146) obtained softening points as high as 80°C for undried wood from rate of twisting measurements that they attributed to the softening of the hemicelluloses, but it is doubtful that these represent T_g values.

The findings of Kelley et al.(55) are particularly interesting for the insight they provide into the mechanical behavior of wood at room temperature. The data clearly show that in completely dry wood, all components are well below their respective T_g. At the fiber saturation point or above, the lignin is still below its T_g, while the hemicelluloses and the noncrystalline portion of cellulose are well above theirs. Some of the components of wet wood are therefore clearly in a rubbery state, which helps to explain the large dependence of mechanical properties on moisture content. The change in T_g is greatest between 0 and 10% moisture content. At 10% moisture content, T_g of hemicellulose appears to be approximately equal to room temperature. This could explain why the tensile strength parallel to grain, which depends more on the cellulose microfibril reinforcement, is less sensitive to moisture content than compression strength, which relies more heavily on the performance of the matrix. Taking published data for four hardwood and seven softwood species (147), we see that the average ratio of strength values at 12% moisture content and in the moisture-saturated condition is 1.22 for tensile strength and 2.01 for compression strength parallel to grain, which indicates almost five times the sensitivity to moisture-content changes in compression as compared to tension.

Almost all strength properties increase as the moisture content is reduced from the fiber saturation point to about 10% moisture content. With further reductions in moisture content, strength properties increase at a reduced rate, and some properties show a maximum at about 6–8% moisture content, as for instance, tensile strength parallel to grain (148). The reduced moisture sensitivity at low moisture contents can be explained on the basis of the effect of moisture on T_g because all components are then in the glassy state. The maximum at relatively low moisture contents has its parallel in the antiplasticization effect in other polymer systems, including cellulosics (149), where small amounts of plasticizer provide some initial strengthening and stiffening of the polymer structure.

2. Elastic and Strength Properties

When Mark (135,99) developed his cell wall model, he found little of the required information on the elastic constants and strength properties of wood constituents. He therefore proceeded to calculate elastic constants for crystalline cellulose from crystal structure. The calculated modulus of elasticity parallel to the chain axis was close to but less than an experimental value of 134 GPa obtained by Sakurada et al. (150), and Mark chose to use the latter in his analysis. Additional approaches to the calculations that included consideration of hydrogen bonds (151,27) yielded a complete set of elastic constants, with an upper bound value for the modulus of elasticity parallel to the chain axis of 246 GPa, i.e., greater than the experimental value. Poisson's ratios, however, were surprisingly small, none of the six values exceeding 0.041 and two of them being extremely small and negative. These calculated values have not found general acceptance, and more recent efforts at developing cell wall models have returned to the use of the experimental value (150) for the modulus parallel to the chains, and either Mark's (99) original calculations (152) or other assumptions (153) for the other constants. As observed by Mark and Gillis (154), the experimental value of 134 GPa (150) is probably too low because the method used leads to underestimation of the true stress level parallel to the microfibrils. Whatever values are used, they apply only to the crystalline portion, and thus they are independent of any adsorbed water.

The properties of the microfibril are generally taken to be those of crystalline cellulose. This may be justified on the basis that the degree of crystallinity of cellulose in wood is rather high. Further support may be derived from the concept of paracrystallinity, suggesting a spectrum of order that may contain very little that could be termed truly amorphous. Thus, it is not unreasonable to assume similar mechanical behavior for both crystalline and noncrystalline phases. This is not true for rheological properties because crystalline regions are not expected to be subject to time-dependent deformations. Short-term creep in wood has been found to decrease with increasing crystallinity (155). Crystal lattice strain increases during creep, but this can be attributed to creep in the non crystalline regions, resulting in a shifting of internal loads to the crystalline regions and consequently larger elastic strains being produced there, rather than creep in the crystal structure itself (156).

Mark (99) assigned some of the hemicelluloses to the framework and the remainder, together with lignin, to the matrix. He assumed the matrix to be isotropic with a single set of elastic constants. By using an experimental value for the modulus of elasticity of lignin of 2.0 GPa (157) and an assumed Poisson's ratio of 0.3, the shear modulus could then be calculated as 0.77 GPa since isotropic materials have only two independent elastic constants. A later calculation based on the energy of C-C bond rotation resulted in a modulus of elasticity value of 1.8 GPa for the amorphous matrix (158). More recent experiments with lignin have yielded values of the same order of magnitude for the modulus of elasticity (159,160) and also showed it to be dependent on moisture content. It exhibits the antiplasticization effect in increasing from 6 GPa in the dry state to 7 GPa at 4% moisture content and then decreases to 2.8 GPa near saturation. The behavior of hemicellulose is similar up to moisture contents in equilibrium with a relative vapor pressure of 0.8, except that the antiplasticization effect was not observed (161). Once the relative vapor pressure of 0.8 was exceeded, however, a transition became apparent as the modulus of elasticity then decreased sharply, more than two orders of magnitude, to about 10 MPa near saturation. This is to be expected based on the effect of moisture on the glass transition temperature of hemicellulose as already discussed above. The moisture dependence of lignin and hemicellulose properties is generally recognized and has been incorporated into the models of Cave (162,153) and Salmén and de Ruvo (152).

Although there is little question that lignin is amorphous and therefore isotropic, the evidence suggests that hemicellulose has some preferred orientation by virtue of its close association with the cellulose microfibril. This would make hemicellulose at least somewhat anisotropic. Mark (99) accounted for this by assigning some of the hemicellulose to the anisotropic framework. Cave (153) treated hemicellulose as a separate component and assumed it to be transversely isotropic, with the modulus of elasticity parallel to the microfibril exceeding the transverse one by a factor of 2; a similar treatment was adopted also by Salmén and de Ruvo (152). Some sets of elastic constants are shown in Table 4. It illustrates the high degree of anisotropy of the cellulose framework as compared to the matrix materials that makes the former such an effective reinforcing material, the moisture sensitivity of hemicellulose and lignin, and some of the different approaches that have been taken. The constants used by Cave (153) were not included because they were not explicitly stated, but appear to be similar to those of Salmén (164).

Little is known of the strength properties of the individual components of wood. Some theoretical calculations have been made for crystalline cellulose that indicate tensile strength values on the order of 7.5 GPa parallel to the chain axis and 1.1 GPa perpendicular to it (99), but no meaningful data are available for the other constituents. Uncertainties about the structure of lignin make theoretical calculations impossible, and the difficulties in extracting lignin in unaltered form would have an even greater effect on attempts at the experimental determination of strength than of elastic properties. Even in manmade composites, the prediction of

Table 4 Elastic Constants of Wood Constituents

Constituent and elastic constant	Mark (158,163)[a]			Salmén (164)[b]	
	High	Medium	Low	Stiff	Soft
Cellulose					
E_1 (GPa)	—	246	—	134	—
E_2 (GPa)	—	20.8	—	27.2	—
G_{12} (GPa)	—	0.21	—	4.4	—
μ_{12}	—	0.037	—	0.1	—
Hemicellulose					
E_1 (GPa)	—	—	—	8	0.02
E_2 (GPa)	—	—	—	4	0.01
G_{12} (GPa)	—	—	—	2	0.005
μ_{12}	—	—	—	0.2	0.2
Lignin					
E (GPa)	7	1.8	0.2	4	0.06
G (GPa)	2.7	0.7	0.08	1.5	0.025
μ	0.3	0.3	0.3	0.33	0.33

[a]In Mark's models, the hemicellulose is divided between the framework (cellulose) and matrix (lignin). Cellulose has been assumed to be transversely isotropic, and E_2, G_{12}, and μ_{12} are average values. Lignin may be of high, medium, or low stiffness, depending on moisture content.

[b]The soft state of hemicellulose or lignin is reached when Tg falls below ambient temperature.

strength of laminated composites requires extensive experimentation with individual laminae (165), which given the present state of the art is not possible for
individual cell wall layers in wood. Thus, there is only limited motivation for
pursuing the question of constituent strength properties at this time.

B. Elastic Constants of Cell Wall Layers

Computation of the elastic constants of reinforced composites from the elastic
properties of the components is a fundamental problem, and many approaches
toward its solution have been used, depending on the nature and configuration of the
reinforcement and other factors (165). An early method widely used in wood cell
wall mechanics was to take a very thin slice parallel to the microfibrils in such a way
that the resulting sheet consisted of alternating strips of framework and matrix (99).
By using a mechanics of materials approach, it can then be readily shown that the
elastic constants are given by the so-called rule of mixtures. This rule takes the form

$$E = P_f E_f + P_m E_m \tag{1}$$

where P is the volume fraction, E is either an elastic constant or its inverse, and the
subscripts f and m refer to the framework and matrix, respectively. That is to say,
the elastic constant of the composite, which may be a modulus of elasticity, shear
modulus, or Poisson's ratio, is calculated as a volume-weighted average of the
constants of the components. Refinements were made by Gillis (166) by allowing
the framework to be embedded in matrix.

Cave (136,167) used Hill's version of the "self-consistent method" (165) to
calculate elastic constants of his layer elements. An approximation to Hill's method
that is widely used today for all types of composites are the Halpin-Tsai equations
(168,165). In contrast to the methods described that all assume continuous fiber
reinforcements, the Halpin-Tsai equations make provisions for accommodating
reinforcements of a wide variety of configurations, including discontinuous fibers
by means of shape factors, and they are easy to use. Thus, Salmén and de Ruvo (152)
used them to model softening of the hemicellulose and noncrystalline fractions of
cellulose when the temperature exceeds their respective glass transition temperatures by assuming discontinuous microfibrils of a length adjusted to the physical
state of the components. Table 5 shows cell wall layer properties calculated by four
different procedures. In each case, the cellulose elastic constants of Mark in Table
4 were used, together with Mark's original values for the matrix, $E = 2$ GPa, $G = 0.77$ GPa, and $\mu = 0.3$ (99). The proportion of framework was 0.531. The modulus
of elasticity parallel to the microfibrils is the same for all methods. For the other
constants, the Gillis equations and either version of the Halpin-Tsai equations give
comparable results, whereas Mark's original method deviates significantly from the
others. Since the modulus of elasticity parallel to the microfibrils depends almost
solely on cellulose stiffness in the chain direction, the choice of elastic constants for

the framework has a far greater effect than the choice of method for calculating the properties of the composite layer.

Barber and Meylan (135) viewed the cell wall as being dominated by the S_2 layer and represented it as a single layer with a constant angle of alignment of the microfibrils. The cellulose microfibrils were embedded in an amorphous matrix of hemicellulose and lignin. Mark (99) made a different disposition in considering part of the hemicellulose to be so closely associated with the cellulose microfibril that it becomes part of the reinforcing framework. The framework was taken as transversely isotropic, i.e., isotropic in the plane perpendicular to the microfibrillar axis. The cell wall layer then becomes orthotropic in the plane of the cell wall with principal axes parallel and perpendicular to the microfibril direction, and it is transversely isotropic. Such a layer has four independent elastic constants: two moduli of elasticity (parallel and perpendicular to the microfibril axis), a shear modulus, and a Poisson's ratio.

In contrast to the division of three components into two phases as used by Mark and others, more recent models have incorporated three separate phases (153,152). In each case, the cellulose microfibrils are embedded in sheaths of hemicellulose, thus forming a carbohydrate complex that exists in tangential lamellae within a cell wall layer. These then either alternate with microlamellae of lignin, as shown in Fig. 5 (153), or the lignin is concentrated in a single lamella in the middle of each layer (152).

C. Cell Wall Models

In recognition of the great importance of framework content and microfibrillar alignment to strength and stiffness, all of the modern cell wall models introduce some scheme of layered structure. The intent in each case is to make a reasonable

Table 5 Cell Wall Layer Elastic Constants

Elastic constant	Mark (99)	Gillis (166)	Halpin-Tsai (168)	(165)
E_1 (GPa)	132	132	132	132
E_2 (GPa)	3.8	6.4	6.0	6.1
G_{12} (GPa)	0.31	0.43	0.47	0.41
μ_{12}	0.0047	0.0044	0.0065	0.0066
μ_{21}	0.16[a]	0.10[b]	0.14[a]	0.14[a]

[a]Calculated from symmetry.
[b]Calculated from Gillis' equations.

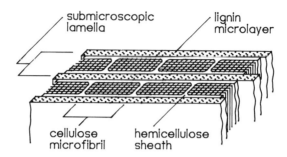

Figure 5 Carbohydrate complex of cellulose microfibrils embedded in hemicellulose sheaths, sandwiched between microlayers of lignin. (From Ref. 153.)

representation of the actual structure that includes the middle lamella (ML), primary wall (P), and the outer (S_1), middle(S_2), and inner (S_3) layers of the secondary wall (Fig. 30, Chap. 3).

1. Plane Stress, Slab Models

Here it is assumed that the typical cell is a thin-walled tube of arbitrary cross section that can be represented by an elemental slab of the wall subjected only to stress in the plane of the wall. The element is a layered composite in which the proportion of framework and the inclination of the microfibrillar angle may vary from one layer to the next. It is subjected only to loads parallel to the cell axis. This means that the *average* stress on the element can only be tensile or compressive parallel to the cell axis. However, additional stresses will, in general, exist within each of the component layers. The reason is found in the orthotropic nature of the layers. Such a layer, in isolation, when stressed in tension at an angle other than 0 or 90° to the reinforcement direction will not only extend in the direction of stress, but will also deform in shear. The reason becomes clear if we imagine an element with reinforcement at 45° to the applied tensile stress. The consequences of the applied stress will be the same for both diagonals of a square element as far as stress is concerned, but the diagonal perpendicular to the reinforcement will extend more than the other diagonal; the element thus becomes diamond-shaped as though it had been subjected to a shear stress. The amount of shear strain will depend on the reinforcement angle, so that different layers will tend to deform differently. They are, however, bonded together and must all deform together, which leads to two results. First, the distribution of stresses within layers will vary from one layer to the next, and the composite slab element of a single cell will deform not only by extension in the cell axis direction but also in shear. A complete cell will therefore not only extend under an axial tensile load, but also twist (169–171).

Mark (99) in his original model placed no restrictions on fiber twisting. His cell wall element was a layered composite consisting of the combined middle lamella and primary wall, the S_1 layer that was divided into two lamellae with microfibril angles of opposite sign to represent the crossed microfibril structure of that layer, and S_2 and S_3 layers, each of which was further subdivided into radial and tangential walls with different microfibril angles. All layers were assumed to undergo the same deformation in the plane of the cell wall, making this a seven-layer composite element.

Since tracheids and other cells in wood are not really free to twist, other authors introduced the concept of shear restraint (136,169,170). In this concept, the basic element combines the walls of two adjacent cells, where in general the mean inclination of the microfibrils will be of approximately equal angles of opposite sign, as illustrated in Fig. 6. This would produce a shear strain under longitudinal stress also of opposite sign in adjacent cell walls, but since they are firmly bonded together, such shear strains are prevented from taking place. Some additional stresses result from this restraint, and the wall becomes stiffer as well. Although strictly speaking, the shear restraint is complete only at the interface between two cells, this approach is more realistic than allowing cells to twist freely within solid wood and has been widely adopted.

Cave's (136) model had the double wall, each half of which consisted of a single layer made up of an infinite number of lamellae with a Gaussian distribution of microfibril angles. This was later expanded to two layers, each with a distribution of microfibril angles, but one with small mean angle representing the S_2 layer and the other with a large mean angle representing a combination of the S_1 and S_3 layers

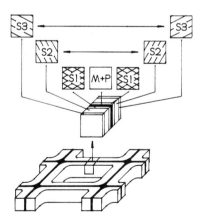

Figure 6 Double cell wall element, illustrating shear restraint of paired S_2 and S_3 layers. (From Ref. 169.)

(153). Although there is evidence that microfibril angles are not constant in a given layer (172), most authors have preferred to use more than two layers and hold the microfibril angle in each layer constant. Salmén (164), for instance, used four layers, ML, P + S_1, S_2, and S_3, whereas Mark and Gillis (170) proposed a "two-wall" model in which radial and tangential walls of tracheids were deformed equally in the direction of the cell axis, but were free to deform independently in the transverse direction.

2. Three-Dimensional Models

Not surprisingly, far fewer attempts have been made to employ three-dimensional models of wood cells. In one of these, the solution given for stress distribution and displacements in a single anisotropic hollow cylinder (173) was extended to a four-layer composite of concentric cylinders (174), but some of the boundary conditions chosen in the analysis led to inconsistencies. A closed-form solution, also for a four-layer composite of concentric anisotropic cylinders, was developed by Gillis and Mark (175) but only applied to problems of shrinking and swelling. Barrett and Schniewind (176) used a finite-element approach to the same problem that gave results for stress distributions very close to values obtained by those by the plane stress, slab element methods described in the previous section. The two-dimensional models were thus shown to be good approximations, except for stresses in the cell wall thickness direction, which can only be calculated from the three-dimensional model but are generally very small.

Although it has been possible to make an effective comparison between two- and three-dimensional models, evaluation of the models in absolute terms has only been possible to a very limited extent. It is extremely difficult to devise critical tests regarding the validity of any of the models, because the required input data such as relative layer volumes, microfibril angles, and relative volumes and distribution of constituents can only be obtained by very painstaking and tedious measurements. Good compendia of cell wall layer dimensions may be available (177), but may not be complemented by composition data for matching material. Furthermore, there are uncertainties regarding constituent properties, as well as simplifying assumptions regarding wood structure, such as cell ends, pits, and ray tissue. Choices between some of the models described must therefore be made mainly based on the reasonableness of the underlying assumptions. There are, however, two important points that speak in favor of using layered systems of oriented, fiber-reinforced polymers as models. One of these is that calculations of stress distributions in individual layers indicate relatively high stresses perpendicular to the microfibrils in the S_1 layer, which agrees well with experimental observations that place failure planes preferentially in the S_1 layer (99,169,170). The other point is that variations in the mechanical properties of wood with changes in microfibril angle in the dominant S_2 layer can be accounted for with a suitable choice of model. Figure 7 shows calculations according to the latest model by Salmén (164) in relation to experimental values of modulus of elasticity parallel to grain of radiata pine by Cave (136).

D. Models of Wood Structure

The cell wall models discussed above can be applied to uniaxial loading parallel to grain, but are not suitable for other types of loading, including uniaxial loads perpendicular to grain. This is where gross anatomic structure rather than the fine structure of the cell wall must be taken into consideration. Loading perpendicular to the grain involves the bending of cell walls that is virtually absent in loading parallel to grain. This has been effectively demonstrated by a model of balsa wood consisting of hexagonal-prismatic cells with pointed end caps (178). The cell walls were given orthotropic material properties based on data of radiata pine (136). Surprisingly, predictions of crushing strength were much closer to experimental values in all three principal directions than the predicted moduli of elasticity. Especially, the stiffness values in the radial and tangential directions were much lower than experimental values; which was attributed to the omission of the septa and end caps of the septate fibers from the analysis (178). Another analysis focused on cell wall bending as the important factor in loading perpendicular to grain by the use of a "triple point" element, representing the juncture of three hexagonal cells as seen on a cross section (179). By using such an element, it was possible to correctly predict the relative magnitude of wood's orthotropic elastic and shear moduli even

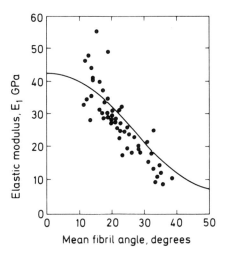

Figure 7.   Modulus of elasticity parallel to grain as a function of S_2 mean fibril angle, showing data by Cave (from Ref. 136) with theoretically calculated curve (From Ref. 164.)

though the cell wall material was assumed to be isotropic. On the other hand, an attempt to model the radial and tangential stiffness of wood based on cell wall properties without provision for the bending of cell walls resulted in estimates that were about one order of magnitude larger than experimental moduli of elasticity (176).

The last and largest-scale level of composite interaction within regular wood structure is found in the radial reinforcement of tissues made up of longitudinally aligned cells by the rays, and in the growth ring structure of alternating bands of earlywood and latewood. A recent, three-dimensional model of wood is composed of alternating layers of tissue with vertically aligned cells and ray tissue (180). Another possibility of modeling the mechanical behavior of whole wood is to consider it as a double laminate, where the earlywood and latewood are each a composite of alternating layers of ray tissue and tissue of longitudinally aligned cells, and wood is then composed of alternating layers of earlywood and latewood. Experiments with California black oak showed that the radial and tangential tensile strengths and moduli of elasticity could be predicted from experimental values for isolated ray tissue, earlywood, and latewood (181).

E. Effect of Extractives on Mechanical Behavior

It is expected that only the extractives located within the cell wall would have any effect on the strength and stiffness of wood, but it is not clear what the mechanism of such an effect would be. It has been suggested that extractives in the cell wall may act either as plasticizers or bulking agents, or alternately as stiffening agents that have an effect principally on compressive strength (182). Experimental evidence is rather limited and not very consistent. The few studies of the effect of extractives on mechanical properties invariably are based on measurements of the total amount of extractives removed, regardless of their original location in either lumen or the cell wall. It is also to be expected that the chemical nature of extractives will be an important factor, leading to great variability from one species to the next.

It has already been pointed out that extractives in the cell wall are thought to occupy sorption sites. Possible mechanisms for influencing mechanical properties are that the extractives (1) act as plasticizers similar to the water they have replaced, (2) act as inert bulking agents, or (3) become part of the matrix. As part of the matrix, extractives should increase compressive strength parallel to grain, and there is some evidence to that effect in some species (182,183). Tensile strength would be expected to be relatively insensitive to any of the above mechanisms, and in fact, a study of 20 species did not show any significant effect of extractives on tensile strength parallel to grain (131). Extractives as plasticizers or as bulking agents could either increase or decrease mechanical properties, depending on their effectiveness relative to the adsorbed water they replace. The only observations of a negative effect of extractives is that under some conditions they lowered shock resistance

(183). Tests of several species in impact bending in an oven-dry condition showed appreciable increases in shock resistance following extraction with alcohol-benzene in four of five species tested, but not all of this increase persisted after wetting and redrying, indicating that at least part of the increase was introduced by the extraction procedure itself rather than being due to the removal of the extractives (184). Luxford (183) had found for redwood that extractives increased compressive strength the most, the modulus of rupture to a lesser extent, whereas the effect on shock resistance was least. More recent work could not detect the significant effect of extractive content on the modulus of rupture in redwood, but a small, positive effect on the modulus of elasticity was observed (185). Thus, the effect of extractives on the mechanical properties of wood is far from clear, but the limited evidence available tends to favor the view that the extractives located within the cell wall may be considered part of the matrix.

IV. SORPTION AND SHRINKAGE CHARACTERISTICS

Individual species of wood show significant differences in water sorption and shrinkage behavior, which must be due to differences in their composition and structure. The major constituents of wood differ in their sorption capacity and presumably also in their shrinkage coefficients. If the sorption capacity of whole wood can be assumed to be a weighted average of the individual component capacities, changes in composition will also be reflected in changes in the sorption of water. The shrinkage of wood that accompanies changes in moisture content will depend not only on the shrinkage coefficients of the constituents, but also on their mechanical interaction. Unless the shrinkage coefficients are the same for all components, there is mutual restraint from one to the next, with the final shape and dimension a function of the relative stiffness and geometry of the constituents as well as their shrinkage coefficients.

A. Water Sorption and Sorption Isotherms

Christensen and Kelsey (186) determined sorption isotherms for *Eucalyptus regnans* wood and its holocellulose, cellulose, hemicellulose, and lignin fractions. Some of the results are shown in Fig. 8 in the form of adsorption isotherms at 25°C. Hemicellulose is clearly the most hygroscopic fraction. It is followed, in order, by holocellulose, whole wood, and lignin. The cellulose sorption isotherm is not shown because it is almost identical with that for whole wood; this is surely a coincidence because the effect of the more highly hygroscopic hemicellulose is balanced by the low hygroscopicity of the lignin. The two types of lignin exhibit rather different sorption behavior, indicating that either one or both have been sufficiently altered in the isolation process to affect sorption characteristics.

In the relative vapor pressure range from 0.1–0.9, the experimentally determined

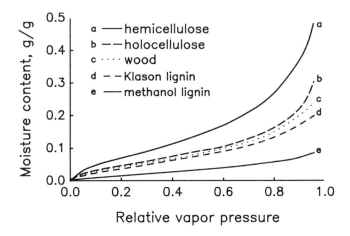

Figure 8 Sorption isotherms for wood and wood constituents. (From Ref. 186.)

sorption capacity relative to whole wood was 0.94 for cellulose and 1.09 for holocellulose. Using these values and the percentage composition of whole wood and holocellulose, Christensen and Kelsey (186) were able to calculate estimates of an in situ relative sorption capacity of 1.56 for hemicellulose and 0.6 for lignin. It follows from these figures that of the total adsorbed water in wood, 47% is held by cellulose, 37% by hemicellulose, and 16% by lignin, even though these fractions represent 50, 24, and 26% of total wood, respectively. Working with spruce and beech, Runkel and Lüthgens (187) had also come to the conclusion that hemicellulose was the most and lignin the least hygroscopic of wood constituents. We can thus expect that species rich in hemicellulose will be more highly hygroscopic, whereas a high lignin content would result in lower overall sorption capacity.

Indeed, tropical woods that tend to be more heavily lignified than temperate woods have been observed to have a lower shrinkage intersection point (which is related to the fiber saturation point) and lower equilibrium moisture content at 0.9 relative vapor pressure when compared at equal specific gravity (188). The differential behavior of wood components also contributes to an explanation of the heat stabilization of wood, because in the course of treatment the polysaccharides are preferentially degraded while leaving the less hygroscopic lignin largely intact (189).

In some species, particularly those with a high content of extractives, the latter can have significant effects on hygroscopicity. California redwood is a good example, where the removal of the heartwood extractives resulted in raising the fiber saturation point from 18 to 25% (190). Increases in the fiber saturation point upon removal of extractives have been observed for a number of different species

(120,191,119). As might be expected, increases tend to be greatest for species with initially low fiber saturation point values, as seen in Table 6. Equilibrium moisture content values are similarly affected, but for the most part, the effect is restricted to the higher relative vapor pressure range (191). This indicates that extractives are not adsorbed directly on sorption sites in competition with water molecules that would otherwise form the monomolecular layer of adsorbed water, but rather act as bulking agents. Additionally, extractives may influence apparent hygroscopicity by acting as inert substances that add to wood weight but not to sorptive capacity (119). The effects of extractives are not limited to water as the sorbate; similar results have been obtained with alcohols and organic acids (121).

Nearn (120) noted that some species, which included teak and black locust, had relatively low values of fiber saturation point after extraction even though the removal of extractives did cause it to increase significantly. Detailed studies of the reasons for the exceptional dimensional stability of teak showed that this was not caused by its extractives, but rather by an unusually low content of hydrolyzable hemicellulose (192,193). A similar observation was made for black locust (193).

B. Shrinkage Behavior

Shrinking and swelling in wood are closely related to the hygroscopicity and sorption behavior of wood, and in general, species that are more hygroscopic also tend to shrink more. Woods such as teak and California redwood that have exceptionally low fiber saturation point values also show exceptionally good

Table 6 Effect of Extraction on Fiber Saturation Point

| Species | Fiber saturation point (%) | | B-A |
	Unextracted (A)	Extracted (B)	
Mora Amarilla	15.4[a]	26.6[a]	11.2
Cupiuba	18.0	26.8	8.8
Manwood	18.6	25.6	7.0
Mahogany	23.0	28.9	5.9
Primavera	27.7	23.7	4.0
Tauary	23.7	26.9	3.2
Hibiscus	24.0	27.0	3.0
Sangre	24.5	29.4	4.9
Copaia	25.9	29.3	3.4

[a]Fitted values at 1.0 relative vapor pressure from adsorption isotherms, taken from Wangaard and Granados (191).

dimensional stability, and treatments such as the removal of extractives that will increase the fiber saturation point also result in higher shrinkage (120,119,193). In the genus *Eucalyptus*, extractive content was positively correlated with collapse shrinkage but negatively correlated with normal volumetric shrinkage (194), where the increase in collapse shrinkage would be due to physical impedance of the flow of moisture and the decrease in normal shrinkage due to a decrease in hygroscopicity. Schroeder (195) compared the volumetric shrinkage of hardwoods and softwoods of similar relative density and found that hardwoods shrank more, which he attributed to their lower lignin content.

As is well known, wood is highly anisotropic in its shrinking and swelling, the total longitudinal shrinkage from green to oven-dry being 0.1–0.3% and the radial and tangential shrinkage being one to two orders of magnitude greater (196). Tangential shrinkage is generally close to twice the radial shrinkage. If the constituents of wood were arranged in the cell wall as a homogeneous blend of hygroscopic gels, we would not only expect the anisotropy of shrinkage to disappear, but also that all woods of a similar chemical composition would show similar shrinkage values. There is, however, a tendency for the volumetric shrinkage of wood to increase with increasing relative density, because cell cavities tend to remain constant in size during shrinking and swelling (196). Woods of higher relative density have thicker cell walls, and if these must shrink and swell about a constant lumen size, thicker walls will result in greater overall dimensional changes. It can be shown that for constant lumen size, volumetric shrinkage S is related to the product of the fiber saturation point M and the relative density based on oven-dry weight and water-saturated volume G:

$$S = MG \tag{2}$$

Chafe (197) examined deviations from this "law" and proposed what he called the R-ratio, which measures the extent of deviation from Eq. (2), as a quality index for wood.

Equation (2) is most likely a coincidental relationship that arises because individual cell wall layers are anisotropic in their shrinkage potential. The source of the anisotropy is again the cellulose microfibril that shrinks little, if at all, in length but is embedded in a matrix that shrinks and swells with the loss or gain of adsorbed moisture. Shrinkage in a layer with a constant microfibrillar angle will therefore be much greater perpendicular than parallel to the microfibrils. In the complete cell, individual cell wall layers interact in such a way as to leave the cell cavity approximately constant in size. For example, the S_1 and S_3 layers can be viewed as giving rise to a hoop effect that restrains the shrinking and swelling of the S_2 layer in cross section (198,199). Quirk (200) found for Douglas-fir that lumen size increased during drying in earlywood tracheids, whereas it decreased in latewood tracheids. Such differences in behavior are not surprising if it is considered that in one study of Douglas-fir the S_2 layer was found to occupy 19% of the total cross section in earlywood tracheids as compared to 70% in latewood (201), which indicates that in this case the S_1 and S_3 layers dominate in the earlywood tracheids and the S_2 layer in the latewood tracheids.

An excellent review of wood shrinkage in relation to structure was published by Kelsey in 1963 (202). It summarizes early attempts at understanding and modeling wood shrinkage in terms of composition and structure. Here we can also distinguish between models of single or double cell wall elements that focus on longitudinal shrinkage, and models on a higher level of organization that deal with the interaction of various tissues.

C. Cell Wall Shrinkage Models

The theoretical treatment of cell wall shrinkage can be achieved by the same types of models used for elasticity prediction and stress analysis in cell walls already discussed. It is merely necessary to introduce into the constitutive equations terms that represent shrinkage (which is a form of strain) of the components as a function of moisture content or vapor pressure change (162,203). This may take the form

$$\varepsilon = S\sigma + \alpha\phi \tag{3}$$

where ε is the strain, S the compliance, σ the stress, α the shrinkage coefficient, and ϕ the change in relative vapor pressure. Equation (3) may be either one-dimensional as shown, or each term can be in matrix form for application to two-dimensional treatments (203).

Actually, the first model considering the cell wall as a fiber-reinforced composite was not concerned with mechanical properties as such, but with the shrinkage of wood (134). The model incorporated for the first time not only the geometric and shrinkage characteristics of the components but also their elastic properties, in order to analyze the mechanical interaction of phases with unequal shrinkage potential. In general, shrinkage models involve the establishment of framework and matrix properties, an assessment of framework and matrix interactions in a layer with constant fiber orientation, and finally an assessment of the combined action of several layers that may differ in composition as well as framework orientation.

1. Component Shrinkage Characteristics

Inasmuch as all wood constituents with the possible exception of extractives are known to be hygroscopic, it may be presumed that all of them will also show some degree of shrinking and swelling. In general, the change in volume can be expected to be approximately equal to the volume of water adsorbed or desorbed. The cellulose microfibril, however, is anisotropic and most unlikely to show much shrinking or swelling along its length. Therefore, Barber and Meylan (134) chose to assume that the framework was unaffected by water sorption and that all shrinking and swelling took place in the matrix. The assumption of a framework that is inert with respect to water sorption was also adopted by Cave (162,204,153).

Neglecting the longitudinal shrinkage of framework appears to be reasonable, but it is questionable whether this can be extended to the transverse direction since according to estimates, 47% of the adsorbed moisture is held by cellulose (186). Barrett et al. (203) assumed that the framework contained 30% adsorbed water at saturation, and that changes in volume would be equal to the volume of water removed. The framework was assumed to be transversely isotropic, the shrinkage coefficient parallel to the microfibril being 1/40 of the transverse coefficient. Finally, shrinkage was assumed to be a linear function of relative vapor pressure.

Barber and Meylan (134) limited themselves to investigating the qualitative relationship of shrinkage to microfibrillar angle, and thus did not adopt any quantitative measure for matrix shrinkage. Barrett et al. (203) assumed that the isotropic matrix, consisting of lignin and some of the hemicellulose, also contained 30% of adsorbed water at saturation. Shrinkage was assumed to be equal to the volume of water removed and the shrinkage coefficient constant over the entire relative vapor pressure range. The value appears to be somewhat high and might be more appropriate for a matrix containing the entire lignin and hemicellulose fractions, since the estimated water content at saturation for lignin is approximately 18%, as compared to 47% for hemicellulose (186). On the other hand, it is more realistic than the 7.5% content of adsorbed water chosen by Cave (153) for the combined lignin and hemicellulose fractions. The figure of 7.5% was taken from the literature for what Cave referred to as "bound water," but was actually the moisture content at which a monomolecular layer of adsorbed water is achieved (205). The assumption by Cave that adsorbed water in additional layers does not participate in shrinking and swelling is not compatible with the known shrinkage behavior of wood, especially in the radial and tangential directions, which takes place over the entire hygroscopic range from the dry state to the fiber saturation point of approximately 28%. Cave also assumed that the change in volume was equal to the volume of water removed or added. His matrix was divided into two phases— namely, hemicellulose sheaths around the microfibrils that do not shrink longitudinally while all the shrinkage takes place transversely in an isotropic manner, and lignin that shrinks isotropically.

2. Cell Wall Layer Shrinkage

The shrinkage of an element of cell wall layer as a fiber-reinforced composite was treated by Barber and Meylan (134) as a problem of restraint on the shrinking (or swelling) matrix by the inert framework. Conceptually, the separate phases can be thought of as shrinking or swelling independently according to their respective shrinkage potential, with subsequent compression applied to those parts that have expanded more and tension to those that have expanded less to bring them back into the integral unit which they were in the beginning. With a swelling matrix and nonswelling fibers, the former will then be in compression and the fibers in tension, and there must be internal equilibrium of forces. The final dimension will then depend on unrestrained shrinkage potential, relative volume, and relative stiffness of the components. Barber and Meylan (134) followed this general approach, but their analysis is not valid because of conflicting assumptions regarding force equilibrium and interrelationships between strains. Barrett et al. (203) applied a similar concept, but followed a more rigorous and general approach using mechanics of materials methods. Several independent sets of assumptions regarding the interaction of the matrix and framework gave identical results and generally were near the lower of upper and lower bounds derived by energy methods. Cave (162, 204,206,207) has developed generalized constitutive equations for the fiber composite layer that include not only the shrinkage component but also allow for changes in the stiffness of the components as shrinkage takes place. As already pointed out, Cave's model includes three phases in a layer: cellulose microfibrils surrounded by hemicellulose sheaths arranged in lamellae that alternate with microlayers of lignin (153). Accordingly, the polysaccharide complex is treated as a fiber- reinforced composite, and considerations of force equilibrium and strain compatibility are used to assemble the parallel lamellae of lignin and polysaccharides into a cell wall layer.

3. Cell Walls as Layered Systems

Once the stiffness and shrinkage characteristics of each layer are known, the behavior of the assembled cell wall can be predicted using equilibrium and strain compatibility conditions. Shear restraint is usually introduced by considering a double cell wall of two adjoining cells. This was included by Barber and Meylan (134) in their model that otherwise represented the wall of one cell by a single layer. Barber (208) later found that the introduction of a hoop layer representing the combined action of S_1 and S_3 made the model more effective in predicting observed shrinkage behavior. Cave proposed both the use of a single layer composed of lamellae with a continuous distribution of microfibrillar angles (206) and another version with a layer representing S_2, a hoop layer representing S_1, S_3, and the carbohydrate fraction of the primary wall, and the true middle lamella (207). Barrett et al. (203) used a model comprised of M + P, S_1, S_2, and S_3.

The focus of shrinkage models has been on longitudinal shrinkage because the latter is to a very great extent dependent only on the shrinkage of the cell wall, especially in conifers. Shrinkage in the radial and tangential directions depends also significantly on anatomic factors that go beyond individual cells. However, differences in composition or structure between radial and tangential walls are also possible factors in transverse shrinkage anisotropy (209,210), and these can be built into any of the models discussed, as was done in fact in one study (208).

Several of the shrinkage models described have been shown to be capable of making qualitative predictions of the longitudinal shrinkage behavior of wood and its relationship to microfibril angle (203,207,211 214). Close quantitative agreement between model shrinkage and empirical data was achieved by Cave (Fig. 9), but only after calibrating the model through adjustment of the input parameters (153,207). Further work is needed on determining the basic parameters of constituent properties, including a consideration of rheological behavior, as well as geometric and structural factors that are still not sufficiently well known.

D. Models of Wood Shrinkage

There is ample evidence, even if not universal acceptance that the shrinkage behavior of wood, especially as it concerns the transverse anisotropy of wood, cannot be explained entirely on the basis of cell wall composition and structure (202,209). The two major factors in transversely anisotropic shrinkage are the restraint of radial shrinkage by rays and the dominance of tangential shrinkage by latewood (in species with pronounced density differences between earlywood and latewood).

If latewood has much greater density than earlywood, the former will also be expected to shrink more than the latter in the transverse plane. In the radial direction, the overall shrinkage will be approximately equal to the volume-weighted average of the earlywood and latewood shrinkages. In the tangential direction, however, the

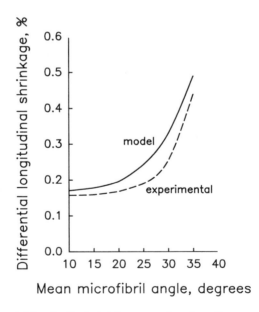

Figure 9   Differential longitudinal shrinkage as a function of mean microfibril angle, comparing Cave's model with a curve fitted to experimental data by Meylan (from Ref. 213). (Adapted from Ref. 207.)

larger shrinkage potential and greater stiffness of latewood will cause it to dominate the overall shrinkage. Models for quantitative predictions of this effect, treating wood as a composite of alternating layers of earlywood and latewood, have been devised by several authors (215,216). Similarly, rays have been found to restrain radial shrinkage because of low shrinkage potential and high stiffness, as compared to the remaining tissues of longitudinally aligned cells. This mechanism can also be modeled as a layered composite, in this case, of ray tissue and longitudinally aligned tissue (217). Both mechanisms can further be combined in the form of a doubly layered composite, the planes of the layers being in the radial-longitudinal plane on one level (ray tissue and longitudinal tissue) and in the tangential-longitudinal plane on the other level (earlywood and latewood). The validity of such a model has been confirmed by the measurement of shrinkage and mechanical properties of isolated tissues in California black oak (181).

V. CONCLUSION

The physical organization of the principal wood constituents—cellulose, hemicellulose, lignin, and extractives—has been examined with special reference to their composite action in determining the overall physical characteristics and behavior of wood. The mechanics of the cell wall have been discussed in terms of composite models designed for the prediction of elasticity and shrinkage and for internal stress analysis. Details of the mathematical treatment of the various models have been purposely omitted, and interested readers are referred to the original literature. Models have been further extended to the composite interaction of wood tissues at a higher level of structural organization.

The discussion has been limited to certain physical and mechanical properties, namely to those thought most important and therefore most intensively studied. There is no reason that the same approaches that were discussed could not be extended to other topics, such as the rheological behavior of wood (204). For example, interesting model studies have been made of dielectric properties (218) and the development of growth stresses in a growing cell (219). It has long been clear that a true understanding of the physical behavior of wood depends on the study of its composite nature, and there is ample opportunity for future developments in this area.

ACKNOWLEDGMENTS

The authors wish to thank Dr. Tore E. Timell, Dr. Richard E. Mark, and Dr. Timothy G. Rials for their valuable comments, and the editors for their patience.

REFERENCES

1. A. J. Stamm, *Wood and Cellulose Science*, New York, Ronald Press, 1964.
2. C. Nägeli and S. Schwendener, *Das Mikroskop, Theorie und Anwendung desselben, zweite verbesserte Auflage*, Wilhelm Engelmann, Leipzig, 1867.
3. A. Sarko, in *Cellulose: Structure, Modification, and Hydrolysis* (R. A. Young and R. M. Rowell, eds.), Wiley, New York, 1986.
4. D. Fengel and G. Wegener, *Wood: Chemistry, Ultrastructure, Reactions*, Walter de Gruyter, Berlin, 1984.
5. D. A. I. Goring and T. E. Timell, *TAPPI*, *45*(6):454 (1962).
6. G. Patscheke and S. Poller, *Cell. Chem. Technol.*, *14*:3 (1980).
7. M. Marx-Figini, *Papier*, *18*(10A):546 (1964).
8. M. Marx-Figini and G. V. Schultz, *Naturwissenschaften*, *53*:466 (1966).
9. M. Marx-Figini and G. V. Schultz, *Biochimica et Biophysica Acta*, *112*:81 (1966).
10. M. Marx-Figini, *Appl. Polym. Symp. No. 37*:157 (1983).
11. M. Shimizu, M. Inoue, and T. Yamashita, Bull. Faculty Agri., Shizuoka Univ., No. 20, p. 37 (as quoted in Ref. 4).
12. A. Sarko and R. Muggli, *Macromolecules*, *7*:486 (1974).
13. K. H. Gardner and J. Blackwell, *Biopolym.*, *13*:1975 (1974).
14. C. Y. Liang and R. H. Marchessault, *J. Polym. Sci.*, *37*:385 (1959).
15. H. A. Swenson, *TAPPI*, *56*(2):106 (1973).
16. R. H. Atalla, *Appl. Polym. Symp. No. 28*:659 (1976).
17. R. H. Atalla, "The Ekman-Days 1981," *International Symposium on Wood and Pulping Chemistry, Stockholm*, SPCI, 1981, Vol. I, p. 57.
18. J. R. Colvin, *Appl. Polym. Symp. No. 37*:25 (1983).
19. K. H. Meyer and L. Misch, *Helvetica Chimica Acta*, *20*:232 (1937).
20. R. H. Marchessault and P. R. Sundararajan, *Advances Carbohyd. Chem. Biochem.*, *33*:387 (1976).
21. P. R. Sundararajan and R. H. Marchessault, *Advances Carbohyd. Chem. Biochem.*, *35*:377 (1978).
22. P. R. Sundararajan and R. H. Marchessault, *Advances Carbohyd. Chem. Biochem.*, *36*:315 (1979).
23. R. H. Atalla and D. L. VanderHart, *Sci.*, *223*:283 (1984).
24. D. L. VanderHart and R. H. Atalla, in *The Structures of Cellulose* (R. H. Atalla, ed.), ACS, Washington, D.C., 1987.
25. R. H. Atalla, in *The Structures of Cellulose* (R. H. Atalla, ed.), ACS, Washington, D.C., 1987.
26. D. H. Krässig, in *Cellulose and Its Derivatives: Chemistry, Biochemistry and Application* (J. F. Kennedy, G. O. Phillips, D. J. Wedlock, and P. A. Williams, eds.), Ellis Horwood, Chichester, 1985.
27. P. P. Gillis, *J. Polym. Sci., Part A-2*, *7*:783 (1969).
28. A. G. Assaf, R. H. Haas, and C. B. Purves, *J. ACS*, *66*:59 (1944).
29. C. C. Conrad and A. G. Scroggie, *Ind. Eng. Chem.*, *37*:592 (1945).
30. V. J. Frilette, J. Hanle, and H. Mark, *J. ACS*, *70*:1107 (1948).

31.   P. H. Hermans and A. Weidinger, *J. Polym. Sci.*, *4*:135 (1949).

32.   J. Mann and H. J. Marrinan, *Trans. Faraday Soc.*, *52*:492 (1956).

33.   A. Basch, T. Wasserman, and M. Lewin, *J. Polym. Sci.*, Polym. Chem. ed., *12*:1143 (1974).

34.   B. E. Dale and G. T. Tsao, *J. Appl. Polym. Sci.*, *27*:1233 (1982).

35.   H. Krässig, *Appl. Polym. Symp. No. 28*:777 (1976).

36.   J. Haase, R. Hosemann, and B. Renwanz, *Cell. Chem. Technol.*, *9*:513 (1975).

37.   R. H. Marchessault and J. A. Howsmon, *Textile Res. J.*, *27*:30 (1957).

38.   G. Jayme and E. Roffael, *Papier*, *24*:614 (1970).

39.   R. Hosemann and S. N. Bagchi, *Direct Analysis of Diffraction by Matter*, North-Holland, Amsterdam, 1962.

40.   R. Hosemann and M. P. Hentschel, *Cell. Chem. Technol.*, *19*:459 (1985).

41.   G. B. Mitra and P. S. Mukherjee, *Polym.*, *21*:1403 (1980).

42.   P. S. Mukherjee and G. B. Mitra, *Polym.*, *24*:525 (1983).

43.   P. K. Ray and S. B. Bandyopadhyay, *J. Appl. Polym. Sci.*, *19*:729 (1975).

44.   M. L. M. El-Osta, R. M. Kellogg, R. O. Foschi, and R. G. Butters, *Wood and Fiber*, *6*:36 (1974).

45.   T. E. Timell, in *The Encyclopedia of Materials Science and Engineering, Vol. 7* (M. B. Bever, ed.), Pergamon, Oxford, 1986.

46.   K. Mühlethaler, in *Cellular Ultrastructure of Woody Plants* (W. A. Côté, ed.), Syracuse Univ. Press, Syracuse, N.Y., 1965.

47.   R. D. Preston, *J. Microscopy*, Part 1, *93*:7 (1970).

48.   J. Sugiyama, H. Harada, Y. Fujiyoshi, and N. Uyeda, *Mokuzai Gakkaishi*, *31*:61 (1985).

49.   J. Sugiyama, H. Harada, Y. Fujiyoshi, and N. Uyeda, *Planta*, *166*:161 (1985).

50.   D. Fengel, *Die Naturwissenschaften*, *61*:31 (1974).

51.   D. Fengel, *Holzforschung*, *32*:37 (1978).

52.   T. Goto, H. Harada, and H. Saiki, *Mokuzai Gakkaishi*, *21*:537 (1975).

53.   F. C. Beall and W. K. Murphey, *Wood and Fiber*, *2*:282 (1970).

54.   P. P. Gillis, *Cell. Chem. Technol.*, *4*:123 (1970).

55.   S. S. Kelley, T. G. Rials, and W. G. Glasser, *J. Mater. Sci.*, *22*:617 (1987).

56.   K. V. Sarkanen and C. H. Ludwig, in *Lignins: Occurrence, Formation, Structure and Reactions* (K. V. Sarkanen and C. H. Ludwig, eds.), Wiley-Interscience, 1971.

57.   H. Nishimura, T. Okano, and I. Asano, *Mokuzai Gakkaishi*, *27*:611 (1981).

58.   D. A. I. Goring, in *Lignins: Occurrence, Formation, Structure and Reactions* (K. V. Sarkanen and C. H. Ludwig, eds.), Wiley-Interscience, 1971.

59.   K. Forss and K.-E. Fremer, *Appl. Polym. Symp. No. 37*:531 (1983).

60.   M. Wayman and T. I. Obiaga, *TAPPI*, *57*(4):123 (1974).

61.   W. G. Glasser and H. R. Glasser, *Paperi ja Puu*, *63*:71 (1981).

62.   O. Faix, *Holzforschung*, *28*:222 (1974).

63.   O. Faix, *Papier*, *30*:(10A):V1 (1976).

64.   H. H. Nimz, D. Robert, O. Faix, and M. Nemr, *Holzforschung*, *35*:16 (1981).

65.   K. V. Sarkanen and H. L. Hergert, in *Lignins: Occurrence, Formation, Structure and Reactions* (K. V. Sarkanen and C. H. Ludwig, eds.), Wiley-Interscience, 1971.

66.   H. L. Hardell, G. J. Leary, M. Stoll, and U. Westermark, *Svensk Papperstidn.*, *83*·44 (1980).

67. J. Gravitis and P. Erins, *Appl. Polym. Symp. No. 37*:421 (1983).
68. U. P. Agarwal and R. H. Atalla, *Planta, 169*:325 (1986).
69. C. Y. Liang, K. H. Basset, E. A. McGinnes, R. H. Marchessault, *TAPPI, 43*(12):1017 (1960).
70. R. H. Marchessault, C. Y. Liang, *J. Polym. Sci., 59*:357 (1962).
71. D. H. Page, F. El-Hosseiny, M. L. Bidmade, and R. Binet, *Appl. Polym. Symp. No. 28*:923 (1976).
72. I. A. Nieduszynski and R. H. Marchessault, *Biopolym., 11*:1335 (1972).
73. S. M. Gabbay, P. R. Sundararajan, and R. H. Marchessault, *Biopolym., 11*:79 (1972).
74. I. Nieduszynski and R. H. Marchessault, *Can. J. Chem., 50*:2130 (1972).
75. R. H. Marchessault, W. Settineri, and W. Winter, *TAPPI, 50*(2):55 (1967).
76. D. Fengel, *Svensk Papperstidn., 70*:70 (1967).
77. D. Fengel, *Cell. Chem. Technol., 13*:279 (1979).
78. G. Annergren and S. A. Rydholm, *Svensk Papperstidn., 62*:737 (1959).
79. W. Czirnich and R. Patt, *Holzforschung, 30*:19 (1976).
80. P. Hoffman and R. Patt, *Holzforschung, 30*:124 (1976).
81. J. W. T. Merewether, *Holzforschung, 11*:65 (1957).
82. Y. Z. Lai and K. V. Sarkanen, in *Lignins: Occurrence, Formation, Structure and Reactions* (K. V. Sarkanen and C. H. Ludwig, eds.), Wiley-Interscience, 1971.
83. E. E. Dickey and L. Roth, Bibliographic Series No. 287, The Institute of Paper Chemistry, Appleton, Wi.,1980.
84. G. J. Leary, *Wood Sci. Technol., 14*:21 (1980).
85. G. J. Leary, D. A. Sawtell, and H. Wong, "The Ekman-Days 1981," *International Symposium on Wood and Pulping Chemistry, Stockholm SPCI,*1981, Vol. I, p. 63.
86. H. H. Brownell, *TAPPI, 53*(7):1278 (1970).
87. J. R. Obst, *TAPPI, 65*(4):109 (1982).
88. Ö. Eriksson, D. A. I. Goring, and B. O. Lindgren, *Wood Sci. Technol., 14*:267 (1980).
89. J. Joseleau and R. Kesraoui, *Holzforschung, 40*:163 (1986).
90. D. Fengel, *Die Naturwissenschaften, 62*:182 (1975).
91. D. Fengel, *Svensk Papperstidn., 79*:24 (1976).
92. B. Kosíková, L. Zákutná, and D. Joniak, *Holzforschung, 32*:15 (1978).
93. B. J. Fergus, A. R. Procter, J. A. N. Scott, and D. A. I. Goring, *Wood Sci. Technol., 3*:117 (1969).
94. B. J. Fergus and D. A. I. Goring, *Holzforschung, 24*:113 (1970).
95. B. J. Fergus and D. A. I. Goring, *Holzforschung, 24*:118 (1970).
96. S. Asunmaa and P. W. Lange, *Svensk Papperstidn., 57*:501 (1954).
97. H. Meyer, *J. Polym. Sci., 51*:11 (1961).
98. D. Fengel, *Wood Sci. Technol., 3*:203 (1969).
99. R. E. Mark, *Cell Wall Mechanics of Tracheids*, Yale Univ. Press, New Haven, Conn., 1967.
100. W. E. Moore, M. Effland, B. Sinha, M. P. Burdick, and C. Schuerch, *TAPPI, 49*(5):206 (1966).
101. O. Sawabe, *Mokuzai Gakkaishi, 26*:641 (1980).
102. J. E. Stone, A. M. Scallan, and P. A. V. Ahlgren, *TAPPI, 54*(9):1527 (1971).
103. A. J. Kerr and D. A. I. Goring, *Wood Sci., 9*:136 (1977).
104. K. Ruel, F. Barnoud, and D. A. I. Goring, *Cell. Chem. Technol., 13*:429 (1979).

105. A. J. Kerr and D. A. I. Goring, *Cell. Chem. Technol.*, 9:563 (1975).
106. R. H. Marchessault, *Symposium on Chemistry and Biochemistry of Lignin, Cellulose and Hemicellulose*, Univ. Grenoble, 1964, p. 287.
107. A. M. Scallan, *Wood Sci.*, 6:266 (1974).
108. J. E. Stone and A. M. Scallan, *Cell. Chem. Technol.*, 2:343 (1968).
109. N. Parameswaran and W. Liese, *Holz als Roh- und Werkstoff*, 40:145 (1982).
110. A. B. Anderson, *J. Inst. Wood Sci.*, (8):14 (1961).
111. A. B. Anderson, *J. Inst. Wood Sci.*, (10):29 (1962).
112. A. B. Anderson and E. Zavarin, *J. Inst. Wood Sci.*, (15):3 (1965).
113. W. E. Hillis, *Wood Sci. Technol.*, 5:272 (1971).
114. C. H. Lee, *Wood and Fiber Sci.*, 18:376 (1986).
115. A. D. Hofstrand, A. A. Moslemi, and J. F. Garcia, *For. Prod. J.*, 34:(2):57 (1984).
116. M. H. Simatupang and X. X. Lu, *Holz als Roh- und Werkstoff*, 43:325 (1985).
117. A. B. Wardrop and G. W. Davies, *Austral. J. Botany*, 6:96 (1958).
118. R. L. Krahmer and W. A. Côté, Jr., *TAPPI*, 46(1):42 (1963).
119. E. T. Choong, *Wood and Fiber*, 1:124 (1969).
120. W. T. Nearn, Bull. 598, School Forestry Series No. 2, Penn. State Univ., Coll. Agri., Agri., Exper. Station, 1955.
121. H. N. Rosen, *Wood Sci.*, 10:151 (1978).
122. H. Tarkow and J. Krueger, *For. Prod. J.*, 11:(5):228 (1961).
123. M. -L. Kuo and D. G. Arganbright, *Holzforschung*, 34:17 (1980).
124. M. -L. Kuo and D. G. Arganbright, *Holzforschung*, 34:41 (1980).
125. T. Kubo and J. Karburagi, *Bull. Exper. Forests, Tokyo Univ. Agri. Technol.*, No. 10 1973, p. 108.
126. A. B. Anderson, E. L. Ellwood, E. Zavarin, and R. W. Erickson, *For. Prod. J.*, 10:(4):212 (1960).
127. C. A. Hart, *For. Prod. J.*, 34:(11/12):45 (1984).
128. A. P. Schniewind, in *The Mechanical Behavior of Wood* (A. P. Schniewind, ed.), Univ. Calif., Berkeley, Calif., 1963.
129. G. Ifju and R. W. Kennedy, *For. Prod. J.*, 12:(5):213 (1962).
130. S. H. Clarke, *Tropical Woods*, 52:1 (1937).
131. R. M. Kellogg and G. Ifju, *For. Prod. J.*, 12:(10):463 (1962).
132. E. Münch, *Flora* (Jena), *N. F. 32*:357 (1938).
133. H. Ziegenspeck, in *Handbuch der Mikroskopie in der Technik*, (H. Freund, ed.), Umschau Verlag, Frankfurt am Main, 1951, Vol. 5, Part 1.
134. N. F. Barber and B. A. Meylan, *Holzforschung*, 18:146 (1964).
135. R. Mark, in *Cellular Ultrastructure of Woody Plants* (W. A. Côté, Jr., ed.), Syracuse Univ. Press, Syracuse, N.Y., 1965.
136. I. D. Cave, *Wood Sci. Technol.*, 2:268 (1968).
137. M. Suzuki, *Mokuzai Gakkaishi*, 15:278 (1969).
138. D. Atack, *Svensk Papperstidn.*, 75:89 (1972).
139. J. Blechschmidt, P. Engert, and M. Stephan, *Wood Sci. Technol.*, 20:263 (1986).
140. H. G. Higgins, G. M. Irvine, V. Puri, and A. B. Wardrop, *Appita*, 32:23 (1978).
141. E. L. Back and N. L. Salmén, *TAPPI*, 65(7):107 (1982).
142. G. M. Irvine, *TAPPI*, 67(5):118 (1984).

143. D. A. I. Goring, *Pulp Pap. Mag. Can. 64*:T517 (1963).
144. N. L. Salmén, "Temperature and Water Induced Softening Behavior of Wood Fiber Based Materials," Ph.D. Diss., Dept. Paper Technol., Royal Inst. Tech., Stockholm, 1982.
145. N. L. Salmén and E. L. Back, *TAPPI, 60*:(12):137 (1977).
146. W. E. Hillis and A. N. Rozsa, *Wood Sci. Technol., 19*:57 (1985).
147. Forest Products Laboratory, *Wood Handbook: Wood as an Engineering Material*, Agri. Handbook No. 72, U.S. Dept. Agri., Washington, D.C., 1974.
148. M. Kufner, *Holz als Roh- und Werkstoff, 36*:435 (1978).
149. J. R. Darby and J. K. Sears, in *Encyclopedia of Polymer Science and Technology* (N. M. Bikales, ed.), Wiley, New York, 1969, Vol. 10.
150. I. Sakurada, Y. Nukushina, and T. Ito, *J. Polym. Sci., 57*:651 (1962).
151. M. A. Jaswon, P. P. Gillis, and R. E. Mark, *Proc. Roy. Soc., London, A., 306*:389 (1968).
152. L. Salmén and A. de Ruvo, *Wood and Fiber Sci., 17*:336 (1985).
153. I. D. Cave, *Wood Sci. Technol., 12*:75 (1978).
154. R. E. Mark and P. P. Gillis, in *Handbook of Physical and Mechanical Testing of Paper and Paperboard* (R. E. Mark and K. Murakami, eds.), Marcel Dekker, New York, 1983, Vol. 1.
155. M. L. M. El-Osta and R. W. Wellwood, *Wood and Fiber, 4*:204 (1972).
156. S. Moriizumi and T. Okano, *Mokuzai Gakkaishi, 24*:1 (1978).
157. P. S. Srinivasan, *Quar. J. Ind. Inst. Sci., 4*:222 (1941).
158. R. E. Mark, in *Theory and Design of Wood and Fiber Composite Materials* (B. A. Jayne, ed.), Syracuse Univ. Press, Syracuse, N.Y. 1972.
159. W. J. Cousins, *Wood Sci. Technol., 10*:9 (1976).
160. W. J. Cousins, *New Zeal. J. For. Sci., 7*:107 (1977).
161. W. J. Cousins, *Wood Sci. Technol., 12*:161 (1978).
162. I. D. Cave, *Wood Sci. Technol., 6*:157 (1972).
163. R. E. Mark, in *The Encyclopedia of Materials Science and Engineering* (M. B. Bever, ed.), Pergamon, Oxford, 1986, Vol. 7.
164. N. L. Salmén, in *Paper, Structure and Properties* (J. A. Bristow and P. Kolseth, eds.), Marcel Dekker, New York, 1986.
165. J. C. Halpin, *Primer on Composite Materials: Analysis (revised)*, Technomic Publishing Co., Lancaster, Pa., 1984.
166. P. P. Gillis, *Fib. Sci. Technol., 2*:193 (1970).
167. I. D. Cave, *Wood Sci. Technol., 3*:40 (1969).
168. J. E. Ashton, J. C. Halpin, and P. H. Petit, *Primer on Composite Materials: Analysis*, Technomic Publishing Co., Stamford, Conn., 1969.
169. A. P. Schniewind and J. D. Barrett, *Wood and Fiber, 1*:205 (1969).
170. R. E. Mark and P. P. Gillis, *Wood and Fiber, 2*:79 (1970).
171. R. E. Mark, J. L. Thorpe, A. J. Angello, R. W. Perkins, and P. P. Gillis, *J. Polym. Sci.*, Part 36:177 (1971).
172. C. E. Dunning, *Wood Sci., 1*:65 (1968).
173. S. G. Lekhnitskii, *Theory of Elasticity of an Anisotropic Elastic Body*, Holden-Day, San Francisco, Calif., 1963.
174. R. C. Tang, *Wood and Fiber, 3*:210 (1972).

175. P. P. Gillis and R. E. Mark, *Cell. Chem. Technol.*, 7:209, (1973).
176. J. D. Barrett and A. P. Schniewind, *Wood and Fiber*, 5:215 (1973).
177. H. Saiki, *Mokuzai Gakkaishi*, 16:244 (1970).
178. K. E. Easterling, R. Harrysson, L. J. Gibson, and M. F. Ashby, *Proc. Roy. Soc. London, A, 383*:31 (1982).
179. P. P. Gillis, *Wood Sci. Technol.*, 6:138 (1972).
180. F. El Amri, "Contribution a la modélisation élastique anisotrope du matériau bois-feuillus et résineux," Ph.D. Thesis, Institut National Polytechnique de Lorraine, Nancy, 1987.
181. A. P. Schniewind, *For. Prod. J.*, 9(10):350 (1959).
182. M. L. M. El-Osta, O. A. Badran, and E. M. A. Ajoung, *Wood Sci.*, 13:225 (1981).
183. R. F. Luxford, *J. Agri. Res.*, 42:801 (1931).
184. J. W. Raczkowski, *Holzforschung und Holzverwertung*, 20:73 (1968).
185. D. G. Arganbright, *Wood and Fiber*, 2:367 (1971).
186. G. N. Christensen and K. E. Kelsey, *Holz als Roh- und Werkstoff*, 17:189 (1959).
187. R. O. H. Runkel and M. Lüthgens, *Holz als Roh- und Werkstoff*, 14:424 (1956).
188. S. H. Clarke and C. B. Pettifor, *Nature*, 145:424 (1940).
189. F. Kollmann and D. Fengel, *Holz als Roh- und Werkstoff*, 23:461 (1965).
190. A. J. Stamm and W. F. Loughborough, *ASME Trans.*, 64:379 (1942).
191. F. F. Wangaard and L. A. Granados, *Wood Sci. Technol.*, 1:253 (1967).
192. A. Burmester and W. E. Wille, *Holz als Roh- und Werkstoff*, 33:147 (1975).
193. A. Burmester, *Holz als Roh- und Werkstoff*, 33:333 (1975).
194. S. C. Chafe, *Wood Sci. Technol.*, 21:27 (1987).
195. H. A. Schroeder, *Wood and Fiber*, 4:20 (1972).
196. C. Skaar, *Water in Wood*, Syracuse Univ. Press, Syracuse, N.Y., 1972.
197. S. C. Chafe, *Wood Sci. Technol.*, 21:131 (1987).
198. W. W. Barkas, "Swelling Stresses in Gels," Dept. Sci. Ind. Res., For. Prod. Res. Spec. Rept. No. 6, London, 1945.
199. A. P. Schniewind, *Holzforschung, 14*:161 (1960).
200. J. T. Quirk, *Wood Fiber Sci.*, 16:115 (1984).
201. A. P. Schniewind, *Holz als Roh- und Werkstoff*, 24:502 (1966).
202. K. E. Kelsey, "A Critical Review of the Relationship Between the Shrinkage and Structure of Wood," Tech. Paper No. 28, Div. For. Prod., Commonwealth Sci. and Indus. Res. Organ., Australia, 1963.
203. J. D. Barrett, A. P. Schniewind, and R. L. Taylor, *Wood Sci.*, 4:178 (1972).
204. I. D. Cave, *J. Microscopy.*, 104, Part 1:47 (1975).
205. H. A. Spalt, *For. Prod. J.*, 8(10):288 (1958).
206. I. D. Cave, *Wood Sci. Technol.*, 6:284 (1972).
207. I. D. Cave, *Wood Sci. Technol.*, 12:127 (1978).
208. N. F. Barber, *Holzforschung*, 22:97 (1968).
209. J. D. Boyd, *Mokuzai Gakkaishi*, 20:473 (1974).
210. J. D. Boyd, *Wood Sci. Technol.*, 11:3 (1977).
211. J. M. Harris and B. A. Meylan, *Holzforschung*, 19:144 (1965).
212. B. A. Meylan, *For. Prod. J.*, 18(4):75 (1968).
213. B. A. Meylan, *Wood Sci. Technol.*, 6:293 (1972).

214.   B. A. Meylan and M. C. Probine, *For. Prod. J.*, *19*(4):30 (1969).
215.   R. E. Pentoney, *J. For. Prod. Res. Soc.*, *3*:27–32, 86 (1953).
216.   A. Ylinen and P. Jumppanen, *Wood Sci. Technol.*, *1*:241 (1967).
217.   D. C. McIntosh, *For. Prod. J.*, *5*(5):355 (1955).
218.   T. Tanaka, M. Norimoto, and T. Yamada, *J. Soc. Mat. Sci. Japan*, *24*:867 (1975).
219.   R. R. Archer, *Wood Sci. Technol.*, *21*:139 (1987).

Index